BAB

Structure and Properties of Dislocations in Semiconductors 1989

Sponsors

The Symposium was largely funded by the NATO Scientific Affairs Division as
an Advanced Research Workshop.

The Institute of Physics (EMAG), The Royal Society, The US Army European
Research Office, The Furukawa Electric Co Ltd, JEOL (UK) Ltd,
The Plessey Co Ltd, Siemens AG

Structure and Properties of Dislocations in Semiconductors 1989

Proceedings of the Sixth International Symposium on the
Structure and Properties of Dislocations in Semiconductors
held at the University of Oxford, 5–8 April 1989

Edited by S G Roberts, D B Holt and P R Wilshaw

Institute of Physics Conference Series Number 104
Institute of Physics, Bristol and New York

CODEN IPHSAC 104 1–470 (1989)

British Library Cataloguing in Publication Data

International Symposium on the Structure and Properties of
    Dislocations in Semiconductors *(6th: 1989: University Of Oxford)*
    Structure and Properties of Dislocations in
    Semiconductors, 1989.
    1. Semiconductors. Structure & physical properties
    I. Title   II. Roberts, S.G.   III. Holt, D.B.   IV. Wilshaw,
    P.R.   V. Series
    537.6'22

    ISBN 0-85498-060-1

Library of Congress Cataloging-in-Publication Data are available

Honorary Editors
        S G Roberts, D B Holt and P R Wilshaw

Published under The Institute of Physics imprint by IOP Publishing Ltd
Techno House, Redcliffe Way, Bristol BS1 6NX, England
335 East 45th Street, New York, NY 10017-3483, USA

Printed in Great Britain by J W Arrowsmith Ltd, Bristol

Contents

Chapter 1: Structure of grain boundaries and dislocations

Chapter 2: Electronic effects of dislocations and associated point defects

Chapter 3: Dislocation mobility

Chapter 4: Dislocations, plasticity and fracture

Chapter 5: Dislocations and device performance

Contents

Preface

The sixth in the series of International Symposia on the Structure and Properties of Dislocations in Semiconductors was held in Oxford University from 5–8 April 1989. The meeting was largely funded by the NATO Scientific Affairs division as an Advanced Research Workshop. Accommodation was provided by Keble College; oral presentations were made in the Department of Nuclear Physics, and poster sessions were a short distance away in the Engineering Science/Metallurgy Common Room. As with previous meetings in the series, the emphasis was on active participation by all those attending; to this end, the number of participants was kept small, and much of the time of the meeting was given over to discussion, either of specific papers or topics, or more generally.

The objective of the Symposium was to bring together international experts on all aspects of dislocations in semiconductors, ranging from fundamental structural, electronic, optical and mechanical properties, to their effects on devices. The conference opened with a session on grain boundaries and dislocations. Elegant high resolution electron microscopy studies are giving a detailed understanding of the interaction between glide dislocations and boundaries, showing the structural rearrangement in the boundary as a result of absorption of dislocations. *In situ* high voltage electron microscope experiments on the transmission and blocking of glide dislocations by grain boundaries in Si bicrystals gave valuable insight into the mechanisms operating.

A review of theoretical models of atomic structure showed tight binding calculations to be in agreement with experiments, for boundaries in Si; this is not surprising since the calculations are guided by the experiments. However, calculations of electronic structure predict absence of midgap states, and of band tails, in disagreement with experiments. It was proposed tentatively that midgap states are due to impurities. The origin of band tails is also still not clear. There is clearly much to be done in establishing theoretical models for carefully assigned and executed experiments. This may be difficult since experimental data on electronic properties of grain boundaries show such boundaries to be electrically inhomogeneous due to disorder.

With regard to electronic properties of dislocations in semiconductors, there are still problems with the unambiguous interpretation of DLTS data. There is now, however, a good understanding of the basis of EBIC contrast and the information which EBIC can provide. The experiments on straight dislocations in Si suggest a low density of states in the gap; the origin of these is still not clear. The photo-luminescence spectra from Ge and Si are now well understood.

For GaAs, good progress has been made in understanding the dislocation-related optical properties in terms of dislocation bonds. Such effects have been shown to contribute to laser degradation. On the other hand, a full understanding of the electronic properties associated with point defects produced by deformation is still lacking.

A number of papers dealt with dislocation mobility—there are now good experimental data on the effects of impurities, including locking by non-electrical impurities, and the effect of electrically active impurities on velocities. Theoretical difficulties still remain in the kink migration energy, the effect of point obstacles, and the pre-exponential factor in the velocity–stress–temperature relationship. Stress-pulse experiments were interpreted in terms of a high secondary Peierls potential and the effect of point defect obstacles. Computer simulation methods showed that the results can also be explained in terms of strong obstacles and a low migration energy. Clearly this needs to be resolved.

The importance of the different behaviour of different partials at low temperatures was brought out by deformation experiments. The electronic doping effect on dislocation velocity is different at low temperatures from that at high temperatures.

Workers on fracture and the brittle–ductile transition in Si are now agreed on the controlling effect of dislocation mobility on the transition. Experiments in which the precursor cracks are introduced by surface indentation, coupled with computer simulations of the dynamics of dislocation emission, showed that in these cases the transition is controlled by pre-existing dislocations. The detailed modelling of other experiments carried out under other conditions has yet to be carried out. These studies are giving an insight into the factors controlling brittle–ductile transitions in general.

The final session of the meeting considered the effects of grown-in and process-induced dislocations in devices. The importance of extended defects on device properties was stressed justifying the continued basic work in this area. This was further underlined by the lack of understanding of defect structures in GaAs, of gettering mechanisms, and of generation of dislocations in strained-layer structures (subjects of considerable importance to potential device applications).

The overall conclusion of the meeting was that this remains an active and important area of research, providing underlying knowledge and understanding to the development of devices. While considerable progress has been made on some topics, the meeting identified directions for future research in a number of areas.

We are grateful to all those who contributed to the success of the meeting, particularly the companies and organisations which provided financial support (without which the meeting would not have been possible), the members of the various organising committees, who gave freely of their time and expertise and the local team who kept the conference running. Lastly, we thank all the participants for their energy in presentation and discussion—and for handing in their manuscripts!

P B Hirsch
S G Roberts

The International Symposia on Structure and Properties of Dislocations in Semiconductors

Year	Venue	Published as
1973	Zakopane (Poland)	*Proc. 4th Int. Summer School on Defects* (Warsaw: Polish Scientific Publishers) 1974
1976	Krynica (Poland)	*Proc. 5th Int. Summer School on Defects* (Warsaw–Łódz: Polish Scientific Publishers) 1976
1978	Hünfeld (FRG)	*Proc. Int. Symp. on Dislocations in Tetrahedrally-Coordinated Semiconductors* (*J. Physique*, Colloque C6) 1979
1983	Aussois (France)	*Proc. Colloq. Int. 'Propriétés et Structure des Dislocations dans les Semiconducteurs'* (*J. Physique*, Colloque C4) 1983
1986	Zvenigorod (USSR)	*Proc. 5th Int. Symp. on the Structure and Properties of Dislocations in Semiconductors* (*Akad. Nauk. Ser. Fiz.* **51** nos 4 & 9) 1987 Published in English translation as: 1987 *Bulletin of the USSR Academy of Sciences, Physical Series* **51** (New York: Allerton Press)

Participants

H Alexander
G R Booker
M Brohl
J Castaing
A Cavallini
P Charsley
D R Clarke
J L Demenet
R Falster
B Farber
J L Farvacque
T Figielski
A George
R Gleichmann
H Gottschalk
P Haasen
M I Heggie
P Hemment
J Heydenreich
P B Hirsch
D B Holt
C J Humphreys
R Jones
C Kisielowski-Kemmerich
B Kolbesen
V Kravchenko
F Louchet
K Maeda

N Maeda
G Michot
E Nadgorny
V I Nikitenko
Yu A Ossipyan
A R Peaker
P Pirouz
B Pohoryles
R C Pond
J Rabier
H Richter
S G Roberts
W Schröter
W Seifert
Yu Shreter
W C Simmons
J Steeds
D J Stirland
K Sumino
A P Sutton
H Teichler
J Thibault-Desseaux
J Vanhellemont
J Werner
P R Wilshaw
P Wurzinger
A Zozime

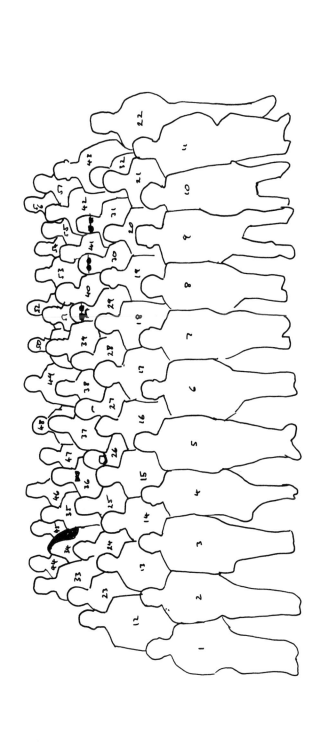

1) P. Pirouz
2) W. Shröter
3) D.B. Holt
4) A. Cavallini
5) P. Haasen
6) P.B. Hirsch
7) H. Alexander
8) J. Castaing
9) K. Sumino
10) V. Kravchenko

11) A. Zozime
12) J.L. Hutchison
13) B. Kolbesen
14) R.C. Newman
15) E. Nadgorny
16) P.R. Wilshaw
17) S.G. Roberts
18) D.R. Clarke
19) M.I. Heggie
20) V.I. Nikitenko

21) K. Maeda
22) A. Umerski
23) Z. Laczik
24) Y. Maeda
25) P. Charsley
26) Yu. Schreter
27) J. Steeds
28) D.J. Stirland
29) D.A. Smith
30) C. Kisielowski-Kemmerich

31) H. Gottschalk
32) J. Rabier
33) J. Czernuszka
34) M. de Couteau
35) C.J. Humphreys
36) A.G. Cullis
37) R. Beanland
38) R.C. Pond
39) F. Louchet
40) J. Thibault-Dessaux

41) R. Jones
42) P.D. Warren
43) T.S. Fell
44) B.Ya. Farber
45) A. George
46) J.L. Demenet
47) C.D. Qin
48) N. Maeda
49) J. van Hellemont
50) J.L. Farvaque

51) G. Michot
52) P. Wurzinger
53) R. Gleichmann
54) W. Seifert
55) A.P. Sutton
56) M. Brohl
57) H. Teichler

Inst. Phys. Conf. Ser. No 104: Chapter 1
Paper presented at Int. Symp. on Struct. Prop. Disloc. Semicond., Oxford, 1989

1

Structure of deformation-induced bulk and grain boundary dislocations in a silicon $\Sigma = 9(122)$ bicrystal. A HREM study

J. Thibault-Desseaux, J.L. Putaux

Département de Recherche Fondamentale/Service de Physique.
Centre d'Etudes Nucléaires. 85X-38041 Grenoble - FRANCE.

ABSTRACT: HREM seems to confirm that the structural unit concept could be applied to describe the GB atomic structure and the bulk dislocation core as well. However the interaction of these defects with point defects leads to structural changes detected by HREM. This might be accounted for to explain mechanical and electrical behaviour of the material.

1. INTRODUCTION

The defects arising from either the growth of silicon crystals or mechanical constraints occurring during the processing greatly affect the properties of the devices. It is therefore important to determine their structure in order to understand and control their behaviour. High resolution electron microscopy (HREM) is one of the powerful means to characterize the structure of the defects. However, HREM investigations are mainly restricted to defects whose structure is periodic along the electron beam, even if attempts could be made to explore the third dimension (Bourret, Rouvière, 1988). The purpose of this paper is to examine the structure of the defects induced during the deformation of an initially perfect $\Sigma=9(122)[011]$ Si tilt bicrystal. The paper will be divided into two main parts, each of which corresponding to the description of two distinct areas of the bicrystal. Firstly, the region close to the grain boundary (GB) will be considered. It is in an advanced state of deformation as confirmed by the existence of various locks which result from the interaction of the gliding dislocations. As a matter of fact, these dislocation configurations look similar to those existing in as-grown subgrain boundaries (SGB). Furthermore, HREM revealed the very early stages of the climb of dissociated dislocations. Secondly, the GB structure will be examined. As deformation induced dislocations enter the GB, the GB undergoes large structural changes. The incoming dislocation decomposes to GB dislocations, which move along the interface by glide and climb, making the stress at the tip of the dislocation pile-up release but impeding the transmission of slip. The structural evolution of the GB will be described.

HREM experiments were carried out on a JEOL 200CX electron microscope with a HR pole piece, the [011] common tilt axis of the bicrystal lying parallel to the optical axis. The deformation conditions were chosen so that the activated slip systems were appropriate for HR observations: the dislocations in both grains could be viewed edge-on (Elkajbaji,et al. 1987).

The deformed samples were provided by A. George and A. Jacques from the Ecole des Mines in Nancy (France). A. George (this conference) will report the studies on the overall plasticity of Si and Ge bicrystals.

2. DISLOCATION CONFIGURATIONS.

The core structure of some dislocations was already investigated by HREM.
It was studied either in as-grown low angle bicrystals (Bourret, Thibault,
Lancon 1983) or in deformed crystals (Olsen et al.,1981, Anstis et al,
1981). It might be argued that the complex configurations found in SGB's
could be far from the ones expected in deformed crystals. However, new
investigations of a deformed high angle bicrystal showed the similarity
between the configurations close to the GB and the ones found in SGB's. In
the following, the structure of these configurations will be reported.

2.1. 60° and screw dislocations.

These are the basic dislocations created by deformation. The structure of
the 60° dislocation (60°D) is shown in Fig.1a. It is dissociated to two
Shockley partials : one 90°D and one 30°D bounding an intrinsic stacking
fault (SF). The screw dislocation would be dissociated to two 30° partials.
The controversy to know whether the partials belong to the glide set or the
shuffle set, will not be detailed here (Louchet, Thibault-Desseaux 1987).
The same 60° D embedded in a SGB, has a similar structure, the SF width depending on the misorientation angle (Bourret, Desseaux, 1979). The main
feature(Fig.1b) is that, at least in projection, the 30° and the 90° partials are made up of well defined structural units (Papon, Petit,1985) that
will be encountered in the description of other dislocation cores. As
already seen by weak beam experiments (Ourmazd et al. 1983), the climb of
the dissociated 60°D was clearly revealed by HREM (Thibault-Desseaux,
Putaux, Kirchner, 1989a).Fig.1b shows the agglomeration of an interstitial
perfect loop on the 90° partial. Unlike Decamps et al.(1983) or Cherns
(1984), one never detected an interstitial Frank loop (with an extrinsic
stacking fault) . This would have involved a a/6<011> stair-rod at the edge
of one intrinsic and one extrinsic stacking faults , both making an obtuse
angle. Climb affected most of the dislocations stopped close to the GB or
blocked in dipoles. This phenomenon is three dimensional and consequences
on bond reconstruction might be pointed out.

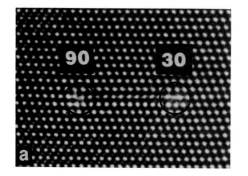

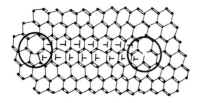

Fig.1 a) 60° dissociated dislocation
b)structural units in the 60°D core,
c)growth of an interstitial perfect
loop on the 90° partial.

2.2. Lomer-Cottrell lock: a/2<011>.

Two 60° dislocations can react and give a Lomer-Cottrell lock (LC). Unlike the theoretical predictions (Korner et al.1979, Fig.2c) this lock was never found asymmetrically dissociated. As asserted by Thibault-Desseaux, Putaux, Kirchner(1989) , this certainly came from the energy of the stair-rod core. As a matter of fact this stair-rod a/6<011> was expected to appear during the agglomeration of the interstitial Frank loop on the 90° partial (§2.1). However, in the LC lock case, the stair-rod would have been at the edge of two intrinsic SF's. The structure of the LC lock is shown on Fig.2a. It looks like the Hornstra model (1959), made up of a five-seven atom ring structural unit called L by Bourret and Bacmann(1987) (Fig.2b).

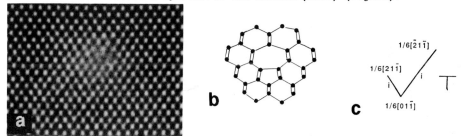

Fig.2 :a) L-C lock, b)core model, c)theoretical dissociation.

2.3. Hirth lock: a<100>.

This lock has been listed by Korner et al.(1979) and was also detected in SGB's . In SGB's the core was always decorated by a large precipitate which looks like hexagonal silicon (Bourret, 1987)(Fig.3a). The same occurred in the core of this deformation induced lock (Fig.3b). It has to be pointed out that platelets of hexagonal Si were also detected in Si deformed under high hydrostatic pressure (Demenet et al.,1987). The precipitation impeded the complete dissociation of the lock; however one "half" of the theoretical dissociation took place, by emitting a glissile 90° Shockley partial and leaving the remaining a/6<411> partial blocked by the precipitate. The same semi-dissociation was also detected in SGB's (Fig.3c). The existence of a <411> strain might have certainly favoured the growth of a <311> interstitial precipitate (Bourret, 1987).

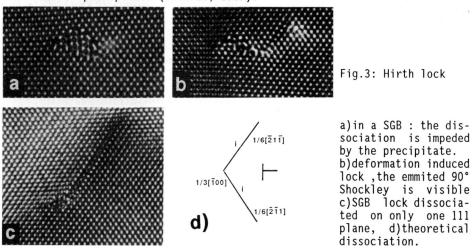

Fig.3: Hirth lock

a)in a SGB : the dissociation is impeded by the precipitate.
b)deformation induced lock ,the emmited 90° Shockley is visible
c)SGB lock dissociated on only one 111 plane, d)theoretical dissociation.

2.4. B2 lock: a<010>.

This lock whose Burgers vector is at 45° with respect to [011] was listed
by Korner et al.(1979). It was detected in deformed Si close to the GB. Its
dissociation (Fig.4a), which involves a a/3<011> stair-rod and two 30°
Shockley partials (with the same screw component), is the theoretical one
(Elkajbaji, Kirchner, Thibault-Desseaux,1988). The structure of the stair-
-rod, at the edge of two intrinsic SF's making an acute angle, is drawn on
Fig.4b. The core is built up with one distorted 5-7 atom ring SU which differs from the L unit as well as from the 90° partial SU (cf fig.1c).

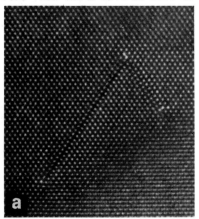

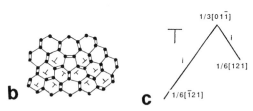

Fig.4 :a) Dissociated B2 lock, b) core
structure of the stair-rod c) theoretical dissociation.

2.5. Antidipole: a/2<211>.

This configuration (fig.5a) might result from the interaction of two 60°
dislocations (with opposite screw component) gliding on parallel 111
planes. Owing to the deformation conditions, this could occur in the tension tests, and this was detected. The two 60° dislocations were stopped by
the GB. In order to release the screw components, the two 30° partials
reacted by emitting a glissile Shockley partial. The resulting configuration is completely sessile (Fig.5c). Two stair-rods have appeared. One of
them is the a/6<011> whose core is thought to be highly energetic (§2.2)
but whose existence in this configuration is probably needed to release the screw stress. Its structure is shown on fig 5.e. It is composed of a
distorted 5-8-5 atom ring unit which might imply a bond reconstruction
along the dislocation line. The other stair-rod is a/3[100] at the edge of
two intrinsic SF's making an obtuse angle. The core is built up with two
different 5-7 SU's : one is a L unit in an uncommon stacking, whereas the
other is the 90° partial SU (Fig.5f). The same antidipole was detected in
SGB's (Fig.5b). The width of the involved SF's is controlled by the
misorientation angle and this configuration completely releases the screw
components of the 60°D's which originally formed the SGB (Bourret,
Desseaux,1979). As in §2.4 a semi-dissociation may occur: one of the
partials would be a/6<411> in the core of which grew a <311> defect,
whereas the core of the second partial a/3<111> might remain dissociated or
not (Fig.5d).

3. DISLOCATIONS IN A HIGH ANGLE GB.

The structure of periodic high angle grain boundaries has been extensively
studied. Geometrical approaches have been developed (Bollmann, 1970, Pond,
Bollmann 1979). The concept of structural units (SU's) has been introduced

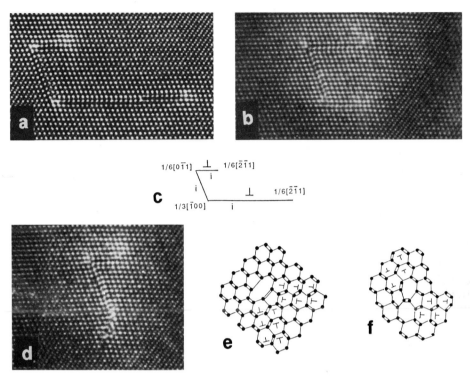

Fig.5 a) Deformation induced antidipole, b) SGB similar antidipole c) configuration, d)SGB semi dissociated antidipole,) a/6<011> and) a/3<100> core structures.

to provide an atomistic description of GB's in coincidence position (Sutton, Vitek 1983). This was applied to <011> Si tilt boundaries (Papon et al.- 1985, Kohyama et al.1986). A misorientation with respect to this special position is accommodated by grain boundary dislocations (GBD's), whose Burgers vectors belong to the DSC (discrete shift complete) lattice. This lattice is made up of linear combinations of the lattice vectors belonging to one and the other grain. The main consequence is that the GB structure remains the same on both sides of a GB defect whose Burgers vector is a DSCL vector. Previous HREM studies were performed on as-grown Ge bicrystals with $\Sigma=27$, $\Sigma=11$, and $\Sigma=3(122)$ <011> tilt GB's (Bourret, Bacmann 1987). They showed that the structure of these GB's could be described with the help of a limited number of SU's and that bond reconstruction is the general rule even if it has to take place along the <011> axis. The secondary dislocations in $\Sigma=11,33,41,51$ [011] as-grown or deformed Ge tilt bicrystals were also studied by HREM (Skrotzki et al. 1988). Owing to a twist component the core atomic structure was difficult to achieve. Nevertheless the Bv's were determined and were found to be non-primitive DSC vectors. Previous studies of the interaction of dislocations with the $\Sigma=9$ Si bicrystal were carried out by X-Rays topography and 1 Mev in-situ experiments (Baillin et al.1987, Jacques et al.1987). The very early steps of the entrance of dislocations into the GB were examined at atomic scale by HREM (Elkajbaji, Thibault- -Desseaux, 1988, Thibault-Desseaux,Putaux,Jacques,Elkajbaji, 1988) : it was shown that the incorporation mechanism involved both glide and climb. In the following, the structure of the residual GBD's and the changes in the GB structure will be presented.

3.1. Isolated GB dislocations.

The 60° dislocation on primary glide planes entered the GB and then decomposed to three GBD dislocations whose Burgers vectors were elementary vectors of the $\Sigma=9$ DSC lattice (Fig.6c). In compression, these GBD's were b_c, b_{30}^1, b_g whereas in tension they were $-b_c$, b_{30}^3, $-b_g$ (the incoming dislocation glides in grain I). As the deformation conditions were compatible, an equal number of dislocations came from grain II and decomposed to either b_c, b_{30}^2, $-b_g$ (compression) or $-b_c$, b_{30}^4, b_g (tension). As a consequence, the shear component induced by the b_g and the b_{30} cancelled. As the strain increased, the number of bulk dislocations into the GB increased, leading to an increasing number of GBD's whose Burgers vector was normal to the GB plane. This induced a variation of the misorientation angle.   While the strain remained relatively low (<1%) and the deformation temperature remained below 850°C, the elementary GBD's could be isolated and their structure determined. The Burgers vector (Bv) of each GBD and the associated step height could be determined as proposed by King and Smith (1980).

3.1.1. b_c and $-b_c$.

These two basic GBD's are shown on Fig.6a,b. The Bv is normal to the GB plane and there is no associated GB step. The b_c structural unit (produced in compression) is a boat-shaped six atom ring (T unit) inserted in the middle of one of the two 5-7 SU's (L-L') which characterize the structure of a $\Sigma=9$ period; L' is related to L by a glide mirror symmetry. The $-b_c$ SU (produced in tension) is composed of two crystal six atom rings (C unit) inserted in a 5-7 SU of the GB. The decomposition of the incoming dislocation requires climb. Consequently, the b_c core might contain jogs and this certainly would play a role on the electronical properties of the GB.

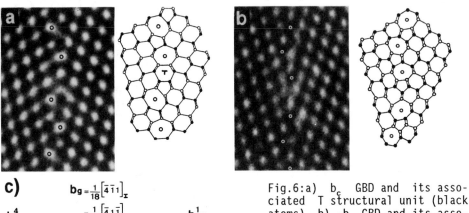

c) $b_g = \frac{1}{18}[4\bar{1}\bar{1}]_I$

$= \frac{1}{18}[\bar{4}1\bar{1}]_{II}$

$b_c = \frac{1}{9}[\bar{1}2\bar{2}]_I$

$= \frac{1}{9}[12\bar{2}]_{II}$

Fig.6:a) b_c GBD and its associated T structural unit (black atoms), b) $-b_c$ GBD and its associated SU (white atoms). c) DSC lattice and elementary DSC dislocation Bv's.

3.1.2. b_{30}^1, b_{30}^2, b_{30}^3, b_{30}^4.

In the same way, these 30° GBD's result from the entrance and the decomposition of a 60°D either in compression or in tension. The dislocations b_{30}^1 and b_{30}^3 are presented on Fig.7a,b. They were produced respectively in compression and in tension by the incorporation of a grain I 60°D. They have an associated step of the lowest height respectively $h= +1.75h_0$ and $h= -1.75h_0$ with $h_0=|b_c|= a/3$. If the 60°D came from grain II, b_{30}^2 ($h=-1.75h_0$) and b_{30}^4 ($h=+1.75h_0$) were left respectively in compression and in tension. The structure of the two latter dislocations could be deduced from the former ones by a mirror symmetry with respect to the GB plane. It has to be pointed out that the structural units describing the core are comparable to the ones appearing in the bulk 30° partial. Moreover, as in the bulk, the reconstruction of the bonds might lead to a doubling of the periodicity along the line. All these dislocations were also observed with an associated step whose height was ± $2.75h_0$. Fig.7c,d shows the corresponding GBD's:b_{30}^1 ($h= -2.75h_0$) and b_{30}^3 ($h=+2.75h_0$). In fact, due to the GB symmetry, only four different structures are needed to depict the eight possible 30° GBD's. As mentioned in the b_c case these dislocations result from the climb decomposition of the incoming dislocations and their presence could lead to electronic effects. Furthermore, as they are linked to a GB step, the GB migrates during their motion along the GB.

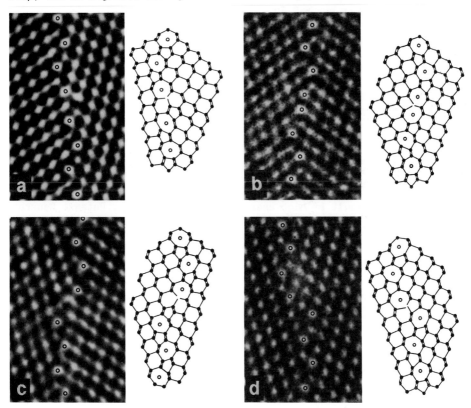

Fig. 7. a,b,c,d : b_{30}^1 ($h=1.75h_0$), b_{30}^2 ($h=-1.75h_0$), b_{30}^1 ($h=-2.75h_0$), b_{30}^2 ($h=2.75h_0$) and their core structure. (black atoms).

3.1.3. b_g, $-b_g$.

These dislocations, whose Bv is parallel to the GB plane, are glissile along the GB. They are characterized by a supplementary 5-7 atom ring (L unit) inserted in the original Σ=9 GB (Fig.8) resulting in the ..LL'LLL'LL'.. sequence. This induces a GB step whose height is $+h_0$ and $-h_0$ respectively for $-b_g$ and b_g. As they move, they can annihilate together or react with other GBD's. Furthermore the motion of b_g produces the GB migration. The dislocation core might be completely reconstructed. However, since it moves, it could react with point defects on the same way as the bulk glissile dislocations could do.

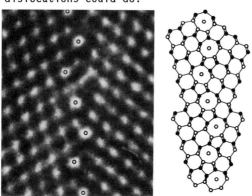

Fig.8 : the glissile b_g GBD and its core structure. (black atoms).

3.1.4. Non elementary DSC dislocations.

At a relatively low temperature (<800°C) complex configurations have been found in the GB. The simplest residues detected in tension and in compression will be described.

In tension, residues stemming from the incomplete decomposition of the incoming 60°D (from grain I) were detected. b_r^3 (h=-1.75h_0) and b_r^4 (h=-2.75h_0) are shown on fig 9a and b. In both cases the core exhibits a special feature which involves T units and L' units (up side down L units). In fact, the core might be described as dissociated to two "incomplete" GBD's linked by a GB stacking fault of high energy (Thibault-Desseaux, Putaux, Bourret, Kirchner, submitted). This SF was never found extended on more than one GB period. These imcomplete dislocations differ from the so-called GB partial dislocations in that they do not separate GB regions of different structures but of the same energy or they do not border GB facets. At high temperature, these configurations decompose by climb to elementary DSCD's. In compression, the complex residues left by an incomplete decomposition of the in-coming dislocations are different from the ones detected in tension conditions. The compact or slightly dissociated configurations of the equivalent residues b_r^1(h=1.75h_0) or b_r^2(h=2.75h_0) have never been detected even at 750°C. For instance, b_r^1 was found decomposed by climb in b_{90}^1 (h=-h_0) and b_{30}^2 (h=2.75h_0). Moreover, b_{90}^1 often exhibits a dissociation in incomplete GB dislocations. The core structure of the global configuration is built up with a combination of T units and L' units (Fig.9c). Besides, the non dissociated core (Fig.9d) exhibits a new SU made of two closely joined up L units. At higher temperature, the complete decomposition of the b_r residue is achieved by climb. It occurs by local atomic rearrangements and always results in elementary GBD's associated with a GB step of the lowest height. One has to emphasize that the "compact" original configurations are always built up with GBD's whose associated step is the highest leading to the idea that the step energy does not govern the configuration stability : at this stage of the dislocation incorporation the energy of the global configuration is certainly the main parameter. However, when the GBD's are very far from each other, they are always linked to the lowest height associated step in order to minimize the surface energy.

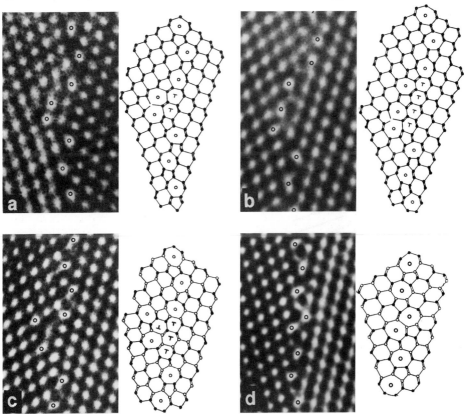

Fig.9: Non elementary GBD's. a) b_r^3 and b) b_r^4 with a dissociated core. c) and d) $b \frac{1}{90}$ respectively with a dissociated and a compact core. (black atoms).

3.2. Structural evolution of the GB.

At temperatures higher than 850°C and at strains higher than 2%, the increasing number of incoming dislocations leads to an appreciable variation in the misorientation angle between the two grains. Large climb occurs, resulting in a secondary dislocation network superimposed on the original GB. In compression, the misorientation angle increases and the original $\Sigma=9$ GB tends to the $\Sigma=3$ GB, whose period structure is composed of two T units. In tension the angle will decrease and the GB tends to the $\Sigma=1$ perfect crystal, whose period along <100> is made up of C units. The main point is that the structure of the defects which only induce a tilt angle variation, might be described by adding simple structural units. Moreover, these are the basic SU's of the nearest coincidence GB's having the lowest energy namely, $\Sigma=3$ and $\Sigma=1$ in our instance (cf. Fig.6a and b).

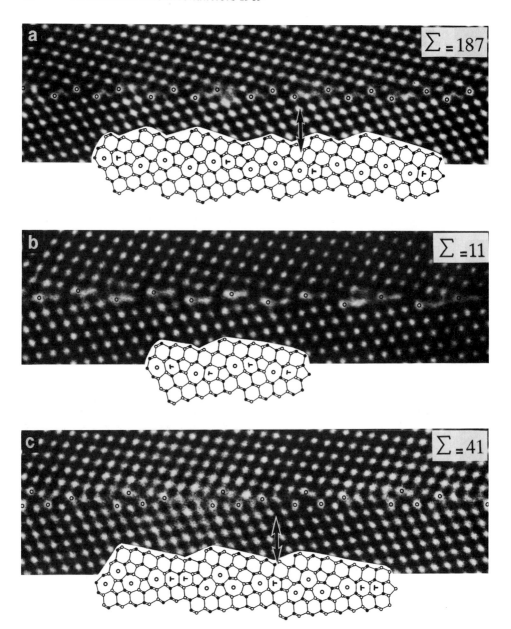

Fig.10 : a) Σ=187 : two different structures, mirror related and separated by a defect corresponding to a missing L unit, b) perfect Σ=11 with a glide mirror symmetry, c) Σ =41 : the arrow indicates one defect corresponding to one missing T unit; the structure exhibits the special SU made of a group of three closely connected L units.

As the angle varies, the GB structure can be related to different Σ values. A 6.8% compression strain was applied to the original $\Sigma=9$ GB ($\theta=38.94°$) and owing to the inhomogeneity of the deformation, different GB's were clearly recognizable on the same sample. Moreover, subgrainboundaries appeared in one or the other grain separating GB domains of different Σ (Thibault--Desseaux, Putaux, Kirchner 1988). We will present here three kinds of Σ : $\Sigma=187$, $\Sigma=11$ and $\Sigma=41$ which contained defects as well.

3.2.1 $\Sigma=187$ (599) <011> - $\theta=42.88°$.

When one T unit is introduced per two $\Sigma=9$ periods, half a period of $\Sigma=187$ is obtained (Fig.10a), then described by the sequence (5-T-7)L'LL'. Since the T unit introduces a a/4[011] shift, the effective $\Sigma=187$ period is the doubled previous sequence. The T unit can be put either in one L or the other L' unit of the original $\Sigma=9$ period. This can lead to two distinct structures of the same GB. Both structures have neither a mirror nor a glide mirror symmetry, nevertheless they are related by a mirror symmetry. From the energetical point of view, both structures are equivalent. As the study of the energy of these configurations has not been completed, one cannot actually decide what is the state of equilibrium (i.e one or the other structure, or a convenient mixture of both), especially as the stress limit conditions are not well known. Nevertheless areas of both types were observed , separated by defects. Fig.10a shows such an example where a L unit is missing in the original sequence (indicated by an arrow). On each side of this defect, one can observe each possible $\Sigma=187$ GB structure : one with T units and the other with T' units.

3.2.2. $\Sigma=11$ (233) <011> - $\theta=50.48°$.

When two T units are introduced per one $\Sigma=9$ period, this makes one $\Sigma=11$ period. This structure (5-T-7)(5-T-7)' was observed in our deformed speci- mens (Fig.10b). It has a glide mirror symmetry as in the $\Sigma=9$ case. As a matter of fact, the structure is not the same as the one detected in an as- -grown bicrystal (Bourret, Bacmann, 1987). Unlike in our case, the period was doubled, with respect to the one detected in the deformed bicrystal. Moreover a second structural unit and the mirror symmetrical equivalent unit were observed : this unit was similar to the one associated with the non dissociated b $_{90}^{1}$ (h=-h $_0$) defect in $\Sigma=9$ (cf Fig.9d).

3.2.3. $\Sigma=41$ (344) <011> - $\theta=55.88°$.

If the GB absorbs more dislocations, the misorientation angle keeps increasing. The GB presented Fig.10c illustrates the fact that the previous examples could not be straightforwardly extrapolated. $\Sigma=41$ could have been obtained by the introduction of two T units per L unit of $\Sigma=9$. The sequence would have been (5-T-T-7)(5-T-T-7)'. Fig.10c shows that the structure actually observed is not the expected one. Owing to a twist component, one has to be very cautious to analyse the HREM pictures. Nevertheless, a simple model of the period may be proposed. It exhibits two structurally distinct parts : one part corresponds to the expected sequence 5-T-T-7 whereas the second one involves a new structural unit made of three closely joined up L units which has to be compared to the additional SU found in the as-grown $\Sigma=11$ GB. Moreover, a defect corresponding to a missing T unit is presented on the image (indicated by an arrow). On both sides of the defect, the GB structure remains unchanged.

4. CONCLUSION.

The structure of the defects introduced by deformation or found in as-grown silicon <011> tilt bicrystals has been studied with the help of HREM. It was shown that the region close to the GB contains locks arising from either glide or climb or even precipitate growth. Furthermore, these locks are similar to the ones found in the SGB's. From this work, we have to emphasize that the incorporated dislocations act as sinks as bulk disloca-tions do but unlike the perfect Σ=9 GB. Though it is illustrated that complex interactions between point defects, dislocations and the GB take place, it turns out that the atomic description in terms of structural units remains valid, at least in projection. As the interaction mechanisms are undoubtly three dimensional, the existence of intrinsic localized point defects cannot be excluded resulting in electronic effects. This work sug-gests also that the defects aligned along the <011> axis would be described by a limited number of SU's. However, it was shown in the deformation indu-ced Σ=41 that the SU's are not necessarily the simplest ones. Bourret and Bacmann came to the same conclusion in the as-grown Σ=11 GB. The interpre-tation we put on these results is that, these configurations might account for local equilibrium conditions which are temperature and internal stress dependent. Though the study of the configuration energy by using suitable atomic potential must be carried out with caution, it is under progress.

Anstis, P.B. Hirsch, C. Humphreys, J. Hutchinson, A. Ourmazd, Inst. Phys. Conf. (1981) 60, 15.
Baillin X., Pelissier J., Bacmann J.J., Jacques A., George A., Phil. Mag. (1987)A55, 143.
Bollmann W, Crystal Defects and Crystalline Interfaces (1970), Springer.
Bourret A., J. Desseaux, Phil. Mag. (1979) A39, 419.
Bourret A.,J. Thibault-Desseaux,F. Lancon Jour. Phys.(1983) 44, C4-15.
Bourret A., J.J. Bacmann, Rev. Phys. Appl. (1987) 22, 563.
Bourret A., Inst. Phys. Conf. (1987) 87, 39.
Bourret, J.L. Rouvière, (1988)Polycrystalline Semiconductors, ed. J.H. Werner, Springer Verlag, Berlin.
Cherns D., Dislocations, ed P. Veyssière, L. Kubin, J. Castaing. (editions du CNRS, Paris).
Decamps B., Cherns D., Condat M., Phil. Mag. (1983) A43, 123.
Demenet J.L., Rabier J., Garem H., Inst. Phys. Conf.(1987) 87, 355.
Elkajbaji,Thibault-Desseaux J.,Martinez-Hernandez M.,Jacques A.,George A., Rev. Phys. Appl. (1987) 22, 569.
Elkajbaji M., J. Thibault-Desseaux, Phil.Mag. (1988)A58, 325.
Elkajbaji M., Kirchner H., Thibault-Desseaux J.,Phil.Mag.(1988) A57,631
Hornstra, Physica, (1959) 25, 409.
Jacques A., George A., Baillin X., Bacmann J.J., Phil. Mag. (1987)A55,165.
King A.H., Smith D.A., Acta Cryst. (1980) A36, 335.
Kohayama M., Yamamoto R., Doyama M., Phil. Stat. Sol. b (1986) 138, 387.
Korner A., Schmid H., Prinz F., Phys. Stat. Sol. a (1979) 51, 613.
Louchet F., Thibault-Desseaux J., Rev. Phys. Appl.(1987) 22, 207.
Olsen A., Spence J., Phil. Mag. (1981) A43, 945.
Ourmazd A., Cherns D., Hirsch P., Inst. Phys. Conf. (1983) 60, 39.
Papon A.M., Petit M., Scripta Met. (1985) 19, 391.
Pond R.C., Bollmann W., Phil. Trans. Roy. Soc. Lon. (1979), 292, 449.
Strotsky W.,Wendt H., Carter C.B., Kohlstedt D.,Phil.Mag.(1988)A57, 383
Sutton A., Vitek V., Phil. Trans. Roy. Soc. Lon. (1983) A309, 1.
Thibault-Desseaux J., J.L. Putaux, A. Jacques, M. Elkajbaji, in Inter-facial Structure, Properties and Design, MRS series (1988) 122,293.
Thibault-Desseaux J.,Putaux J.L.,Kirchner H., Phil. Mag. (1989a, accepted)
Thibault-Desseaux J., Putaux J.L., Kirchner H., (1988)in Point, Extented and Surface Defects in Semiconductors, ed. Cavallini, Schroter, Benedek, Plenum Pub. (to be published).
Thibault-Desseaux J.,Putaux J.L.,Bourret A.,Kirchner H., Jour. de Phys., (1989b,submitted).

Inst. Phys. Conf. Ser. No 104: Chapter 1
Paper presented at Int. Symp. on Struct. Prop. Disloc. Semicond., Oxford, 1989

Models of the atomic and electronic structures of grain boundaries in silicon

Adrian P. Sutton

Department of Metallurgy and Science of Materials, Oxford University, Parks Road, Oxford, OX1 3PH.

ABSTRACT: An overview of recent attempts to model the atomic and electronic structures of grain boundaries in intrinsic silicon is presented. The methods that have been used are assessed critically and illustrated. The calculated electronic structures of various twin boundaries and stacking faults are discussed and compared with experimental measurements. A theoretical analysis is given of band tails at grain boundaries, and it is suggested that dopant atoms may be the cause of the experimentally observed band tails.

1. INTRODUCTION

High resolution electron microscopy observations in Si and Ge (see Bourret and Bacmann (1985) for a review) have revealed that the structural width of the grain boundary region is between 0.5 and 1 nm. It follows that models which treat the boundary as an amorphous or liquid like layer are inappropriate since all atoms within the layer are within 1 or 2 nearest neighbour distances from crystalline regions. Thus, the atomic environment in a (clean) grain boundary represents a compromise between the equally strong ordering tendencies of the nearby misoriented crystals. At certain misorientations there is a common periodicity in the two crystalline surfaces at the grain boundary. The grain boundary itself may then assume a periodic structure with a periodicity that is some multiple of the common periodicity in the two surfaces. These misorientations are usually described as coincident or commensurate. At other misorientations there is no periodicity in the boundary plane and it turns out that the boundary is then quasiperiodic (Gratias and Thalal (1988), Sutton (1988a), Sutton (1989a)). From the point of view of constructing atomistic models of grain boundaries the coincident grain boundaries are particularly convenient since they admit the use of periodic boundary conditions parallel to the boundary plane. By studying boundaries with longer and longer periods we can gain insight into the structures of the quasiperiodic or incommensurate boundaries, as described by Sutton (1988a). Grain boundaries in metals have been modelled extensively in this way with considerable success (Sutton and Vitek (1983a–c), Sutton (1984), (1988b), (1989b)). By contrast there has been relatively little modelling of the atomic structures of grain boundaries in covalently bonded semiconductors. The main reason for this is the lack of confidence felt in satisfactorily describing covalent bonding in the distorted atomic environment of a grain boundary. Two distinct approaches have been followed to treat bonding and interatomic forces in covalent semiconductors.

First there are models that use classical interatomic potentials, for example Phillpot and Wolf (1988) used the potential of Stillinger and Weber (1985) to model (110) and (111) twist boundaries in Si. These models have the advantages of simplicity and computational speed. Their weakness is that they are not grounded in quantum mechanics so that as their limitations become evident one cannot systematically improve the potentials. Thus there will be situations where any particular potential will work and others where it will not and one does not always know in advance whether the particular application one has in mind is going to succeed. For example, the Stillinger–Weber potential fails to predict the π–bonded chain structure for the (2×1) (111) surface of Si or the tilting of [1$\bar{1}$0] chains on the (1×1) (110) surface of Si as the lowest energy structures. The formation of π bonds on the (111) surface and the Jahn–Teller effect which tilts the chains on the (110) surface are examples of quantum phenomena that are beyond the capability of a simple classical potential. At a grain boundary, however, one might be inclined to be more optimistic since the requirement that the bonding is tetrahedral on *both* sides of the boundary region is a much stronger constraint than exists at a free surface. As will be shown in section 4 the Stillinger–Weber potential is able to predict the structures of certain grain boundaries in Si that have been studied by tight binding. However from a preliminary study, reported in section 3, it appears that the Stillinger–Weber potential fails to predict the structure of the simplest of all (001) twist boundaries. This suggests that models of grain boundaries in covalent semiconductors that are based on classical potentials are not always reliable.

The second approach is to solve Schrodinger's equation, in one form or another, and minimize the total (electronic and ionic) potential energy of the boundary. This is a much more ambitious undertaking and two schemes have been developed and applied to grain boundaries in covalent semiconductors. The first (Payne, Bristowe and Joannopoulos (1987); Tarnow, Bristowe, Joannopoulos and Payne (1989)) is based on the local density approximation of electronic density functional theory and the molecular dynamics method of Car and Parrinello (1985). This work is reviewed briefly in section 2. The second scheme is based on a tight binding description of bonding and is reviewed in section 3.

The rest of the paper is organized as follows. In section 4 we summarise the results of the atomistic models and compare them with experimental observations. In particular we focus on band tails and mid–gap states. In section 5 we give an analysis of the absence of band tails in the models and an explanation for the experimental observation of band tails at semiconductor grain boundaries is offered.

2.   LOCAL DENSITY CALCULATIONS

In the Car–Parrinello method one solves for both the electronic wave functions and the ion positions by molecular dynamics. Starting from a Lagrangian, in which the potential felt by an electron is the usual Kohn–Sham potential of density functional theory (Kohn and Sham (1965)), equations of motion for both the electron wave functions and the ionic coordinates are derived. Temperature enters the formalism through the real kinetic energies of the ions and fictitious kinetic energies assigned to the electronic wave functions in the equations of motion. In the applications to grain boundaries (Payne et al. (1987), Tarnow et al. (1989)) the real and fictitious kinetic energies are quenched from the beginning of the simulation and there is no "simulated annealing" at an elevated temperature. Thus there is the same risk as in a static simulation (where there is no kinetic energy) of getting trapped in local minima in the energy surface. For this reason a variety of starting configurations are tried in order to map out the energy surface. The more complete study was carried out by Tarnow et al. and we shall concentrate on that here.

Tarnow et al. (1989) modelled the (001) twist boundary in Ge that is obtained by a rotation about [001] of $2\tan^{-1}(1/3)$. This produces a periodic structure on the (001) boundary plane with a unit cell that is bounded by 1/2[310] and 1/2[$\bar{1}$30] vectors. There are 5 atoms per atomic layer in the unit cell. A superlattice of grain boundaries with equal misorientations but alternating signs was generated in order to impose three dimensional periodic boundary conditions. In this "supercell" geometry two (004) layers on either side of each boundary were relaxed and two further layers were inserted, between each bicrystal slab, in which the atoms were fixed at the perfect crystal positions. There are two distinct locations for the boundary plane in (001) twist boundaries in diamond lattices, differing by one (004) atomic plane. Tarnow et al.'s study was confined to one particular location of the boundary plane. As in all crystalline materials an important mode of grain boundary relaxation is the rigid body translation **t** of one crystal with respect to the other parallel to the boundary plane. Only in the limit of no periodicity in the boundary plane is **t** of no importance, although the boundary expansion remains important. For entirely geometrical reasons **t** may be defined uniquely within a square cell in the boundary plane bounded by $1/10[310]$ and $1/10[\bar{1}30]$ vectors. With the origin at the centre of this cell Tarnow et al. found that the lowest energy structure has a translation state of $\mathbf{t} = 1/20[\bar{1}30] \pm 3/100[\bar{1}30]$, an expansion of 0.1 ± 0.05 A and an energy of 482 ± 100 mJ/m^2. The relaxed atom positions were considered to be accurate to within 0.04A. (The *imprecision* for these numbers arises from the use of a small supercell, the k–point sampling, the finiteness of the basis set and the finite mesh used in the Fourier transform grid; there may be *inaccuracies* (which are difficult to quantify) arising from the local density approximation and the use of a local pseudopotential for Ge.)

The most interesting result from this study is the atomic structure of the lowest energy configuration shown in fig.1.

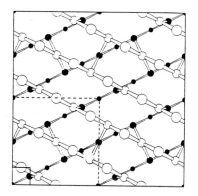 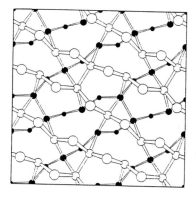

Fig.1. Atomic positions in planes normal to the [001] direction for 2 layers above (open circles) and below (full circles) an unrelaxed (left) and relaxed (right) (001) twist boundary in Ge. The boundary unit cell is delineated by broken lines. Atoms are connected by bonds if their separation is within 15% of the ideal bond length. From Tarnow et al. (1989).

Of the 10 atoms in each unit cell immediately adjacent to the geometrical boundary plane, which passes between two atomic layers, there are two five–fold and two three–fold coordinated atoms, the remaining six being four–fold coordinated. Moreover the symmetry of the structure is lower than would be obtained by simply rigidly translating two diamond lattices by $1/20[\bar{1}30]$. Jahn–Teller like distortions have taken place in the boundary plane which result in a mixture of dimerised and undimerised atoms and a lowering of the symmetry of the structure and of the grain boundary energy. This suggests that a simple classical potential such as that of Stillinger and Weber would not be able to correctly predict the structure of this boundary. In a preliminary study we have tested this conjecture and found that there is a relaxed structure with $\mathbf{t} = 0.96/20[\bar{1}30] + 0.05/20[310]$, with an expansion of 0.093A and an energy of 1343mJ/m^2. The structure is the nearest we have found to the configuration given by Tarnow et al. and of the 10 atoms nearest to the boundary plane two are three–fold and four are five fold coordinated (using the same bonding criterion as Tarnow et al.). However, a lower energy structure was found with $t = 1/10[210]$, an expansion of 0.244A, an energy of 1192 mJ/m^2 and four five–fold but no three–fold coordinated atoms. According to Tarnow et al. this translation state corresponds to a saddle point configuration, not a local energy minimum. We conclude that there appears to be a marked disagreement between the predictions of the Stillinger–Weber potential and the local density calculations for this grain boundary.

3.   TIGHT BINDING CALCULATIONS

The tight binding method has its roots in the development of quantum chemistry in the 1930's and was formulated for the solid state in a pioneering paper by Slater and Koster in 1954. The basic ansatz is to express the valence electron states as linear combinations of atomic like orbitals (LCAO). The method is semi–empirical because rather than evaluating integrals involving products of two orbitals and an atom centred potential these integrals, which are matrix elements of the electronic Hamiltonian, are regarded as parameters to be empirically fitted to the energy bands of the crystal in question. In the case of Si with a minimal sp^3 basis set the parameters are the four fundamental bond (or hopping or transfer or resonance) integrals between the s and p orbitals on neighbouring atoms and the s–p splitting energy difference. It is straightforward to set up a Hamiltonian matrix for any assembly of Si atoms and solve Schrodinger's equation either in k–space or real space. The *angular* dependence of bond energy and hence interatomic force results from the fundamental transformation properties of the p–orbitals under rotations and it reflects directly the quantum mechanical nature of the atomic orbitals. The *radial* dependence of the interatomic force results from an assumption, proposed by Harrison (1980), that the bond integrals decay as the inverse square of the bond length. The electronic contribution to the energy is augmented with a repulsive pair potential in order to ensure mechanical stability of the crystal. Sutton, Finnis, Pettifor and Ohta (1988) have given a detailed analysis of the tight binding model, including the physical meaning of the pair potential, that is based on the variational principle of density functional theory. The stability of the diamond cubic structure, as predicted by the tight binding model, relative to the hexagonal diamond, β–tin and close packed structures has been studied by Paxton, Sutton and Nex (1987) and Paxton (1988). The application of the model to grain boundaries in silicon was described briefly in Paxton and Sutton (1988) and in more detail by Paxton and Sutton (1989). Using a tight binding approach the following information is obtained: the relaxed atomic structure, the bond orders and local densities of states, and the hybridization state of each atom in the boundary. The weaknesses of the method are its empirical nature and the assumption of transferability of the Hamiltonian parameters from the perfect crystal to the grain boundary environment. Furthermore there is not a well defined self–consistency algorithm for the electronic density matrix although in narrow band gap semiconductors (i.e. narrow with respect to the total width of the valence and conduction bands) the imposition of local charge neutrality is a reasonable approximation to the full self–consistent solution (Sutton et al. (1988)). It turns out in silicon that imposing local charge neutrality does not affect either the atomic or electronic structures of the boundary significantly (Paxton and Sutton (1989)).

Thomson and Chadi (1984) carried out the first tight binding study of a grain boundary in Si using the same k–space method that Chadi pioneered for surfaces of Si (Chadi (1979)). They modelled the (221) twin, which is a second order twin and occurs frequently in polycrystalline Si. Like all the boundaries discussed in this section the boundary had first been observed by high resolution electron microscopy (in this case by Krivanek, Isoda and Kobayashi (1977)) and this provided the motivation to study it theoretically. From the structural point of view the boundary is not very exotic: it has a (1×1) periodicity, i.e. the periodic cell is 1/2[11$\bar{4}$] by 1/2[1$\bar{1}$0], all atoms in the boundary region are four–fold coordinated with the boundary region consisting of an alternating sequence of 5 and 7 membered rings and the boundary plane is a mirror glide plane. This boundary structure can probably be reproduced by any classical interatomic potential for Si which favours tetrahedral bonding, as Phillpot and Wolf(1988) have shown for the Stillinger–Weber potential. Thomson and Chadi sampled the electronic states of the boundary at 8 points in the irreducible part of the Brillouin zone. The interesting result of their paper is that *there are no electronic states in the minimum gap.* The minimum gap lies between the maximum of the valence band and the minimum in the conduction band. The work of Thomson and Chadi was confirmed by DiVincenzo, Alerhand, Schluter and Wilkins (1986) who modelled the same boundary with both local density functional theory and tight binding. The equilibrium structure of the boundary was obtained by tight binding and then DiVincenzo et al. carried out a local density pseudopotential calculation to study the electronic structure. They showed that there are localised states in the gap emanating from both the conduction and valence bands but, owing to the dispersion of the bands, these states do not lie in the minimum gap (see fig.2). Thus in the density of states associated with the boundary *there are no band tails in the gap but sharp (square root) band edges.*

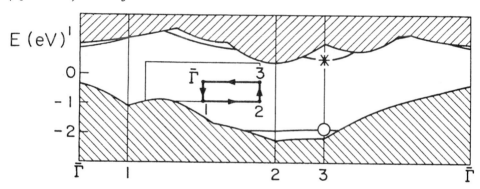

Fig.2. The projected energy bands for the relaxed (221) twin boundary as calculated in the local density approximation by DiVincenzo et al. (1986). The bulk crystal bands are hatched. The inset shows the interfacial Brillouin zone and the wave vectors that have been plotted. Note the absence of states in the gap below the conduction band minimum and above the valence band maximum.

Paxton and Sutton (1988,1989) modelled the (111) intrinsic stacking fault and the (111), (112) and (310) twins in Si with tight binding, all of which have been observed by electron microscopy. Independently, Kohyama, Yamamoto, Ebata and Kinoshita (1988a) and Kohyama, Yamamoto, Watanabe, Ebata and Kinoshita (1988b) modelled the (310) and (112) twins in Si. The same tight binding Hamiltonian parameters for Si were used in these two studies but whereas the Schrodinger equation was solved in k–space by Kohyama et al. it was solved in real space with the recursion method by Paxton and Sutton. Two further differences were (i) the pair potentials (ii) Paxton and Sutton imposed local charge neutrality as a form of self–consistency whereas Kohyama et al. did not attempt any form of self–consistency. Despite these differences the agreement between the calculated structures and energies of the (112) and (310) twins was excellent. The (112) twin is a particularly interesting grain boundary. It was shown by Bourret and Bacmann (1986), using high resolution electron microscopy and electron diffraction in Ge, that the boundary has a cm(2×2) structure, i.e. double the normal crystal periodicity along the [11$\bar{1}$] and [1$\bar{1}$0] directions in the boundary plane.

This observation came as a considerable surprise and was not predicted by any previous attempts to model the boundary with classical potentials (see Paxton and Sutton (1989) for a review). Perhaps the most important point to note is that the assumption that the boundary periodicity is always the same as the common crystal periodicity in the boundary plane because in reality it may be some integer multiple of this basic periodicity. This is a familiar phenomenon at surfaces of metals, semiconductors and insulators and it may well turn out to be equally widespread at grain boundaries. The relaxed structure of the $cm(2\times2)$ is shown in fig.3.

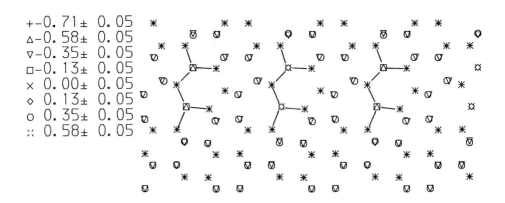

Fig.3. Relaxed structure of the $cm(2\times2)$ (112) twin in Si. The boundary is seen in projection along the tilt axis $[1\bar{1}0]$, and the boundary plane goes from left to right across the figure through the middle of the three delineated clusters. Atoms represented by '△' and '□' have pinched together along the projection direction and bonded as have atoms represented by '✕' and '◇'. The key gives the distance of the atoms above the plane of projection in units of the lattice parameter. Atoms within ±0.05 of these positions are represented by the same symbol.

The reason for the doubling of the period along $[1\bar{1}0]$ is clear: just as in the 30° partial dislocation alternate pairs of atoms that were separated by $1/2[1\bar{1}0]$ vectors pinch together and bond. In this way tetrahedral bonding is maintained in the boundary. The period is then doubled along $[11\bar{1}]$ by the sense in which the pinching together along $[1\bar{1}0]$ takes place switching from one $[11\bar{1}]$ period to the next. The *reason* for the switching of the reconstruction along $[1\bar{1}0]$ is unclear. If there were no switching the boundary symmetry would be $pm(1\times2)$. Paxton and Sutton modelled the $pm(1\times2)$ boundary and were unable to distinguish its energy from that of the $cm(2\times2)$ structure.

Owing to the relatively low energy resolution of the recursion method Paxton and Sutton were unable to determine whether band tails exist at the band edges of the boundaries they studied. However, Kohyama et al. (1988a, 1988b) plotted the projected band structures and showed that, as in the (221) twin, while there are localised states emanating from the valence and conduction bands they are buried under the band edges in the density of states of the boundary. Thus there are *no band tails in the density of states of the (112) and (310) twins*. However, a state was found in the minimum gap in the simulation of a high energy reconstruction of the (112) twin containing an atom with a dangling bond (Paxton and Sutton (1989)). The dangling bond gave rise to a prominent mid–gap state. The energy of this structure was more than 3 times that of the $cm(2\times2)$ structure and it has never been observed experimentally.

I have applied the Stillinger–Weber potential to the (310) and (112) twins to check whether its predictions agree with those of tight binding. Five metastable structures were found for the (310) twin, assuming the (1×1) periodicity, and the lowest energy structure was indeed the one found by tight binding and observed experimentally by Bacmann, Papon, Petit and Silvestre (1985). Paxton and Sutton (1989) considered five proposed reconstructions for the (112) twin and these were all found to be metastable with the Stillinger–Weber potential and tight binding. The ranking of the energies of the five structures was the same although the energies of any particular structure differed by up to a factor of two. The main differences in the structures were in the boundary expansions, which varied by up to 0.25A; otherwise the atomic positions were the same to within 0.05A. The energies of the $cm(2\times2)$ and $pm(1\times2)$ structures were 684 and 698 mJ/m^2 with the Stillinger–Weber potential and 340 ± 30 and 350 ± 30 mJ/m^2 with tight binding. These energy differences are too small to explain why the $pm(1\times2)$ structure is not observed experimentally. Nevertheless, we conclude that the Stillinger–Weber potential predicts the same energy ranking of metastable structures of the (310) and (112) twins as tight binding.

The intrinsic and extrinsic (111) stacking faults have been modelled by a number of groups (see Chou, Cohen and Louie (1985) and refs. therein). The surprising feature about these calculations is the existence of a localized state, associated with both types of fault, at about 0.1eV above the maximum of the valence band at the centre of the Brillouin zone. Both tight binding (Mattheiss and Patel (1981)) and local density calculations (Chou et al (1985)) obtain this result. The state is occupied and doubly degenerate and it has been observed using photoluminescence by Weber and Alexander (1983). There are also states pulled out of the conduction band but they do not lie below the minimum of the band. In view of the structural similarity of the (111) twin to the stacking faults it would seem quite likely that it displays the same state. At the (111) stacking faults and the twin there is virtually no variation in the bond lengths (less than 1%) or bond angles, but there are dihedral angle changes of 60°. As Paxton and Sutton (1989) have demonstrated the local density of states at atoms in the intrinsic fault and the (111) twin resemble closely the density of states in the wurtzite phase. It is suggested that the localised states, above the valence band and below the conduction band, that have been calculated at the stacking faults are features of two embedded layers of wurtzite phase in the diamond cubic structure rather than states that have been created by "structural disorder" in the faults. Similar states may be found in the polytypes of SiC.

4. SUMMARY OF RESULTS AND COMPARISON WITH EXPERIMENT

Atomic structures: The (111), (221), (310) and (112) twins in Si have been modelled by tight binding and the lowest energy structures obtained agree with available measurements of the translation states, expansions and direct observations of the boundary atomic structures in Si and Ge by electron microscopy (see Paxton and Sutton (1989)). The same structures are predicted by the Stillinger–Weber potential. The (001) twist boundary modelled by Tarnow et al. (1989) has not been studied experimentally. The Stillinger–Weber potential appears, from a preliminary study, not to predict the same structure as the local density calculations for this boundary.

Mid–gap electronic states and band tails: No states were found inside the minimum gap of any of the low energy structures of the (221), (112) or (310) twins. A prominent mid–gap state was found associated with a dangling bond in a *high energy* boundary structure. Comparison with experiment is particularly thorny because the models treat intrinsic silicon (i.e. containing no impurities at all) whereas experimental samples contain dopant atoms and possibly other impurities such as oxygen. The experimental results have been summarized clearly by Queisser and Werner (1988). Those authors report that there is almost always a *continuum of mid–gap states* associated with grain boundaries in bicrystals and polycrystals as seen by electrical measurements such as I–V characteristics, capacitance transients, photoconductance, temperature dependent zero–bias conductance, admittance spectroscopy, current transients and noise experiments. Long–range potential fluctuations within the boundary plane are caused by the charged midgap states and this allows them to be detected experimentally. Although no conclusive evidence has been found it is very likely that the midgap states are caused by segregated impurities such as oxygen. This is consistent with the absence of midgap states in the tight binding models noted above. Secondly Queisser and Werner (1988) report the observations of *exponential band tails* in all cases where it was possible to measure the density of states over a wide energy range. Werner (1989) has speculated that band tailing should occur at any grain boundary. He argues that it arises from the short range potential fluctuations in the boundary due to the misfit in the boundary region and it is therefore an intrinsic property of the boundary that would exist even in the absence of all impurities. This proposal is discussed in section 5.

5.   ON BAND TAILS AT GRAIN BOUNDARIES

Werner's argument for band tails at grain boundaries (Werner (1989)) goes as follows: in the grain boundary there is disorder in the bond lengths, angles and dihedral angles which result in short wavelength fluctuations in the electronic potential on a length scale of the lattice constant. These fluctuations are assumed to localise free carriers in the same way as they do in amorphous silicon and MOS interfaces and then cause the band tails. Thus the origin of the tails is seen as an intrinsic property of the grain boundary and any boundary exhibiting distorted bond lengths, angles and dihedral angles should show band tails.

In the (221), (112) and (310) twins simulated by tight binding there are bond length distortions of up to 6%, bond angle distortions of up to 26% and even larger distortions of the dihedral angles. These distortions give rise to perturbations in the off–diagonal Hamiltonian matrix elements (the hopping integrals) and also to the site–diagonal Hamiltonian matrix elements in the work of Paxton and Sutton (1989) through the requirement of local charge neutrality. These variations in the Hamiltonian matrix elements have a wavelength of the order of the interatomic spacing and they correspond to Werner's short wavelength fluctuating potential. While it is true that there are localised states emanating from the valence and conduction bands into the gap these states do not penetrate the minimum gap in the tight binding models. The reason for this, as is evident in fig.2, is that the states appear at only certain parallel wave vectors outside the bulk bands. Clearly if they appeared at all parallel wave vectors in the gap then they would give rise to some structure in the density of states in the gap. This raises two related questions. Why do the localised states that are found in the tight binding models appear at only certain parallel wave vectors and not others and is there some condition that has to be satisfied, such as a minimum bond length/angle distortion, before localised states appear at all parallel wave vectors? Before we address these questions we point out a tight binding model of a completely disordered interface between two dissimilar crystalline materials A and B and which *does not* give rise to band tails even though a three dimensional random alloy of A and B *does* display band tails. The point of this model is that it highlights the important role of the adjoining crystals at an interface in determining the structure of the density of states at the band edges. We shall see that the adjoining crystals have an equally important role for the density of states at the band edges in grain boundaries.

Consider first a disordered alloy of A and B atoms on a regular crystal lattice. It is assumed that the hopping integrals between neighbouring atoms are independent of whether they are between A or B atoms, but the site diagonal Hamiltonian matrix elements are ϵ_A or ϵ_B depending on whether the site is occupied by an A or a B atom.

Thus this tight binding model shows site diagonal disorder only. What is the electronic energy spectrum of the disordered alloy compared with the spectra of the pure A and pure B crystals? Thouless (1970) proved that the spectrum of the disordered alloy is contained in the union of the spectra of the pure A and B crystals. The proof is based on the mini–max theorem for Hermitian matrices. Having obtained bounds on the spectrum of the disordered alloy we still do not know whether these bounds are reached. There is an argument due to Lifshitz (1964) which shows that the bounds are reached and goes as follows. In a disordered alloy there is always some finite probability of finding a large cluster containing only A or only B atoms, so that there are always states in the alloy close to the states of the pure constituents. Clearly the density of states near the bounds should be vanishingly small since they correspond to large statistical fluctuations. Thus there are tails in the density of states of the disordered alloy which just reach the bounds on the spectrum. These tails are known as "Lifshitz tails".

Now consider a bicrystal of pure A and pure B crystals. We allow intermixing to take place across the interface but *the width of the disordered region at the interface is assumed to be small in comparison to the sizes of the adjoining A and B crystals.* We assume the same Hamiltonian applies to the interface as to the disordered alloy, i.e. we assume site diagonal disorder only in the interface. What is the spectrum of the bicrystal? Thouless's theorem still applies. The key difference with the three dimensional random alloy is that we no longer invoke Lifshitz's argument to establish that the bounds are reached since there are pure A and pure B crystals sufficiently far from the interface. Furthermore, the density of states at the bounds of the spectrum is not vanishingly small because the quantity of pure A and pure B material far exceeds (by assumption) the quantity of disordered material at the interface. *We conclude that an interface with site diagonal disorder only does not display band tails.* We should emphasize though that the "site diagonal disorder" does not mean that the diagonal matrix elements may assume any value but only ϵ_A or ϵ_B. I have not managed to generalize this simple argument to the more realistic case of off–diagonal disorder, which exists at a grain boundary. But it is instructive because it shows that in order to determine whether band tails exist at an interface one has to consider both the interface *and* the adjoining crystals. This point reappears in a somewhat different guise below.

Let us return to the question of why, in the tight binding and local density calculations, localised states associated with the grain boundary were pulled out of the bands only at certain parallel wave vectors but not others. In the following analysis it will be shown that whether a state is pulled out of a band at a particular parallel wave vector is dependent on both the perturbation to the bulk crystalline Hamiltonian that exists at the grain boundary *and* the electronic scattering properties of the adjoining crystals. We use the scattering theoretic formalism, as described by Pollmann and Pantelides (1980). The formal scattering properties of the adjoining crystals are conveniently encapsulated in their electronic Greens functions. Let G_I^0 and G_{II}^0 be the Greens functions of the infinite perfect crystals I and II. That is, $G_I^0(E) = (E-H_I^0)^{-1}$ and $G_{II}^0(E) = (E-H_{II}^0)^{-1}$ where H_I^0 and H_{II}^0 are the Hamiltonians of the infinite perfect crystals I and II. We imagine that the grain boundary is created by (i) orienting the two crystals in space at the required boundary misorientation, (ii) cleaving the two infinite crystals into four semi–infinite crystals on the appropriate boundary plane, (iii) bonding together one semi–infinite crystal of I to one of II to form a bicrystal, (iv) allowing the bicrystal to relax and finally (v) discarding the remaining two semi–infinite crystals. As shown by Pollmann and Pantelides (1980) steps (ii–iv) may be represented by a single perturbation V which does not extend very far into the adjoining crystals. The Greens function G of the relaxed bicrystal may be obtained from Dyson's equation using the perturbation V and the projection, G^O, of G^O_I and G^O_{II} onto the Hilbert space spanned by the bicrystal: $G = (1 - G^O V)^{-1} G^O$. The condition for a localised state to occur at an energy E in the gap is $Det(1 - G^O(E)V) = 0$. We see that the existence of band tail states in the gap is determined both by the perturbation V representing the creation and relaxation of the grain boundary *and* the Greens functions of the adjoining perfect crystals.

For the periodic grain boundary structures that have been modelled by tight binding and local density calculations the wave vector, q, parallel to the boundary plane is a good quantum number. Thus G^O is diagonal in a representation of two dimensional Bloch sums (called "layer orbitals" by Pollmann and Pantelides (1980)) parallel to the boundary plane. We denote a layer orbital by $<r|1,j,\alpha,q>$ where

$$<r|1,j,\alpha,q> = \frac{1}{\sqrt{N}_2} \sum_R \exp(iq.R) \, f_\alpha(r - \nu_1 - \tau_j - R)$$

Here N_2 is the number of 2–D unit cells parallel to the boundary (which is infinite), f_α is an atomic like–orbital, 1 is a layer index, τ_j is the position of the j'th atom in the 2–D unit cell in layer 1, ν_1 is the position of layer 1 relative to some arbitrary origin in the bicrystal and $\{R\}$ is the (infinite) set of 2–D lattice translation vectors parallel to the boundary. Because the perturbation V extends only over a finite number of layers the non–zero elements of V at any particular q in the layer orbital representation are small in number compared to the (infinite) size of the Hilbert space of the bicrystal. Let the non–zero elements of V be grouped into the matrix $V_{11}(q)$. Then the condition for a state to appear in the gap at energy E and wave vector q is

$$Det \left(1_{11} - G^O_{11}(E,q) V_{11}(q) \right) = 0$$

The size of this determinant is the same as that of the matrix $V_{11}(q)$. We can now see why states can be pulled out of the bands at certain wave vectors and not others. It depends on *two* quantities: the wave vector resolved perturbation $V_{11}(q)$ and the wave vector resolved Green's function of the adjoining perfect crystals $G^O_{11}(E,q)$. Since both of these quantities are wave vector dependent it is quite plausible that the above determinantal condition can be satisfied at certain wave vectors and not others.

Armed with this understanding of the absence of band tails in the models let us reconsider Werner's argument for band tails. The argument concentrates on the perturbation V_{11} and essentially ignores the fact that this perturbation has been applied to two crystalline media. That is, the argument does not distinguish between the perturbation being applied to an amorphous medium or a crystalline medium: hence the parallel that is drawn with amorphous silicon and MOS interfaces. What has been argued here is that the existence or otherwise of a localised state in a grain boundary depends on the perturbation V_{11} *and* the response of the adjoining crystals to that perturbation, which is both energy and wave vector dependent.

We cannot rule out the possibility that the grain boundaries that have been modelled to date are rather idealised even though they are fully relaxed. That is, they do not contain intrinsic structural defects such as steps, facet junctions, continuous curvature of the boundary plane, point defects or dislocations run in from the adjoining crystals. It may be that in a more realistic model, containing such structural defects, the perturbation V_{11} is such as to produce band tails.

But there is another possibile explanation for the experimentally observed tails that does not appear to have been considered: segregated dopant atoms. The *experimental evidence for the tails* (Werner and Peisl (1985)) comes from dc–conductivity and ac–admittance spectroscopy measurements on polysilicon films doped up to $10^{18} cm^{-3}$ with As or B by ion implantation. These measurements reveal exponential band tails extending from both the valence and conduction band edges. Grovenor, Batson, Smith and Wong (1984) measured the extent of segregation of As to grain boundaries in Si at temperatures between 700 and 1000C. They found an average binding energy of the As to the boundaries of 0.65eV and a boundary saturation limit of between 10 and 15 at.%. Rose and Gronsky (1982) measured an average binding energy of P to grain boundaries in Si of 0.33eV and a similar saturation limit. Grovenor et al. (1984) point out that segregation of dopant to grain boundaries has two effects: it increases the resistivity of the grain interiors by depleting the dopant concentration there and it can affect the electrical characteristics of the boundaries. Grovenor (1985) showed that the segregated dopant increases the boundary potential barrier very significantly. We speculate that the segregated dopant atoms introduce a range of states in the gap with energies that depend on the binding of electrons to the donor atoms in the various atomic environments they have segregated to. It is suggested that these states introduced by segregated dopant atoms exist and that they are the origin of the band tail states measured by Werner and Peisl (1985). This may be tested by finding out whether the measured band tails change with temperature since the extent of equilibrium segregation is temperature dependent.

ACKNOWLEDGEMENTS

I am grateful to Dr. Juergen Werner for sending me a preprint of Werner (1989) and to Dr. M. Kohyama for sending me a preprint of Kohyama et al. (1988b). The support of the Royal Society is gratefully acknowledged.

REFERENCES

Bacmann J.J., Papon A.M., Petit M. and Silvestre G. 1985, *Phil. Mag. A*, **51**, 697.
Bourret A. and Bacmann J.J. 1985, *Surf. Sci.*, **162**, 495.
Bourret A. and Bacmann J.J 1986, *Proc. JIMIS–4, Trans. Japan Inst. Metals, Suppl.*, 125.
Car R. and Parrinello M. 1985, *Phys. Rev. Lett.*, **55**, 2471.
Chadi D.J. 1979, *J. Vac. Sci. Technol.*, **16**, 1290.
Chou M.Y., Cohen M.L. and Louie S.G. 1985, *Phys. Rev. B*, **32**, 7979.
DiVincenzo D.P., Alerhand O.L., Schluter M. and Wilkins J.W. 1986, *Phys. Rev. Lett.*, **56**, 1925.
Gratias D. and Thalal A. 1988, *Phil. Mag. Letts.*, **57**, 63.
Grovenor C.R.M., Batson P.E., Smith D.A. and Wong C. 1984, *Phil. Mag. A*, **50**, 409.
Grovenor C.R.M. 1985, *J. Phys. C: sol. stat. phys.*, **18**, 4079.
Harrison W.A. 1980, *Electronic Structure and the Properties of Solids*, Freeman, San Francisco.
Kohn W. and Sham L.J. 1965, *Phys. Rev.*, **140**, A1133.

Kohyama M., Yamamoto R., Ebata Y. and Kinoshita M. 1988a, *J. Phys. C: sol. stat. phys.*, **21**, 3205.

Kohyama M., Yamamoto R., Watanabe Y., Ebata Y. and Kinoshita M. 1988b, *J. Phys. C: sol. stat. phys.*, **21**, L695.

Krivanek O.L., Isoda S. and Kobayashi K. 1977, *Phil. Mag.*, **36**, 931.

Lifshitz I.M. 1964, *Adv. Phys.*, **13**, 483.

Mattheiss L.F. and Patel J.R. 1981, *Phys. Rev. B*, **23**, 5384.

Paxton A.T., Sutton A.P. and Nex C.M.M. 1987, *J. Phys. C: sol. stat. phys.*, **20**, L263.

Paxton A.T. 1988, *Phil. Mag. B*, **58**, 603.

Paxton A.T. and Sutton A.P. 1988, *J. Phys. C: sol. stat. phys.*, **21**, L481.

Paxton A.T. and Sutton A.P. 1989, *Acta Metall.*, to appear.

Payne M.C., Bristowe P.D. and Joannopoulos J.D. 1987, *Phys. Rev. Lett.*, **58**, 1348.

Phillpot S.R. and Wolf D. 1988, *Mat. Res. Soc. Symp. Proc.*, **122**, 103.

Pollmann J. and Pantelides S.T. 1980, *Phys. Rev. B*, **21**, 709.

Queisser H.J. and Werner J.H. 1988, *Mat. Res. Soc. Symp. Proc.* **106**, 53.

Rose J.M. and Gronsky R. 1982, *Appl. Phys. Lett.*, **41**, 993.

Slater J.C. and Koster G.F. 1954, *Phys. Rev.*, **94**, 1498.

Stillinger F.H. and Weber T.A. 1985, *Phys. Rev. B*, **31**, 5262.

Sutton A.P. and Vitek V. 1983a, *Phil. Trans. Roy. Soc.*, **A309**, 1; 1983b, *ibid*, **A309**, 37; 1983c, *ibid*, **A309**, 55.

Sutton A.P. 1984, *Int. Met. Revs.*, **29**, 377.

Sutton A.P., Finnis M.W., Pettifor D.G. and Ohta Y. 1988, *J. Phys. C: sol. stat. phys.*, **21**, 35.

Sutton A.P. 1988a, *Acta Metall.*, **36**, 1291.

Sutton A.P. 1988b, *Mat. Res. Soc. Symp. Proc.*, **122**, 81.

Sutton A.P. 1989a, *Proc. Modulated Stuctures, Polytypes and Quasicrystals '88*, ed. D. Pandey, *Phase Transitions*, to appear.

Sutton A.P. 1989b, *Phil. Mag. Letts.*, **59**, 53.

Tarnow E., Bristowe P.D., Joannopoulos J.D. and Payne M.C. 1989, *J. Phys.: Condens. Matter*, **1**, 327.

Thomson R.E. and Chadi D.J. 1984, *Phys. Rev. B*, **29**, 889.

Thouless D.J. 1970, *J. Phys. C: sol. stat. phys.*, **3**, 1559.

Weber E.R. and Alexander H. 1983, *J. Phys. Paris (Colloq.)* C **4**, 319.

Werner J.H. 1989, in *Polycrystalline Semiconductors '88*, Springer, Heidelberg, in press.

Werner J.H. and Peisl M. 1985, *Mat. Res. Symp. Proc.* **46**, 575.

Inst. Phys. Conf. Ser. No 104: Chapter 1
Paper presented at Int. Symp. on Struct. Prop. Disloc. Semicond., Oxford, 1989

Symmetry and crystallography: implications for structure

R C Pond
Department of Materials Science & Engineering, The University of Liverpool, P.O. Box 147, Liverpool L69 3BX, U.K.

ABSTRACT: The crystallographic origin of interfacial line-discontinuities is shown to be the breaking of crystal symmetry. A general formulation for the characteristion of admissible discontinuities is derived, and is shown to encompass interfacial steps, facet junctions, four types of dislocations, disclinations and dispirations. The crystallographic prerequisites for the occurrence of such discontinuities is outlined in terms of the generalisation of the concept of misfit, called bicrystal disconnexion. In addition, the implications of this theory regarding the formation of domains in epitaxial layers, and epitaxial growth on vicinal and rough substrates is discussed.

1. INTRODUCTION

Interfaces in semiconducting materials are hosts to a diversity of linear features such as dislocations and steps, and an improved understanding of these configurations is very desirable. Ideally, we would wish to know their atomic and electronic structures in order to appreciate their roles in physical processes. The present work is a contribution towards this goal, and is concerned with the geometrical character of linear discontinuities. As the title suggests, the theory to be presented is based on the consideration of symmetry and crystallography. The present work takes into account the total spacegroup symmetry of each of the adjacent crystals, and this leads to a fuller understanding of the crystallographic origin of interfacial defects, showing, for example, that the range of potential defects is considerably broader than previously recognised.

The cornerstone of the present theory is an equation which can be used to predict the geometrical character of all linear discontinuities which are admissible in a given interface, and which separate energetically degenerate regions of that interface. All the parameters in this equation are geometrical operators, but it is helpful to distinguish between symmetry operators of the adjacent crystals, and another operator which defines the relation between the coordinate frames of these crystals. The presence of these two types of operators reflects the fact that the treatment is based on abstract symmetry theory, but only has predictive capability regarding the nature of allowed defects in a given interface when the relationship between the adjacent crystals' coordinate frames is known. The crystallographic operator which specifies this relationship can, at present, only be determined reliably by experimental means, i.e. by crystallographic investigation of the actual bicrystal in question.

As indicated above, the foundation of this theory of interfacial defects is based on abstract symmetry theory, and is described in section 2. The underlying principle is symmetry compensation, which means that if the symmetry at one structural level is reduced, a multiplicity of variant configurations (inter-related by the suppressed symmetry operations) arises at another structural level. In the present context, the formation of a bicrystal generally suppresses symmetry initially present in the component crystals, and hence a multiplicity of variant bicrystals arises. The coexistence of such variant bicrystals is the fundamental origin of interfacial defects, and in section 3 the concept that a defect is the discontinuity which arises between a pair of coexisting variants is described.

In section 4 we outline the characterisation and classification of admissible discontinuities. The most convenient procedure for this purpose, in the author's view, is to adopt the well known method of Volterra. By this means it is shown that discontinuities may have defect character, i.e. dislocation, disclination or dispiration character, but discontinuities without defect character, i.e. steps, and facet junctions, can also arise.

In addition to the defects outlined above, it is also important to be able to consider interfacial line defects from which extended defects, such as stacking-faults or domain boundaries, emanate into one or both of the adjacent crystals. This topic can be readily incorporated into the theory.

The implications regarding the structure of homojunction and heterojunction semiconductor bicrystals arising from the present theory are reviewed in sections 4, 5 and 6. Firstly, the deeper understanding of the origin of defects provided by the present approach in terms of symmetry breaking leads to a better appreciation of the possible occurrence of discontinuities in the various types of interfaces. The second aspect to be discussed is concerned with domain formation in epilayers, which can arise either during growth or subsequently by a process such as ordering. The final topic is epitaxial growth on vicinal and rough surfaces. Such surfaces are thought to comprise stepped low-index surfaces, and, according to the present work, the growing overlayer will accommodate these surface features by transforming them into interfacial discontinuities. The resulting array of interfacial defects may cause the orientation of the overlayer to differ from the nominal relationship, and this may in turn have implications regarding the evolution of defect structure in the epilayer.

2. SYMMETRY BREAKING AND INTERFACIAL DISCONTINUITIES

Invariances, or in other words symmetry, in physical laws imply the existence of conservation principles. In the case of geometrical symmetry the associated conservation principle is often referred to as the principle of symmetry compensation (see Shubnikov and Koptsik (1977) for example). According to this principle, if symmetry is suppressed at one structural level, it 'reappears' or is conserved at another level. This notion is familiar to materials' scientists in their analysis, for example, of phase transformations where a daughter phase arises having lower symmetry than that of its parent. All the symmetry operators which leave the parent phase invariant, but which are not exhibited by the daughter phase, are conserved in the sense that they inter-relate variant daughters. Thus, a composite arrangement of all the daughter variants can exhibit the same symmetry as that of the parent, and the initial symmetry is thereby conserved.

In the present context we imagine the formation of a bicrystal from two component crystals, which we refer to as being black, (μ), and white, (λ). Initially the white crystal exhibits all the symmetry operations in its spacegroup, $\Phi(\lambda)$, and similarly the black crystal exhibits the operations in its spacegroup, $\Phi(\mu)$. Now we imagine that surfaces are prepared on the white and black crystals, and that these are brought together so that a bicrystal is formed. In general, the bicrystal is a composite black/white object which exhibits lower symmetry than that of the separate component crystals; in fact, according to Pond and Vlachavas (1983) the bicrystal can be assigned a spacegroup designated $\Phi(b)$. Moreover, all the operators which belong to $\Phi(\lambda)$ or $\Phi(\mu)$, but which do not also belong to $\Phi(b)$ must, according to the principle of symmetry compensation, inter-relate variant bicrystals. Thus a multiplicity of crystallographically equivalent, and hence energetically degenerate, bicrystals exists, and elegant group theoretical methods can be used to establish the number of distinct bicrystals and the symmetry operations which inter-relate them (Shubnikov and Koptsik (1977)).

As a simple illustration we consider fig.1(a) which is a schematic representation of the formation of a $NiSi_2$:Si bicrystal. For simplicity we take the orientation of the unit cells of the two crystals to adopt parallel orientations, and to take up a relative position such that they have an atomic site in coincidence. Moreover, we also consider the lattice parameters to be identical. Regarding the NiSi$_2$ crystal as being white, and the Si as black, we have $\Phi(\lambda) = Fm\bar{3}m$ and $\Phi(\mu) = F\bar{d}\bar{3}m$, and for a bicrystal having an (001) interface as shown in fig.1(a), we also have $\Phi(b) = p2mm$. Clearly, many symmetry operations belonging to $\Phi(\lambda)$ and $\Phi(\mu)$ have been suppressed, and therefore a multiplicity of variant bicrystals arises. One of the suppressed operations, which was exhibited by the NiSi$_2$ before creation of the bicrystal, is 4_z^+ (i.e. rotation by 90° anticlockwise about [001]). This operation relates the bicrystal variant depicted in fig.1(a) to that shown in fig.1(b)), and it can be appreciated that these two objects are energetically degenerate since all corresponding interatomic spacings and environments are equivalent in the two objects.

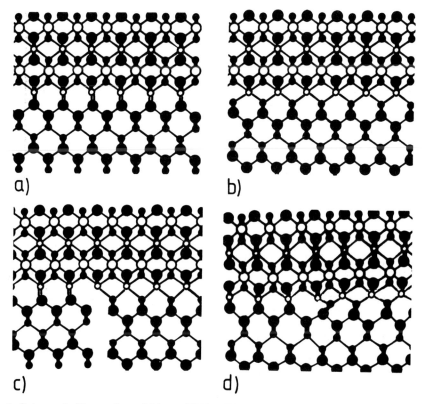

Fig.1. Schematic illustration of (a) on (001) NiSi$_2$:Si bicrystal, (b) a variant bicrystal, (c) undistorted co-existing variants, and (d) the formation of a 'frustrated-symmetry' interfacial dislocation with $\underline{b} = 1/4[111]$ separating completely contiguous variants.

The discussion above shows that, in general, a multiplicity of variant bicrystals exists. It is now necessary to show that interfacial discontinuities are the topological features which arise and separate variant bicrystals when these latter coexist contiguously. For the present we only indicate the nature of the defect which arises in the particular case where the variants shown in fig.1(a) and (b) coexist, and leave the formal characterisation of admissible discontinuities in general until the next section. The two variants shown in fig.1a) and (b) can be brought together so that either the NiSi$_2$ or the Si crystals can be joined contiguously without distortion, but it is not possible to join both simultaneously without distortion. If we choose arbitrarily to join the two NiSi$_2$ crystals without distortion, it can be seen in fig.1(c) that the two Si crystals are relatively displaced by 1/4 [111]. In other words, in order to join both crystals contiguously, a dislocation with Burgers vector $\underline{b}$ = 1/4[111] must be introduced, as depicted schematically in fig.1(d). It can be seen that the dislocation core resides in the interface, and the defect separates energetically degenerate regions of interface. We also note that this defect could, in principle, move along the interface by a combination of glide and climb, but could not move into either crystal without the concomitant introduction of an extended fault.

3. CHARACTERISATION AND CLASSIFICATION OF DISCONTINUITIES

In the previous section the formation of an interfacial dislocation in NiSi$_2$:Si was described. It can be seen that the procedure necessary to bring together the complementary Si surfaces in fig.1(c) in order to create the defect depicted in fig.1(d) is analogous to the Volterra (1907) method for introducing defects into a medium. In the present section we adopt an equivalent but slightly different procedure for creating interfacial defects. This has the advantage of facilitating the visualisation of a general mathematical formulation for characterising admissible defects in any given interface. Based on this general formulation, it is then possible to classify the diversity of possible defects into a small number of distinct categories. Fig.2(a) shows the formation of an initial interface by bringing together chosen black and white crystal surfaces, and allowing them to bond. This imaginary process fixes the relative orientation of the two crystals, expressed mathematically by the matrix $\underline{P}$ relating the black and white coordinate frames, and their relative position designated by the vector $\underline{p}$ which is the displacement, expressed in the white coordinate frame, of the black crystal away from some chosen origin configuration. For the sake of conciseness, we use the notation for augmented matrices set out in the International Tables for Crystallography edited by Hahn (1983), and refer to the matrix pair $(\underline{P},\underline{p})$ as P. A line defect of interest can be introduced into this initial interface by making a cut along the interface, then removing or adding further material to both the crystals in order to prepare new surfaces, and subsequently bonding together these latter surfaces. Now, because the black and white spaces represent crystalline materials, it is possible to expose new black and white surfaces which are crystallographically equivalent to the initial ones. In other words, if the unit vector normal to the initial interface is designated $\underline{n}$ (taken to be pointing into the white crystal and defined in the white coordinate frame), the unit normal to the new white surface is $W(\lambda)_j\underline{n}$, where $W(\lambda)_j = (\underline{W}(\lambda)_j, \underline{w}(\lambda)_j)$ is the jth symmetry operation in the white crystal's spacegroup; and $\underline{W}(\lambda)_j$ is the rotation, mirror or inversion part, and $\underline{w}(\lambda)_j$ the translation part. Similar expressions can be written relating the initial and new black surfaces.

When the new black and white surfaces are brought together they may bond to form an interface which is crystallographically equivalent, and hence energetically degenerate, to the initial interface. In other words, two variant bicrystals are made to coexist, and an interfacial discontinuity separates the contiguous black and white crystals. Following Volterra (1907), the discontinuity is characterised by the operation required to bring the new surfaces together. Let this operation, which would bring the new black onto the new white surface, be designated Q_{ij}. This compound operation is made up of the inverse of the black symmetry operation $W(\mu)_i^{-1}$, which can be regarded as the operation necessary to recover the initial black surface from the new one, followed by the white operation $W(\lambda)_j$, which takes the latter surface onto the new white one. If we choose to express this compound operation in the white crystal's coordinate frame, the inverse of the black symmetry operation must be re-expressed as $P\,W(\mu)_i^{-1}P^{-1}$. Thus the total compound operation is given by

$$Q_{ij} = W(\lambda)_j\,P\,W(\mu)_i^{-1}P^{-1} \tag{1}$$

We also note that defect character, as defined by expression (1), is independent of the line direction of a defect in question. Furthermore, although there generally exists a multiplicity of descriptions of the transformation, P , for a given interface, the set of defects predicted by expression (1) is independent of this choice.

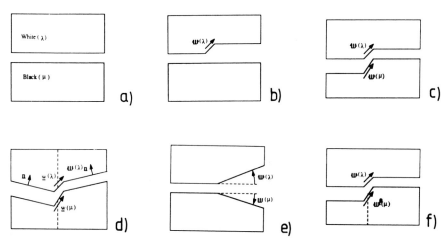

Fig.2. Schematic illustration showing the formation of discontinuities in interfaces; the black and white crystal surfaces are shown before being bonded to form (a) the initial interface, (b) a white crystal dislocation, (c) a 'broken-translation' dislocation, (d) a 'frustrated-symmetry' dislocation, (e) an interfacial disclination, and (f) an interfacial discontinuity delineating the intersection of a domain boundary with the interface in the special case where the interface orientation remains invariant.

The operations defined by expression (1) characterise admissible discontinuities in a specified interface. Since the morphological aspects and defect character of potential defects have been discussed at length elsewhere, Pond (1985 and 1989), we only outline the four principal categories in the present work. Discontinuities in the first category exhibit no defect character, and correspond to the operation Q_{ij} being the identity, (I, o). This situation can only arise when the two crystals exhibit some coincident symmetry operations, $W(c)_k$ (see Table).Provided a particular coincident symmetry operation does not also belong to the spacegroup of the bicrystal in question, $\Phi(b)$, then the discontinuity is an interfacial step when $W(c)_k$ is a translation $(I, t(c)_k)$, and a facet junction otherwise. Such discontinuities can be visualised as arising when complementary stepped or facetted black and white crystal surfaces are brought together. The step heights and facet angles associated with these defects are given in the Table.

When Q_{ij} corresponds to a displacement, (I, q_{ij}), the corresponding discontinuity is an interfacial dislocation, with Burgers vector $b_{ij} = q_{ij}$, and examples of the formation of such defects are illustrated schematically in fig. 2(b), (c) and (d). It is convenient to subdivide this category of defects into four classes which reflect their crystallographic origins. The Burgers vectors, morphological features, and the special crystallographic requirements necessary for their existence are listed in the Table. Dislocations in two of these classes, namely crystal and broken-translation-symmetry dislocations illustrated in fig.2(b) and (c) respectively have been investigated extensively in the past, but observations of defects in the other two classes have only been presented relatively recently.

In the case where Q_{ij} is a proper rotation, (Q_{ij}, o), the defect is a disclination, and this is shown schematically in fig.2(e). The final category of discontinuities corresponds to the case where Q_{ij} is a combination of rotation and displacement (Q_{ij}, q_{ij}). These defects are

TABLE

GEOMETRICAL CHARACTERISTICS OF DISCONTINUITIES IN INTERPHASE INTERFACES

Discontinuity	Special Crystallographic Requirements	Burgers Vector $\underline{b}_{ij}$	Morphological Aspects s : step (height h) sf : combined step and facet junction
Steps	$\underline{t}(\lambda)_j = \underline{P}\ \underline{t}(\mu)_i = \underline{t}(c)_k$	-	s, $h = \underline{n}.\underline{t}(c)_k$
Facet junctions	$\underline{W}(\lambda)_j = \underline{P}\ \underline{W}(\mu)_i\underline{P}^{-1} = \underline{W}(c)_k$ $\underline{w}(\lambda)_j = \underline{P}\ \underline{w}(\mu)_i = \underline{w}(c)_k$ $\underline{W}(c)\underline{p} = \underline{p}$	-	sf, $h = \underline{n}.\underline{w}(c)_k$
Crystal dislocations (e.g. white)	None	$\underline{t}(\lambda)_j$	s, $h = {}^1/_2\underline{n}.\underline{t}(\lambda)_j$
Broken-translation-symmetry dislocations	$\underline{t}(\lambda)_j \neq \underline{P}\ \underline{t}(\mu)_i$	$\underline{t}(\lambda)_j - \underline{P}\ \underline{t}(\mu)_i$	s, $h = {}^1/_2\underline{n}.(\underline{t}(\lambda)_j + \underline{P}\underline{t}(\mu)_i)$
Frustrated-symmetry dislocations	$\underline{W}(\lambda)_j = \underline{P}\ \underline{W}(\mu)_i\underline{P}^{-1}$ $\underline{w}(\lambda)_j \neq \underline{P}\ \underline{w}(\mu)_i$	$\underline{w}(\lambda)_j - \underline{P}\ \underline{w}(\mu)_i + (\underline{W}(\lambda)_j - \underline{I})\underline{p}$	i) $\underline{W}(\lambda)\underline{n} = \underline{n}$ s, $h = {}^1/_2\underline{n}.(\underline{w}(\lambda)_j + \underline{P}\ \underline{w}(\mu)_i)$ ii) $\underline{W}(\lambda)\underline{n} \neq \underline{n}$ sf, $h = {}^1/_2\underline{n}.(\underline{w}(\lambda)_j + \underline{P}\ \underline{w}(\mu)_i)$
Broken-coincident symmetry dislocations	$\underline{W}(\lambda)_j = \underline{P}\ \underline{W}(\mu)_i\underline{P}^{-1} = \underline{W}(c)_q$ $\underline{w}(\lambda)_j + \underline{t}(\lambda)_\ell = \underline{P}\ (\underline{w}(\mu)_i + \underline{t}(\mu)_k) = \underline{w}(c)_q$	$(\underline{W}(c) - \underline{I})\underline{p}$	i) $\underline{W}(c)\underline{n} = \underline{n}$ s, $h = \underline{n}.\underline{w}(c)_q$ ii) $\underline{W}(c)\underline{n} \neq \underline{n}$ sf, $h = \underline{n}.\underline{w}(c)_q$

known as dispirations, Harris (1970), and an example has been observed experimentally by Bourret et al. (1986), and analysed crystallographically by Pond (1985).

4. OCCURRENCE OF INTERFACIAL DEFECTS

Expression (1) clearly shows that the factor governing the range of defects that can arise in an interface is the extent to which crystal symmetry has been broken in the manufacture of the bicrystal. Symmetry may be broken because either the white and black crystals exhibit different spacegroups at the outset, or the relative orientation and position specified by P causes the misorientation or misalignment of similar symmetry elements in the two crystals, or a combination of both. In other words, the number of admissible defects depends on the combinations of broken symmetry operations $^\lambda W(\lambda)_i$ and $^\lambda W(\mu)_i$ (i.e. those not belonging to $\Phi(b)$) which can be substituted into expression (1) and which lead to proper compound operations, Q_{ij}. We refer to the extent to which the two crystals exhibit coincident symmetry as bicrystal connexion, and conversely, we can regard the extent of disconnexion as the generalisation of the concept of misfit. We can imagine the extreme cases of bicrystal connection as follows. In the case where the white and black spacegroups are identical, and the crystals have identical lattice parameters, orientation and no relative displacement (e.g.P = (I,o)), no interfacial defects can arise other than the dislocations, disclinations and dispirations which can also exist in either crystal. Thus, although the interface may be a structural and/or chemical discontinuity in this bicrystal, it does not present a topological discontinuity as far as crystal defects are concerned. However, if these adjacent crystals become disconnected to some degree, as can occur even spontaneously, for example if relaxation in the above situation leads to an interfacial displacement field which breaks symmetry, additional defects can then arise in the interface but cannot propagate into the adjacent crystals. At the other extreme, the interface in a completely disconnected bicrystal can be host to a wide variety of interfacial defects, including defects that can exist in the adjacent crystals, but these latter will not be able to penetrate from one crystal to the other without defect interactions occurring in the interface.

It is well known that interfacial dislocations can arise when some of the translation vectors of the adjacent crystals are different. In the present framework we designate such dislocations 'broken-translation-symmetry dislocations', and their formation by the Volterra process is indicated schematically in fig.2(c) and their geometrical parameters are given in the Table. Dislocations of this type have been referred to as misfit, Matthews (1975), coherency, Olsen and Cohen (1979), twinning, Christian and Crocker (1980) and dsc dislocations, Bollmann (1970 & 1982), and arrays of such defects have been observed to accommodate misfit in interfaces which are found in semiconductor devices. Such dislocations can arise in interfaces irrespective of the interfacial orientation, $\underline{n}$, and their Burgers vectors and associated step heights are independent of any relative displacement $\underline{p}$. It is interesting to note that dislocations can also arise, even if the translation symmetry of the adjacent crystals is identical. Such dislocations are designated 'frustrated-symmetry' dislocations (see Table) and arise when mirror-glide planes or screw-rotation axes are orientationally aligned in the two crystals, but have different intrinsic glide components, as is illustrated schematically in fig.2(d). As can be seen from the Table, the Burgers vectors of these defects depend on the difference between the intrinsic glide components, and also on the relative displacement $\underline{p}$. The dislocation in the $NiSi_2$:Si interface illustrated in fig.1(d) is an example of this type of dislocation, and can be regarded as arising due to, for example, the alignment of a four-fold screw axis in Si with a four-fold axis in $NiSi_2$, or the alignment of diamond-mirror-glide planes parallel to {100} in Si with their ordinary mirror counterparts in $NiSi_2$. In general, dislocations in this class will delineate the intersection of two interfacial facets with orientations related by the orthogonal part of the frustrated symmetry elements. In other words the two variant bicrystals, and hence the orientation of their interfaces, are related by the mirror or rotation part of the frustrated symmetry operation, see fig.2(d). Thus, such dislocations only arise on planar interfaces in the special instances where $\underline{n}$ is left invariant by the frustrated symmetry element. In the case of $NiSi_2$:Si for example, Pond (1989) has shown that this subset of special interfacial orientations is {hko}, i.e. where at least one of the Miller indices is zero, and this is completely consistent with experimental observations, Pond and Cherns (1985). The early stages of the formation of such dislocations has been studied in some detail for {100} interfaces by Batstone et al. (1988).

As was mentioned above, interfacial dislocations can also arise as a consequence of spontaneous bicrystal disconnection. We now consider briefly an example in a GaAs:Al interface, where the defect arises because of the spontaneous breaking of coincident mirror symmetry in the two crystals by translational misalignment due to relative displacement, $\underline{p}$. In the Table, defects of this type are designated 'broken-coincident-symmetry' dislocations, and a schematic illustration of an edge dislocation of this class in a GaAs:Al interface is shown in fig.3. The Burgers vector is equal to 1/4 [1$\bar{1}$0] in this case, and it can be seen that the defect has no associated step at its core and separates energetically degenerate regions of interface (i.e. separates continguous bicrystal variants) which have structures related by the broken mirror indicated in the figure. Defects of this type have been observed experimentally by Kiely (private communication), and occurred in arrays which accommodated misfit in small islands.

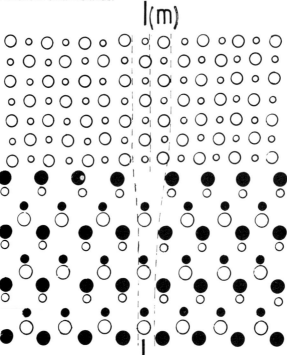

Fig.3. Schematic illustration of a dislocation with b = 1/4[1$\bar{1}$0] in edge orientation in an (001) interface between GaAs and Al. The interfacial structures on either side of the defect are related by the broken (coincident) (1$\bar{1}$0) mirror indicated by the thick vertical line. The upper crystal is also displaced by 1/8[110] parallel to the projection direction.

According to the geometrical treatment introduced earlier, disclinations can arise in interfaces when point symmetry operations in the adjacent crystals are orientationally misaligned. This can be viewed as the point symmetry counterpart of the concept of translational misfit, and for a more detailed account the interested reader should consult Pond (1989). The formation of an interfacial disclination by the Volterra process is illustrated schematically in fig.2(e). Disclinations are only likely to be plausible defects when their angle of rotation is relatively small, or when one or both of the component crystals in the bicrystal has small dimensions. Their possible occurrence in epitaxially grown bicrystals will be outlined later.

5. INTERACTION OF DOMAIN BOUNDARIES WITH INTERFACES

The line of intersection between an extended crystal defect, such as a domain boundary, and an interface will be delineated by an interfacial discontinuity, and the character of such perturbations can be established using the present framework. In this section we first consider briefly the characterisation of extended defects in the bulk of single crystals, and subsequently discuss interfacial defects. The distinctive sub-classes of interfacial defects are exactly the same as described in the previous section, except that in the present case domain boundaries emanate from the defect into one or both of the adjacent crystals. Finally, we illustrate an example of the nucleation of inversion domain disorder in epitaxial GaAs films grown on Ge substrates.

Domains can arise in crystals which are either (i) non-holosymmetric, or (ii) are daughter crystals obtained as a result of a phase transformation from a parent structure with higher symmetry. Adjacent domains in such crystals are related by symmetry operations which do not belong to the crystal's own spacegroup, but which belong to that of the crystal's lattice in case (i), or to that of the parent crystal in case (ii). We shall refer to such operations as exchange operations, and designate them $W^e(\lambda)$ and $W^e(\mu)$. As an illustration we consider briefly the instance of inversion domains in crystals having the zinc blende structure. It is helpful to regard this structure as having been derived from the diamond structure; in the parent structure the two atoms forming the crystallographic basis of the material are the same atomic species, but in zinc blende the two are distinct. Now whereas the parent structure exhibits the forty eight point symmetry operations in the spacegroup $Fd\bar{3}m$, zinc blende exhibits only the twenty four in the spacegroup $F\bar{4}3m$ (which is a subgroup of $Fd\bar{3}m$). In other words, the twenty four operations which belong to $Fd\bar{3}m$ but not to $F\bar{4}3m$ are the exchange operations, W^e, in the present case. All of the latter operations interchange one type of zinc blende domain into the other, i.e. they interchange the species in the material's atomic basis.

The formation, by means of the Volterra-like process, of an interfacial defect associated with a domain boundary is illustrated in fig.1(f). In the case where the domain boundary is located in the black crystal for example, the initial and new black surfaces will be related by the exchange operation $W^e(\mu)_i$ which also characterises the fault. It follows therefore that interfacial defects in such cases are characterised by the operation Q_{ij}, as defined by expression (1), except that the black operation is $W^e(\mu)_i$ rather than a true symmetry operation of the black crystal, $W(\mu)_k$. Similarly, when the domain boundary is located in the white crystal, admissible operations Q_{ij}, i.e. where the compound operation is proper, are given by $W^e(\lambda)_j P W^e(\mu)_i^{-1} P^{-1}$. In the case where domain boundaries emanate from the line defect in the interface into both crystals, the white and black operations to be substituted into equation (1) will both be exchange operations. Bearing in mind these additional qualifications, the geometrical properties of interfacial defects associated with domain boundaries are otherwise exactly the same as is summarised in the Table for the case of dislocations.

We now consider the relatively straight-forward and illustrative case of GaAs layers grown epitaxially on Ge substrates, where the misfit is very small. Ge has the diamond structure with spacegroup $Fd\bar{3}m$, and GaAs has the zinc blende structure which has the spacegroup $F\bar{4}3m$. Epitaxial layers of GaAs grow with their unit cells oriented parallel to those in the substrate, and with virtually the identical lattice parameter, i.e. P can be taken as the identity. Thus the symmetry operations of the Ge crystals which are not also exhibited by GaAs correspond to the set of exchange operations described above. It follows, according to equation (1), that interfacial discontinuities which delineate the intersection of domains with an interface can arise which have no dislocation or disclination character, i.e.

$$Q_{ij} = W^e(\lambda)_j W(\mu)_i^{-1} = (\underline{w}^e(\lambda)_j, \underline{w}^e(\lambda)_j)(\underline{W}(\mu)_i, \underline{w}(\mu)_i)^{-1} = (\underline{I}, \underline{o})$$

where GaAs is taken to be the white crystal, and $W(\mu)_i$ is the Ge symmetry operation coincident with the GaAs exchange operation. (It is possible that the exchange operation relating adjacent domains actually includes a small supplementary displacement which arises because the Ga-Ga or As-As bonds at the fault do not have the same length as Ga-As bonds. It would be expected that such additional displacements would lead to dislocation character arising at the interfacial defect.)

We shall assume that the given substrate surface is planar, although possibly stepped, and demonstrate that we are able to predict whether or not inversion domains can be nucleated in an epilayer grown on this substrate. The presence of such domains can only arise if $W(\mu)_i$ leaves the orientation of the substrate normal invariant. By inspection of the appropriate twenty four operations it can be seen that this is only possible for {hko} substrate surfaces, i.e. those for which at least one of the Miller indices is zero. Moreover, a step of height $1/2\underline{n}.(\underline{w}^e(\lambda)_j + \underline{w}(\mu)_i) = \underline{n}.\underline{w}^e(\lambda)_j$ will be associated with the discontinuity. A schematic illustration of the nucleation of an inversion domain on (001) substrates is shown in fig.4. It is seen that contiguous growth of the epilayer across the demi-stepped substrate, in a manner which leads to degenerate interfacial structures on either side of the step, can only occur if an inversion domain is nucleated. The initial and new substrate surfaces to the left and right of the demi-step respectively, are related by, for example, the four-fold screw axis normal to the substrate. Similarly, the initial and new GaAs surfaces to be bonded to the substrate surfaces, are also related by this screw-rotation, which is an exchange operation for this crystal. The above discussion of the nucleation of inversion domain disorder in GaAs grown on Ge substrates is consistent with the experimental observations of Pond et al. (1985). Domains have also been observed to form as a consequence of ordering in a variety of epitaxial systems, and a crystallographic analysis of the interaction of such domains with the substrate/epilayer interface, using group theoretical methods, has been presented by Pond (1986).

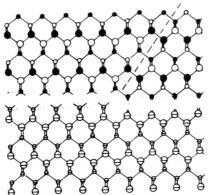

Fig.4. Schematic illustration of the nucleation of inversion domains in GaAs grown on a demistepped (001) Ge substrate. The projection direction in the diagram is [110].

6. EPITAXIAL GROWTH ON ROUGH AND VICINAL SURFACES

The schematic illustrations of the formation of interfacial defects depicted in fig.2 clearly show that features on a substrate surface, such as steps or facet junctions, will be transformed into interfacial defects following deposition of an epitaxial layer. Thus, it is to be expected that the accommodation by the overlayer of regular and irregular terraces of steps, as are thought to exist on vicinal and rough surfaces respectively, will influence the state of strain in the epilayer and hence the subsequent evolution of defect structures.

The simplest situation to consider is a regular terrace of steps on the substrate surface. Following the growth of a misfitting epitaxial layer, these features may delineate the core of 'broken-translation' or 'frustrated-symmetry' interfacial dislocations. Moreover, this resulting array of defects may contribute to the accommodation of in-plane misfit. In addition, if the individual dislocations have components of their Burgers vectors perpendicular to the interface, designated $\underline{b}'$, the configuration will cause the overlayer to be misorientated by a mechanism analogous to a low-angle tilt boundary. Using Frank's formula we can estimate the misorientation θ_m to be equal to $(1-h_e/h_s)\,\theta_v$, where h_e and h_s are the step heights of the surface steps on the epitaxial and substrate surfaces before bonding respectively, and θ_v is the vicinal angle. This expression has been found to be in good agreement with experimental measurements for the case of Si grown on sapphire (α-Al_2O_3), Pond et al. (1987), and CdTe grown on sapphire Aindow (1989), for vicinal angles up to about 2°. However, for larger vicinal angles there are some notable examples which do not conform to this simple model, and it is suggested that these discrepancies arise due to the presence in these interfaces of 60° type crystal dislocations which contribute to the overall normal component of the Burgers vector content of the interface, Beanland and Pond (1989).

The presence of small epilayer tilts as described above may cause the misorientation of point symmetry elements in adjacent crystals, and this is the crystallographic prerequisite for the formation of disclinations. Physically, it would be necessary for terraces of opposite inclination (i.e. related by the point symmetry element in question) to be present as indicated in fig.5. Such a configuration requires the overlayer to tilt in opposite directions on either side of the central apex, and this is formally equivalent to a disclination. In this way disclinations with quite small rotation angles can be envisaged, and it should be kept in mind that the elastic energy stored around such defects increases dramatically as the epilayer thickness increases.

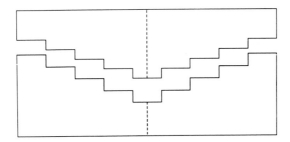

Fig.5. Schematic illustration of the formation of an interfacial disclination by epitaxial accommodation of substrate surface roughness. After bonding, the array of interfacial dislocation dipoles formed at the steps is equivalent to a disclination.

The situation depicted in fig.5, i.e. where the substrate exhibits only a single apex, is somewhat improbable. It is more likely that the substrate surface exhibits a multiplicity of 'hills and valleys' having a spectrum of amplitudes and wavelengths. The accommodation of such configurations could be modelled by arrays of disclination dipoles, and further work is necessary to clarify this point. One interesting possibility might arise if the 'up' steps in fig.5 are crystallographically distinct from the 'down' steps. It would then follow that the normal component, $\underline{b}'$, of the defects resulting from accommodation of the 'up' steps would be different from that of the 'down' steps. As a consequence, the normal component of the total Burgers vector content due to these step configurations might be finite, and hence the overlayer would exhibit a misorientation, even when epitaxial growth took place on a substrate surface which was, on average, precisely the nominal low-index surface. It has been suggested by Aindow (1989) that, in this manner, surface roughness may explain the observation that Si films grown on sapphire are misoriented about an axis parallel to [$2\bar{1}\bar{1}0$]

even when grown on exact ($01\bar{1}2$) substrates, but are not correspondingly offset about the axis parallel to [$01\bar{1}1$]. This latter axis lies parallel to a mirror-glide plane in the sapphire, and hence 'up' and 'down' steps related by this symmetry element are equivalent, and the presence of a diamond-mirror-glide plane in the Si nominally parallel to this ensures equivalent accommodation by the overlayer. On the other hand, no symmetry element is present in the sapphire relating 'up' and 'down' steps parallel to [$2\bar{1}\bar{1}0$].

The presence of small misorientations between epilayer and substrate may influence the evolution of defect microstructures. For example, it has been known for a long time that in the growth of (001) Si on ($01\bar{1}2$) sapphire the extent of twinning can be quite different in the four possible twin planes. This twin population anisotropy appears to be directly correlated to the misorientation present, Aindow (1989), but the mechanism governing this behaviour has not yet been identified in detail. Misorientations may also affect the magnitude and symmetry of the forces which act to nucleate or move dislocations on the various glide planes in an epitaxial layer, and this may, in turn, influence the stress relief mechanisms in semicoherent epilayers.

REFERENCES

Aindow M, Ph.D. Thesis, University of Liverpool (1989).
Batstone J L, Gibson J M, Tung R T and Levi A F J, Applied Phys. Lett. 52 (10)1988,828.
Beanland R and Pond R C, these proceedings (1989).
Bollmann W, Crystal Defects and Crystalline Interfaces (Springer, Berlin, 1970).
Bollman W, Crystal Lattices, Interfaces, Matrices (Published privately by W. Bollmann, Geneva, 1982).
Bourret A, Lasalmonie A and Naka S, Scripta Met. 20 (1986) 861.
Christian J W and Crocker A G, Dislocations and Lattice Transformations, in: Dislocations in Solids, Vol.3, ed. F R N Nabarro (North-Holland, Amsterdam, 1980) 165.
Hahn T (ed.), International Tables for Crystallography (De. Reidel Pub. Co., Dordrecht, 1983.
Harris W F, Phil. Mag. 22 (1970) 949.
Matthews J W (ed.), Epitaxial Growth (Academic Press, New York, 1975).
Olson G B and Cohen M, Acta Metall. 27 (1979) 1007.
Pond R C Interfaces and Dislocations, in Dislocations and Properties of Real Materials (Inst. of Metals, London, 1985) 71.
Mat. Res. Soc. Symp. Proc. 58 (1986) 3; 'Interfacial Dislocations', in 'Dislocations and Solids, 8, ed. F R N Nabarro, North-Holland,ch. 38, (1989).
Pond R C and Cherns D, Surf. Sci. 152/3 (1985) 1197.
Pond R C, Gowers J P and Joyce B A, Surf. Sci. 152/3 (1985) 1191.
Pond R C, J. Cryst. Growth 79 (1986) 946.
Pond R C, Aindow M, Dineen and Peters T, Inst. Phys. Conf. Ser. No. 87 (1987) 181.
Shubnikov A V and Koptsik V A, Symmetry in Science and Art (Plenum, New York, 1977).
Volterra V, Annls. Scent. Ec. Norm. Sup. Paris 24 (1907) 401.

Inst. Phys. Conf. Ser. No 104: Chapter 1
Paper presented at Int. Symp. on Struct. Prop. Disloc. Semicond., Oxford, 1989

37

Incipient dislocation dipoles in silicon

Malcolm I Heggie

Department of Physics, The University, Exeter EX4 4QL, UK

Abstract.   Recent computer simulations have shown that incipient dislocation dipoles arise spontaneously in silicon near its melting point. Such dipoles start as the smallest-conceivable faulted edge glide dipoles, embracing only two atoms of stacking fault, and they have formation energies comparable with, or less than, that of the vacancy. The paper is in three parts, dealing respectively with the role of incipient dipoles in melting, creep and self-diffusion. Theoretical estimates for the formation energy of incipient dipoles vary from 1 to 5 eV, but 3 eV is the most credible estimate.

1. INCIPIENT DIPOLES IN MELTING

Computer simulation has a significant role to play in areas where microscopy is relatively powerless, for example in determining the structure of very small defects and in understanding dynamic processes, such as melting. The incentive of a "microscopic understanding" of materials has driven the development of new, better interatomic potentials and more realistic simulation methods. One of the first studies to exhibit both improvements was a study of melting in silicon by Stillinger and Weber (1985). They employed a potential that gave a reasonable description of both the diamond structure and liquid silicon, whereas former, simple, valence force potentials tended to give reasonable descriptions of atomic configurations close to the diamond structure alone. They also used the technique of molecular dynamics to reveal the dynamic properties of the system, whereas most structural studies had previously used static techniques of energy minimisation, like the conjugate gradient method. At high temperatures it was difficult to discern the underlying structure in their 216 atom cell, so they looked at structures that arose when the system was quenched, using steepest descents to find the nearest local potential minimum. In so doing they found the lowest energy structural excitation of the diamond structure - lower in energy than a vacancy–interstitial pair. The new defect had a formation energy of about 2.9 eV according to their potential and it did not involve broken bonds. It is this defect that has been identified as an "incipient dislocation dipole" and related to processes other than melting (Heggie, 1988).

Figure 1 shows two slices through such a defect, one a perspective view of a slice whose normal is $(1,-1,0)$ and the other a projection of a slice whose normal is $(1,1,1)$. It is a point, not a line, defect.

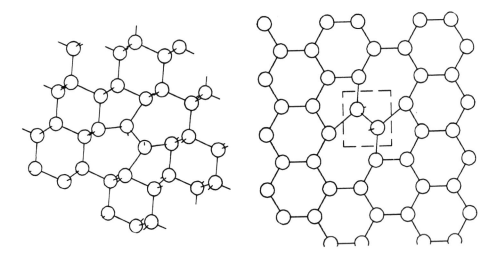

Figure 1. Incipient dipole, left in perspective almost along the (1,-1,0) direction and right in projection parallel to the (1,1,1) direction. In each case a slice only two atoms thick is shown.

The lack of dangling bonds probably accounts for the fact that this defect is lower in energy than a vacancy and this property, together with Stillinger and Weber's simulations, seems to reinforce a dislocation theory of melting (Nabarro, 1967) rather than a point defect theory.

A simplified model in two dimensions due to Kosterlitz and Thouless (1972) allows that dislocation dipoles are nucleated by thermal activation and when the temperature is high enough (*i.e.* at the melting point T_m) these dipoles expand until the mutual attraction of the dislocations in a dipole disappears by virtue of shielding by other dislocations. The resulting heavily dislocated solid will not be stable against shear and could be considered to be liquid-like. In fact the notional process of introducing many incipient dipoles into diamond silicon has been used by Wooten and Weaire (1985) to generate models of amorphous silicon. Clearly, this melting model fails by not taking into account the fact that silicon in the liquid has a mean coordination number of six, as opposed to four in the solid. And, in a sense because of this, it does not describe the true nature of melting which is an accidental degeneracy in the free energy of the liquid and the solid at the melting temperature. However, there may be reason to believe that a dislocation theory could be applicable to Stillinger and Weber's simulation, which did exhibit substantial superheating (estimated as 25% of T_m) due to the lack of a liquid/solid interface for smooth initiation of melting. In other words, melting in their case was attendant on mechanical instability of the solid as is invoked in dislocation theory.

2. INCIPIENT DIPOLES IN CREEP

The relationship between the incipient dipole and the 90° partial and the 90° faulted glide dipole can be seen in figures 2 and 3. Along the axis of a 90° partial there are contiguous pentagonal and heptagonal rings, which may be indentified as

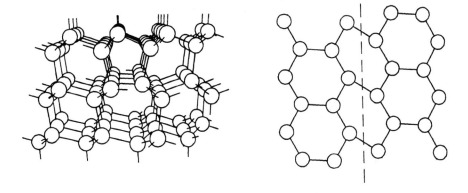

Figure 2. 90° partial dislocation, left in perspective almost along the (1,-1,0) direction and right in projection parallel to the (1,1,1) direction. In the latter case a slice only two atoms thick is shown.

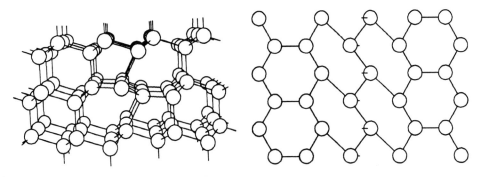

Figure 3. 90° faulted dipole, left in perspective close to the (1,-1,0) direction and right in projection parallel to the (1,1,1) direction. In the latter case a slice only two atoms thick is shown.

the cores of the disclination dipole that gives rise to the dislocation. They are in evidence in all three figures.

The localized nature of the incipient dipole makes comparison of the glide plane diagrams difficult. In effect the incipient dipole is bounded by kinks on the 90° partial. Figure 4 illustrates a kink pair on a 90° partial, showing the close similarity in structure with the incipient dipole. In fact many elasticity theory treatments of kink-kink interaction describe a kink pair as a combination of a dipole and a straight dislocation.

The energetics of these defects have been studied by using an interatomic potential due to Tersoff (1988a) similar in spirit to the Stillinger-Weber potential. The

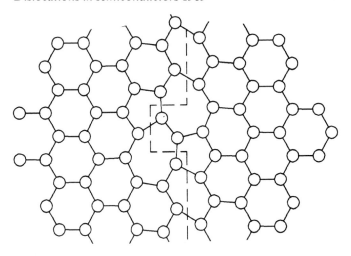

Figure 4. Kink pair on a 90° partial dislocation in projection parallel to (1,1,1) direction (the glide plane). A slice of only two atoms thick is shown and the kinked dislocation line is picked out with a broken line.

Tersoff potential attempts to describe silicon in many different coordinations and its successes and failures have been described elsewhere (Heggie and Jones, 1987, Heggie 1988). Here we note that its most significant shortcoming is in the shear modulus because the potential had only one elastic modulus in its fitting data - the bulk modulus. Indications are that relaxed defects involving shear have energies under- estimated by a factor of 2 to 3 (Heggie, 1988). The potential is used here to give orders of magnitude and to give an energy order to similar defects. It should be noted that the potential has lately been modified to improve the elastic constants, but at the expense of other features (Tersoff, 1988b).

Using clusters of some 200 silicon atoms and the Tersoff (1988a) potential, the incipient dipole was found to have an energy of 1.0 eV (Heggie, 1988). There are many ways in which the incipient dipole might extend. An exploration of one of the ways - a length-wise extension using the two-atom rotation model of Jones (1980) - gave a formation energy for the longer defect of about 2.5 eV. Looking at the fully extended line defects and quoting the energies per repeat distance (of 3.8 Å) , the 90° partial has a core energy of 0.44 eV (Heggie and Jones, 1987) and the faulted edge dipole of figure 3 has an energy of 0.9 eV. If the latter is widened by one primitive lattice vector its energy increases to 1.7 eV. These formation energies are high. Ignoring entropy effects and making approximate allowance for underestimation of energy in the Tersoff potential these results indicate that these dipoles could only be nucleated at a reasonable rate under the action of applied stresses of the order of 0.1μ (shear modulus), *i.e.* the theoretical shear strength of the material. In this context the suggestion that the appearance of these dipoles could contribute to creep (Heggie, 1988) seems unlikely and the dipoles observed to form at high temperatures arise more probably from mutual trapping of dislocations or from jog dragging as suggested by A. George (private communication). Spontaneous dipole creation at high temperatures would require high internal stresses (when the applied stress might be only 2MPa), high entropy (although a calculation of the vibrational entropy in the harmonic approximation of an incipient dipole yielded only 1 k_B) and, possibly, the presence of defects/impurities which could lower dipole formation and migration energies.

3. INCIPIENT DIPOLES IN SELF DIFFUSION

The concerted exchange of a pair of atoms in silicon was thought to be energetically unfavourable for self-diffusion compared with point defect diffusion until the remarkable *ab initio* calculations of Pandey (1986). These showed that a route for the exchange that consisted of three consecutive rotations of the atom pair about their mutual bond centre need only cost about 4.3 eV, in good agreement with the experimental activation energy for self-diffusion of 4.1 to 5.1 eV.

Each atom in the exchange has three bonds to neighbouring, non-exchanging atoms and each of these bonds is parallel to one such bond on the other exchange atom. The three rotations successively break these parallel bond pairs (by stretching) and remake them so that each exchange atom changes one of its neighbours for one of the other atom's neighbours in each rotation step. The mechanism of each step is identical to Jones' model of kink pair nucleation and migration (Jones, 1980). The first step generates an incipient dipole, the second switches the glide plane of the dipole and the third removes the dipole. It is during the second step that the structure is most distorted and the highest saddle point arises. "Ideal" concerted exchange is an approximation to this process which holds all but the two exchanging atoms rigid and maintains the latter's mutual bond centre and length at the crystalline values. Pandey's calculations give the ideal saddle point an energy of 5.4 eV. Considerations, such as relaxation and basis size dependence, lower the ideal energy to about 4.3 eV. The incipient dipole energy is about 4 eV in the ideal case and, if the same considerations apply to it, then its energy from these *ab initio* calculations coincides with Stillinger and Weber's estimate of about 2.9 eV.

However, Pandey's calculations relate to a plane wave basis and a supercell of 54 atoms. Attempts to reproduce his results in a real space basis and a cluster of up to 44 atoms of silicon (with a hydrogen-saturated surface *i.e.* $Si_{44}H_{42}$) have failed. The ideal saddle point energy in a Si_8H_{18} cluster is about 7.3 eV (Heggie, 1989) and it has remained constant in further calculations on larger clusters with different basis sizes in collaboration with P. Briddon and R. Jones. These same calculations give an "ideal" incipient dipole formation energy of 5 eV (*c.f.* Pandey 4 eV). One possible cause of the discrepancy is the interaction of the concerted exchange event in the supercell with its images in neighbouring cells. To test this a long cluster was constructed ($Si_{38}H_{54}$) which allowed two concerted exchange events to occur at opposite ends separated by the same distance as there was between the event and its closest image in Pandey's supercell. The result was that the saddle point energies of both events were almost exactly additive, with no energy lowering interaction. The reconciliation of cluster and supercell calculations is not yet resolved. The difference is most probably due to the large energy gaps (about 4 eV) and isolated defect levels in the former contrasting with the narrow gaps and defect bands up to 1 eV wide in the latter.

4. CONCLUSIONS

The most probable value for the formation energy of an incipient dipole is about 3 eV. The lowest value of 1 eV is due to the poorness of the Tersoff (1988a) potential and the highest values of 4 to 5 eV relate to "ideal" unrelaxed structures. Its vibrational formation entropy is low - of the order of 1 k_B. It appears likely to arise in melting and self-diffusion, but is unlikely to contribute significantly to creep through spontaneous nucleation and expansion. On the other hand it is very easy to nucleate an incipient dipole on a straight dislocation (equivalent to kink pair nucleation which costs about 0.15 eV according to Heggie and Jones 1987). This could mean that concerted exchange, which requires incipient dipole formation, is greatly facilitated in a dislocation core and the core might "melt" (in the sense of rapid atomic interchange) and change its behaviour just below T_m.

Acknowledgments

I would like to thank the SERC, UKAEA, CEGB and the University of Exeter for financial support and R. Jones, P. Briddon, G.M.S. Lister, A. Umerski, P. Fairbrother, A. George and F. Louchet for helpful discussions and private communications.

References

Heggie M I 1988 *Phil. Mag. Letters* **58** 75
Heggie M I 1989 *Molecular Simulation* to be published
Heggie M I and Jones R 1987 *Inst. Phys. Conf. Ser. No.***87** 367
Jones R 1980 *Phil. Mag. B* **48** 379
Kosterlitz J M and Thouless D J 1972 *J. Phys. C* **6** 1181
Nabarro F R N 1967 *"Theory of Crystal Dislocations"* (Oxford: Pergamon) 688
Pandey K C 1986 *Phys. Rev. Lett.* **57** 2287
Stillinger F H and Weber T A 1985 *Phys. Rev. B* **31** 5262
Tersoff J 1988a *Phys. Rev. B* **38** 9902
Tersoff J 1988b *Phys. Rev. B* **37** 6991
Wooten F and Weaire D 1985 *Phys. Rev. Lett.* **54** 1392

Inst. Phys. Conf. Ser. No 104: Chapter 1
Paper presented at Int. Symp. on Struct. Prop. Disloc. Semicond., Oxford, 1989

Interaction of impurities with dislocation cores in silicon

MI Heggie, R Jones, GMS Lister and A Umerski

Department of Physics, University of Exeter, Stocker Road, Exeter. EX4 4QL

ABSTRACT: The interaction of electrically active impurities like phosphorus with dislocations in silicon is examined using a cluster method with local density functional pseudopotential theory.This method is capable of predicting structural properties such as bond lengths and angles to within a few percent. We describe two states of phosphorus at a dislocation core. The first in which phosphorus is four fold coordinated and acts as a normal donor, and the second when the phosphorus atom is three fold coordinated and is electronically inactive. We find the three fold coordinated state to be more stable.

1. INTRODUCTION

Dislocations are known to act as segregation sites for impurities in silicon. These segregated impurities often profoundly affect the mechanical and electrical properties of the material. For example, O strongly locks dislocations rendering them immobile whilst introducing donor states (Sumino, Yonenaga, Yusa, 1980,Yonenaga, Sumino,1985, Sato, Sumino, 1985). N and P also lock dislocations, but with N atoms at dislocations appearing to be electrically inactive (Kamiyama, Sumino, 1985, Sumino, Imai, 1983). This may also be true for P atoms. H removes epr activity associated with plastically deformed Si but increases the photoluminescence observed around .9eV, possibly by suppressing non-radiative recombination processes (Kveder, Ossipyan, 1985). Its effect on the dislocation mobility is however not known. The electrical activity of plastically deformed Si is not understood. It seems that most workers accept that the bulk of dislocation core dangling bonds are removed by reconstruction (Hirsch 1985) leading to electrically inactive dislocations. The electrical effects associated with plastic deformation namely epr, photoluminescence ,DLTS effects, and the change in carrier concentrations are then attributed to either unusually stable clusters of point defects like vacancies and/or point defects at dislocations eg kinks, jogs or solitons (Weber, Alexander, 1983, Heggie, Jones, 1983). One hypothesis, due to Heggie (1987), considers that chemical donors and acceptors can be passivated by the dislocation core. For example, a P atom owes its donor character in crystalline Si to being able to act as a four fold coordinated substitutional defect. However on reaching a dislocation it could take up a lower energy structure by forming three bonds with neighbouring Si atoms. This leaves a lone pair of electrons occupying a state low in the gap or in the valence band. This is believed to occur in a-Si and leads to a loss of the electrical activity of P (Street,1982). Clearly such a large structural change induced by P could explain the strong locking effect upon the dislocation velocity. When a dislocation advances, the actual atoms comprising the core are forever changing, it is simply a switch of bonds that is taking place. Thus a three fold coordinated P atom would have to revert to a four fold coordinated one, at a cost in energy, and acts accordingly as a drag on the dislocation. The gettering of P atoms to the dislocation would also lead to a lowering of the free electron density as each P atom becomes passivated. Thus the dislocations in plastically deformed n-Si:P act as acceptors without having any acceptor levels in the energy gap. This mechanism could only lead to a change from n-Si to p-Si if the concentration of chemical acceptors, that are not themselves passivated, dominated that of the remaining electrically active P atoms in compensated material.

We describe here the results of accurate quantum mechanical calculations to investigate these hypotheses quantitatively. We use a self-consistent local density functional pseudopotential scheme which has shown itself reliable in calculating structures and dynamical properties of semiconductors and their defects. For example the bond lengths and vibrational modes of small molecules such as H_2O and C_2 are given to within a few per cent (Jones, Sayyash, 1986). The bond lengths of Si, diamond and crystalline P are also given to within 2%, whilst bond stretching and bending force constants yield phonon dispersion curves to within 10% of the observed values (Jones 1987, 1988, Briddon, Jones, 1989).

The method requires one to embed the dislocation and the impurity atoms in a large cluster (~50-100 atoms) whose surface is saturated with H's. These serve to tie up dangling bonds which would otherwise give rise to levels in the gap making them difficult to distinguish from defect levels. Unfortunately these H atoms have bonding and anti-bonding states, which couple strongly with states at the top of the valence band and at the bottom of the conduction band, and thus artificially enlarge the gap. For example the band gap in the 56 atom cluster $Si_{26}H_{30}$ is 4.5eV. Nevertherless as we shall see, it is possible to distinguish deep from shallow levels.

2. METHOD

We use a basis of s-,p_x-,p_y- and p_z- Gaussian orbitals to describe the wave function of each state and a second basis of s-Gaussians to expand the charge density. These enable all the necessary integrals to be carried out analytically. The gaussians are centred on both atomic and bond centred sites. Generally most atoms were treated with a wave function basis set consisting of a fixed linear combination of Gaussian orbitals, however atoms of special interest, eg in the dislocation core, were treated with a basis set of four independent s and p orbitals. The total energy of the cluster is then found and the force acting on each atom evaluated semi-analytically. An inner cluster of atoms is then allowed to move in response to these forces, and the relaxed structure found using a conjugate gradient algorithm. Generally the inner cluster consists of atoms of special interest and their nearest neighbours. The method is described in detail in Jones (1988).

3. THE 90 DEGREE PARTIAL IN SILICON

The common 60 degree dislocations in silicon, lying on [111] planes parallel to <110> directions, are known to dissociate into 90 and 30 degree partials. It is believed that these lie on the closely spaced [111] planes and are called the glide set of dislocations, to distinguish them from the shuffle set, which lie on the widely spaced planes (Hirsch 1985). Theoretical work (Marklund 1983, Hirsch 1979, Jones 1979, Heggie and Jones 1987) suggests that both these partials can reconstruct to eliminate dangling bonds. Here we examine the interaction of P impurities with 90 degree partials (similar results are expected for 30 degree partials). Clearly in order to change the bonding of a phosphorus atom, the reconstruction itself must be 'undone'. This is equivalent to creating a pair of anti-phase boundaries or solitons (Fig.1) as discussed by Heggie and Jones (1983). A 57 atom cluster $Si_{26}H_{31}$ (Fig.1) containing one soliton was relaxed as described above. The soliton level came around mid-gap at $E_v+1.5$ eV. It must be remembered that the gap in this cluster is artificially enlarged to 3.6eV because of the H termination. The soliton's bond lengths were 3% shorter than a normal one because of the increased sp^2 hybridisation. It is known for example that the ideal vacancy in Si expands outwards (Bachelet, Baraff, Schluter 1981) to reduce the back bond length on its nearest Si atoms. The reconstructed bonds have lengths about 3% longer than normal Si-Si bonds.

We then investigated the affinity of a P atom with the soliton. Fig.2a shows the slide plane, of the 60 atom cluster $Si_{26}PH_{33}$ in which the P has normal coordination. The relaxed structure has a filled level around mid-gap at $E_v+1.3$ eV showing that the P atom has donated its electron to the dangling bond on the soliton. The P donor level is close to the conduction band edge. The P-Si bonds are 2.2-2.3Å. We then exchanged the P atom with the Si atom having the dangling bond (Fig.2b) and relaxed the cluster once again. The energy reduced by about 1eV and there were no energy levels in the gap. The P-Si bonds have length about 2.2Å. This shows that the 3-fold coordinated P atom is much more stable than the 4-fold coordinated one.

We have also made preliminary investigations into the structure of a pair of P atoms at a dislocation core. An 82 atom cluster $Si_{38}P_2H_{42}$ containing a pair of P atoms along the core separated by a normal Si-Si bond, was relaxed. There was a clear tendency for the four fold coordination of the P atoms to be destroyed, but we have not yet unambiguously identified the relaxed state.

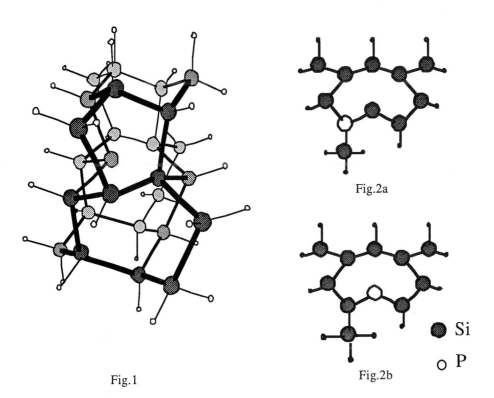

Fig.2a

Fig.1

Fig.2b

● Si

○ P

4. CONCLUSION

Dislocations have a major influence upon the structure of impurities. In the case of P, the tendency to create a three fold coordinated structure appears very strong. This can explain the pronounced locking effect of P upon the dislocation mobility. It also suggests that P atoms at dislocations will be electrically inactive. We hope to present further results on these structures in a future paper.

5. AKNOWLEDGMENTS

We would like to thank P.R Briddon for many useful discussions.

6. REFERENCES

Bachelet GB, Baraff GA, Schluter M, 1981, Phys. Rev. B24, 4736.
Briddon P, Jones R, 1990, J Phys C: Solid St. Phys, submitted.
Heggie M, Jones R, 1983, Phil Mag, B48, 365.
Heggie M, Jones R, 1987, Institute of Phys. Conf. Series, 87, 367.
Heggie M , 1987, Izvestiya Akademii Nauk. Fizicheskaya Seriya, 51, 693.
Hirsch PB, 1985, Mat. Sci. and Technology, 1, 666.
Hirsch PB, 1979, J de Physique, 40, C6-27.
Jones R, Sayyash A, 1986, J Phys.C: Solid State Phys. 19, L653.
Jones R, 1988, J Phys. C: Solid St. Phys, 21, 5735.
Jones R, 1988, J Phys. C: Solid St. Phys, 20, L271.
Jones R, 1979, J de Physique, 40, C6-33.
Kamiyama H, Sumino K, 1985, Yamada Conf. on Dislocations in Solids, edited by H
Suzuki, T Nonomiya, K Sumino, S Takeuchi, University of Tokyo Press, 1985. p 399.
Kveder VV, Ossipyan A, 1985, Yamada Conf. on Dislocations in Solids, edited by H
Suzuki, T Nonomiya, K Sumino, S Takeuchi, University of Tokyo Press, 1985. p 395.
Marklund S, 1983, J de Physique,44, C4-25.
Sato M, Sumino K, , 1985 Yamada Conf. on Dislocations in Solids, edited by H Suzuki, T
Nonomiya, K Sumino, S Takeuchi, University of Tokyo Press, 1985. p 391 .
Street RA, 1982, Phys. Rev. Lett., 49, 1187.
Sumino K,Harada H, Yonenaga I, 1980, Jpn, J Appl. Phys. 19, L49.
Sumino k, Imai M, 1983, Phil Mag., A47, 753.
Weber ER, Alexander H, 1983, J de Physique, C4, C4-319.
Yonenaga I, Sumino K, 1985,Yamada Conf. on Dislocations in Solids, edited by H Suzuki,
T Nonomiya, K Sumino, S Takeuchi, University of Tokyo Press, 1985. p 385.

Inst. Phys. Conf. Ser. No 104: Chapter 1
Paper presented at Int. Symp. on Struct. Prop. Disloc. Semicond., Oxford, 1989

On the shuffle–glide controversy

F Louchet* and J Thibault**

*INPG - LPTCM, ENSEEG, BP75 38402 Saint Martin d'Héres Cedex, France
**DRF/SPg/S, CENG, BP 85 X 38041 Grenoble Cedex, France

Different structures of partial dislocations in elementary (ESC) and compound (CSC) semiconductors are proposed on the basis of simple criteria such as atomic spacing, bond distortion and possible reconstruction of dangling bonds, and discussed in the light of recent HREM observations (Bourret et al 1983) and core energy calculations. It turns out that partial dislocations probably have complex cores, with an average glide character, but with a number of shuffle vacancy (S_v) and/or shuffle interstitial (S_i) sites. The 30° partial is likely to have a reconstructed glide core (in ESC), some isolated S_v sites and some S_i sites arranged in 3 different types of zig-zag chains, the most frequent lying parallel to the stacking fault plane. In contrast, S_i sites in the 90° partial are quite unlikely, since they would lead to a very compact core. HREM images (Bourret et al 1983) are consistent with a pure glide 90° core, but some S_v sites would certainly not be detectable and could be distributed along the core.

The dangling bonds associated with these shuffle sites can account for EPR results (Alexander 1979, Weber and Alexander 1979) and EBIC contrasts (Ourmazd et al 1981). The consequences for dislocation mobilities are also of interest, in so far as S_v sites can facilitate kink nucleation and migration. Indeed, the motion of dissociated dislocations through the propagation of S_v kinks (i.e. kink-vacancy complexes) satisfies two apparently incompatible requirements:

 (i) the dissociated dislocation must slip in the stacking fault plane, the GLIDE plane.
(ii) the lattice friction is minimum when slip occurs in a SHUFFLE plane which minimizes the "distortion ratio": b/d_{hkl}.

The particular proposed structure of S_i sites in the 30° partial has been used by one of the authors in order to explain the reduction of dislocation density in GaAs by In doping, through the incorporation of In atoms in 30° cores (Louchet 1988).

A much more detailed discussion of these complex core structures can be found in Louchet and Thibault-Desseaux (1987).

Alexander H, 1979 J. Phys. Coll. 40 C6 1
Bourret A, Thibault-Desseaux J and Lancon F 1983 J. Phys. Coll. 44 C4 15
Louchet F 1988 J. Physique 49 1219
Louchet F and Thibault-Desseaux J 1987 Revue. Phys. Appl. 22 207
Ourmazd A, Weber E, Gottschalk H, Booker G R and Alexander H 1981 Inst.
  Phys. Conf. 60 63
Weber E and Alexander H 1979 J. Phys. Coll. 40 C6 101

Extended Abstract only

Inst. Phys. Conf. Ser. No 104: Chapter 1
Paper presented at Int. Symp. on Struct. Prop. Disloc. Semicond., Oxford, 1989

49

Dislocation mechanisms for twinning and polytypic transformations in semiconductors

P. Pirouz

Department of Materials Science and Engineering, Case Western Reserve University, Cleveland, OHIO 44106, U.S.A.

ABSTRACT: A dislocation model recently proposed for twin formation in cubic semiconductors is extended to explain polytypic transformations in semiconductors with a low stacking fault energy and a high Peierls stress, e.g. SiC.

1. INTRODUCTION

Twinning is the predominant mode of deformation in both elemental and compound semiconductors at low temperatures. In many compound semi-conductors, specially of the II–VI type, the structure is wurtzite at the growth temperatures but tranforms to the cubic zincblende on cooling to room temperature. In addition to the zincblende (3C) and the wurtzite (2H) structure, SiC occurs in over one hundred other structures. These are known as polytypes and the only difference between them is the ordering sequence along an axis normal to the polar layers of Si_4C (or C_4Si) co–ordination tetrahedra. As Jepps and Page (1983) have noted in a review of polytypic transformations in SiC, the conditions under which one polytype becomes unstable with respect to another is ill–defined. This paper presents a unified dislocation model to explain twinning and polytypic transformations in semiconductors.

2. DISLOCATION VELOCITY AND FRANK–READ SOURCE

The critical resolved shear stress necessary to activate a Frank–Read source is approximately:

$$\tau^*_g \simeq \mu b/L \qquad \ldots(1)$$

where L is the length of the source dislocation segment which is pinned, b is its Burgers vector and μ is the shear modulus of the crystal. Consider-ing a source segment of 1 μm, then in SiC [μ=180 GPa, b=0.31 nm (Fujita et al., 1987)], τ_g=55.8 MPa while in Si [μ=68 GPa, b=0.384 nm (George and Rabier, 1987)], τ^*_g=26 MPa. Equation (1) ignores the Peierls stress which is high in tetrahedrally bonded semiconductors. It also really gives the critical stress needed to activate the first loop. After this the inter-action of the source dislocation segment with the surrounding loops needs to be taken into account.

Consider Fig. 1 where a Frank–Read source has generated n dislocation loops and the source dislocation, XY, is ready to generate the (n+1)st loop. The segment XY, of course, expands by the action of the resolved shear stress, τ ($>\tau^*_g$), on the slip plane on which the dislocation lies. At the same time, however, the segment XY interacts with all the n loops surrounding it; these, of course have the same Burgers vector, b, as the segment XY.

Interaction between straight, elliptical, or circular dislocation loops has been considered in detail by Eshelby, Frank, Li, Nabarro, and others. However, for our purposes, we consider a very approximate treatment of the situation in the following.

(a)

Fig. 1. The dislocation loops generated by a Frank-Read source and surrounding it, (a) plan view, (b) side view.

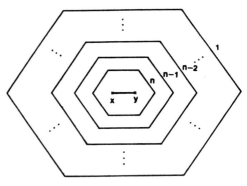

(b)

For simplicity, suppose that the segment XY is a screw dislocation and consider the effect of just the innermost loop (Fig. 2).

Fig. 2. Interaction of a source dislocation with the n'th loop surrounding it.

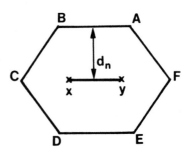

Further assume that there are only interactions with segments AB and DE parallel to XY. Segments AB and DE have, respectively, similar and opposite line directions as XY. Thus, there is an attractive interaction between DE and XY and a repulsive interaction between AB and XY. The magnitude of both the forces on XY arising out of these interactions is given by:

$$F/L = \mu b^2 / 2\pi d_n$$

where d_n is the "radius" of the n'th loop (i.e. the distance between the source segment XY and the segment AB of the n'th loop) as shown in Fig. 2. Hence the total stress acting on XY is given by:

$$\tau_T = \tau - 2\mu b/2\pi d_n$$

In the presence of other loops:

$$\tau_T = \tau - (\mu b/\pi) \sum_{j=1}^{n} 1/d_j$$

The surrounding loops are, of course, expanding in the presence of the external shear stress, i.e. d_j in the above equation is a function of time: $d_j(t)$. The configuration of the loops around the segment XY will be in the form of an inverse pile-up (Fig. 1). However, as a first approximation, it may be assumed that the loops expand with equal velocity and that they are equidistant from each other, i.e. $d_j(t) = (n-j+1)d_n(t)$, then:

$$\tau_T = \tau - (\mu b/\pi d_n)\sum_{k=1}^{n} 1/k$$

where the summation on the right hand side is just the harmonic series. Now, the segment XY will not expand if $\tau_T < \tau_g$. This condition holds for an innermost loop radius, d_{nc}, where:

$$d_{nc} = [\mu b/\pi(\tau-\tau_g)]\sum_{k=1}^{n} 1/k \qquad \ldots(2)$$

So the source will stay inactive while $d_n(t) < d_{nc}$ and starts operating only when the innermost loop has expanded far enough that $d_n(t) > d_{nc}$. For the source to become inactive, d_{nc} should of course be greater than the initial radius of the innermost loop, d_i, which is of the order of L. Thus, assuming an applied stress of 30 MPa on a 1 μm source dislocation, and taking d_i to be equal to 1 μm, after three loops have been generated, $d_{nc} = 3.81$ μm. Thus, while the third loop expands from an initial radius of 1 μm to a critical radius of $d_{nc} = 3.81$ μm, the back stress of the first three loops will stop the operation of the Frank-Read source in Si. Once the loop has expanded to a radius greater than 3.81 μm, the source will generate the fourth loop.

In semiconductors, dislocation motion is a thermally activated process and so is a strong fuction of temperature. At high temperatures, the dislocation mobility is large and the dislocation loops expand rapidly as soon as they form. Hence $d_n(t)$ exceeds d_{nc} very quickly and the source can be active practically continuously. On the other hand, at lower temperatures, the source may be inactive for a long duration until the surrounding loops have moved sufficiently away from XY to allow its activation. Dislocation velocity in semiconductors may be expressed as:

$$v = v_o (\tau/\tau_o)^m \exp(-E/k_B T) \qquad \ldots(3)$$

where τ_o is a unit shear stress ($= 1$ Pa), m is the stress exponent and E is the activation energy for dislocation motion. The time, t_c, taken for the innermost dislocation loop to expand from its initial radius, d_i, to the critical radius, d_{nc}, is given by:

$$t_c = [(1/v_o)(\tau_o/\tau)^m \exp(E/k_B T) \{[\mu b/\pi(\tau-\tau_g)\sum_{k=1}^{n} 1/k] - d_i\} \qquad \ldots(4)$$

Hence, the time that the source dislocation for the Frank-Read source is inactive is strongly temperature dependent and its operation slows down as more and more lopps are generated. Although some parameters on dislocation mobility in 6H-SiC were obtained by Maeda et al. (1986), (e.g. m=1.1 and E=3.4 eV), actual dislocation velocity measurements have not been carried out in this material and thus not all the parameters necessary for estimating t_c are available. For Si, however, considerable data exists on dislocation velocity. Thus, using the data of Imai and Sumino (1983) for screw dislocations in intrinsic Si (m≈1, $v_o = 3.5 \times 10^{-2}$ ms^{-1}, E=2.35 eV in the range 876-1046 K), and the same values of applied stress ($\tau = 30$ MPa) and source length (1 μm) as before, the dwell time, t_c, after three loops is given by:

$$t_c = 9.624 \times 10^{-15} \exp(28200/T)$$

where t_c is in seconds and T is in degrees Kelvin. Hence, at T=600°C, $t_c \approx 286$ s while at 800°C, the dwell time decreases to 0.7 s.

Although the dwell time in SiC cannot be estimated because of a lack of hard data on dislocation velocity, it is expected to be much larger than in, say, Si because of the very large activation energy in this material.

2. TWINNING AND DISLOCATION VELOCITY

According to the dislocation mechanism proposed by Pirouz (1987, 1989) twinning in semiconductors arises because of the different mobility of partial dislocations in these materials, particularly at intermediate temperatures. In this model, a pinned segment of screw dislocation is dissociated into two 30° partial dislocations with Burgers vectors b_l (for the leading partial) and b_t (for the trailing partial). In semiconductors, the mobility of these partials is different and the difference is a function of temperature, generally increasing with decreasing temperature. Suppose that the leading partial has a higher mobility than the trailing partial dislocation. It is then possible that, over a certain temperature range, the leading partial has a sufficiently greater mobility than the trailing partial that it forms a complete loop while the trailing partial lags behind. The loop thus formed is faulted and is surrounded by the partial dislocation b_t. After the formation of the fault, the two partials recombine to re-form the screw dislocation. This, however, cannot dissociate on the same glide plane and repeat the process because of the formation of a high energy fault with an AA stacking. However, if there exists a resolved shear stress on a cross-slip plane, the screw dislocation cross-glides to the next {111} slip plane where it is free to dissociate and form a faulted loop. Repetition of this process n times (i.e. over n parallel planes) forms a twin of thickness nd_{111} as shown in Fig. 3. The process will stop when, for example, the local resolved shear stress on the primary glide plane falls below the critical stress for twin formation, τ_t. In this case, the screw dislocation may continue its glide on the cross-slip plane without switching back to the primary glide plane. In this way the source twinning dislocation moves away from the twinned region although it leaves a perfect dislocation in the twin. The configuration of the source screw dislocation relative to the leading partial dislocations is shown in profile in Fig. 3.

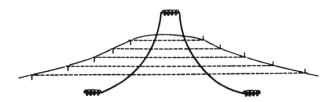

Fig. 3. Formation of a twin by the sequential cross-slip of a screw dislocation, its dissciation and formation of faulted loops on neighboring {111} primary glide planes.

The thickness of the twin depends on the stress on the cross-slip plane together with the mobility of the screw dislocation on the cross-slip plane which is a function of temperature. The width of the twin, on the other hand, depends on the expansion of the faulted loops on the primary glide plane and is thus a function of the resolved shear stress on this plane and the mobility of the leading partial dislocation. The latter is, of course, also a function of temperature. Hence, the pre-requisites for twinning are:

  i) A pinned segment of screw dislocation dissociated on the primary glide plane,
 ii) A temperature range where the mobility of the leading partial dislocation is appreciably greater than that of the trailing partial,

iii) A relatively large resolved shear stress, τ, on the primary glide plane ($>\tau_t$),
  iv) Sufficient resolved shear stress on a cross-slip plane.

The stress required for twinning, τ_t^*, must be sufficient to overcome the stress, τ_1, for the bending of the leading partial dislocation fully through the pinning points, and the stress, τ_2^*, required to overcome the stacking fault energy, γ. τ_1 is given by Eqn. (1) with the Burgers vector of the partial dislocation, i.e. $\tau_1 = \mu b_1/L$. τ_2^*, is given by:

$$\tau_2^* = \gamma/b_1 \qquad \qquad \ldots(5)$$

For instance for a 1 μm length of pinned dislocation in Si, where $\gamma \approx 60$ mJ/m^2, $\tau_1 \approx 15.1$ MPa and $\tau_2 \approx 271$ MPa, while for the same dislocation segment in SiC, assuming that 3C-SiC has the same stacking fault energy as 6H-SiC, $\gamma \approx 2.5$ mJ/m^2 (Maeda et al. 1988), $\tau_1 \approx 32$ MPa and $\tau_2 \approx 14$ MPa. Hence, for this segment of pinned dislocation, $\tau_t = 286$ MPa for Si and $\tau_t = 46$ MPa for SiC, i.e. the critical resolved shear stress for twinning is more than six times larger in Si than in SiC. It should be stressed that in the above treatment, the Peierls stress has not been taken into account. This can be very important and, in particular, will tend to keep the screw dislocation in the <110> Peierls valley and oppose it s bending out of the valley. Hence, cross-slipping of the screw dislocation would be easier which is important for the mechanism of polytypic transformation discussed in the next section.

4. CUBIC → NON-CUBIC POLYTYPIC TRANSFORMATIONS

For the same reason that a Frank-Read source may stop operating for a while because of interactions with the surrounding loops, a twin source may also become inoperative when it has cross-slipped to the next glide-plane. This is despite the fact that the loops are not on the same slip plane but are located on neighboring parallel planes. Hence, if a twin of thickness n has formed, the back stress of the n partial dislocations on the source dislocation - now on the (n+1)st {111} plane - may lower the resolved shear stress to below the critical value for the operation of the source. As before, the source dislocation can wait for the surrounding loops to expand beyond d_{nc} before it starts operating. A second possibility, however, arises in this case. If the resolved shear stress on the cross-slip plane is sufficiently large, the screw dislocation may skip the (n+1)st plane and slip to the (n+2)nd plane, or generally skip the next m planes to the (n+m)th plane, where it is sufficiently far from the influence of the first n loops so that it can get back on the primary glide plane and start its operation of forming a faulted loop. The source can now start a new sequence, i.e. form n faulted loops on neighboring planes until it becomes inoperative again due to the back stress of the n loops already formed, cross-slip m planes to the (n+m)th plane where the interactions are sufficiently weak that the operation can re-start. In this way a crystal with a stacking repeat of (n+m) or 3(n+m) will form. Thus, in the case of a cubic crystal of low stacking fault energy with a repeat sequence of 3, (i.e. 3C), different polytypes can form as in the following table.

This operation depends on a relatively long dwell time, a low stacking fault energy, and the ease of cross-slip by the screw dislocation on the cross-slip plane. The numbers n and m are determined by the resolved shear stress on the primary glide plane, resolved shear stress on the cross-slip plane, mobility of the leading partial dislocation and mobility of the screw dislocation. Both the mobility of the leading partial dislocation and that of the screw dislocation are of course temperature dependent although their activation energy are presumably different. Hence, a variety of polytypes can be envisaged depending on the stress distribution

in the crystal and the range of temperaure which affects the relative
mobilities of the leading partial and the screw dislocation.

n (Number of faulted planes on the primary glide plane)	m (Number of {111} planes cross-slipped by the screw dislocation)	Polytype formed from 3C
n	0	Twin
1	1	2H
2	1	9R
2	2	4H
3	1	12H
3	2	15R
3	3	6H
.	.	.
.	.	.
.	.	.

In addition, electrically active impurities, e.g. acceptors such as Al and
B, or donors such as N and P, are expected to influence the formation of a
polytype because they are known to affect the dislocation mobility in
tetrahedrally-bonded semiconductors. This is well-known in the polytypic
transformations of SiC (Jepps and Page, 1983).

5. NON-CUBIC → CUBIC TRANSFORMATIONS

Formation of the cubic 3C polytype by the present mechanism is simpler.
During its cross-slip, the screw dislocation comes across only a few {111}
primary glide planes where it is able to dissociate without forming a
high-energy stacking sequence of the type AA, BB, or CC. Dissociation and
formation of a faulted loop on all these available primary glide planes
transforms the non-cubic polytype to the cubic 3C phase. This applies to
all the non-cubic polytpes and implies that the transformation of one
non-cubic polytype to another passes through the intermediate stage of 3C
formation.

ACKNOWLEDGEMENT

The author would like to thank Dr. P. M. Hazzledine for very helpful
comments on the paper.

REFERENCES

S. Fujita, K. Maeda, and S. Hyodo, Philos. Mag. A 55, 203 (1987).
A. George and J. Rabier, Revue Phys. Appl. 22, 941 (1987).
M. Imai and K. Sumino, Philos. Mag. A, 47, 599 (1983).
N. W. Jepps and T. F. Page, J. Crystal Growth & Charac., 7, 259 (1983).
K. Maeda, S, Fujita, and K. Suzuki, Proc. Vth Intn. Symp. 'Structure and
    Properties of Dislocation in Semiconductors', Zvenigorod, U.S.S.R.
    March 17-22, (1986).
K. Maeda, K. Suzuki, S. Fujita, M. Ichihara, and S. Hyodo, Philos. Mag. A
    57, 573 (1988).
P. Pirouz, Scripa Metall. 21, 1463 (1987).
P. Pirouz, Scripta Metall. 23, 401 (1989).

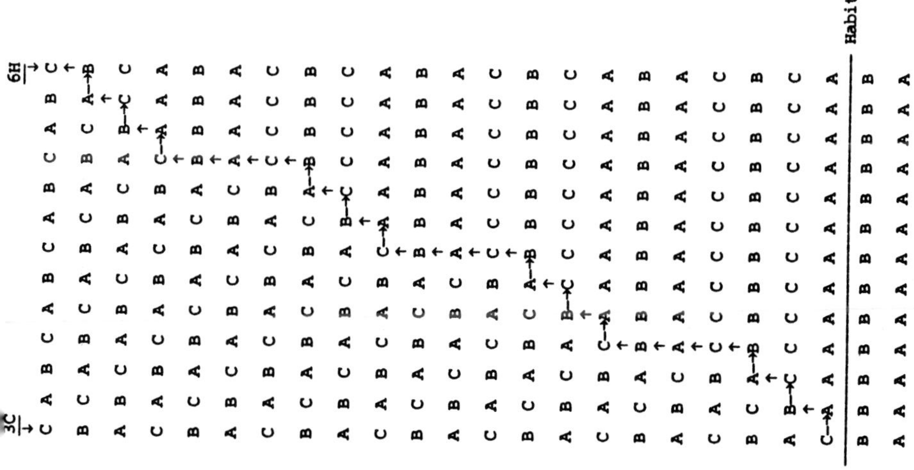

Fig. 4(b). 3C→6H transformation.

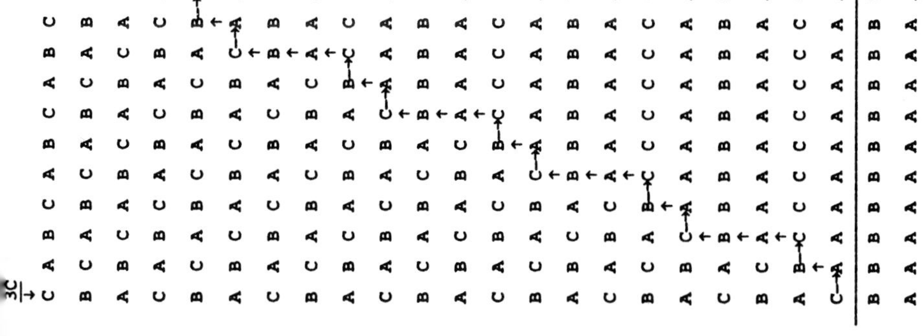

Fig. 4(a). 3C→4H transformation.

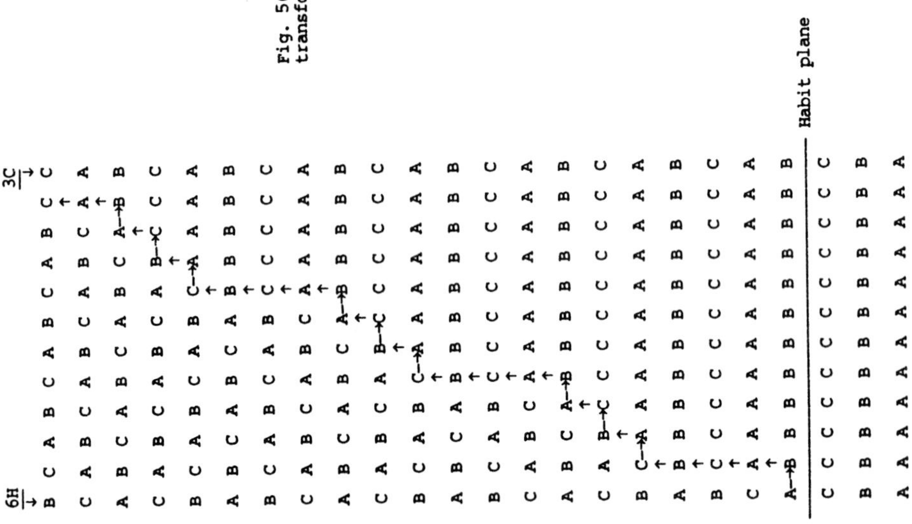

Fig. 5(b). 6H→3C transformation.

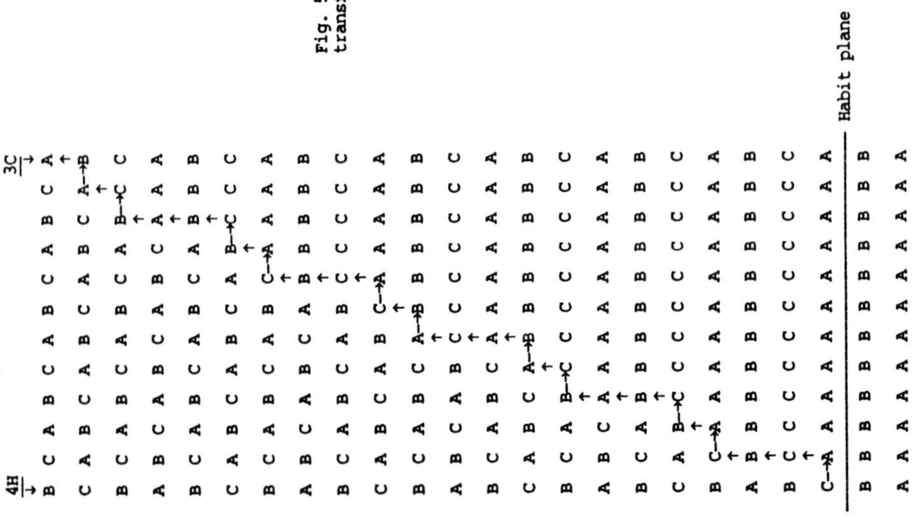

Fig. 5(a). 4H→3C transformation

Habit plane

Habit plane

Inst. Phys. Conf. Ser. No 104: Chapter 1
Paper presented at Int. Symp. on Struct. Prop. Disloc. Semicond., Oxford, 1989

Reconstructed dislocations and broken bond defects

H Teichler

Institut für Metallphysik der Universität Göttingen, and Sonderforschungsbereich 126, D-3400 Göttingen, F.R. Germany

ABSTRACT: Dangling bonds in dislocation cores are treated within an LCAO
approach. For 90°-partial dislocations in Ge and Si the energy
difference between reconstructed and unreconstructed configurations is
calculated. For the reconstruction defect of the 90° - partial in Si
formation energy and electronic structure are estimated.

1. INTRODUCTION

The present contribution is concerned with the properties and the
theoretical description of broken bonds *viz.* unpaired 'dangling' bonds
(db's) in the core of dislocations in Ge and Si. This is a rather urgent
question since reliable treatments of atomic arrangements with db's are
needed to clarify whether db carrying defects in the dislocations may
account for the observed electrical activity of the latter.

According to the early proposal of Shockley (1953) rows of db's in the
perfect dislocations are the source of the strong electrical activity of
the dislocations in semiconductors. In the last years, however, energy
estimates by a number of authors (e.g., Marklund 1979, Jones 1979, Altmann
Lapiccirella and Lodge 1983) give strong evidence that in Si and Ge the
predominating 30°- and 90°-partial dislocations do occur in reconstructed
configurations with pairwise rebonding of the core db's rather than in the
broken bond geometry. At the same time electron theoretical studies
revealed that the ideal reconstructed configurations do not provide deep
electron levels in the band gap (Chelikowsky 1982, Veth and Teichler 1984)
or - at most - some bound states in the uppermost part of the gap (Marklund
1979, Jones 1979, Lodge, Altmann, Lapiccirella and Tomassini 1984) -besides
the deformation potential states (Teichler 1975). In this situation it is
tempting to ascribe the richness of the observed deep levels at
dislocations (see, e.g., the recent reviews by Labusch and Schröter (1980)
or Alexander (1986)) to defects in the dislocation cores, which raises the
question of the formation energy, atomic arrangement and electronic level
structure of such defects.

The *atomic arrangement* in deformed lattices without broken bonds can be
easily modelled by use of phenomenological interatomic potentials. Some
care must be taken when estimating *formation energies* from these
potentials, since for this purpose the phenomenological interaction must be
able to correctly describe the energy of long wavelength deformations
(elastic stress field) and of short wavelength perturbations as found in

defect cores, (see, e.g. our recent discussion on that point (Teichler 1989)). Also in the presence of db's the atomic arrangement can be simulated by use of the phenomenological interactions, with db's modelled in the 'virtual atom' approach of Marklund (1979), provided the db's do not introduce noticeable interactions with neighbouring atoms. The main problem for defects with db's is the estimation of the formation energies, since it demands knowledge of the db self-energy, a quantity so far not well known.

In the following we discuss how to treat db's within an LCAO approach. We consider in particular the case of db's with marked residual interactions, as is the case in the core of the 'unreconstructed' 90°-partials in Si and Ge, and the reconstruction defect in the 'reconstructed' 90°-partial in Si. Some details of our model will be described in Section 2 of the paper. Section 3 is devoted to the db's in the unreconstructed partials and Section 4 deals with the reconstruction defect.

2. BOND ORBITAL MODEL FOR SATURATED COVALENT BONDS

Here we sketch some basic features of the bond orbital approach (BOA) proposed by Harrison (1980) for the case of compatible deformations without db's. In this approach the structure dependent energy of the system is modelled by a phenomenological short-ranged repulsive interatomic potential and an attractive bond-oriented part deduced from electron theory. It relies on an independent-particle picture for the electrons with the understanding that some net effects of the Coulomb interaction are incorporated in the repulsive potential. The BOA models the Bloch waves of the valence bands by linear combinations of 'bond orbitals' $|j\rangle$ which are normalized, symmetric combinations of the two sp^3-hybrids in each bond

$$|j\rangle = (\ |h_j^{\alpha}\rangle + |h_j^{\beta}\rangle\ )\ /\ \sqrt{\{2\cdot(1+S_j)\}}\ , \qquad (2.1)$$

$S_j = \langle h_j^{\alpha}|h_j^{\beta}\rangle$. For compatible deformations with saturated covalent bonds the one-particle contribution to the electronic energy can be expressed as

$$E_{el} = \sum_j \langle j|\ H\ |j\rangle\ , \qquad (2.2)$$

where H includes kinetic and potential energy of the electrons. According to Harrison's (1980) approach the repulsive core-core interaction (and further corrections) are taken into account by a short - ranged interatomic pair potential, $W(r)$, which we (for $r \leq r_0$) model by

$$W(r) = \sum_{\ell=0}^{3} w_{\ell} \cdot (\ r/d_0 - 1\ )^{\ell}\ . \qquad (2.3)$$

The parameters w_{ℓ} are selected so that the total energy (E_{el} and repulsive interactions) describes well the equilibrium distance d_0, bulk modulus, pressure dependence of the bulk modulus, and the energy difference E_0 between an isolated atom with sp^3 configuration and the ideal lattice energy. (For $r > r_0$, $r_0 = 1.15\ d_0$ in Si, $1.06\ d_0$ in Ge, we use the form $W(r) = A \cdot \exp(-B \cdot r)$, with A and B from the condition that $W(r)$ and its derivative are continuous at r_0.) E_0 is the sum of cohesion energy and the $s^2 p^2 \to s p^3$ promotion energy and has a value of 5.84 eV per bond in Si (5.96 eV in Ge) (Harrison 1980). Due to the short-range nature of the interaction this construction fixes the bond breaking energy (per bond) for situations without residual interactions or relaxations to E_0 and thus predicts for the ideal db self energy the value $E_{db}^0 = E_0/2$ (2.92 eV in Si, 2.98 eV in Ge).

Estimation of E_{db}, the true db self energy including residual interactions and electronic relaxations, and of the bond deformation energies demands a model for the atomic orbitals and potentials. In our calculation the atomic s- and p-orbitals and the atomic potentials are described by Gaussians with

parameters adopted from Louie's (1980) paper. (Within this model the w_ℓ in
(2.2) for $\ell=0$ to 3 have the values 5.07, -26.04, 43.39, -53.02 eV per bond
in Si and 3.89, -26.05, 46.43, -86.15 eV in Ge.)

Application of this model to arbitrary deformations deserves an additional
detail. At atom α the $|h_j^\alpha>$ and the s- and p-orbitals $|\mu^\alpha>$ are related by

$$|h_j^\alpha> = \sum_\mu v_{j\mu}^\alpha \ |\mu^{\alpha j}> \qquad (2.4)$$

In the ideal lattice one has $v_{j\mu}^\alpha = \pm 1/2$ where $(v_{jx}^\alpha, v_{jy}^\alpha, v_{jz}^\alpha)$ describes the
orientation of the hybrid and points towards the neighbouring atoms. In a
complete approximation-free treatment the energy will be independent of the
choice of the v^α. However, approximat treatments like the Heitler-London
approach used here depend on the selected v^α and optimized values have to
be deduced for them *via* Ritz's principle by considering the v^α as
additional variation parameters in the energy minimization procedure. We
have taken into account orthogonal transformations with $\{v^\alpha\} \in SO(4)$.

3. DANGLING BONDS IN UNRECONSTRUCTED 90°-PARTIAL DISLOCATIONS

The db's in the core of the 90°-partials
provide an interesting system to show how
db's can be incorporated in the BOA. In
the diamond lattice the 90°-partials of
glide set geometry are periodic along the
dislocation line with periodicity $b=a/\sqrt{2}$
and contain two rows of threefold
coordinated atoms (below named core atom
rows (c.a.r.'s)). In the unreconstructed
configuration (unrec.c.) the two c.a.r.'s
are mutually displaced by b/2 whereas the
reconstructed configuration (rec.c.)
involves an additional symmetry breaking
shift u (Marklund 1979) as shown in Fig.1.

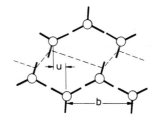

Fig.1. Atomic arrangement and
sp-hybrids of core atoms in
90°-partial dislocations in Ge
and Si with periodicity b
along the dislocation line

The rec.c. shows a pairwise rebonding of the db's across the dislocation
core, in the unrec.c. no *pairwise* reconstruction is possible. Nevertheless,
even in the unrec.c. there is a marked interaction between db's on
different c.a.r.'s as can be inferred from the drastic splitting of the db
energy levels in the LCAO treatment of the dislocation energy bands by
Teichler and Marheine (1987). In accordance with this finding the BOA has
been generalized to apply to the unrec.c. as well as to the rec.c. (and
intermediate cases) by Schulz and Teichler (in preparation) and Schulz
(1987). In this approach the rows of db's at the c.a.r.'s are described by
'Bloch-Wannier' functions (Brown 1967), i.e. Bloch-wave like linear
combinations of the db hybrids,

$$|k,q> = N^{-1/2} \sum_\alpha |h_q^\alpha> \exp\{ikR_q^\alpha\} \quad , \qquad (3.1)$$

(q labels the two c.a.r.'s, R_q^α denotes the position of the atom α on the
c.a.r. q). The BOA is applied to these k-dependent functions and yields in
the simplest case as bonding orbitals the symmetric combination

$$|k,b> = (\ |k,1> + |k,2> \ ) \ / \ \sqrt{\{2\cdot(1 + S(k))\}} \ , \qquad (3.2)$$

with $S(k)=<k,1|k,2>$ and bonding energy $E(k,b) = <k,b|H|k,b>$. (In practice,
due to the stacking fault attached to one of the two c.a.r.'s the bonding
energy $E(k,b)$ has to be determined as the bonding root of the 2nd order
secular equation in the $|k,1>$, $|k,2>$ subspace.) The electronic binding
energy between the two c.a.r.'s is obtained by integrating $E(k,b)$ in the

interval $-\pi/b < k < \pi/b$.

The structure dependent part of $E(k,b)$ is primarily determined by the energy matrix elements and overlap integrals between one hybrid and its neighbours on the opposing c.a.r., as indicated by the dashed lines in Fig.1. In the unrec.c. (u=0) the atomic arrangement has an additional mirror symmetry and left-hand and right-hand couplings are identical. Finite u breaks this symmetry, makes one set of couplings dominant, and leads to a smooth transition into a pairwise reconstructed covalent bond geometry, where the latter is reached once the minor couplings are negligible compared to the major ones.

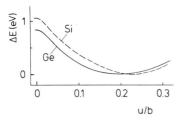

Fig.2. Change of the total energy of 90°-partials in Ge and Si with symmetry breaking displacement u

This approach is applied here to the 90° partials in Si and Ge (a preliminary treatment for Si was given by Schulz, 1987). Energy and atomic structure are estimated by energy minimization treating the c.a.r.'s (and their back bonds) in the above indicated BOA and by describing the residual interactions via Keating's (1966) model. Fig.2 displays the resulting energy (per periodicity length b) as a function of u with the energy zero put to the rec.c., which is characterized by u=0.23 b in Si (0.19 b in Ge). The rec.c. indeed corresponds to a bond reconstruction geometry in the above mentioned sense, as may be seen from the overlap-integrals S. For a perfect bond in Si we find $S_{0}=0.61$. In the rec.c. the short bond across the core (Fig.1) gives $S_{1}=0.55$, whereas the overlap between the next nearest neighbour hybrids across the core is $S_{2}=0.045$.

In the unrec.c. the overlap between neighbouring hybrids on different c.a.r.'s is $S_{1}^{*}=0.21$. According to the S values the core bonds in the unrec. c. are between db's and complete covalent bonds. This is also found when comparing energies: according to Keating model calculations of Veth (1983) the 'lattice' energy of the rec.c. is larger by about 1.6 eV in Si (1.5 eV in Ge) than that of the unrec.c. (neglecting any c.a.r. interaction). Interpretation of our values for the total energy difference (including the c.a.r. interactions) of Fig.2 in terms of these lattice energy differences plus additional db self energies, E_{db}, yields $E_{db}=1.33$ eV in Si (1.16 eV in Ge). This is much less than the ideal db self energy $E_{db}^{0} \approx 2.95$ eV. It, indeed, shows that the db's in the unrec.c. regain an appreciable amount of bond breaking energy through residual interactions (and electronic relaxation).

4. RECONSTRUCTION DEFECTS ON RECONSTRUCTED 90°-PARTIALS

A situation with an isolated db is found in the reconstruction defects (RCD's) of the 90°-partials proposed e.g., by Hirsch (1980) and sketched in Fig.3. Atomic arrangement and 'lattice' energy (neglecting db self energies) have been deduced by Veth (1983) within Keating's (1966) model in the 'virtual atom' approach (Marklund, 1979). From treating a periodic array of RCD's (with RCD distance 3.5 b) he predicts a 'lattice' energy of 0.4 eV per RCD in Si (0.3 eV in Ge).

We here consider one isolated RCD. For this case we find from Keating's (1966) model a 'lattice' energy of about 0.7 eV in Si - 0.6 eV in Ge. (The difference to Veth's values may be due to attractive RCD interactions in his treatment.) In the spirit of the generalized BOA the db self energy can be estimated by taking into account the interaction between the db hybrid h_o and its most important neighbours $(h_1, ..., h_4$ of Fig.3). Due to symmetry the Hamiltonian in this 5-dimensional subspace decays into one 3-dim. and two 1-dim. irreducible representations, where h_o enters only the 3-dim. part. Use of LCAO parameters as in the BOA and atomic structure from energy minimization within Keating's model shows that h_o markedly couples only to the combination

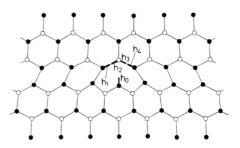

Fig.3.Atomic arrangement and significant sp-hybrids of a reconstruction defect in the $90°$-partials

$$|s> = (|h_1> + |h_2> + |h_3> + |h_4>) / 2 \quad , \qquad (4.1)$$

i.e. the symmetric combination of bonding orbitals in the opposing bonds, whereas the antisymmetric combination of bonding orbitals and the anti-bonding orbitals are not affected by h_o. In the 2-dim. h_o-s subspace the h_o-s-coupling has two effects: a) it lowers the energy of the bonding eigenstate (which has a dominating amplitude on the $|s>$ state and is twofold occupied in the neutral case) below that of a genuine covalent bond and b) it raises the energy of the anti-bonding solution (with dominating amplitude on the $|h_o>$ state and single occupancy in the neutral case) above that of an uncoupled db. Our LCAO parameters yield that for RCD's in Si due to a) the total electronic energy is lowered by 1.65 eV, whereas due to b) the db level is raised by 0.75 eV. The repulsive interaction between the db atom and its opposing partners turns out negligibly small.

With these data, by considering a pair of infinitely far separated RCD's, we can estimate the db self energy for RCD's on $90°$ partials in Si as

$$E_{db} = E_{db}^o - 1.65 \text{ eV} + 0.75 \text{ eV} \approx 2.0 \text{ eV} \quad .$$

So far the electronic relaxation processes in the back bonds of the db atom and effects of the residual h_o-s force on the atomic positions have been neglected.They may reduce the apparent value E_{db} where, however, reductions below about 1.3 eV seem to be rather improbable. Thus our estimate of the 'lattice' energy of an RCD and our E_{db} value indicate that the formation energy for an RCD on a $90°$-partial in Si should be of the order of 2 eV or even more.

Proper location of the db level in the band gap needs knowledge of its position under neglect of the db coupling to the opposing atoms, since we here have estimated only the corresponding db level shift. This position can be taken from the tight binding recursion treatment of db's by Teichler and Veth (1983) which just neglects this coupling and which predicts a db level position 0.02 eV above the valence band edge E_v. The here deduced level shift thus locates the db level at about 0.77 eV above E_v.

A more detailed determination of the level location needs, however, incorporation of band structure effects. Preliminary estimates of these effects have been obtained by simulating the residual crystal in the Bethe-Lattice approach (e.g., Davidovich, Koiller, Osorio and Robbins, 1988) with valence and conduction band attached to each bond and by coupling the $|s>$-state to the valence band and the $|h_o>$-state to valence and conduction band. For valence and conduction band of equal width of 11 eV, band gap of

1.2 eV,and semielliptic model density of states for the bands this approach yields a slight lowering of the db level location to 0.75 eV above E_v. Fig.4. displays the local density of states $N(E)$ obtained for the $|h_0\rangle$- (solid line) and $|s\rangle$-state (broken line) by including their mutual coupling and the coupling to the (Bethe-) lattice. The $|h_0\rangle$ state has nearly equal weight, $\int dE\ N(E)$,in valence and conduction band and is found with 0.355 weight in the bound state at 0.75 eV. The $|s\rangle$ state appears with weight 0.09 in the bound state and has minor weight in the conduction band.

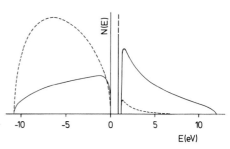

Fig.4. Local density of states for $|h_0\rangle$ (solid line) and $|s\rangle$ state (broken line) with bound state 0.75 eV above E_v.

In the neutral RCD the bound level at 0.75 eV is singly occupied. Occupation by a second electron introduces additional Coulomb energy which will raise the level to a value near the conduction band edge. This means that in n-material negative charging of the RCD has only a minor effect on the apparent formation energy of the RCD. In p-material, however, positively charged RCD's may exist with an apparent formation energy lowered by about 0.75 eV, where this lowering is a consequence of the db-coupling to the opposing covalent bonds.

REFERENCES

Alexander H 1986 *Dislocations in Solids* Vol.7 ed F R N Nabarro (Amsterdam: North Holland) p 118
Altmann S L, Lapiccirella A and Lodge K W 1984 *Int.J.Quant.Chem.* 23 1057
Brown R A 1967 *Phys. Rev.* 156 889
Chelikowsky J R 1982 *Phys. Rev. Lett.* 49 1569
Davidovich M A, Koiller B, Osorio R and Robbins M 1988 *Phys.Rev.* B 38 10524
Harrison W A 1980 *Electronic Structure and the Properties of Solids* (San Franzisco: Freeman)
Hirsch P B 1980 *J. Microscopy* 118 3
Jones R 1979 *J. Physique Suppl.* C6 - 33
Keating P N 1966 *Phys. Rev.* 145 637
Labusch R and Schröter W 1980 *Dislocations in Solids* Vol.5 ed F R N Nabarro (Amsterdam: North Holland) p 128
Lodge K W, Altmann S L, Lapiccirella A and Tomassini N 1984 *Phil.Mag.*B49 41
Louie S G 1980 *Phys. Rev.* B 22 1933
Marklund S 1979 *phys. stat. sol. (b)* 92 83
Schockley W 1953 *Phys. Rev.* 91 228
Schulz A 1987 *Thesis* Univ. Göttingen (the paper contains a numerical error in the energy of the unreconstr. disloc. which is corrected here)
Teichler H 1975 *Lattice Defects in Semiconductors* ed F A Huntley Inst.Phys. Conf.Ser.No 23 (London Bristol: The Institute of Physics) p 374; 1989 *Polycrystalline Semiconductors* ed J H Werner, H P Strunk and H J Möller, Springer Proc.Physics (Heidelberg: Springer) in press
Teichler H and Marheine C 1987 *Izv. Akad. Nauk. SSSR* 51 663
Teichler H and Veth H 1983 *J. Physique Suppl.* C4 - 93
Veth H 1983 *Thesis* Univ. Göttingen
Veth H and Teichler H 1984 *Phil. Mag.* B 49 379

Inst. Phys. Conf. Ser. No 104: Chapter 2
Paper presented at Int. Symp. on Struct. Prop. Disloc. Semicond., Oxford, 1989

63

Interface states at silicon grain boundaries

Jürgen H Werner

Max-Planck-Institut für Festkörperforschung,
D-7000 Stuttgart 80, Federal Republic of Germany

ABSTRACT: This paper reviews results from the analysis of interface states at Si grain boundaries; continua at midgap and asymmetric band tail states are found. There is strong experimental and theoretical evidence that defect states at midgap originate from the segregation of impurities. No systematic correlation between crystallographic structure and electrical activity has been found. The intrinsic band tail states are compared to tail states in amorphous silicon and at SiO_2/Si interfaces. It is speculated that such tails and their asymmetry arise from carrier localization due to intrinsic structural disorder.

1. INTRODUCTION

The electronic properties of silicon grain boundaries are dominated by the crystallographic misfit at the boundary between the misoriented grains. The misfit results in *intrinsic* lattice defects such as primary and secondary dislocations, dangling bonds, and fluctuations of bond lengths and angles, as well as in *extrinsic* electronic defects due to a preferential segregation of impurities and doping atoms at the grain boundary. These intrinsic and extrinsic defects introduce states into the forbidden gap which act as traps and recombination centers as discussed by Queisser (1983), Grovenor (1985), Seager (1985), Werner (1985), and Queisser and Werner (1988).

The defect states at silicon grain boundaries trap preferentially majority carriers. This effect causes a *depletion* region with a potential barrier impeding majority carrier current flow in n-type as well as in p-type material. No boundaries with *accumulation* and a barrier in the minority carrier band have been reported for polycrystalline Si. The barrier in the majority carrier band at Si boundaries is, on the other hand, not high enough to cause *inversion*, i.e. a *minority* carrier concentration at the boundary which is higher than the majority carrier concentration within the grains. Here the results for Si are in stark contrast to those of Uchida et. al. (1987) on n-type Ge, Herrmann et. al. (1984) on p-type InSb, and Grabecki et. al. (1984) on p-type $Hg_{1-x}Mn_xTe$ who investigated the properties of a two-dimensional minority carrier gas at inversion boundaries. No theory yet exists which explains *why* these materials display inversion boundaries whereas one always encounters depletion in Si. There might be some intrinsic mechanism at work which is related to the energy distribution of interface states and/or the different band structures of the single crystalline materials. This scientific problem (for Si) has been *experimentally* investigated by measuring the energy distribution of

interface traps giving rise to the charges at the boundary which are
responsible for depletion. Data for the energy distribution of defect
states at grain boundaries of semiconductors other than Si are scarce.

Part 2 of the present paper lists some characterization techniques which
were applied to electronic states at silicon grain boundaries; we summarize
results for the energy distribution of interface states. Most techniques
yield midgap continua and additional band tail states. Part 3 discusses
extrinsic grain boundary behavior and Part 4 interfacial inhomogeneities.
Part 5 concentrates on a comparison (Werner (1989)) of experimentally
observed band tails at silicon grain boundaries with band tailing in
amorphous silicon and at SiO_2/Si interfaces. It is speculated here that
band tailing in disordered silicon is an intrinsic effect and is a
consequence of Anderson localization of carriers within short wavelength
potential fluctuations due to disorder.

2. ANALYSIS TECHNIQUES

Figure 1 shows a band diagram of the
typical depletion region around an
electrically active grain boundary in
n-type Si: The prevailing negative
charge within the interface plane
establishes a potential barrier which
impedes the current flow of electrons
as majority carriers. The potential
well in the minority carrier band
acts as a preferential sink which
enhances hole recombination.

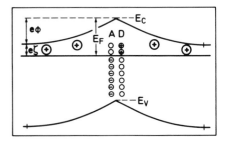

Fig.1: A charged boundary in
n-type Si. Negative interface
charge in acceptor states prevails
and is compensated by positive
donors in the space charge regions.

Most of the characterization techniques of grain boundary states rely
on the analysis of electrical measurements; little information about
the optical properties of grain
boundaries is available. Pike and
Seager (1979) and Seager and Pike
(1979) proposed the first deconvolution scheme for the energy distribution
of traps at Si boundaries on the basis of current/voltage characteristics
of single grain boundaries in bicrystals. *Bicrystal* techniques based on
capacitance transients were later used by Seager (1982) and Broniatowski
and Bourgoin (1982). Infrared photoconductance was applied by Werner,
Jantsch and Queisser (1982), and temperature dependent zero bias
conductance by Werner (1983, 1985). Werner and Strunk (1982) and Werner
(1983, 1985) showed later that admittance spectroscopy is particularly
suitable to analyze potential fluctuations within the grain boundary plane
which are due to the random spatial distribution of charged grain boundary
defects. Similar information can be obtained from current transients
(Stützler et al. (1985)) and noise experiments which were analyzed by
Madenach and Werner (1985, 1988) and Madenach, Werner and Stützler (1985).
Petermann (1988) and Petermann and Haasen (1989) applied the admittance
technique of Werner (1983) to bicrystals in float-zone material. Jackson,
Johnson and Biegelsen (1983) and Harbeke (1985) investigated traps in
fine-grained polycrystalline silicon films by optical absorption. Thermally
activated conductance was used by Hirae, Hirose and Osaka (1980) and later
by de Graaff, Huybers and de Groot (1982). Taylor and Tong (1984) as well
as Werner and Peisl (1985) used similar techniques in combination with
admittance spectroscopy; a field effect method was recently applied by

Fortunato et al. (1988). Conclusions about the energy distribution of interface states were also drawn by modeling the conductivity of polycrystalline films (Peisl and Wieder (1983)) and the behavior of thin-film transistors (Faughnan (1987) and Dimitriadis et al. (1988)).

The application of some of the mentioned techniques to bicrystals suffers from the fact that transport properties across single grain boundaries are not completely understood as recently pointed out by Queisser and Werner (1988). The interpretation of experiments using large voltage pulses (for example DLTS) is particularly difficult. Despite these objections there are two common features in the results for the density of states (DOS) at silicon grain boundaries in bicrystals as well as in fine-grained films. Almost all experiments yielded *midgap continua* in the energy distribution of electronic defects, and secondly, whenever it was possible to measure the DOS over a wide energy range, then *band tails* as shown in Fig.2 for fine-crystalline films were found. The midgap continua seem to have an extrinsic origin, whereas the appearance of such tails is a direct consequence of the disorder within the boundary plane and therefore an *intrinsic* effect as discussed in Part 5.

3. EXTRINSIC GRAIN BOUNDARY BEHAVIOR

There is strong evidence that midgap states at silicon grain boundaries are dominated by *extrinsic* properties. Most of the bicrystals studied are, for example, prepared from Czochralski-grown silicon which contains usually up to $1 \times 10^{18} cm^{-3}$ oxygen. This concentration corresponds to an areal density of $10^{12}-10^{13} cm^{-2}$. It is known from experiments on single crystalline Si that the electrical activity of oxygen depends critically on annealing conditions such as time, temperature and thermal history. Annealing temperatures around 450°C, which are usually used in order to anneal contacts for electrical measurements, are, for example, typical for the activation of so-called thermal donors. (For a review about the role of oxygen in Si see, for example, the book edited by Mikkelsen et al. (1986)). At higher temperature other defects, such as "new thermal donors" are known to form in single crystals. One can therefore understand that the activity of silicon grain boundaries containing a high concentration of oxygen depends critically on thermal history, as reported by Kazmerski and Russel (1982), Stützler and Queisser (1986), Yang (1987) and Kazmerski (1988).

Bicrystals with low oxygen content were investigated by Petermann (1988) and Petermann and Haasen (1989). These authors found electrical activity on float-zone grain boundaries in the as-grown state (without anneal). They applied the I/V-method of Pike and Seager (1979) as well as the doping method, the thermal method and admittance spectroscopy of Werner (1983, 1985) to their bicrystals and found also midgap continua. It is not clear if these continua were due to intrinsic lattice defects or if they were the result of impurity contamination; the impurity content was not investigated in detail. However, the oxygen content of such float-zone bicrystals should be lower than in Czochralski material.

It seems that at most investigated Si grain boundaries the basic dislocation network which is necessary to accommodate the misfit between differently oriented grains is indeed free of charges as proposed by Werner and Strunk (1982) and later reviewed by Queisser (1983). Impurities as well as disturbances of the basic dislocation networks (secondary dislocations, kinks and jogs etc.) are more likely to cause interface charges instead of the reconstructed primary dislocation lines. Activity of secondary dislocations at $\Sigma=5$ bicrystal boundaries was recently, for example, reported by Maurice and Colliex (1989). Geometrical arguments demonstrate also that the contribution of primary dislocations to grain boundary charge must be small: Measured areal grain boundary charges in Si amount to $10^{10}cm^{-2}$ to $10^{12}cm^{-2}$, the distances between the point charges range therefore between 10 and 100nm whereas the distance between primary dislocation lines is usually smaller. These intrinsic linear lattice defects, with bond angles and lengths which deviate from the values of the bulk material, are more likely to result in band tails similarly to the tailing in amorphous Si and at the Si/SiO_2-interface than to charged midgap states.

Extensive studies of the influence of impurities on the electrical activity of grain boundaries were recently reported (Möller, Strunk, and Werner (1989)). Martinuzzi (1989) investigated, for example, the activation and passivation of Σ=9 and Σ=13 boundaries in artificially grown Czochralski bicrystals as well as boundaries in coarse-grained Si for photovoltaic applications. He showed that the assumption of an intrinsic origin of trap centers at midgap did not agree with experiment. The basic dislocation network per se did not show recombination activity; impurities which are gettered at the dislocation network are instead responsible! Models which ascribe grain boundary charges solely to dangling bonds must thus be rejected (Queisser and Werner (1988)). Martinuzzi states: "If there is a tendency of higher electrical activity when misfit is larger at grain boundaries, this activity has essentially an extrinsic origin, due to a direct impurity segregation or to a segregation assisted by dislocations."

Pizzini et al. (1989) investigated the influence of oxygen, carbon and boron segregation on the recombination velocities of Si grain boundaries. Carbon and boron induced strong recombination activity, while simultaneous oxygen and carbon segregation caused a variety of effects. Aucouturier et al. (1989) found strong influence of oxygen and copper precipitates on the electrical activity of Σ=25 boundaries: The boundary in their *as-grown* bicrystals with a <001> tilt axis and an angle of 16.26° (yielding a [710] boundary plane) was electrically inactive. Annealing at temperatures between 450-950°C was necessary to obtain electrical activity. Other samples cut from the same bicrystal were investigated by Hamet and Nouet (1989). These authors explained the electrical activity as observed in DLTS experiments as due to copper, and, in addition, to nickel silicides within the grain boundary plane. Ihlal and Nouet (1989) made similar observations in EBIC measurements on Σ=13 and Σ=25 bicrystals. The Σ=25 boundary was also investigated by Maurice and Colliex (1989), who confirmed the electrical activity of copper and nickel related precipitates. All these investigations showed that the crystallographic structure of grain boundaries is of less significance; it has no *direct* influence on electrical activity. These findings are also in agreement with recent work of Bary and Nouet (1988) on Σ=9 boundaries.

It is interesting to compare these experimental results with recent theoretical work of Kohyama et al. (1988) and Paxton and Sutton (1989). Kohyama et al. examined the structural and electronic properties of a Σ=5 symmetrical tilt boundary around <001> with a semi-empirical tight binding model using a supercell method. Similar calculations were earlier performed by DiVincenzo et al. (1986) for the <011> symmetrical tilt boundary. In both cases there were no deep centers within the fundamental gap because of the absence of broken bonds at these boundaries. The calculations revealed localized boundary states at the band edges (usually termed as band tail states) and inside the valence band. These states were ascribed to structural disorder such as odd-membered rings and bond length and bond angle variations as is also found for amorphous Si (see discussion below). However, Kohyama et al. found that the energy of these localized states lies outside the fundamental gap. They predicted electrical inactivity for their perfectly periodic grain boundary.

Paxton and Sutton (1989) computed recently the atomic and electronic structure of three symmetrical tilt grain boundaries in Si. In those boundaries which contained only four-fold coordinated atoms (no broken bonds) they found no deep states in the forbidden gap. Deep states were found for three-fold coordinated atoms. The resolution did not allow to decide whether band tails appear at the edges of valence and conduction band. Results of these calculations are discussed in the contribution of A. P. Sutton which is presented in this volume.

A synthesis of presently known results from experiments *and* theory leads to the hypothesis that silicon grain boundaries with perfect periodicity without any broken bonds and impurity contamination should be electrically inactive. At least they should not contain deep levels in the forbidden gap. There seem no data available which contradict such a conclusion. However, from experiment alone one can not conclude that grain boundaries without any impurities do not display any charge. The *complete* absence of electrical activity of the basic dislocation network has not been conclusively demonstrated. Such experiments would need to use extremely clean and crystallographically well-defined bicrystal boundaries with a perfect periodicity over macroscopic dimensions (of the order of millimeters) in order to apply most of the presently available characterization methods. One usually judges the perfection of bicrystals with the help of high resolution transmission electron microscopy (Bourret and Rouviere (1989)) and one is therefore only able to check periodicity over distances of the order of several ten nanometers. Most parts of the grain boundary which determine the macroscopic electrical properties cannot be tested with respect to structural perfection. New techniques which use the scanning tunneling microscope as recently reported by Kazmerski (1989) might be a better tool to investigate the influence of single dislocation lines on the electrical activity of grain boundaries.

Theory and experiment are in disagreement on the existence of tail states at grain boundaries. There were no tails found for the perfect periodic grain boundaries which are assumed in theory, whereas experiments revealed tail states as discussed below.

4. INTERFACIAL INHOMOGENEITIES

If one wants to compare experiments on bicrystals to theories which are based on the assumption of perfect periodicity one is left with the problem of how to demonstrate the homogeneity of the boundary under investigation. Three different length scales can be distinguished here: Non-homogeneities on the scale of the

   i) width of the space charge region around the grain boundary (of the order of $1 \mu m$);
   ii) electrostatic fluctuations on a length scale of distances between charges (10-100nm);
   iii) short wave length potential fluctuations due to bond disorder (a nanometer scale).

Grain boundaries which show non-linear current/voltage-curves characteristics together with similar results for the band bending evaluated from current/voltage and capacitance/voltage are homogeneous on scale i) (Queisser and Werner (1988)). Inhomogeneities on scale ii) can be detected by admittance and noise spectroscopy as shown by Werner (1983, 1985) and Madenach and Werner (1985, 1988). Inhomogeneities on scale iii), which are here proposed to give rise to band tails are not directly accessible by electrical measurements.

5. BAND TAILING

Midgap states at Si grain boundaries are strongly influenced by impurities. The measured *band tail* states seem, on the other hand, to have an intrinsic origin: Werner and Peisl (1985) proposed that tails in fine-crystalline films are the result of carrier localization within short wavelength potential fluctuations, which are characteristic for the intrinsic disorder within the grain boundary planes. One could indeed imagine that the statistical fluctuations of the bond lengths and angles at grain boundaries result in fluctuations of the potential on a length scale of inter-atomic distances; these quantum well fluctuations should then immobilize carriers. The behavior of grain boundaries resembles, therefore, a two-dimensional amorphous material. Band tails which were experimentally detected in specific samples should consequently occur at *any* grain boundary. The degree of disorder determines just the slope of the exponential band tails: Tails in fine-crystalline films are expected to decay deep into the forbidden gap. Tails in relatively well-ordered bicrystal boundaries should be restricted to the band edges as, for example, shown in Fig.4.

Band tails due to disorder occur not only at grain boundaries but also in amorphous Si and at the Si/SiO$_2$-interfaces, and in all these systems the tails are asymmetric. It is here proposed that these tails and their asymmetry are the result of carrier localization and the different effective masses of electrons and holes.

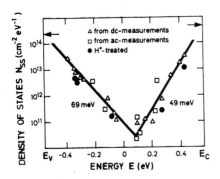

Fig.2: Tails at grain boundaries in fine-grained silicon films. The valence band tail has an inverse slope $E_{ov}=69$meV (see Eq.(1)). The conduction band tail shows $E_{oc}=49$meV. The different slopes result in an asymmetry in the density of states and shift the minimum towards the conduction band edge. Similar behavior is known from MOS-interfaces (Fig.3), from amorphous silicon and from bicrystal boundaries (Fig.4).

5.1 Tail states at MOS-interfaces

Traps at the SiO$_2$/Si-interface of MOS-structures have been the subject of innumerable investigations during the last twenty years. The density of interface states is continuous and consists of a U-shaped exponential distribution of band tails as in Fig.3 and additional midgap states. Nicollian and Brews (1982), Laughlin, Joannopoulos and Chali (1978) and Johnson, Biegelsen and Moyer (1980) discussed various models for the origin of these tails. There seems to be general agreement that tails originate from a disordered silicon region with local strain and a distribution of bond angles and bond lengths at the interface. Laughlin, Joannopoulos and Chadi (1987) found, for example, that at an ideal Si/SiO_2-interface without broken or distorted bonds there are no states within the forbidden gap. Broken bonds result in midgap states. Distortions of the Si-O-Si angle create a tail at the conduction band whereas the valence band tail stems from distortions of Si-Si bonds.

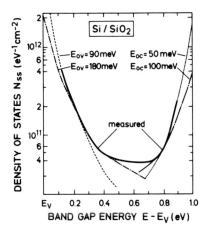

Fig.3: Measured U-shaped tail states of Johnson, Biegelsen and Moyer (1980) at the SiO_2/Si interface are explainable by Anderson localization of free carriers within disorder induced potential fluctuations. Singh and Madhukar (1981a,b) used a fit to Eq.(1) with n=3/2 and obtained E_{ov}=90-180meV for the valence band tail and E_{oc}=50-100meV for the conduction band tail. Our fit to Eq.(1) with n=1 yields E_{ov}=90-100meV, E_{oc}=60-70meV.

Singh and Madhukar (1981a,b) ascribed the tails at SiO_2/Si interfaces to Anderson localization of carriers within disorder induced potential fluctuations; tail states are thus created below the mobility edge. These authors described the density of states N_{SS} at energy E from the band edge E_b by

$$N_{ss}(E) = N_o \exp\left\{-\left(E_b - E\right)/E_o\right\}^n \tag{1}$$

with n=3/2. The asymmetry in the density of states which stems from a steeper behavior of the conduction band tail (E_{oc}=50-100meV) than at the valence band edge (E_{ov}=90-180meV) was not explicitly explained. If we fit the data in Fig.3 with Eq.(1) and use n=1 (instead of n=3/2) then we obtain E_{ov}=90-100meV and E_{oc}=60-70meV.

5.2 Band tails in Amorphous Silicon

Winer and Ley (1987) and Winer, Hirabayashi and Ley (1988) recently measured the *intrinsic* tails in amorphous silicon. They found an intrinsic valence band tail which was exponential over more than three orders of magnitude. A fit of their data to Eq.(1) with n=1 yields E_{ov}=45-50meV. Similar measurements in a phosphorus-doped sample yielded the conduction band tail with E_{oc}=35meV. The asymmetry in the density of states due to a valence band tail which decays deeper into the gap than the conduction band tail is thus also found in amorphous silicon.

Joannopoulos (1977) used a Bethe lattice model and proposed earlier that the *valence* band tail in a:Si stems from bond angle fluctuations (which are more important than bond length distortions) in agreement with the results for the MOS-interface. Singh (1981) made use of a tight-binding scheme for a:Si and investigated quantitative disorder (variations of bond lengths, angles, and bond rotations) as well as topological disorder (ring statistics). Quantitative disorder affects the valence band edge much more than the conduction band edge. *Dihedral disorder* (rotations around bonds) is primarily responsible for the shift of the valence band edge; bond-angle and bond-length disorder are of inferior significance and result in smearing of the tails due to dihedral disorder. Singh predicted total energy fluctuations around 0.5eV near the top of the valence band and 0.2eV for the conduction band. Topological disorder mainly influences the conduction band edge.

5.3 Tails at Silicon Bicrystals

We found the asymmetry of band tailing also at single grain boundaries in artificially grown silicon bicrystals. Figure 4 shows the density of states as measured by Werner (1983, 1985) and Madenach and Werner (1985, 1988) by different methods at a [110]-tilt boundary. The data for E<0.55eV were obtained by admittance and noise spectroscopy as well as from the analysis of current/voltage-curves. The results for E>0.55eV stem from the photocapacitances of Werner, Jantsch and Queisser (1982) which revealed for E>0.95eV an exponential behavior following Eq.(1) with n=1 and an E_0-value of E_{OC}=11-12meV for the conduction band tail. The dashed line in Fig.4 for E<0.4eV represents a valence band tail with E_{OV}=29meV, which was found by Madenach, Werner and Stützler (1985) in a sample with different orientation. This result of a deeper tail at the valence band edge than at the conduction band edge is consistent with the results for fine-crystalline films (Fig.2), amorphous Si, and Si/SiO$_2$-interfaces (Fig.3).

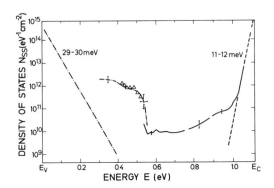

Fig.4: Band tails at bicrystal boundaries. The conduction band tail has an inverse slope E_{oc}=11-12meV (after Madenach and Werner (1988)). Dash-dotted line represents the valence band tail of a different sample with E_{ov}=29-30meV. The slopes are smaller than for polycrystalline films (Fig.2) and Si/SiO$_2$ (Fig.3).

Tail states at the conduction band edge are expected to behave like acceptors. They are therefore electrically neutral under thermodynamic equilibrium conditions for the case of n-type Si with the Fermi level located around midgap. Similar arguments hold for the donor-like states at the valence band edge. The existence of such tail states in bicrystal boundaries can then only be inferred from non-equilibrium measurements which allow the charging of the defect states close to the band edges. Werner, Jantsch and Queisser (1982) used, for example, infrared photoconductivity to charge the acceptor-like tail states at the conduction band edge with electrons which are excited from the valence band. The donor-like bicrystal tail states at the valence band edge in Fig.4 could be detected by Madenach and Werner (1988) because in that particular bicrystal there were insufficient states at midgap to pin the Fermi level there.

5.4 Theories of Band Tailing

There is presently no theory which specifically describes band tailing at grain boundaries. In fact, the MOS-interface seems to be the only interface at all that has been so far considered in band tail theory. There are, however, general theories available which model band tailing due to *charge localization within potential fluctuations* in any d-dimensional physical system. If one speculates that the disorder of bond lengths, angles and dihedral angles at grain boundaries, Si/SiO_2-interfaces and in a:Si results in short wavelength potential fluctuations on a length scale of the lattice constant then these quantum well fluctuations could localize free carriers. (Similar fluctuations could also be introduced by impurities because they distort the periodicity of the lattice even when they are uncharged.) Band tailing theories, recently reviewed by Sa-Yakanit and Glyde (1987) as well as by Cohen et al. (1988), which *assume* charge localization in potential fluctuations seem therefore appropriate for the description of band tailing in the disordered, real physical systems discussed here.

Band tailing in a three-dimensional system (d=3) due to potential fluctuations with correlation length L was first described by Halperin and Lax (1966). Their model predicts n=1/2 for fluctuations with L=0 and n=2 for the semi-classic limit with L=∞. Soukoulis, Cohen, and Economou (1984) showed that tails which follow Eq.(1) with n=1 could be understood within the models of Halperin and Lax if one included the effects of disorder on the energy of localization. Sritakool, Sa-yakanit, and Glyde (1986) confirmed analytically later that the *original* Halperin and Lax model already explains the occurrence of strictly exponential tails with n=1 in a:Si if one takes L=0.2nm and $W^2=(1eV)^2$ which are typical values for amorphous silicon. Similar results were recently obtained by Bacalis, Economou and Cohen (1988).

According to Soukoulis, Cohen and Economou (1984) in a three-dimensional description one obtains for the inverse slope E_0 of the band tail

$$E_o = \frac{\pi}{2} \frac{W^2 L^2 m_x}{h} \qquad (2)$$

for carriers of mass m_x which are localized in Gaussian potential fluctuations with variance W^2 and correlation length L. Large effective masses m_x, strong potential fluctuations (large values for W^2) and large correlation lengths L result thus in large E_0-values and deep tails in the forbidden gap.

The influence of disorder on the potential fluctuations at the conduction and valence band can be different. The values for correlation length L and variance W^2 are generally not equal for the two band edges. It is therefore interesting to note that the different effective masses m_x of electrons and holes explain the experimentally observed asymmetry of tails in disordered Si.

5.5 A Proposal for the Origin of Asymmetric Band Tails

We suggest that the asymmetry in the density of states of polycrystalline films (Figs.2), at the Si/SiO_2 interface (Fig.3), in amorphous silicon, and at bicrystal boundaries (Fig.4) may be understood on the basis of Eq.(2): For *equal* values of L and W^2 for the potential fluctuations at the two band edges we predict a ratio $R=E_{0v}/E_{0c}=1.7$ for the tails just by the different relative (density of states) masses of 0.55 and 0.33 for holes and electrons in silicon. This value R>1 is indeed compatible with R=1.4 for fine-grained films in Fig.2. For the states at the Si/SiO_2 interface in Fig.3 one obtains R=1.3-1.6 from the experimental data. The data of Winer and Ley (1987) and Winer, Hirabayashi and Ley (1988) for amorphous Si yield also R=1.4. The conduction band tail and the valence band tail in Fig.4 stem from two different bicrystals with different orientation and yield a ratio R=2.4-2.7; the result is therefore also compatible with the prediction R>1 which follows from this simple model which is exclusively based on different effective masses.

Asymmetric band tails at grain boundaries, MOS-interfaces, in amorphous Si, and at bicrystals seem therefore to be an intrinsic property of silicon. The tails could originate from charge localization within disorder induced potential fluctuations. The larger hole mass implies that valence band tails extend deeper into the forbidden gap of silicon than conduction band tails. This asymmetry is eventually enhanced by the stronger distortion of the valence band edge due to dihedral disorder as proposed by Singh (1981). The enhancement could in Eq.(2) be interpreted in terms of larger W^2-values for fluctuations at the valence band edge.

5.6 The Different Length Scales of Potential Fluctuations

It is here emphasized that the short wavelength, *disorder* induced potential fluctuations which cause carrier localization with the consequence of band tailing should have typical correlation lengths of the order of interatomic distances, i.e. around a few nm. The postulation of such potential fluctuations at grain boundaries allows a possible explanation of the measured exponential behavior of the density of interface states as shown in Figs.2,4. In addition, within such a picture the asymmetry of the density of defect states in amorphous Si, polycrystalline Si, and at the SiO_2/Si-interface can be consistently explained. The short wavelength potential fluctuations are, however, not directly accessible by experiment.

Long wavelength potential fluctuations at grain boundaries as discussed in Part 4 are, on the other hand, more directly accessible. These *electrostatic* fluctuations stem from the statistical spatial distribution of *charged* defects (impurities, dangling bonds, charged tail states) with separations of 10-100nm for typical grain boundary charges of $10^{10}-10^{12}cm^{-2}$. The electrostatic fluctuations modulate the mean carrier concentration at grain boundaries and result in a *spatial* distribution of time constants for capture/emission processes, an effect which is *directly* observable in admittances (Werner 1983, 1985) or noise spectroscopy (Madenach and Werner 1985, 1988)).

CONCLUSIONS

The electronic properties of Si grain boundaries are strongly influenced by extrinsic effects such as contamination and impurity segregation. The sensitive dependence of the recombination activity of grain boundaries on thermal history indicates that oxygen plays a major role. Fast diffusing transition metals such as copper and nickel, and their silicides, were found at bicrystal boundaries. Extreme care has therefore to be taken in sample preparation and handling if one searches for intrinsic electrical activity of dangling bonds etc. in the dislocation network at Si grain boundaries.

Theory predicts that certain perfectly periodic bicrystal boundaries should neither contain midgap states nor tail states. The energy of localized band tail states was predicted to be outside the fundamental gap. The results of theory are therefore in contradiction to measured tail states at grain boundaries which were free of measurable amounts of segregation (Werner and Strunk (1982)). The discrepancy might originate from the fact that grain boundaries which were studied in experiment contained much stronger and *non-periodic* lattice disturbances than were assumed in the theories, which were applied to perfectly *periodic* grain boundaries.

ACKNOWLEDGMENTS

The author gratefully acknowledges most valuable collaboration with Armin Madenach, Hans Queisser, Martin Peisl and Horst Strunk. It is a pleasure to thank Tony Paxton and Adrian Sutton for several open minded, scientifically interesting and fruitful discussions. I am obliged to Andrew Blakers for a critical reading of the manuscript.

REFERENCES

Bacalis N, Economou E N, and Cohen M H 1988 Phys. Rev. B37, 2714
Bary A and Nouet G 1988 J. Appl. Phys. 63, 435
Bourret A and Rouviere J L 1989 in *Polycrystalline Semiconductors - Grain Boundaries and Interfaces*, Springer Proc. in Phys. 35 (in press)
Broniatowski A and Bourgoin J C 1982, Phys. Rev. Lett. 48, 424
Cohen M H, Chou M Y, Economou E N, John S, and Soukoulis C M 1988 IBM J. Res. Develop. 32, 82
Dimitriadis C A, Papadimitriou L, Stoemenos J, and Economou N A 1988 J. Appl. Phys. 63, 1104
DiVincenzo D P, Alerhand O L, Schlüter M and Wilkins J W 1986 Phys. Rev. Lett. 56, 1925
Faughnan B 1987 Appl. Phys. Lett. 50, 290
Fortunato G, Meakin D B, Migliorato P, and Le Comber P G 1988 Phil. Mag. B 57, 573
Halperin B I and Lax M 1966 Phys. Rev. 148, 722
Hamet J F and Nouet G 1989 in *Polycrystalline Semiconductors - Grain Boundaries and Interfaces*, Springer Proc. in Phys. 35 (in press)
Harbeke G 1985, in *Polycrystalline Semiconductors - Physical Properties and Applications*, G. Harbeke ed., (Springer Berlin), p. 156
Herrmann R, Kraak W, and Glinski M 1984 Physica Status Solidi B 125, K85
Hirae S, Hirose M, and Osaka Y 1980 J. Appl. Phys. 51, 1043
de Graaff H C, Huybers N, and de Groot J C 1982 Solid State Electron. 25, 67
Grabecki G, Dietl T, Sobkowicz P, Kossut J, and Zawadski W 1984 Appl. Phys. Lett. 45, 1214
Grovenor C R 1985 J. Phys. C: Solid State Phys. 18, 4079
Ihlal A and Nouet G 1989 in *Polycrystalline Semiconductors - Grain Boundaries and Interfaces*, Springer Proc. in Phys. 35 (in press)
Jackson W B, Johnson N M, and Biegelsen D K 1983, Appl. Phys. Lett. 43, 195
Joannopoulos J D 1977 Phys. Rev. B16, 2764
Johnson N M, Biegelsen D K, and Moyer M D 1980 in *The Physics of MOS Insulators*, edited by G. Lucovsky, ST. Pantelides, and G.F.L. Galeener, (Pergamon, New York), p.311
Kazmerski L L and Russel P E 1982 J. Phys. Colloque (Paris) 43, C1-171
Kazmerski L L 1988 Mat. Res. Soc. Symp. Proc. 106, 199
Kazmerski L L 1989 in *Polycrystalline Semiconductors - Grain Boundaries and Interfaces*, Springer Proc. in Phys. 35 (in press)

Kohyama M, Yamamoto R, Ebata Y, and Kinoshita M 1988 J. Phys. C:
   Solid State Phys. 21, 3205
Laughlin R B, Joannopoulos J D, and Chali D J 1978 in *The Physics of SiO₂
   and its Interfaces*, edited by ST. Pantilides (Pergamon, New York), p.321
Madenach A J and Werner J H 1985 Phys. Rev. Lett. 55, 1212
Madenach A J, Werner J H, and Stützler F J 1985, in *Proc. 18th
   Photovoltaics Spec. Conf.*,(IEEE, Las Vegas), p.1080
Madenach A J and Werner J H 1988 Phys. Rev. B 38, 13150
Martinuzzi S 1989 in *Polycrystalline Semiconductors - Grain
   Boundaries and Interfaces*, Springer Proc. in Phys. 35 (in press)
Maurice J L and Colliex C 1989 in *Polycrystalline Semiconductors - Grain
   Boundaries and Interfaces*, Springer Proc. in Phys. 35 (in press)
Mikkelsen J C, Pearton S J, Corbett J W, and Pennycook S J 1986
   Oxygen, Carbon, Hydrogen and Nitrogen in Crystalline Silicon
   Mat. Res. Soc. Symp. Proc. 59 (Materials Research Society, Pittsburgh)
Möller H J, Strunk H P and Werner J H 1989 eds. *Polycrystalline Semi-
   conductors - Grain Boundaries and Interfaces*,
   Springer Proc. in Phys. 35 (Springer Verlag, Heidelberg) in press
Nicollian E H and Brews J R 1982 *MOS Physics and Technology*,
   (John Wiley and Sons, New York) chapter 16
Paxton A T and Sutton A P 1989 Acta Met. (in press)
Peisl M and Wieder A W 1983 IEEE Trans. Electr. Dev. ED-30, 1792
Petermann G 1988 Phys. Stat. Sol. (a) 106, 535
Petermann G and Haasen P 1989 in *Polycrystalline Semiconductors - Grain
   Boundaries and Interfaces*, Springer Proc. in Phys. 35, (in press)
Pike G E and Seager C H 1979 J. Appl. Phys. 50, 3414
Pizzini S, Borsani F, Sandrinelli A, Narducci D, Anderle M and Canteri R
   1989 in *Polycrystalline Semiconductors - Grain Boundaries and
   Interfaces*, Springer Proc. in Phys. 35, (in press)
Queisser H J 1983 Mat. Res. Soc. Symp. Proc. 14, 323
Queisser H J and Werner J H 1988 Mat. Res. Soc. Symp. Proc. 106, 53
Sa-Yakanit V and Glyde H R 1987 Comm. Cond. Mat. Phys. 13, 35
Seager C H and Pike G E 1979 Appl. Phys. Lett. 35, 709
Seager C H 1982 Appl. Phys. Lett. 40, 471
Seager C H 1985, Ann. Rev. Mat. Sci. 15, 271
Singh J 1981 Phys. Rev. B23, 2156
Singh J and Madhukar A 1981a J. Vac. Sci. Technol. 19 437
Singh J and Madhukar A 1981b Appl. Phys. Lett. 38, 884
Soukoulis C M, Cohen M H, and Economou E N 1984 Phys. Rev. Lett. 53, 616
Sritakool W, Sa-yakanit V, and Glyde H R 1986 Phys. Rev. B33, 1199
Stützler F J, Madenach A J, Werner J H, Lu Y C, and Queisser H J 1985
   Proc. 4th Internat. Conf. on *Grain Boundary Structure and Related
   Phenomena*, J. Jap. Inst. Metals, Sendai 1985, page 1005
Stützler F J and Queisser H J 1986 J. Appl. Phys. 60, 3910
Taylor D M and Tong D W 1984 J. Appl. Phys. 56, 1881
Uchida S, Remenyi G, Landwehr G 1987 in *High Magnetic Fields,*
   G.Landwehr, ed. (Springer, Heidelberg), p. 353
Werner J H and Strunk H 1982 J. Phys. Colloque (Paris) 43, C1-89
Werner J H, Jantsch W and Queisser H J 1982 Solid State Comm. 42, 415
Werner J H 1983 Thesis, University of Stuttgart (unpublished)
Werner J H 1985 in *Polycrystalline Semiconductors - Physical Properties
   and Applications*, G. Harbeke ed., (Springer Berlin 1985), p. 77
Werner J H and Peisl M 1985 Phys. Rev. B31, 6881
Werner J H 1989 in *Polycrystalline Semiconductors - Grain Boundaries
   and Interfaces*, Springer Proc. in Phys. 35 (in press)
Winer K and Ley L 1987 Phys. Rev. B36, 6072
Winer K, Hirabayashi I, and Ley L 1988 Phys. Rev. Lett. 60, 2697
Yang C 1987 Appl. Phys. Lett. 51, 112

Inst. Phys. Conf. Ser. No 104: Chapter 2
Paper presented at Int. Symp. on Struct. Prop. Disloc. Semicond., Oxford, 1989   75

Capacitance transient spectroscopy of dislocations in semiconductors

W. Schröter, I. Queisser and J. Kronewitz

IV. Physikalisches Institut der Universität Göttingen and
Sonderforschungsbereich 126, Bunsenstr.11-15, D-34 Göttingen
West Germany

Abstract: Dislocations or point defect clouds surrounding
them are represented in transient capacitance spectra by
broadened lines and characteristics, which are significantly
different from those of isolated point defects. The results
of computer simulations are used to show to what extent
these differences result from modification in the capture
and emission rate or from the spectral distribution of
localized states at dislocations.

1. Introduction

Despite of considerable efforts, the characterization of
single dislocation types in semiconductors by their electrical parameters is still an open problem.

For this purpose experimental methods with sufficient
spatial resolution to study single dislocation have been
developed - like the EBIC-mode (electron beam induced
current) of the scanning electron microcopy (Wilshaw
(1988)), the IRBIC-method (infrared beam induced current)
(Cavallini (1988)) or Scanning DLTS (deep level transient
spectroscopy) (Breitenstein (1988)). The present state of
these efforts will be presented in several lectures at this
conference.

An alternative way to characterize single dislocation types
is to apply conventional DLTS to an array of dislocations,
that are of a single type. The first problem, that has to be
solved going this way, is to generate a large number of
dislocations of a single type. To demonstrate that this part
of the problem is now solved let me refer to a recent
investigation of α- and β-dislocations in InP by Zozime et
al (1989). These authors placed a narrow Schottky contact
between two linear arrays of hardness indentations. They
have demonstrated that, if the {100}-surface is chosen,
capacitance transient spectra of α- and β-dislocations can
be measured separately. Details of this work are presented
elsewhere in these proceedings.

The second, still unsolved problem is to develop models for
the electron capture and emission at a dislocation, that
allow to simulate and analyze the observed DLTS-characteristics. In this paper the existing models will be
briefly described together with the DLTS-characteristics
which according to computer simulation are associated with
them. It is our aim to find the ingredients of a realistic
model that accounts for all observed characteristics and
allows to determine the electrical parameters of the defect.

2. The Simulation Problem

To simulate by computer the capacitance transient spectrum
of a defect the solution of the rate equation, describing
the capture and emission of electrons or holes at the
defect, has to be inserted into the solution of Poisson's
equation, relating a charge variation within the space
charge region to the capacitance change ΔC of the Schottky
contact.

The solution of the Poisson's equation is easily obtained,
if the density of defects N_T is small compared to the doping
concentration N_D (see e.g. Pons (1984)):

$$\frac{\Delta C}{C_0} = - \frac{N_T}{N_D} \cdot \frac{1}{W_0^2} \cdot \int_0^l \Delta f_T(x,t) \cdot x \; dx \tag{1}$$

ΔC is registered by a lock-in or box-car technique during
the emission period, when the Schottky contact is under
reverse bias V_b. $\Delta f_T(x,t)$ is the deviation of the occupation
probability of the defect from its stationary value in the
space charge region under those conditions. C_0 is the capacitance, W_0 is the width of the space charge region at
reverse bias V_B, and $l > W_0$.

$\Delta f_T(x,t)$ is obtained as the solution of the rate equation:

$$\frac{\partial f_T(x,t)}{\partial t} = r_c - r_e \tag{2}$$

with r_c and r_e being the capture and emission rate of the
defect, respectively. Equation (2) has to be solved for the
capture period and for the emission period separately, using
the final value of Δf_T of the previous period as the
entering value for the next period. This procedure has to be
repeated, until the entering value for the emission period
becomes stationary.

The electrical parameters of the defect enter into DLTS
through the capture and emission rate. For isolated point
defects one has:

$$r_c = \sigma_n \cdot n \cdot v_{th} \cdot (1 - f_T) \quad , \quad r_e = e_n \cdot f_T \tag{3}$$

σ_n and v_{th} are the cross section for electron capture at the
defect and mean thermal velocity of electrons in the conduction band, respectively.

Applying the principle of detailed balance one obtains for the emission rate:

$$e_n = \sigma_n \cdot v_{th} \cdot N_c \cdot \exp(-E_T/k_B T) \qquad (4)$$

N_c is the effective density of states in the conduction band, E_T the free enthalpy of electron emission from the defect to the conduction band:

$$E_T = \Delta H_T - T \cdot \Delta S_T \qquad (5)$$

The electrical parameters of a point defect that can be obtained from DLTS measurements are:

$$\sigma_n, \quad \Delta H_T, \quad \Delta S_T$$

Dislocations and point defect clouds surrounding them are extended multi-electron defects. They have a long range electric and elastic potential, which modifies capture and possibly emission of electrons at these defects. Furthermore they are characterized by a density of states.
Assuming that the electron or hole density transferred to these defects during DLTS is small compared to the doping concentration we apply equation (1) and (2) to calculate ΔC, but modify the capture and emission rate.

3. Modified Electron Capture

3.1. Model

Due to the interaction between charge carriers occupying the extended defect and between these trapped carriers and free charge carriers an electrostatic potential, superimposed to the elastic potential, is built up around the defect. This potential modifies the capture rate at the defect:

$$r_c = \sigma_n \cdot n \cdot v_{th} \cdot (1 - f_T) \cdot \exp(-E_e/k_B T) \qquad (6)$$

where E_e defines the shift on the energy scale of the defect states with occupation f_T. In a first approximation one obtains $E_e = \alpha \cdot f_T$ (Labusch and Schröter (1980)). It is assumed in this model that the emission in the space charge region remains unaffected.

The electrical parameters of the extended defect within this approximation are:

$$\sigma_n, \quad \Delta H_T, \quad \Delta S_T, \quad \alpha \qquad (7)$$

3.2. Numerical Results

Computer simulations on the basis of equations (1), (2) and (6) have been performed first by Schmalz et al (1987) and also by us (Queisser (1988)) under the assumption, that the free electron distribution instantaneously follows the voltage variations, which are applied to switch from the

capture to the emission period and vice versa. This means
that the experimental conditions have to be chosen so that
the dielectric relaxation time $T_d = \varepsilon\varepsilon_o/e \cdot n \cdot \mu_n$ (μ_n free
electron mobility, ε dielectric constant) has to be small
compared to the time constant for the capture of electrons
at the defect $T_c = 1/\sigma_n \cdot v_{th} \cdot n$.

We have chosen an acceptor-type defect with prescribed
values of σ_n, ΔH_T and ΔS_T and have investigated the
influence of the coupling constant α on its DLTS
characteristics. The value for the free electron density n
is chosen as $4 \cdot 10^{15}$ cm^{-3}. Please note that $\alpha = 0$ for an
isolated point defect and that it is expected to lie between
1 and 5 eV for a row of dangling bonds in the core of a 60°-
dislocation (Labusch and Schröter (1980)).

The results of the computer simulation are as follows:

i. For $\alpha = 0$, i.e. an isolated point defect, the occupation
probability f_T in the space charge region changes from 0 to
1 within a Debye length L_D at the position, $W_i - L_i$ (i=0,1),
at which the Fermi level E_F crosses the defect level E_T.
With increasing α the transition region broadens and the
maximum value of f_T becomes increasingly smaller than 1
(see fig. 1).

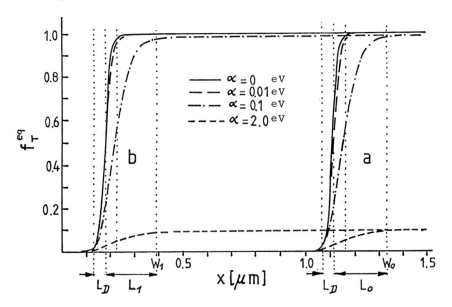

Fig. 1. Calculated depth profiles of the equilibrium
occupation f_T^{eq} in the depletion layer at full bias voltage
(a) and at reduced bias (b) for different values of the
coupling constant α. W_o, W_1 depletion layer width at full
bias and reduced bias respectively, L_D Debye length.
The defect level crosses the Fermi level at $W_o - L_o$ and $W_1 - L_1$
respectively. Defect parameters: $\Delta H_T = 0.29$ eV, $\Delta S_T = 1.3$ k_B,
$\sigma_n = 1.5 \cdot 10^{-16}$ cm^2, doping concentration $N_D = 4 \cdot 10^{15}$ cm^{-3}.
Temperature T=163 K.

ii. The shape of the DLTS-line is independent of α and remains that of an isolated point defect. Correspondingly the capacitance transient during the emission period is $\Delta C(t) = \Delta C(0) \cdot \exp(-t/\tau_e)$ with $(\tau_e)^{-1} = e_n$ being the emission rate.

iii. The capture characteristics, i.e. the line amplitude with the filling pulse time $\Delta C_m(t_p)$ shows a sensitive dependence on the coupling constant. For a point defect $(\alpha = 0$ eV) $\Delta C_m(t_p) = \Delta C(0) \cdot \{1 - \exp(-t_p \cdot \sigma_n \cdot v_{th} \cdot n)\}$. With increasing α this behaviour changes to $\Delta C(t_p) \sim \ln(t_p)$. In the t_p-range, which is easily accessible to measurements $(10^{-7} - 10^{-3}$ s) this logarithmic dependence on t_p is found throughout the t_p-range for $\alpha > 0.2$ eV (see fig.2). An approximate analytic expression has been derived from a solution of the rate equation (for the validity range see Schröter and Seibt (1983)) and by approximating the difference in the equilibrium occupation between emission and capture period by a rectangular function (Queisser (1988)):

$$\frac{\Delta C_m}{C_0} = \frac{N_T}{N_D} \cdot \frac{\beta}{2} \cdot \frac{k_B T}{\alpha} \cdot \ln\left(\frac{t_p \cdot \alpha}{\tau_c \cdot k_B T}\right) \quad , \quad \beta = 0.7 \qquad (8)$$
for the parameters indicated

The dotted lines in figure 2   show $\Delta C_m(t_p)$ according to this approximative description.

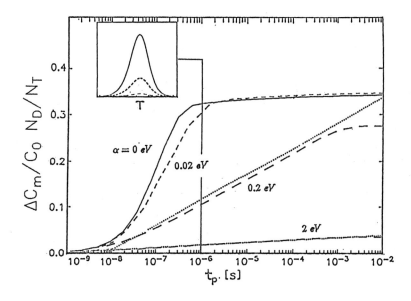

Fig. 2. Normalized maximum DLTS signal height ($\sim \Delta C_m$) as a function of filling time t_p for different values of the coupling constant α. The inserted figure shows the DLTS spectra at $t_p = 10^{-6}$ s. Dotted lines: approximation $\Delta C_m \sim \ln(t_p)$.

4. Modified electron capture and emission

4.1. Model

As a further correction to the rate equation of linear
defects a modification of the emission term has been proposed and discussed (Ferenczi and Dosza (1981), Zdansky and
Thuc Hien (1984a und 1984b), Queisser (1988)):

$$r_e = e_n \cdot f_T \cdot \exp(\alpha_e \cdot f_T /kT) \qquad (9)$$

This correction implies a lowering of the effective free
enthalpy of electron emission and could result from
tunneling through the potential barrier surrounding the line
defect. As will be seen below this model yields quite
unusual DLTS-characteristics.

4.2. Numerical Results

Using the same values for σ_n, ΔH_T and ΔS_T as in the simulation of the previous model and in addition $\alpha = 2$ eV, simulations on the basis of the modified capture and emission
rate yield the folowing results (Queisser (1988)):

i.   With increasing α_e the DLTS-line is shifted to lower
      temperatures (see fig.3).

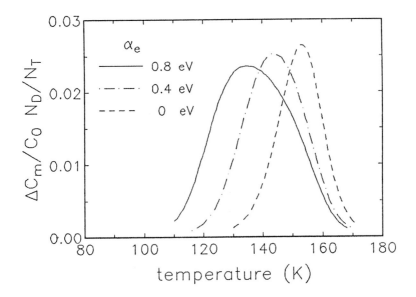

Fig.3. Simulated DLTS spectra for different values of the
effective emission barrier $\Delta H_{eff} = \Delta H_T - \alpha_e \cdot f_T$ calculated
with coupling constant $\alpha=2$eV, pulse length $t_P =10^{-4}$s, lock-in
fequency $f_m =10$Hz (rate window $e_n (T_{max})=23.5$s^{-1} for $\alpha_e =0$ eV).
Defect parameters as in Fig.1.

ii. The capture characteristic remains the same as that of the first model, i.e. $\Delta C_m(t_p) \sim \ln(t_p)$.

iii. With increasing α_e the line broadens; the slope on the low-temperature side remains nearly unchanged, the high-temperature side becomes flatter (see fig.3). If one calculates an emission rate from the lock-in frequency in the way, used for point defects, the Arrhenius plot of e_n/T^2 versus $1/T$ yields for increasing α_e significantly increased values for σ_n and lower values for ΔH_T. Both effects decrease for increasing values of the coupling constant α, which limits the maximum occupation probability.

iv. For $\alpha_e > 0$ eV the DLTS-line shifts to lower temperature with increasing t_p, so σ_n as derived from the $e_n/T^2(1/T)$-plot, increases significantly with increasing t_p.

5. Discussion

The two models, which have been described above, do not account for all the DLTS-characteristics, which are observed for extended defects.

The modification of the emission rate in the second model leads to a dependence of the line position from the filling pulse time t_p, a prediction, which can be easily checked in experiments. In the data published so far for dislocation-related defects this dependence has not been found. Recent, yet unpublished DLTS-results for $NiSi_2$-platelets in silicon indicate this dependence, however (Gnauert (1989)).

In the following we confine our discussion to the first model, which as a main result predicts a logarithmic dependence of the line amplitude ΔC_m on t_p within a wide range of t_p. To our knowledge this characteristic has indeed been observed for all electronic states localized at disloca-tions, but it is combined with a symmetric or asymmetric line broadening, which the model does not reproduce.

The analysis of experimental data in terms of $\Delta C_m(t_p)$ according to equation (8) allows to determine two parameter combinations: $\alpha \sigma_n$ and N_T/α, where N_T is the total density of localized states at the dislocation. To determine α and from that N_T, σ_n has to be determined with good precision. For point defects the emission characteristic yields ΔH_T and $\sigma_n{}^* = \sigma_n \cdot \exp(\Delta S_T/k_B)$. ΔS_T is obtained from the analysis of $\Delta C_m(V_p)$, where V_p is the pulse voltage during the capture period.

For extended defects the emission characteristic cannot be analyzed in the same way as for point defects, since the DLTS-lines are now broadened and the transients are non-exponential.

Any further progress in this problem requires an interpretation of the line broadening and a model to simulate it.

5.1. Line Broadening

5.1.1. Point defects

The computer simulations have shown that the modification of the capture rate has only negligible influence on the line shape and the transient. Both remain the same as for a point defect. On the other hand, both are significantly modified, when the emission rate is corrected for a possible electron tunneling through the potential barrier surrounding the dislocation. But this correction leads to a dependence of the line position on the filling time t_P, which is not confirmed by the experimental data for dislocation-related defects.

This means that apparently tunneling does not contribute to emission at dislocations in the space charge region. But the model shows that modification of the emission rate are needed to explain line broadening and non-exponential transients.

Omling et al have proposed that a distribution of activation enthalpies E_{T1} for electron emission could be responsible for line broadening and non-exponential transients, i.e.

$$C(t) = C_0 \cdot \int_0^\infty g(E_{T1}) \cdot \exp(-e_n(E_{T1}) \cdot t) \cdot dE_{T1} \qquad (10)$$

They developped this model to explain DLTS-data for the EL 2-centre in $GaAs_{1-x}P_x$, which showed symmetrically broadened lines and non-exponential transients in the case of x>0. A distribution of E_{T1} for this point defect could result from alloy fluctuation in its neighbourhood. The authors have shown that all the characteristics are explained by assuming a Gaussian distribution function for $g(E_{T1})$.

5.1.2. Point Defect Clouds

The strain field of a dislocation with edge component produces a region under compression and another one under dilatation, i. e. with enlarged and reduced band gap. Point defect clouds surrounding the dislocation should therefore be associated with a distribution of E_T-values.

There is good evidence that some of the broadened lines arising with the introduction of dislocations have to be attributed to point defect clouds. In a few cases such a line is well separated from other lines, so that its shape can be well compared with model simulations.

Omling et al (1985) have obtained a symmetrically broadened DLTS-line (in their notation: the B-line) in silicon by plastic deformation at 650°C. They analyzed the measured emission characteristic and the line shape assuming again a Gaussian distribution for the E_{T1}. Using for both evaluations the same center of the distribution, viz. $\Delta H_T = 0.29 \text{eV}$, they derive $\sigma_n{}^* = \sigma_n \cdot \exp(\Delta S_T /k_B)$ from the emission characteristic: $\sigma_n{}^* = 1.3 \cdot 10^{-15} \text{cm}^2$ and from the fit to the DLTS-line: $\sigma_n{}^* = 1.5 \cdot 10^{-16} \text{cm}^2$.

5.1.3. Dislocations

Isolated point defects with an E_T-distribution become fully occupied during the capture period, if $t_p > 1/(\sigma_n \cdot n \cdot v_{th})$. The measured transient is then obtained by a summation of the individual tansients, which are obtained independently as solutions of the rate equation inserting the individual E_{Ti}.

For point defect clouds or dislocation core states the calculation becomes much more complex, since the potential barrier limits capture and thereby reduces occupation. Since emission occurs only from occupied sites, it is not expected that the emssion characteristic gives some direct picture of the E_T-distribution or density of states of the extended defect.

What is the reason then that the procedure of Omling et al (1985) applied to point defect clouds in silicon gave to some extent consistent results. One possibility is that the value of α for this defect is rather small ($\alpha < 0.1 \text{eV}$). Due to the uncertainty of their σ_n-values this possibilty cannot be checked at the moment. The other one is that even for large α (>0.5 eV) the complex problem might have some simple approximate solution.

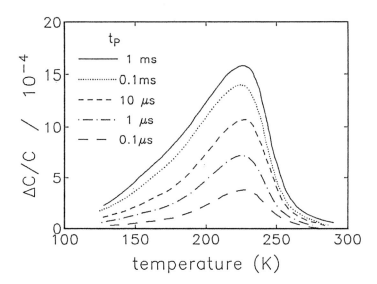

Fig.4. DLTS spectra of dislocation induced defects in n-type silicon, deformed at 680°C (annealed 1h at 800°C) with one dominating dislocation type (60°). Dislocation density $N_d = 4 \cdot 10^6 \text{cm}^{-2}$, doping concentration $N_D = 4.5 \cdot 10^{14} \text{cm}^{-3}$, lock-in frequency $f_m = 17 \text{Hz}$.

Figure 4. shows DLTS-data obtained for 60⁰-dislocations in silicon (Kronewitz (1987). The DLTS-lines, measured for different capture times t_p, are asymmetrically broadened. Both sides deviate in their shape from that of an isolated point defect: the high-temperature side is significantly flatter than for a point defect and the low-temperature side is extremely broadened. If one normalizes them to their amplitude, their high-temperature sides almost coincide, whereas the low-temperature sides change with t_p. This could mean that at least the high-temperature side reflects in a direct way some part of the density of states.

References

Breitenstein O 1988 Proc. on the International Workshop on
         Beam Injection Assessment of
         Defects in Semiconductors
         Paris 1988
Cavallini A 1988 Proc. on the International Workshop on Beam
         Injection Assessment of Defects in Semiconductors
         Paris 1988
Ferenczi G and Dosza L 1981 Crystal Res. and Technology 16
         203
Gnauert U 1989 Diploma thesis Göttingen, to be published
Labusch R and Schröter W 1980 in "Dislocation in Solids",
         edited by F.N.R. Nabarro (North Holland Publishing
         Company), pp 127 - 192
Omling P, Samuelson L and Grimmeiss H G 1983 J. Appl. Phys.
         54 5117
Omling P, Weber E R, Montelius L, Alexander H and Michel J
         1985 Phys. Rev. B 32 6571
Pons D 1984 J. Appl. Phys. 55 3644
Queisser I 1988 Diploma thesis Göttingen, to be published
Schmalz K, Tittelbach-Helmrich K and Richter H 1987 phys.
         stat. sol. (a) 99 K97
Schröter W and Seibt M 1983 Journal de Physique C4-329
Wilshaw P 1989 Proc. Intl. Workshop on Point, Surface and
         Extended Defects, Erice, Sicily, 1989. ed. Benedek G,
         Schroter W, Cavallini A, Plenum Press, in press.
Zdansky K and Thuc Hien N T 1984 phys. stat. sol. (a) 81 353
         and 85 219
Zozime A and Schröter W 1989 Phil. Mag. in print

Acknowledgement

We gratefully acknowledge the assistance of Dr.K.Ahlborn for the computer simulations and M.Schrader for the preparation of the figures.

Inst. Phys. Conf. Ser. No 104: Chapter 2
Paper presented at Int. Symp. on Struct. Prop. Disloc. Semicond., Oxford, 1989

85

The electronic properties of dislocations in silicon

P.R.Wilshaw and T.S.Fell

Department of Metallurgy and Science of Materials, Parks Road,
Oxford OX1 3PH, United Kingdom.

ABSTRACT: The electrical properties of dislocations are considered in terms of their charge and resulting Coulomb potential. The process of carrier recombination is described. Experimental results are presented which indicate dislocation kinks in silicon to be reconstructed.

1. INTRODUCTION

The study of the electrical properties of dislocations in semiconductors can be divided into two parts. The first is to try to understand the relationship between the experimentally measured electrical properties of dislocations and the electronic states in the band gap that may be associated with them. These states are characterised by such parameters as their position within the band gap and their cross-section for carrier capture. The overall properties of the dislocation are then determined by the spatial distribution of these states and the way they interact when in close proximity. It is the understanding of these effects which allows data gained from techniques such as DLTS, EPR, EBIC, microwave conductivity, photoluminescence and others to be interpreted in terms of the electronic states associated with dislocations.

The other aspect of the study of the electrical properties of dislocations is to try to understand on a microscopic basis the particular features of the structure of dislocations and their surroundings that leads to the electronic states in the band gap and hence to the electrical activity observed. This is an intriguing problem which although still in need of further research is already shedding light on the interaction between dislocations and point defects and impurities. In the first sections of this paper the relationship between individual, discreet electronic states and the overall measured electrical properties of the dislocation at which they are located is discussed. In the final section of this work experimental results are presented and the cause of the observed activity considered.

2. DISLOCATION CHARGE AT EQUILIBRIUM

The key factor which distinguishes the electrical behaviour of dislocations and other extended defects such as grain boundaries, precipitates or point defect clusters from that of isolated point defects is that they are associated with a long range Coulomb potential when the band gap states are charged. This effect, which is due to the many electron nature of the defects concerned, does not occur for isolated point defect centres, but has a profound influence on the dislocation properties observed. For example the capture of both minority and majority carriers and the ability of dislocations to scatter carriers is normally dominated by the Coulomb potential. The equilibrium occupation of the centres as detected by DLTS, EPR and other techniques can also be controlled by the Coulomb potential (and vica versa). Thus in considering the electrical properties of dislocations it is vital that the Coulomb potential surrounding them is taken into account and hence because they are inter-related the occupation of the states themselves.

In considering the Coulomb potential induced by a dislocation the case of a dislocation present in n-type silicon which introduces acceptor states into the gap will be used. However many parts of this treatment are also applicable to other extended defects and also to defects present in p-type material provided the states then considered are donor states. Figure 1 shows the band diagram for such a defect where, in this case, the states lie at a position, when uncharged, deeper than the Fermi level. Some of the states are occupied which due to their small separation, which in this model is assumed to be less than λ_D the Debye length, can thus be considered to give rise to a uniform line charge, Q per unit length. This increases the energy of electrons in the vicinity of the dislocation and is represented on a band diagram by the conduction and valence bands bending up. It is normally assumed that the spatial extent of the wave functions associated with the defect states is small compared to the long range electrostatic potential which normally extends in the region of 0.1μm or more. Thus the entire defect level is considered to be rigidly shifted by the electrostatic potential of its charge by an amount ϕ.

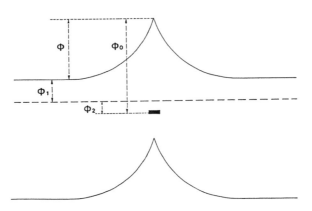

Figure 1. The band diagram for a negatively charged dislocation.

The amount of band bending at a dislocation in 10^{15}cm^{-3} doped silicon is shown in figure 2 as a function of its charge according to the relation derived by Masut et al (1982):

$$\phi = \frac{Q}{2\pi\epsilon\epsilon_0}\frac{(\ln(\lambda_D Q) - 0.5)}{q} \tag{1}$$

where λ_D is the Debye length. From this it can be seen that even very small line concentrations of charge give rise to relatively large amounts of band bending. For example, in this case only 200 electronic charges evenly distributed along a 1μm segment of dislocation give rise to a band bending of ~0.14eV. When considering the concentrations of impurities normally present in semiconductors it is clear that if only a small fraction of this segregates to a small number of dislocations where it is then electrically active, significant band bending will result.

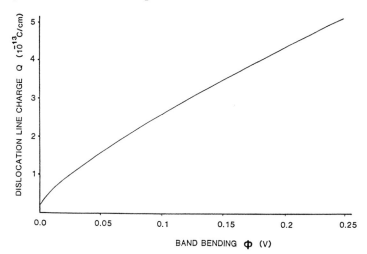

Figure 2. The Coulomb potential of a dislocation plotted as a function of its line charge.

2.1 Dislocations With a Single Energy Level

A full solution for the occupation of dislocation states with a single energy level requires the numerical solution of equation (1) and the Fermi function for the relevant values of temperature, doping concentration, position of the dislocation energy level and the line concentration of dislocation states. Values can be obtained for the band bending ϕ, occupation probability of a given state f, and line charge Q, see Schroter and Labusch (1969) for a similar treatment. In the following, for simplicity, such a detailed analysis will not be used. Instead, so that the different regimes of dislocation occupation which occur can be identified the approximation will be made that a band gap state is occupied if below the Fermi level, is empty if above it, and is partly filled only if very close to the Fermi level. In this way three separate cases of dislocation occupation and hence band bending and overall dislocation behaviour can be identified.

i) $\phi_0 \leq \phi_1$, the dislocation state is above the Fermi level and the dislocation is uncharged.

ii) $\phi_0 > \phi_1$ and N_D sufficiently large that the band bending produced by the filled states is sufficient to make the dislocation states intersect the Fermi level. In this case the dislocation states are only partially occupied and are "pinned" to the Fermi level.
iii) $\phi_0 > \phi_1$ and N_D small so that all the dislocation states are full but the band bending this induces is insufficient to make them intersect the Fermi level.

In the first case the dislocation would show very little electrical activity, the dislocation state would not be detected by DLTS and it would act only as a weak recombination centre for minority carriers and hence might not be detected in EBIC measurements.

In the second case the dislocation level is pinned to the Fermi level and not all of the available states are occupied. Moreover their occupation varies with temperature as the position of the Fermi level changes and hence the dislocation charge changes. Under these conditions the state would be detected by DLTS but measurements of the concentration of states made in the usual way by observing the maximum change in specimen capacitance would be in error since they would record only the concentration of occupied states rather than the total number of states available for occupation. In addition since the concentration of occupied states changes as a function of temperature the DLTS peak observed would be distorted from that of a normal dislocation peak. Recombination at the dislocation would decrease with increasing temperature as the dislocation charge is reduced which in turn reduces its efficiency for capturing minority carriers.

In the third case all the dislocation states are occupied and the dislocation charge is independent of temperature. Under these conditions DLTS measurements can give an accurate measure of states present at dislocations since all can be filled during the filling pulse. A detailed analysis of carrier recombination at dislocations shows that in this regime recombination, as measured by EBIC contrast, is now approximately independent of temperature (Wilshaw et al. 1989).

It is important to note that the behaviour of dislocations, due to the associated band bending, is much more complicated than for isolated point defects. For example a given type of dislocation state may cause the dislocation to exhibit all three regimes of behaviour described above, depending on the temperature, doping concentration and line concentration of defect states present. Figure 3 is plotted using the approximations described above to show the charge on two dislocations with a concentration of states of 2.10^6 and 3.10^6 at a position in the gap $\phi_0 = 0.27$eV present in 10^{15}cm^{-3} doped silicon. All three regimes are observed in each case but the temperature of transition between regime 2 and 3 is seen to depend on the concentration of states present.

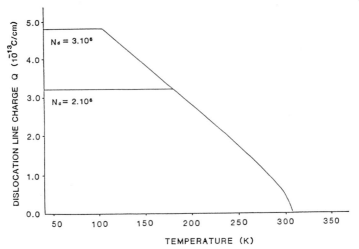

Figure 3. Dislocation charge as a function of temperature and concentration of states. See text for details.

2.2 Dislocations With Multiple Energy Levels

It is likely that in many practical situations dislocations or other extended defects will be associated with more than one type of defect state which could be caused for example by different kinds of impurity atoms, point defects or possibly dislocation jogs or kinks. In such a situation there will be two or more energy levels closely localised in space and the band bending produced by the charging of each state will be experienced by all the others also located at that particular point on the defect. In such a situation it is possible for the occupation of one or more deep levels to be sufficiently large that the band bending so caused will be sufficient to raise the shallower states above the Fermi level. Thus in this case, shallow states which would be charged when present at a dislocation by themselves will be rendered neutral by the presence of deeper levels. The detection of such neutral states would then be impossible using normal DLTS or carrier recombination techniques such as EBIC, and depending on the details of the state, EPR.

In this way it is possible for a relatively small number of deep states at dislocations to mask the presence of a very much larger number of shallower levels. This effect is similar to the process of compensation that occurs in a bulk semiconductor when the deep level concentration is of the same order or larger than the doping concentration. However it is important to note that only a very small number of deep level states e.g. ~$200\mu m^{-1}$ is required to produce this effect at dislocations and thus it is likely to happen even if the number of impurity or other deep levels is so small that if uniformly distributed throughout the material they would be equivalent to a concentration many orders of magnitude below the doping concentration.

The corollary of this effect is that if DLTS or other measurements show, for example, two types of defect state present in a single specimen in circumstances such that the band bending caused by the deepest state would leave the other uncharged then the different types of state are necessarily spatially isolated from each other. Thus an analysis of DLTS data to obtain the barrier heights associated with DLTS peaks can be used to give information as to whether the defect states concerned are located close to each other or whether they must be spatially isolated and hence probably associated with different defect types.

2.3 Compound Semiconductors

The analysis of band bending for a dislocation in a compound semiconductor is further complicated over that for elemental materials due to the partly ionic character of the compounds which results in each atomic species carrying a net electrostatic charge. Thus for those dislocations which contain an excess of one atomic species over another there will be a net electrostatic line charge Q, irrespective of whether they are associated with any defect states in the band gap. For relatively highly ionic materials such as the II-VI compounds this effect is expected to be very large and the band bending so induced will be sufficient to make the valence band at the dislocation approach the Fermi level in the case of a negatively charged dislocation and the conduction band approach the Fermi level for a positively charged dislocation. In such a situation the dislocation charge would be screened not just by the repulsion of free majority carriers as described by equation (1) but increasingly by the carriers contained in the inversion or accumulation region now produced by the increased band bending around the dislocation. Thus for compound semiconductors the relation between the occupation of dislocation states,

Fermi level position and band bending which is true for elemental semiconductors will not be valid. In addition it is expected that the inversion or accumulation region produced around dislocations in the more ionic materials will produce effects not observed in elemental semiconductors such as substantial dislocation conductivity and a non-logarithmic dependence of occupation on the length of the filling pulse in a DLTS measurement. The latter because the band bending is no longer approximately proportional to the dislocation level occupation as is the case with elemental semiconductors where equation (1) is valid.

3. CARRIER RECOMBINATION AT DISLOCATIONS

In the present work full details of the recombination process which have been given elsewhere (Wilshaw et al 1989) are not described, instead the dependence of recombination on the Coulomb potential surrounding the dislocations is emphasised. This demonstrates the importance of this effect on the electrical activity of dislocations. The effect of the Coulomb potential on other aspects of dislocation behaviour has been considered by other authors, for example in relation to DLTS measurements (Schroter 1989, Omling et al 1985), photoconductivity measurements (Figielski 1978) and Hall effect measurements (Schroter and Labusch 1969).

Recombination of an electron hole pair via a dislocation level such as that shown in figure 1 can be considered to take place in four steps:

(1) Diffusion of a hole to the space charge region.

(2) Capture of the hole into the bound hole states at the top of the valence band which are produced by the electrostatic potential well of the space charge region.

(3) Transition of the hole to the primary dislocation level.

(4) Capture of an electron to the primary dislocation level via the repulsive Coulomb potential barrier of the space charge region.

Step 1 is not a "bottleneck" for recombination at dislocations in silicon (Wight et al 1981) for which the presence of dislocations only weakly perturbs the surrounding minority carrier distribution. It is to be expected that step 3 is fast because once a hole has become trapped in the bound hole states it is spatially located at the same position in the crystal as the primary dislocation states and thus the probability of this transition is likely to be high.

Thus for recombination at dislocations it is step 2, the capture of holes into the bound hole states and step 4 the excitation of electrons over the Coulomb potential barrier which control recombination. In the analysis used here all other steps in the recombination process are potentially sufficiently fast that their rates adjust to fit the limiting processes of electron and hole capture. Thus the detailed nature of the dislocation levels in the band gap and the nature of the transitions to and between them are not limiting in the overall process of recombination which can be described by considering only both hole and electron capture.

3.1 Hole Capture

Once holes have diffused to the space charge region their capture into the bound hole states may be modelled by a reduced lifetime τ'. The hole capture rate per unit dislocation length J_h is related to τ' by $J_h = \pi r_d^2 \Delta P/\tau'$ where r_d is the radius of the space charge cylinder and ΔP the excess minority carrier concentration. The radius of the space charge region is related to the dislocation line charge according to $r_d^2 = Q/\pi n_0 q$ and hence:

$$J_h = Q\Delta P/n_0 q\tau' \qquad (2)$$

Thus it can be seen that the recombination rate is directly proportional to Q.

In many practical cases recombination at dislocations is measured using the EBIC technique and hence in the following recombination will be discussed in terms of the EBIC dislocation contrast C, defined as $C = (I_b - I_d)/I_b$ where I_d and I_b are the EBIC signal collected when the incident electron beam is respectively at and away from the defect. At this point it is usual to introduce the concept of the dislocation recombination strength γ defined by Donolato as $\gamma = J_h/\Delta PD_h$. Donolato (1978) has shown that:

$$C = A\gamma \qquad (3)$$

where A is a constant dependent on certain specimen parameters. Thus a relation between dislocation recombination as measured by EBIC contrast and the dislocation line charge is obtained : $C = AQ/D_h n_0 q\tau'$. The reduced minority carrier lifetime τ' depends on the cascade capture of holes into the potential well surrounding the dislocation. This has been treated theoretically by Sokolova (1970) who finds that τ' varies as $T^{1.5}$. Experiment has shown that D_h varies as $T^{-1.4}$ and consequently the product $\tau'D_h$ is closely independent of temperature allowing the EBIC contrast to be expressed as $C \sim BQ/n_0 q$ where B is a constant. Thus although the absolute magnitude of Q cannot be directly obtained from measurements of the dislocation contrast because, in general, the value of the constant of proportionality B, is not accurately known, the changes in the dislocation contrast will reflect changes in the dislocation line charge Q.

3.2 Electron Capture

The net electron capture rate per unit dislocation length J_e, over the Coulomb potential barrier is determined by detailed balance :

$$J_e = C_e N_d [(1-f)n_0 \exp(-q\phi/kT) - fN_C \exp(-q\phi_0/kT)] \qquad (4)$$

where C_e is the probability of the transition of an electron from the conduction band to an empty dislocation level, N_d is the number of states per unit dislocation length, N_C is the effective density of states in the conduction band and f is the occupancy factor of the dislocation states. The expression for J_e consists of two terms. The first represents capture of electrons to the dislocation states by thermal excitation over the Coulomb barrier, the second describes the thermal re-emission of electrons form the dislocation level back into the conduction band. From this expression it can be seen that the value of J_e determines the value of ϕ. For example when a dislocation is at equilibrium, in the absence of an excess minority carrier concentration, the net electron capture rate J_e is zero and hence the capture and re-emission terms of equation (4) must be equal. Under these conditions it is easy to show that the occupation f, of the dislocation states is that predicted by the Fermi function when the states are a distance $q\phi_2$ away from the Fermi level. Hence the value for ϕ in this case is the equilibrium value for the Coulomb potential at the dislocation in the absence of recombination, ie. when the occupation f is the equilibrium value as determined by the Fermi function. If the minority carrier concentration is now increased, J_e is also increased from zero and the electron capture term becomes larger than the thermal re-emission term and hence the values of ϕ and f are decreased. In an EBIC experiment the minority carrier concentration is increased by increasing the incident electron beam current.

It has already been shown that EBIC contrast is a measure of Q and hence ϕ and thus the above analysis shows that if the beam current is increased in an EBIC experiment so that the dislocation charge moves from equilibrium then the EBIC contrast will decrease as the dislocation charge is decreased. This has been verified experimentally (Wilshaw et al 1989). Alternatively the capture of electrons can be viewed as a straight forward thermally activated process with an activation energy qϕ when the thermal re-emission term is negligible. However it is noted that ϕ is not necessarily constant but depends on the recombination taking place according to equation (4). If the electron beam current is altered during an EBIC experiment hence changing J_e such that the EBIC contrast from a dislocation is maintained constant then it follows that the dislocation charge and hence Coulomb potential ϕ, are also maintained constant. Thus if the recombination current at a dislocation, which is equal to J_e, is measured as function of temperature whilst constant contrast is maintained then an Arrhenius plot of $\ln(J_e)$ versus 1/T will yield the activation energy qϕ for electron capture according to equation (4). In this case the activation energy measured will be that due to the Coulomb potential ϕ of the dislocation corresponding to that particular line charge Q which gives rise to the value of contrast, for that particular dislocation, which was held constant throughout the experiment.

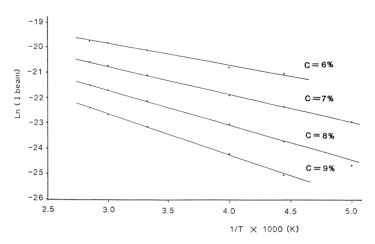

Figure 4. Arrhenius plots of J_e versus 1/T for a screw dislocation in silicon for various values of EBIC contrast.

Figure 4 shows Arrhenius plots for a screw dislocation in n-type 10^{15} silicon for various values of contrast. From these measurements of ϕ it can be seen that the band bending decreases as the contrast of the dislocation decreases. Moreover since the dislocation contrast is directly proportional to Q a plot of values of ϕ versus contrast obtained from these Arrhenius plots, when suitably scaled, should coincide with the theoretical ϕ versus

Q relation shown in figure 2. Figure 5 shows values of ϕ versus contrast for the dislocation shown in figure 4 and also for a sixty degree dislocation, superimposed on a theoretical plot of dislocation line charge versus Coulomb potential. The only fitting parameter used in this figure is the scaling parameter A' such that $C = A'Q$. A' is different for each dislocation because they lie at different depths in the specimen so altering the constant of proportionality between C and γ in equation (3).

The above analysis has demonstrated the dependence of recombination at dislocations on the Coulomb barrier surrounding them and also how the amount of recombination can in turn alter the size of the Coulomb potential. In addition it demonstrates that if sufficient EBIC contrast measurements are made on a given dislocation, then the EBIC contrast of that dislocation can be "calibrated" in terms of its line charge and barrier height.

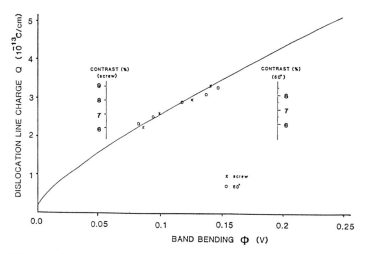

Figure 5. Values of band bending from Arrhenius plots for a screw and 60° dislocation superimposed on the theoretical curve of figure 2.

3.3 The Use of Recombination at Dislocations to Determine the Electronic Properties of the Defect States

Section 2.1 showed how the equilibrium dislocation line charge varied as a function of temperature depending on defect parameters such as the concentration of defect states N_d and their position in the gap. In regime 1 the equilibrium dislocation line charge $Q = N_d q$ and is independent of temperature whereas in regime 2 the states are pinned to the Fermi level and hence $\phi_0 = \phi + \phi_1$ and the line charge decreases with temperature. The previous section has shown how EBIC measurements of recombination at dislocations can be used to measure the Coulomb potential and line charge. Thus by progressively reducing the beam current used to observe a dislocation in an EBIC experiment, its line charge and hence EBIC contrast increases until the equilibrium value is approached, see figure 6. Further reduction of the beam current then leaves the dislocation charge unchanged at its equilibrium level. If the equilibrium charge is measured as a function of temperature its behaviour will show whether regime 1 or 2 pertains and hence whether N_d or ϕ_0 can be obtained from the data. However in either case one of the important dislocation parameters is deduced. This procedure has been followed for deformation induced dislocations in silicon which are expected to be relatively free from impurity decoration. In each case regime 1 was found to describe the activity observed and so values of N_d could be deduced. The results so obtained are described in the following section.

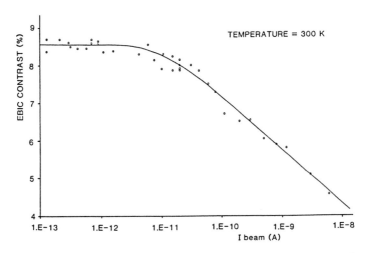

Figure 6. EBIC contrast versus beam current for a screw dislocation produced by deformation at ~950°C and 420°C. This shows the contrast approaching its equilibrium value as the beam current is reduced.

4. EXPERIMENTAL

EBIC measurements have been performed using a system similar to one described previously (Wilshaw et al 1983) which is now based around a Philips 505 SEM with LaB_6 gun. In previous work (Wilshaw et al 1989) measurements have been made of dislocations produced by deformation of float zone, swirl free, n-type $10^{15}cm^{-3}$ silicon under two stage compression along a [213] axis at 850°C or ~950°C and then 420°C. Specimens deformed both by Prof. Alexander and coworkers in Cologne and by the present authors have been investigated with a total of 49 dislocations having been studied. In the present work the EBIC experiments have been extended to dislocations produced by deformation at ~950°C and 650°C for which DLTS experiments on similar specimens have shown greater activity than for those deformed at 420°C, see for example Weber and Alexander (1983) and Omling et al (1985). The deformation procedure used in the present work is as follows.

4.1 Specimen Deformation

Ingots of float zone, swirl free silicon, 7mm long were deformed along a [213] axis at ~950°C and 650°C. The pre-deformation procedure is carried out initially by compressing the specimens at 500μm per minute until a resolved sheer stress of 15MPa is achieved at which point deformation is continued at a constant strain rate of 50μm per minute until the specimens have deformed by 10 to 15μm. This process takes the silicon ingot over its upper yield point and induces the nucleation of many dislocations at the crystal surfaces. In addition the high stress is sufficient to activate a few dislocation sources of unknown character within the bulk of the specimens. It was found that these remain inactive if the deformation is carried out at a low strain rate. After pre-deformation the load is reduced to a resolved shear stress of 3MPa and the specimens are then cooled.

At least 500µm of material is then polished away from all the crystal surfaces, thus removing the surface nucleated dislocations and so leaving the surfaces damage free. The specimens are then remounted in the deformation apparatus and deformed at 650°C at a resolved shear stress of 30MPa. The dislocation sources activated in the bulk of the specimen during pre-deformation now generate many concentric hexagonal loops on (111) planes, whilst very few dislocations are generated at the specimen surfaces. In this way it is possible to generate complete dislocation loops entirely within the bulk of a crystal, see for example figure 7 which shows an EBIC image from a specimen cut parallel to the (111) plane on which the dislocation loops are lying.

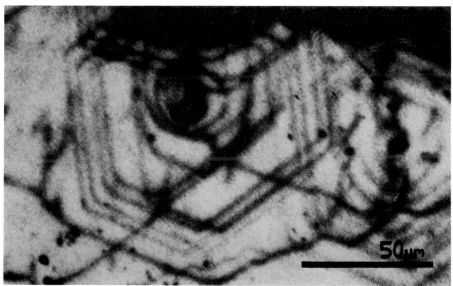

Fig. 7 EBIC micrograph of dislocations produced by pre-deformation at ~950°C followed by futher deformation at 650°C.

4.2 EBIC Results and Comparison With DLTS

The EBIC technique has been used to investigate individual dislocations produced mainly by deformation at 650°C such as those shown in figure 7. EBIC measurements on these dislocations show an EBIC contrast of up to ~2% and a preliminary analysis of the behaviour of this contrast and hence their charge as a function of temperature and incident beam current showed them to have a line concentration of states N_d ~$5.10^5 cm^{-1}$, lying deeper in the band gap than ~0.3eV. Because of the range of temperatures and doping used for these experiments states shallower than ~0.3eV could not be detected as they would be uncharged.

At first sight this small number of active sites is surprising since DLTS measurements of dislocations produced at 650°C generally show a larger concentration of states, for example Omling et al (1985) find a line concentration of ~$4.10^6 cm^{-1}$ for a dislocation density of $5.10^7 cm^{-2}$. However the EBIC results are consistent with the DLTS results provided two factors are taken into account.

i) DLTS measurements show that the relative proportion of the dominant DLTS trap in 650°C material at high dislocation densities, the DLTS D-line, decreases as the dislocation density is reduced. In the present work the dislocations studied are produced in concentric loops and generally do not interact with each other. Thus for the individual dislocations studied here the effective dislocation density may be close to zero so yielding the concentration of the D-line states also close to zero.

ii) One of the major sources of the DLTS activity is the B-line which is due to states 0.29eV below the conduction band edge. However this state would not be observed by the present EBIC measurements which are only sensitive to states deeper than ~0.3eV, which contribute to the dislocation charge.

Thus the EBIC results obtained here are in good agreement with DLTS if the activity observed is correlated to centres other than the B and D-lines, that is to the centres that generate the DLTS C-line. In addition the results are consistent with the absence of the state causing the DLTS D-line on the dislocations produced by the deformation procedure described above.

4.3 EBIC Results From Curved Dislocations.

Hexagonal dislocation loops produced by deformation at 650°C and with a resolved shear stress of 30MPa have significant segments of curved dislocations at the corners where the radius of curvature is typically 0.5μm (Gottschalk 1983). The high spatial resolution of the EBIC technique can be exploited to make measurements of the line concentration of charged states along these short segments of dislocation. Measurements made in this way on the dislocations described in the present work, show no increase in the number of charged states at curved segments compared to the ~5.10^5cm^{-1} found along the adjacent straight dislocation segments. Since the concentration of kinks at the curved dislocations is much higher than this, these results imply that, for the specimens studied here which are deformed at 650°C, kinks do not introduce charged states into the band gap and hence if they introduce any acceptor states into the band gap these must be within 0.3eV of the conduction band edge. Such a result implies that the dislocation bonds at kinks are largely reconstructed.

5.REFERENCES

Donolato C 1978 Optik **52** 19
Figielski T 1978 Solid State Electron **21** 1403
Gottschalk H 1983 J. Physique **44** C4-69
Masut R, Penchina C M and Farvacque J L 1982 J. Appl. Phys. **53** 4694
Omling P, Weber E R, Montelius L, Alexander H and Michel J 1985
    Phys. Rev. B **32** 6571
Schroter W, Queisser I and Kronewitz J 1989 Published in these
    proceedings.
Schroter W and Labusch R 1969 Phys. Stat. Sol. **36** 359
Sokolova E B 1970 Sov. Phys. Semiconductors **3** 1266
Weber E R and Alexander H 1983 J. Physique **44** C4-319
Wight D R, Blenkinsop I D, Harding W and Hamilton B 1981
    Phys. Rev. B **23** 5495
Wilshaw P R, Fell T S and Booker G R 1989 NATO ASI Series
    "Point, Surface and Extended Defects", Erice, Sicily,
    Eds. Benedeck G, Cavallini A and Schroter W, Plenum Press, in press.
Wilshaw P R, Ourmazd A and Booker G R 1983 J. Physique **44** C4-445

Inst. Phys. Conf. Ser. No 104: Chapter 2
Paper presented at Int. Symp. on Struct. Prop. Disloc. Semicond., Oxford, 1989

97

Combined application of SEM-CL and SEM-EBIC for the investigation of compound semiconductors

J Schreiber and W Hergert

Sektion Physik der Martin-Luther-Universität Halle-Wittenberg, Postfach, Halle, DDR-4010, GDR

ABSTRACT: For SEM investigations of III-V and II-VI compound semi-conductors the conception of combined CL and EBIC measurements is favoured because of simultaneous availability of information on electrical and optical properties. Thus, this SEM mode is an adequate method for quantitative examinations of the recombination activity of dislocations. The theoretical description as well as experimental results of combined CL and EBIC experiments are discussed.

1. INTRODUCTION

For SEM investigations of III-V and II-VI compound semiconductors with a gap > 1eV the conception of combined CL and EBIC measurements is favoured because of simultaneous availability of information and microscopic electrical and optical properties of the material. SEM-EBIC mode has been extensively used to reveal the presence of electrically active dislocations. In recent years the SEM-CL mode became interesting in connection with investigations of optical properties of dislocations (Brümmer et al 1974, Schreiber 1978). Thus, the combined CL/EBIC technique is an adequate method for quantitative examinations of the recombination activity of dislocations. In order to deduce quantitative information and to relate it to the recombination process of individual defects a theoretical model for an interpretation of the CL and EBIC data is necessary (Pasemann 1987). An advantage of the combined CL/EBIC technique is its ability to yield additional and more reliable data from the crystal matrix and individual defects.

The paper deals with the theoretical fundamentals of the calculation of luminescence and induced current signals from semiconducting samples and of the description of defect contrasts in the CL and EBIC experiments. Some experimental results of CL and EBIC investigations of matrix properties and at dislocations in samples of III-V and II-VI compound semiconductors are presented. The electrically and optically active recombination behaviour at the dislocation is considered by comparing the defect contrasts in the CL and EBIC modes.

2. THEORETICAL FUNDAMENTALS

A system of partial differential equations, which contains
(i) the continuity equation for the excess minority carrier in the homogeneous extrinsic semiconductor,
(ii) the transport equation for the radiation field,
represents the starting point for the calculation of induced currents and

luminescence signals from semiconducting samples (v Roos,1983, 1985, Hergert 1988b). The coupling between these equations describes the effect of reabsorption of recombination radiation (RRR-effect) (Akamatsu et al 1981, Koch et al 1987, Rossin et al 1986). This effect will be neglected throughout the paper.

If we consider the case of a stationary generation of electron-hole pairs, we can calculate the flux of radiation just outside the sample per wavelength interval $d\lambda$ (Hergert 1988b)

$$\frac{df}{d\lambda} = S(\lambda) \int_0^{\theta_c} \sin\theta \; d\theta \int_{\Omega_s} d^3\underline{r} \quad \frac{\delta p(\underline{r})}{\tau_r} \exp\left[\frac{-\alpha(\lambda)\,z}{\cos\theta}\right] \tag{1}$$

(δp-excess minority carrier density, τ_r-radiative recombination lifetime, θ_c-critical angle of total reflection, α-absorption coefficient, Ω_s-sample volume). The flux of radiation is proportional to the spectral distribution function $S(\omega)$ and proportional to a function, which depends on the radiative recombination rate of the excess minority carriers and the optical loss corresponding to absorption in the sample and total reflection at the surface.

In CL experiments the recombination radiation leaving the sample is collected by an optical system of collection efficiency $A_{opt}(\lambda)$. A resolutionfunction $A_{spec}(\lambda)$ corresponds to the spectral decomposition of the recombination radiation in a spectrometer. The photomultiplier converts the optical signal in an electrical one with the efficiency $A_{PM}(\lambda)$. The last step is an electrical amplification of the signal ($A_{amp}(\lambda)$). Therefore, the electrical signal $I_{exp}^{CL}(U_b,\lambda)$ which is recorded as a function of the beam voltage of the SEM and the wavelength λ of the recombination radiation, is given by a convolution of the flux of radiation corresponding to (1) with the different measuring instrument functions

$$I_{exp}^{CL}(U_b,\lambda) = \int_0^\infty d\lambda' \; S(\lambda') \; I^{CL}(U_b,\lambda') \; A_{opt}(\lambda') \; A_{spec}(\lambda-\lambda') \; A_{PM}(\lambda') \; A_{amp}(\lambda') \tag{2}$$

$I^{CL}(U_b,\lambda)$ is given by

$$I^{CL}(U_b,\lambda) = 2\pi \int_0^{\theta_c} \sin\theta \; d\theta \int_{\Omega_s} d^3\underline{r} \quad \frac{\delta p(U_b,\underline{r})}{\tau_r} \exp\left[\frac{-\alpha(\lambda)\,z}{\cos\theta}\right] \tag{3}$$

The quantity $I^{CL}(U_b,\lambda)$ plays an important role in the discussion of luminescence experiments. The excess minority carrier density depends on the beam voltage U_b of the SEM, because of the beam voltage dependence of the generation rate of electron-hole pairs $g(U_b,\underline{r})$.

Two different kinds of experiments are important:

- spectral resolved experiments
  The resolution function of the spectrometer is given by a Dirac δ-function ($A_{spec}(\lambda) = \delta(\lambda - \lambda_o)$). Then it follows from (2) that

$$I_{exp}^{CL} (U_b,\lambda_0) \;\approx\; S(\lambda_0)\; I^{CL}(U_b,\lambda_0) \tag{4}$$

– spectral integral experiments

A special wavelength region $\lambda_\ell < \lambda < \lambda_u$ is selected. Therefore, the resolution function of the spectrometer is given by

$$A_{spec}(\lambda) \;=\; A_{spec}^0 \; [\![\; \theta(\lambda-\lambda_\ell) - \theta(\lambda-\lambda_u) \;]\!] \tag{5}$$

Corresponding to the mean-value theorem we get

$$I_{exp}^{CL}(\tilde{\lambda}) \;\approx\; S(\tilde{\lambda})\; I^{CL}(U_b,\tilde{\lambda}) \tag{6}$$

with $\lambda_\ell < \tilde{\lambda} < \lambda_u$. All wavelength dependent quantities are represented by "effective quantities". For instance it is only possible to get an "effective absorption coefficient" $\alpha(\tilde{\lambda})$ from such luminescence experiments.

The induced currents can be calculated from the excess minority carrier density δp directly. The EBIC signal at a pn-junction in the depth z_B parallel to the surface is given by

$$I^{EBIC}(U_b) \;=\; -q\; D_p \int_{-\infty}^{\infty} \int_{-\infty}^{\infty} dx\; dy\; \frac{\partial}{\partial z}\; \delta p(U_b,\underline{r})|_{z\,=\,z_B} \tag{7}$$

For combined EBIC and CL experiments we use the sample geometries given in Fig. 1. In our universal description the quantities $I^{EBIC}(U_b)$ and $I^{CL}(U_b,\lambda)$ are given by

$$I^{EBIC}(U_b) \;=\; -q\; G_0 \left[\; \Phi(\tfrac{1}{L}, z_m+z_d, \infty;\, U_b\,) \right. $$
$$\left. +\; \Phi(\,0,\, z_m,\, z_m+z_d;\, U_b\,) \right] \tag{8}$$

and
$$I^{Cl}(U_b,\lambda) \;=\; \frac{2\pi}{\tau_r} \int_{0}^{\theta_c} \sin\theta\; d\theta\; F(\hat{\alpha})$$

with $\hat{\alpha} = \alpha(\lambda)/\cos\theta$ and

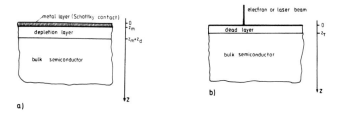

Fig. 1. Schematic view of the sample geometries under consideration.

$$F(\hat{\alpha}) = \frac{\tau\, G_0\, e^{-\alpha z_T}}{(1-(\hat{\alpha}L)^2)}\left(\Phi(\,\hat{\alpha},\, z_T,\, \infty;\, U_b\,) - \frac{\hat{\alpha}L+S}{1+S}\,\Phi(\,\tfrac{1}{L},\, z_T,\, \infty;\, U_b\,) \right) \qquad (9)$$

The thickness z_m of the Schottky contact and a depletion layer of thickness z_d are taken into account in the calculation of the EBIC signal. In the calculation of the CL signal a dead layer of thickness z_T, that means a layer at the surface from which no recombination radiation arises, is taken into account. Φ is the so called "universal function". Φ is given by

$$\Phi(x,\, z_1,\, z_2;\, U_b) = \int_{z_1}^{z_2} dz\, g_z(z;\, U_b)\, e^{-x(z-z_1)} \qquad (10)$$

Because of the rotational symmetry of the samples only the depth distribution $g_z\,(z;\, U_b)$ of the generation rate enters the calculation of the signals.

Fig. 2 shows the experimental CL signal $I_{exp}^{CL}(U_b,\, \lambda)$ as a function of the beam voltage U_b for a fixed wavelength λ of the recombination radiation, detected in the experiment, in a schematic representation ($U_b I_b$ = const.). Typical properties of such an experimental curve are the occurrence of a maximum at a certain beam voltage U_b^{max} and the behaviour of the signal for low beam voltages (Hergert et al 1987a,b, Koch et al 1988). It is possible to get information about the reduced surface recombination velocity S from the limit $U_b \to 0$ as indicated in Fig. 2. The beam voltage position U_b^{max} yields information about α if L is known from a corresponding EBIC experiment.

For a fixed value of L, U_b^{max} increases with decreasing value of the absorption coefficient. Theoretical considerations show that if the diffusion length L is nearly 1μm and U_b^{max} changes from 20kV to 40kV in the wavelength region under consideration, α changes by more than order of magnitude (Hergert et al 1987b, Hildebrandt et al 1988).

Crystal defects in semiconducting samples result in local changes of the recombination properties of the sample. The total recombination lifetime τ as well as the radiative recombination lifetime τ_r can be different in a corresponding defect region (cf. Fig. 3):

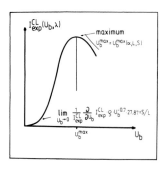

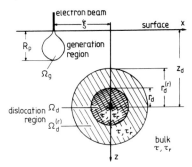

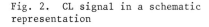

Fig. 2. CL signal in a schematic representation

Fig 3. Schematic illustration of a contrast experiment at a surface parallel dislocation

$$\tau(\underline{r}) = \begin{cases} \tau' & \text{within } \Omega_d \\ \tau & \text{elsewhere} \end{cases} \qquad \tau_r(\underline{r}) = \begin{cases} \tau_r' & \text{within } \Omega_d^{(r)} \\ \tau_r & \text{elsewhere} \end{cases} \quad (11)$$

Detailed theoretical considerations of the defect contrast in EBIC and CL experiments are given by Hergert (1988), Pasemann (1987), Pasemann et al (1986). The CL contrast can be calculated from

$$C_{Cl} = \left\{ -\frac{\tau/\tau' - 1}{\tau_r \, I_0^{CL}} \int_{\Omega_d} d^3r' \; \delta p(\underline{r}_g, \underline{r}') \; j_{CL}(\underline{r}') \right\}$$

$$+ \left\{ \frac{\tau_r/\tau_r' - 1}{\tau_r \, I_0^{CL}} \int_{\Omega_d^{(r)}} d^3\underline{r}' \; \delta p(\underline{r}_g, \underline{r}') \; A(z') \right\} \qquad (12)$$

$$= C_{CL}^{(E)} + C_{CL}^{(C)}$$

($\underline{r}_s$ - position of the generation region). I_0^{CL} is the CL signal far away from the defect. j_{CL} corresponds to the CL signal from point source excitation and is given by

$$j_{CL}(\underline{r}') = \frac{1}{4\pi L^2} \int_{\Omega_s} d^3\underline{r} \; G(\underline{r}', \underline{r}) \; A(z) \qquad (13)$$

A(z) describes the optical loss, i.e. the common effect of internal absorption of the luminescence and total reflection at the surface, where the transmission coefficient of the surface is assumed to be constant for all angles less than the critical angle θ_c. The first term in (12) results from a change in the total lifetime, like the EBIC contrast too. Therefore this term is called "EBIC-like" CL contrast. The change of τ_r in the defect region results in a typical contrast mechanism in the luminescence signals. Therefore the second term in (12) is called "CL-specific" CL contrast (cf Hergert 1988, Pasemann 1987, Pasemann et al 1986).

If the change of τ_r is unimportant in the material under consideration, combined EBIC and CL contrast experiments can provide information about the depth position of the defect below the surface of the sample. Theoretical considerations (Paseman et al 1986) have shown that the contrast ratio C_{EBIC}/C_{CL} is determined by the semiconductor parameters of the homogeneous semiconductor and the depth position z_d of the defect:

$$\frac{C_{EBIC}}{C_{CL}} = \frac{\tau_r \, I_0^{CL}}{\tau \, I_0^{EBIC}} \; F(z_d) \qquad (14)$$

The depth position z_d of the defect enters the theoretical known function $F(z_d)$.

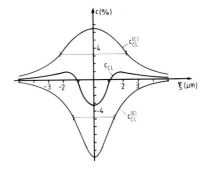

Fig. 4. Model calculation of dark-bright stripe contrast (L = 3µm, S→∞, R_p = .5µm, α = 0µm^{-1}, z_d = 2µm, $\Gamma_d^P(r)$ = 2µm, γ = 3, $(\tau_r/\tau_r' - 1)$ = .15, r_d = .2µm)

If the "CL-specific" CL contrast $C_{CL}^{(C)}$ is important, dark-bright contrasts are possible for $C_{CL}^{(E)}$ and $C_{CL}^{(C)}$ have different signs and different full widths at half maximum. Fig. 4 shows a model calculation for a surface parallel dislocation. The superposition of the two CL contrast contributions gives a dark-bright stripe.

3. EXPERIMENTAL METHOD

In performing combined CL and EBIC examinations one has to realise the simultaneous measurement of CL and EBIC signals in the SEM apparatus. A schematic representation of the experimental setup used is given in Fig. 5. The quantitative CL and EBIC experiments carried out under conditions of low carrier injection made demands for high sensitivity and stability of the complete equipment (better than 1% required). The registration of the CL signals was performed by means of the highly efficient light collection system. To minimize background signals the photomultiplier was cooled. Optical filters provided for quasi-monochromatic CL observations. Both the CL and EBIC signal were measured using a lock-in technique. The quantitative investigations were made by beam voltage dependent measurements in the range 2kV to 45kV for a constant excitation power of $\leq 20\mu W$ (Hergert et al 1987b). Because of the relatively low excitation level the beam current had to be measured with high accurancy. The temperature of the samples could be varied from 300K to 80K.

Combined CL and EBIC studies are widely applied to optoelectronic compound semiconductor materials and devices. The experimental results presented here were obtained on GaAs, $GaAs_{1-x}P_x$, and various III-V specimens. Sample preparation and Schottky contact formation were carried out using standard procedures.

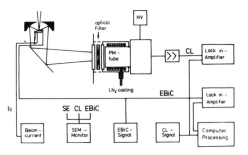

Fig. 5 Scheme of experimental arrangement for combined SEM-CL/EBIC

4. RESULTS AND DISCUSSION

4.1 Determination of semiconductor matrix parameters.

The EBIC technique is well established for the determination of the electrical semiconductor parameters (L, v_s). Only recently has it been shown that the CL mode is able to yield quantitative information about the optical parameters like absorption coefficient α or quantum yield η in homogeneous samples (Hergert et al 1987a, Hildebrandt et al 1988, Koch et al 1988). In our recent work the CL experiments were extended to spectral resolved measurements (Hildebrandt et al 1988).

Fig. 6 shows the result of a wavelength resolved CL experiment on $GaAs_{1-x}P_x$. The curves of the monochromatic CL signal vs. beam voltage give evidence of the shift $U_b^{max}(\lambda)$ mentioned above. From the analysis of the experimental data, by fitting the theoretical voltage dependent CL signal using carrier diffusion length values derived from the EBIC

measurement, the optical absorption
coefficient $\alpha(\lambda)$ in the relevant
wavelength region is obtained.

The determination of the semiconductor parameters like
diffusion length L, absorption
coefficient α etc. of the homogeneous semiconducting sample
with combined EBIC and CL
experiments is a necessary
supposition for a quantitative
study of the properties of
crystal defects with the SEM.

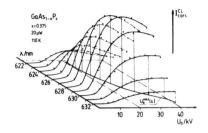

Fig. 6. Monochromatic CL signal
dependence on beam voltage U_b

4.2 Study of crystal defects

Considerations of the defect properties start with the representation of
the defect configuration by CL and EBIC micrographs. Dislocations occur
as point or comma and line shaped image contours depending on defect
geometry (Schreiber 1978). The nature of the defect contrast is
determined by local carrier recombination processes. In Fig. 7 a, b CL and
EBIC images from (001) surface of GaAsP-VPE layer specimens are shown:

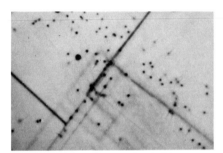
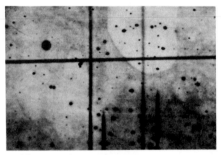

Fig. 7 a) CL image of misfit
dislocations in GaAsP-VPE-layer

b) EBIC image (Au Schottky
contact) of GaAsP-VPE-layer

dislocations are easily identified by comparing the CL and EBIC maps with
etch pit distributions in the surface area investigated. It is possible
to distinguish dislocations which lie parallel to the surface and perpendicular or inclined dislocations which intersect the surface. The crystallographic directions of the dislocation lines are <110>, <001> and <112>.
The system of line-shaped CL and EBIC contrasts belongs to a 3-dimensional
misfit dislocation configuration caused by an x-gradient of 1-3µm depth below
the surface. The dislocation lines are situated in the (001) plane and are
oriented along [110] and [1$\bar{1}$0]. Their burgers vectors are mostly of the type
$\underline{b}$ = a/2<101>, which means they can occur as edge, 60°, and screw type dislocations. Both the CL mode and EBIC mode depict dislocations in a similar manner.
This proves that dislocations act as recombination centres quenching the
luminescence. Some differences in the contrasts for several misfit lines are
seen, which are mainly due to different defect depth. Along the lines one
finds a nearly homogeneous contrast. (Superpositions of contrasts of various
defects are excluded). A striking feature of the contrast at a singular point,
like the 90° knee of a dislocation line, cannot be established.

In the EBIC picture (Fig. 7 b) very small dark spots are recognised.
Their density is two or three orders of magnitude higher than the density
of dot shaped dislocation contrasts. This fact hints at the presence of
micro-defects localized in the depletion layer of the Schottky contact.
The contrasts of micro-defects are absent in the CL image because no
luminescence emission results from the depletion region.

Fig. 8 a, b shows the CL and EBIC micrographs of a GaAs-IRED device
specimen with a pn-junction several microns below the surface. The
pictures illustrate the recombination active defects in the near surface
p-region of the device and demonstrate that remarkably different contrasts
in the CL and EBIC micrographs can arise from the same defect, as is
exemplified by dislocation 2. Such pronounced differences of the defect
contrast in the CL and EBIC mode might be for a variety of reasons, for
instance the defect activity as a radiative or non-radiative centre and
the different responses of the CL and EBIC signals to the defect, depending
on its position with respect to the EBIC charge collection plane. These

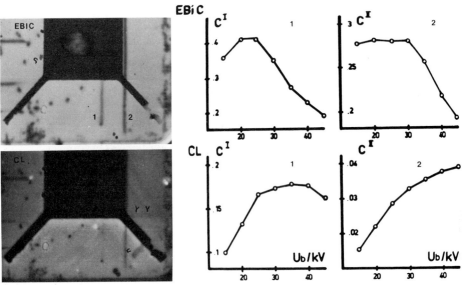

Fig. 8 CL and EBIC image   Fig. 9 Plots of CL and EBIC contrasts vs.
of GaAs device sample   beam voltage U for the line defects 1 and 2

contrast behaviours may be discussed within the framework of the model given
by Pasemann et al (1986). The CL and EBIC contrast values vs. U_b of the
two line defects seen in Fig. 8 are shown in Fig. 9. It is found that the
CL and EBIC contrast of the same defect change independently. However, if
one considers the quotients C_{EBIC}/C_{CL} then a clear correlation between
C_{EBIC} and C_{CL} is established as shown in Fig. 10. The EBIC to CL
contrast ratio was analysed theoretically using relation (14). We
obtained the qualitative correspondence of this experimentally measured
contrast ratio vs. U_b with that expected theoretically, given the
assumption of different depth positions of dislocations 1 and 2. In
our case, dislocation 2 lies deeper (2 μm) than dislocation 1 (1 μm).

On the other hand, the two dislocations would be expected to be in a similar state of decoration since they have interacted with point defects during the thermal treatment in the manufacturing process. Therefore it seems reasonable that they possess nearly equal recombination behaviour.

The strong interaction between the dislocations and point defects leads to a higher degree of decoration state (Schreiber 1978) from which implicitly arises the Halo contrast. An example is given in the CL micrograph of Fig. 11. Halo-contrasts result from local changes of the recombination properties in the vicinity of the dislocation as discussed above by $\tau_r' < \tau_r$. A correlation with a changed point defect configuration may be supported by spectroscopical findings.

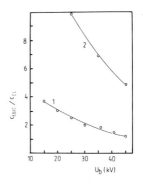

Fig. 10 Contrast ratio C_{EBIC}/C_{CL} of line defect 1 and 2

Due to the interaction with point defects and geometrical factors influencing the dislocation contrasts in CL and EBIC pictures, serious difficulties exist in estimating absolute and relative recombination activity from defect contrast measurements. For that purpose the values of defect strength have to be deduced. Our relevant results obtained from quantitative contrast measurements at dislocations in n-GaAs bulk samples will be published elsewhere. It could be shown that in bulk samples of several compound semiconductors (GaAs, InP, GaP, CdTe, CdS) freshly produced dislocations (introduced at 300K) also act as recombination centres. Fig. 12 shows the CL image of a GaAs (111) specimen; contours of dislocation half loops lying in surface-parallel glide planes are seen. The dislocation half loops have extended in the stress field of a surface scratch.

Kinematical in-situ CL observations give evidence of stimulated motion of the dislocations (REDM effect after Maeda et al 1983). It is an interesting result that stimulated dislocation movement

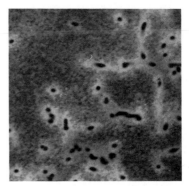

Fig. 11 CL Halo contrasts at dislocations in GaAs bulk sample

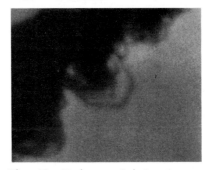

Fig. 12 CL image of induced dislocation loops on (111) GaAs

was observed also at low temperature (down to 80K). The fresh defects
have relatively high mobilities, thus Cottrell clouds should not occur.
Therefore, the influences of extrinsic point defects on the local carrier
recombination at the dislocations should be weak. Point defect centres
produced by moving dislocations must, if they exist, move together with
the dislocations or build up traces behind the moving defects (Tarbaev et
al 1988).

The first SEM-CL observations of the glide motion of dislocations were
successfully conducted on CdS samples (Brümmer et al 1974). In this
material, at temperatures below 130K, the dislocations exhibit a real
bright CL contrast caused by the defect bound luminescence emission
(λ = 508 nm) (Schreiber 1978). A detailed discussion of the nature of
this dislocation induced radiative recombination process is given in
papers by Grin et al (1979) and Ossipyan et al (1987).

Material	$j_{th} \cdot A/m^2$	v,nm/s (j = 200 A/m^2)
GaP:S (1 · 10^{17} cm^{-3}	0.1	3 200
GaAs:Si (2 · 10^{18} cm^{-3})	1	60
InP:Sn (7 · 10^{17} cm^{-3})	100	800

Table 1. Threshold values of REDM and dislocation velocity values for
j = 200 A/m^2 for III-VI semiconductors

The REDM effect for dislocations occur in diverse semiconductor materials
under different conditions (Schreiber et al 1988). In Table 1 REDM
threshold values of beam current density and the dislocation glide
velocities observed on GaAs, GaP and InP under equal excitation density
are compiled. In III-V materials only real dark CL contrasts at movable
dislocations are found. It might be concluded that non-radiative
recombination via dislocation core states (probably at kink sites) takes
place which may promote the dislocation motion.

5. CONCLUDING REMARKS

The combined application of SEM CL and SEM EBIC investigations was
considered. The experimental CL and EBIC measurements carried out
in dependence on beam voltage were treated within the framework of a
theoretical description for CL and EBIC signals. Thus quantitative
information on electrical and optical semiconductor parameters and
quantities of defects could be derived. The analysis of combined CL
and EBIC defect contrast investigations provided the depth position
of dislocations below the surface investigated. From CL and EBIC
contrast behaviour first conclusions on the recombination activity of
grown-in and induced dislocations showing REDM effect were obtained.

ACKNOWLEDGEMENTS

The authors wish to acknowledge collaboration with Dr H S Leipner and
Dr S Hildebrandt in the experimental and theoretical studies. We also
wish to thank these colleagues for their contributions to the technical
preparation of the paper.

REFERENCES

Akamatsu B, Hénoc J and Hénoc P 1981 J. Appl. Phys. 52 7245
Brümmer O,and Schreiber J 1974 Kristall und Technik 9 817
Grin V F, Salkov E A, Tarbaev N I and Shepelski G A 1979 Fiz.tverd.Tel.
    21 1690
Hergert W, Reck P, Pasemann L and Schreiber J 1987a phys. stat. sol. (a)
    101 611
Hergert W and Hildebrandt S 1987b phys. stat. sol. (a) 103 K145
Hergert W and Hildebrandt S 1988a phys. stat. sol. (a) 109 625
Hergert W 1988b Thesis (Halle)
Hildebrandt S, Schreiber J, Hergert W and Petrov V I 1988 phys. stat. sol.
    (a) 110 283
Koch F and Oelgart G 1987 phys. stat. sol. (a) 104 931
Koch F, Hergert W, Oelgart G and Puhlmann N 1988 phys. stat. sol. (a)
    109 261
Maeda K and Takeuchi S 1983 J. Physique 44 C4/175
Ossipyan Yu A, Negri V D and Bulenkov N A 1987 Izv. Akad. Nauk SSSR, Ser.
    Fiz. 51 1458
Pasemann L and Hergert W 1986 Ultramicroscopy 19 15
Passmann L 1987 Thesis (Leipzig)
v Roos O 1983 J. Appl. Phys. 54 1390
v Roos O 1985 J. Appl. Phys. 57 2196
Rossin V V and Sidorov V G 1986 phys. stat. sol. (a) 95 15
Schreiber J 1978 Thesis (Halle)
Schreiber J and Leipner H S 1988 Sov. Phys. – Quantum Electronics 15 2304
Tarbaev N I, Schreiber J and Shepelski G A 1988 phys. stat. sol. (a) 110 97

− Note: All markers given in the figures represent 10 μm.

Inst. Phys. Conf. Ser. No 104: Chapter 2
Paper presented at Int. Symp. on Struct. Prop. Disloc. Semicond., Oxford, 1989

109

Electrical and optical phenomena of II–VI semiconductors associated with dislocations

Yu.A.Ossipyan

Institute of Solid State Physics, USSR Academy of Sciences, Chernogolovka, Moscow district, 142432, USSR

ABSTRACT: A review of numerous studies, pioneered by the discovery of the photoplastic effect is presented. Moving dislocations in II-VI compounds, strongly interacting with point-defect electrons, were shown to form their own anomalously great electric charge, whose presence stipulates the numerous physical phenomena.

Among the extensive researches into the interaction of dislocations with electrons in semiconductors, the papers devoted to II-VI compounds are of particular importance. High plasticity at room temperatures (and in some cases up to liquid helium temperatures) and a wide forbidden band allow studying the dynamical interaction of moving dislocations with an electron subsystem in the conditions when electron-hole pairs are weakly excited by the thermal generation, that is, when the electron subsystem is still rather sensitive to the excitations induced by luminescence, electrical fields, injection of nonequilibrium carriers and even by the dislocation motion. Under these unique conditions the interaction of electron and dislocation subsystems of II-VI compounds gives rise to a whole number of remarkable physical phenomena. The first of the latter is the photoplastic effect (PPE), discovered by Ossipyan and Savchenko in 1968 (Fig.1). The light inducing interband electron transitions turned out to cause an essential increase in the plastic flow stress . The discovery of the PPE in CdS stimulated numerous investigations devoted to plastic properties of II-VI compounds, the dislocation structures of these materials and to the effect of immobile and moving dislocations on the electrical and optical properties. The results of these papers, in which a whole series of new physical effects were discovered, have been summarized in the review by Ossipyan, Petrenko, Zaretskii and Whitworth (1986). Here we present in brief the fundamental experiments,models and theoretical results.

1. DISLOCATION CHARGES AND CURRENTS

The dislocations moving in II-VI compounds under plastic deformation transfer an anomalously large electrical charge, almost as large as an electron, for an interatomic distance.

Fig. 1. Compression diagram of a CdS crystal at T= 360 K and $\dot{\varepsilon}$ = 1.6x10^{-5}s^{-1} showing the enhancement of stress $\Delta\tau$ PPE produced by illumination (Ossipyan and Savchenko 1968).

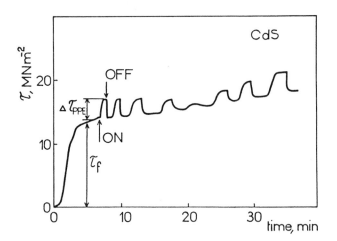

Fig. 2. Schematic arrangement for measuring the dislocation current I_d when a ZnSe crystal is deformed plastically under compressive stress σ. In this case the Se(g) dislocations are more mobile than Zn(g) and the current is predominantly due to them. Also shown are typical results for ZnS of σ and I_d versus time, for deformation rates of 20,16,12,8,4 and 2 μm min^{-1} (Ossipyan and Petrenko 1975).

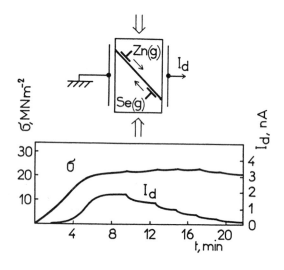

The motion of such charges is responsible for the arising of dislocation currents (Fig.2) and a considerable difference in potentials (Ossipyan and Petrenko 1975). The nature of the dislocation charge in II-VI compounds was studied in the papers by Ossipyan and Petrenko (1975), Kirichenko et al (1978), Petrenko and Whitworth (1980), Petrenko (1982). It was found that this charge is thermodynamically nonequilibrium and can be determined by equalizing the two electron fluxes: the electrons trapped on a deep dislocation level E_d from the point centers, "picked up" by a moving dislocation and a flux of electrons tunneling through the Coulomb barrier from E_d into the conduction band. In this case, before tunneling the electron may undergo a thermoactivated transfer to a state with higher energy. As a result, the linear density of the dislocation charge q becomes dependent on the depth and capacity of the dislocation level, the electron concentration at deep centers and on the velocity of dislocations and temperature, whereas the experimental study of the dependence of q on these parameters provides information on both the dislocation and electron subsystems of crystal. The processes of electron exchange between a moving dislocation, point centers and forbidden bands show up most vividly in the phenomenon of dynamic excitation of electric conductivity in the motion of dislocations (Zaretskii and Petrenko 1983) (Fig. 3). As one may see in the figure, in the region of plastic

Fig. 3. Time dependence of stress τ, Hall mobility μ_H and calculated free-electron concentration n during plastic deformation of a weakly illuminated ZnSe crystal. T= 295 K, strain rate =3 x 10^{-5}s^{-1} (Zaretskii and Petrenko 1983).

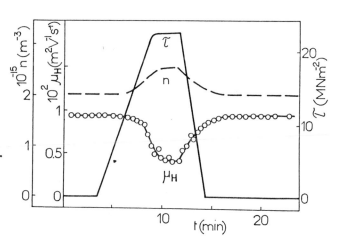

flow the electron concentration of conductivity n increases (first of all). This is due to the transfer of electrons by a charged dislocation or directly from deep point centers to the conduction band (thermal-field ionization), or through level E_d. An increase in the number of charged ionized point defects and the growing charge of dislocations cause an increase in the electron scattering, that is, a decrease in their mobility, μ. The same process of ionization by a moving charged dislocation of point centers defines the arising of continuous (in time) deformation luminescence (Fig.4). The

Fig. 4. Spectrum of the pulsed luminescence produced by the deformation of ZnS: Mn crystals (Bredikhin and Shmurak 1977).

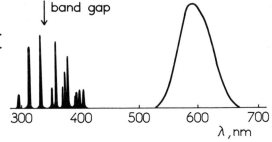

section of the line spectrum depicted in this figure is related to pulsed luminescence of the surface, giving light when there is an electrical breakdown between the sections of the crystal with the dislocation charge carried out (Bredikhin and Shmurak 1977). In case the deformation occurs in vacuum, the emergence of electron-saturated dislocations onto the surface may be accompanied by the electron emission (Poletaev and Shmurak 1981). The charges of dislocations in some II-VI compounds are so large (in ZnS, for instance) that, by applying comparatively small external electrical fields to them, one may change essentially the conditions of plastic deformation (Fig.5). The study of the mechanism of motion of charged

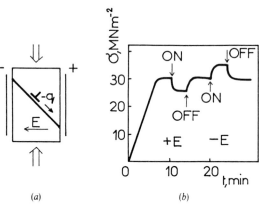

Fig. 5. (a) Application of an electric field E to a crystal during plastic deformation in compression at constant strain rate. Mobile dislocations are assumed to be of the type shown and to be negatively charged. (b) Typical stress-time graph for a ZnS sample measuring 4 x 3 x 1 mm deformed at $h_p=10$ μm min^{-1} with an applied potential difference of 2 kV (Ossipyan and Petrenko 1975).

dislocations in II-VI compounds at about room temperatures carried out by the methods of thermo-activated analysis, electron microscopy in-situ, etc., showed that this is a Peierls' mechanism, and the height of the Peierls' relief and, therefore, the plastic flow **stress** depend on the linear density of dislocation charge (Fig.6). Such a dependence is likely

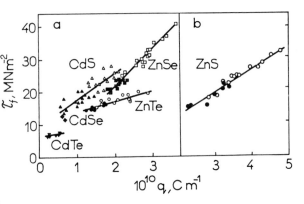

Fig. 6. Collected results of measurements of the flow stress τ_f and dislocation charge q for different II-VI compounds. Solid points are for measurements on different samples in darkness, and open points are obtained under varying levels of illumination. Dislocation charges in (a) have been calculated assuming that perfect dislocations are moving, but in (b) for ZnS partial dislocations have been assumed to move (Petrenko and Whitworth 1980)

to be a result of the interaction of a charged dislocation line with the series of like-charged ions of the crystal lattice (Ossipyan and Petrenko 1978). The model of the dynamical formation of the moving dislocation charge together with the model of the Coulomb interaction of charged dislocations and lattice ions allowed elucidating such phenomena as the PPE and the related one, the injection-plastic effect (Ossipyan and Petrenko 1973), observed under excitation of the electron system of the crystal by the injection of electrons and holes from the contacts during plastic deformation (Fig.7).

Fig. 7. (a) The injection-plastic effect in a crystal of ZnSe 300 μm thick deformed at a constant rate in four-point bending. The graph shows the required force f as a function of time. At the points indicated a potential difference of ± 3 kV is applied to In-$\overline{\text{Hg}}$ electrodes on the large surfaces. The potential difference is removed at the points marked OFF. Temperature = 370 K. (b) Comparison of the effect

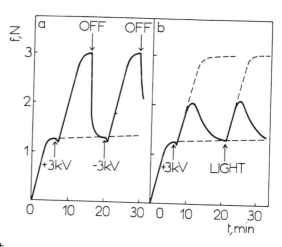

of a 3 kV pulse as in (a) with that of an appropriate level of illumination. The broken line is copied from (a) (Zaretskii et al 1978).

In accordance with the model proposed by Ossipyan and Petrenko (1978), the excitation of the crystal electron system by illumination or injection causes a change in the magnitude of the dislocation charge, q, and, consequently, in the height of the Peierls' relief for its motion.

2. LOW TEMPERATURE DEFORMATION

Successful researches into the low-temperature deformation of the II-VI compounds marked a new stage in studying the electron-dislocation interaction in these compounds. The improved quality of single crystals, high accuracy in sample treatment and perfecting of the deformation procedure made it possible to decrease the plastic flow temperature down to 4.2 K (Negriy and Ossipyan 1978). This presented two qualitatively new possibilities at least. First, due to enhanced luminescence quantum efficiency this rather informative method (luminescence) can be used for studying the interactions of electrons with moving dislocations, that is, in situ. Second, a decrease in temperature allowed suppressing the processes of dislocation decoration with point defects, that is, it allowed examining a "purer" object. In these conditions the method of low temperature photo- or cathodoluminescence (PL and CL) is fairly efficient. The use of scanning electron micros- copes operating in the CL regime permits studying the inhomo- geneities of radiation distribution on the excited surface and relating them with various defects of the crystal struc- ture. One of the variants of this method is that of space- resolved photoluminescence. It has successfully used for ana- lyzing the plastic deformation of II-VI crystals in the tem- perature range (300-4.2 K) on their optic properties within the exciton and edge regions of spectra (Negriy and Ossipyan 1978,1979). In the range of wavelengths λ =505-510 nm at 4.2 K the crystals deformed by basis and prismatic systems exhibit radiation (Fig.8) induced by different type disloca- tions introduced into the crys-

tal. The examination of the effect of the temperature on the optical spectrum of dislocation emission at low temperature showed instabi- lity of the defects resul- ting from plastic deforma- tion, which destruct under long exposure at T=300 and heating the crystal. Most remarkable results were ob- tained while studying the low-temperature plastic de- formation of CdS crystals (Negriy and Ossipyan 1982). The method of space-resolved photoluminescence applied revealed generation and moti- on of separate dislocations.

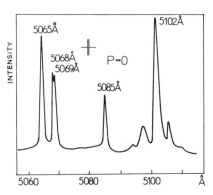

Fig. 8.   Dislocation photo- emission spectrum.

Dislocations on moving in glide planes were shown to genera- te specific crystal lattice defects which are effective cen- ters of radiative recombination with a characteristic emis- sion spectrum and are displayed as luminous tracks passed through the crystal. To elucidate the space structure of ra- diating defects and their symmetry the effect of uniaxial lo- ad on the emission spectrum was investigated. Splitting of all the spectrum lines into six components under loading along the common $[1, \mathcal{E}_1 ; 1+\mathcal{E}_1, \mathcal{E}_2]$ type direction testifi- es of the C_s group symmetry of the arising centers. The stu- dy of the radiation polarization made in different geometri- es of crystal deformation revealed a new characteristic fea- ture - cooperative behaviour of defects (Ossipyan and Negriy 1987). Due to this property a system of the defects, located in traces of the dislocations moving in crystal, in the pola- rized radiation falls into domains characterized by two ori- entations of the vector of the light wave. As a result of excitation of the crystal electron subsystem the domain stru- cture of dislocation traces becomes unstable which shows up in nonstationary behaviour of the polarized dislocation emis- sion in time and space (Fig.9). To account for a whole num- ber of optical defects arising in the process of dislocation motion proposed were a model of configurational defects and the mechanism of their formation. The optical study perform-

ed provides evidence that the system of defects resulting from plastic deformation possesses certain
ordering related to collective
interaction of defects. An investigation of this interaction will
elucidate not only the electron
structure of the defects but their
role in the processes of plastic deformation of crystals as well.

3. SPACE RESOLVED DISLOCATION SPECTROSCOPY

Due to localized electron states
observed in the region of dislocation core and elastic stresses
exhibited around dislocations, the
latter form in the crystal extended inhomogeneities of dielectric
constant which may affect essentially the propagation of light in
plastically deformed crystals.
Bands of glide in CdS and other
II-VI crystals show light focussing and channeling. This is defined by the fact that the inhomogeneities of dielectric permeability give rise to formation of localized light waves in

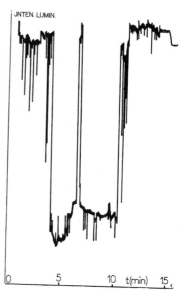

Fig. 9. Fragment of
the time dependence of
the intensity of pola-
rized radiation from
one of the domains of
the dislocation trace
at 4.2 K.

the neighbourhood of the dislocation core. They are sensitive to spectral and space singularities of the permeability,
which are related to resonance frequencies of localized electron states, and, therefore, the spectroscopy of the dislocation light focussing permits studying the energy spectra
of the electron states on dislocations at rather small their
average concentrations. We have thus revealed deep centers
on dislocations in CdS at the concentration of about 10^{13}
cm^{-3}, which two orders of magnitude is better than the sensitivity of the ordinary absorption spectroscopy. Formation of
a quasiperiodic lattice of bands of glide, its period being
commensurate with the wavelength of light, is a remarkable
feature of the developed dislocation structure in plastically deformed II-VI crystals (Emelin et al 1983). This gives
rise to a whole number of diffraction phenomena which occur
under emission of light in plastically deformed crystals,
which resemble effects of the X-ray diffraction on crystal
atomic lattice. For instance, the distribution of the light
field in the crystal and the intensities of the straight
light beam and the diffused light are oscillating functions
of the light direction. In this case when the Bragg condition is fulfilled the light intensity is distributed periodically inside the crystal. And provided the light wavelength
corresponds to the optical absorption on dislocations, the
dislocation absorption decreases, that is, the intensity of
the light passed through the crystal grows. Such oscillations of transmission in plastically deformed cadmium sulfide

are depicted in Fig.10: at angles of 0° and 9°, corresponding here to the Bragg condition, the property of dislocation
absorption is not exhibited in the 560 nm region, whereas at
other angles it is clearly seen. Fig.11 shows spectral oscillations of the
intensity of
light scatte-
red by bands of
glide in a simi-
lar CdS crystal
at a fixed 45°
angle between
the bands of gli-
de and the light
flux.

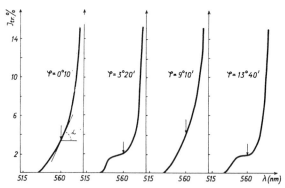

4. ULTRASONIC
SPECTROSCOPY

Fig. 10.   Transmission spectra of prismatic dislocated CdS at various angles
of light incidence. The light polarization is parallel to the glide bands.
a) φ =0°10'; b) 3°20'; c) 9°10';
d) 13°20'.

Finally, outstan-
ding results have
been obtained re-
cently when the
processes of plas-
tic deformation
of II-VI compou-
nds were studied
by the method of
ultrasonic spect-
roscopy. For instance, signals of acoustic emission
(AE) were observed under
plastic deformation of II-VI
crystals. The signals were
in the form of discrete ite-
rated trains of several tenth
of μsec in length (Robsman
and Shikhsaidov 1988a). The
spectral density of AE signal power, calculated on
the computer, has a number
of distinct resonance maxima (Fig.12). Additional experiments on the effect of
light and changing the loading showed that time intervals of AE trains coincide
with those when plastic flow
takes place. This fact suggests that the resonance
peaks of the AE spectral
distribution correspond to
the oscillatory processes
occurring upon motion of
dislocations. Investigation of the pulsed ultrasonic
(US) effect on the plastic flow of II-VI crystals showed

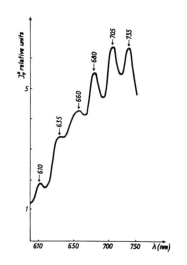

Fig. 11.   Secondary beam intensity oscillations in the
red part of the spectrum.

Fig. 12.
Spectral
density of
AE signals
under plas-
tic defor-
mation of
ZnSe crys-
tals.

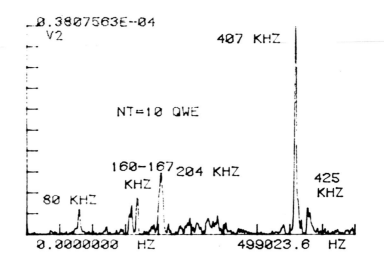

Fig. 13. De-
pendence of
the magnitude
of the softe-
ning effect
on the rate
of the ZnSe
crystal de-
formation.
1 - 25 kHz
2 - 100 kHz

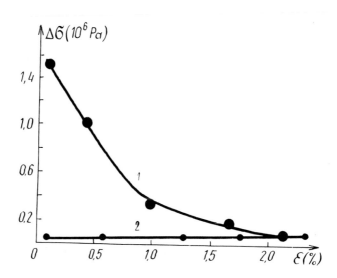

that the softening of the sample observed is essentially dependent on the carrier frequency of the US pulse (Robsman
and Shikhsaidov 1988b). At the initial stage of plastic flow
the magnitude of softening under the action of the pulse
with the carrier frequency of 25 kHz (curve 1, Fig.13) much
exceeds the analogous value for the pulse with the 100 kHz
frequency (curve 2). In this case the amplitudes of pulses
are equal. These results support the assumption that resonance oscillatory processes in the motion of dislocations in
II-VI crystals are very essential. Further researches into
these phenomena will provide more detailed information on

the microscopic processes of the kinetics of dislocations.

ACKNOWLEDGEMENTS

The author is grateful to Dr. N.V.Klassen, Dr. V.D.Negriy, Prof. V.F.Petrenko and Dr. M.Sh.Shikhsaidov for their help in the preparation of the manuscript.

REFERENCES

Bredikhin S I and Shmurak S Z 1977 Soviet Phys. JETP 46 768
Emelin V Ya, Klassen N V, Negriy V D and Ossipyan Yu A 1983
     J. de Physique Colloque C4 44 No9 pC4-125
Kirichenko L G, Petrenko V F and Uimin G V 1978 Soviet Phys.
     JETP 47 389
Negriy V D and Ossipyan Yu A 1978 Fizika tverd. Tela 20 744
Negriy V D and Ossipyan Yu A 1979 Phys.Stat.Sol.(a) 55 583
Negriy V D and Ossipyan Yu A 1982 JETF Lett. 35 598
Ossipyan Yu A and Negriy V D 1987 Inst.Phys.Conf.Ser. No87:
     Section 4 p 333
Ossipyan Yu A and Petrenko V F 1973 JETP Lett. 17 399
Ossipyan Yu A and Petrenko V F 1975 Soviet Phys.JETP 42 695
Ossipyan Yu A and Petrenko V F 1978 Soviet Phys.JETP 48 147
Ossipyan Yu A, Petrenko V F, Zaretskii A V and Whitworth R W
     1986 Advances in Physics 35 No2 pp 115-188
Ossipyan Yu A and Savchenko I B 1968 JETP Lett. 7 100
Poletaiev A V and Shmurak S Z 1981 Pis'ma Zh. tekhnich. Fiz.
     7 1352
Petrenko V F 1982 Thesis, Doctor of Science, Chernogolovka,
     Institute of Solid State Physics
Petrenko V F and Whitworth R W 1980 Phil.Mag.A 41 681
Robsman V A and Shikhsaidov M Sh 1988a Fizika tverd. Tela
     30 No8 p 2437
Robsman V A and Shikhsaidov M Sh 1988b Fizika tverd. Tela
     30 No11 p 3329
Zaretskii A V and Petrenko V F 1983 Soviet Phys. Solid St.
     25 300

Inst. Phys. Conf. Ser. No 104: Chapter 2
Paper presented at Int. Symp. on Struct. Prop. Disloc. Semicond., Oxford, 1989

119

Optical properties of dislocations in germanium crystals

Yu S Lelikov, Yu T Rebane and Yu G Shreter

A F Yoffe Physico-Technical Institute, Academy of Sciences, Leningrad, USSR

ABSTRACT: A sharp line in the photoluminescence (PL) of Ge with grown-in dislocations is observed. The line is interpreted as being due to the 1d-dislocation exciton (DE). The binding energy of the DE is determined as $\sim$ 3 meV and its theoretical estimation is made. The series of lines in the PL spectra of the Ge and Si crystals with non-equilibrium dislocations are explained. The kinetics and the mechanisms of the carrier recombination are analysed.

1. INTRODUCTION

The luminescence of Ge crystals with dislocations has been studied for more than thirty years (Newman 1957, Gippius and Vavilov 1966, Kolyubakin et al 1984, and others). In most experiments, crystals with a dislocation density > 10^6 cm^{-2} were used, and spectra with several broad lines were observed (width > 10 meV). There have been various speculations about the nature of these lines. Gippius and Vavilov (1966) ascribed the main line at $\sim$ 0.5 eV to internal transitions in some centres. Mergel and Labusch (1977) showed that the transitions are direct and supposed that they take place between the valence band and the 1d-band split off by the deformation field from the bottom of the c-band. Kolyubakin et al (1987) supposed that the transitions occur between the 3d-conduction band and a 1d-dangling bond band.

In this paper we report results of PL-investigation of Ge crystals with a low dislocation density of 10^3 - 10^5 cm^{-2}. Using a sensitive photodetector, we could observe in such crystals a sharp photoluminescence (PL) line at 513 meV (width $\sim$ 1 meV) and a background band in the range of 500-600 meV. Typical PL spectra are shown in Figure 1. In part 2 a detailed investigation of the sharp PL line behaviour under temperature variation is presented and a conclusion is made that this line is due to annihilation of a dislocation exciton (DE) formed by electron and hole in one-dimensional bands, split off by the deformation field from the c- and v-bands. The corresponding split-off energies are found to be equal to $\sim$ 150 meV and $\sim$ 80 meV. Further theoretical investigation has shown that the DE can be described in the effective mass approximation by using the zero-range potential model for the electron-hole interaction. The DE must be localised near the 90°-partial dislocation, because its deformation field is stronger than that of the 30°-partial dislocation (Winter 1977). The experimental binding energy of the DE is determined as $\sim$ 3 meV, which coincides with the theoretical estimation. In part 3 the origin of line series in PL spectra of crystals containing non-equilibrium dislocations is considered. Such

series were observed in Si by Sauer et
al (1986) and in Ge by Izotov et al
(1988). It will be shown that the
line series can be explained on the
basis of the DE-model taking into
account the dissociation of 60°
dislocations. The line positions
in the series are determined by the
shifts of the DE energy levels in the
deformation field of the 30° partial
dislocation. The discrete shift
values are related to the discrete
set of the stacking fault widths.
Part 4 is devoted to an investigation
of the origin of the background in the
PL spectra and to luminescence kinetics.
It will be shown that the broad background band is due to hot luminescence
from the dislocation segments with non-
equilibrium stacking fault widths.
Estimations of carrier life and energy
relaxation times in the 1d-dislocation
bands are made and a condition for hot
luminescence observation is found.
Additionally a system of rate equations
is developed, which allow description
of experimental dependences of the PL
sharp line intensity on temperature,
excitation level and dislocation density.

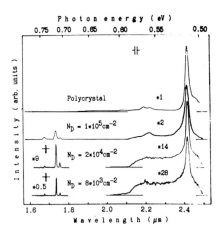

Fig. 1.   The PL spectra of Ge
with different densities of
grown-in dislocations.
Excitation with 1.06 μm Nd-YAG
laser light. Power density
$P = 0.5$ W/cm^2.

In additional, the PL spectra of a polycrystal are studied and it is shown
that in some cases additional peaks may appear in the background, which are
linked with specific features in the distribution function of stacking
fault widths.

2. ONE-DIMENSIONAL EXCITON

The possibility of existence of Mott-
type DE in semiconductors was discussed
by Emtage (1967) and Bajokin et al
(1982). In these papers the screening
of the electron-hole interaction by
dangling bonds of the dislocation core
was neglected. But, as will be shown
below, the screening has a substantial
effect on the exciton binding energy.
Recently, we reported the first
observation of a DE-line in PL of
Ge crystals with a low dislocation
density (Lelikov et al 1988). The
DE-spectra are shown in Figure 2.

2.1 DE-line position

The position of the line maxima for
the DE and free exciton (FE) behave
in a similar way with variation of

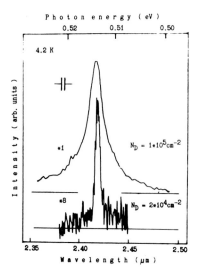

Fig. 2. PL-line of DE.

temperature (Figure 3). By comparing the curves for excitons with the temperature dependence of the forbidden band width, the binding energy of the excitons may be measured and thus we obtain $\approx$ 3 meV(DE) and $\approx$ 4 meV(FE).

2.2 DE-line intensity

With increasing temperature the DE-line intensity decreases with activation energies of $\approx$ 2 meV in the range of 10-30 K and $\approx$ 80 meV above 70K (Figure 4). The first gives the binding energy of DE and the second determines the binding energy of the hole in the 1d-dislocation band E_{v1}. The binding energy of the electron in the 1d-dislocation band E_{c1} can be obtained by subtraction:

$$E_{c1} = E_g - E_{v1} - h\omega = 150 \text{ meV}.$$

The displacement of the activation curve for the FE to lower temperatures (Figure 4) is due to the effectively smaller concentration of carriers in the 3-d bands compared with that in the 1-d bands.

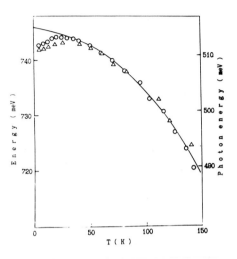

Fig. 3. The dependence of DE-line position on T. o – DE (right-hand scale). Δ – FE + LA-phonon; solid line – energy gap (left-hand scale).

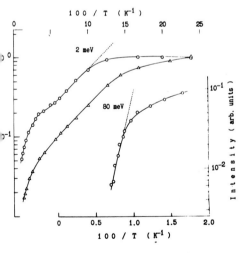

Fig. 4. The dependence of DE-line intensity on 1/T.
o – DE, Δ – FE

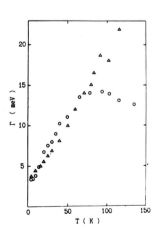

Fig. 5. The dependence of DE-line width on T.
o – DE, Δ – FE

2.3 DE-line width

The saturation of the line width with increasing temperature (Figure 5) is due to the 1-d character of dislocation bands. Indeed, without being broadened the line would be infinitely sharp at all temperatures because the density of states is $\sim E^{1/2}$, and has a singularity at E = 0. Taking into account an inhomogeneous broadening we obtain a finite value for the line width, $\Gamma \approx 10$ meV, at high temperatures (> 50 K), and a linear temperature dependence of $\Gamma \sim 2$ kT at low temperatures. The behaviour of the FE-line width differs from that of the 1d-exciton and is linear at all the temperatures studied.

2.4 Theoretical model

The analysis of the band structure of the Ge crystal, of the dissociation of 60°-dislocations and of the dangling bonds in the dislocation core reveals the following features of DE:

- to a first approximation, the DE can be described by using the effective mass approach, since the DE-line position varies with temperature in the same way as the energy gap in Ge (Figure 3);
- the electrons and holes, bound by the deformation field of dissociated dislocations, are localized at a 90°-partial dislocation (Winter 1978) and therefore the DE must be localized there too;
- the effective mass of carriers along the dislocation is determined by the light components of the anisotropic bulk masses (Winter 1977), and there-fore the reduced mass of the DE μ is small and equal to 0.03 m_0;
- the zero-range potential approach can be used to describe the electron-hole interaction.

The latter is linked with the fact that there is a conducting chain of dangling bonds in the core of a 90°-partial dislocation (Kveder et al 1985). This chain screens the electron-hole Coulomb interaction at distances of more than $r_t \approx 30\text{Å}$ where r_t is the maximum in transverse size of the electron and hole wave functions (Winter 1977). Thus, the electron-hole interaction potential is like a rectangular potential wall of depth $U_0 \approx e^2/\varkappa \ r_t$ and of width 1 $\approx r_t$. Since the non-equality $\hbar^2/\mu l^2 >> U_0$ is satisfied, the zero-range potential approach can be used (Demkov and Ostrovskii 1975) which gives for the DE binding energy a theoretical value of $E_{DE} = \mu e^4/\varkappa^2 \hbar^2 = 3$ meV. It is interesting to note that E_{DE} does not depend on r_t and the DE longitudinal size of $r_l \approx 300\text{Å}$ is more than ten times that of r_t. Therefore considering the DE as one-dimensional exciton is a good approximation. Thus the DE in Ge may be visualized as a "cigar" pierced through by a metallic string of dangling bonds.

3. ON THE ORIGIN OF THE LINE SERIES

Recently series of lines have been discovered in the PL spectra of Ge and Si containing non-equilibrium dislocations (Sauer et al 1986, Izotov et al 1988). The line positions have been described by empirical formulae. For Si, Sauer et al (1986) proposed:

$$E_n = E(D4) - const \ (14-n)^2 \qquad (1)$$

where E(D4) = 997 meV is the energy of D4-line, and n is an integer from 1 to 11. Another version has been proposed by Izotov et al (1988):

$$E_n = E^* - \alpha/\beta^n \qquad (2)$$

where $E^* = 550$ meV, $\alpha = 141$ meV, $\beta = 1.18$ for Ge and $E^* = 1013$ meV, $\alpha = 264$ meV, $\beta = 1.13$ for Si; n is a positive integer. The series disappears after annealing except for one line E_0, whose intensity increases. For Ge this line lies at 513 meV, and for Si at 997 meV. There are presently no satisfactory theoretical explanations of the empirical formulae and the series phenomena proper.

3.1 Non-equilibrium dislocations

It is known that 60°-dislocations in Ge and Si dissociate into 90°- and 30°-partial dislocations connected by a stacking fault ribbon of width r. In specimens prepared by the two-stage deformation method (Wessel and Alexander 1977) there may exist dislocations whose stacking fault width is larger and/or smaller than that of equilibrium dislocations (Alexander et al 1980, Gottschalk 1979). The distance between partial dislocations takes only discrete values

$$r_n = r_o + na_{SF} \qquad (3)$$

where n is positive or negative integer and the minimal step na_{SF} is determined by the structure of the stacking fault 2d-lattice and is linked with the volume lattice constant a by the relation: $n_{SF} = (3/8)^{1/2}a$. Following Sauer et al (1986), we suppose that discrete values of r results in the series of lines E_n. After annealing, only dislocations with equilibrium stacking fault widths $r_o = 65$Å (Si), $r_o = 50$Å (Ge) (Cockayne and Hons 1979) survive in the crystal, and therefore the line E should correspond to r_o.

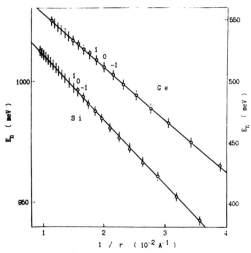

Fig. 6. Peak positions of series in PL-spectra plotted against reciprocal stacking fault widths for Ge and Si. The experimental data of Sauer et al (1986) for Si (left-hand scale) and of Ozotov et al (1988) for Ge (right-hand scale) are shown as 'o'. | are discrete values of stacking fault widths, solid line is the theory, the numbers at lines correspond to n in (3).

3.2 Influence of 30°-partial dislocation on DE energy

In part 2 we have shown that the E_0 line in Ge is linked with DE localized
at the 90°-partial. We suppose that in Si the E_0 (D4)-line is the PL also
of DE at the 90°-partial. The influence of the 30°-partial deformation
field on the DE energy may be considered as a perturbation, whose magnitude
is proportional to $1/r_n$, as the deformation field decays as $1/r$. Thus, the
energy levels of DE at dislocations with different stacking fault widths are:

$$E_n = E_\infty - A/r_n \qquad (4)$$

E_∞ is DE energy when $r \to \infty$, the constant $A = f \cdot D \cdot b \cdot /2\pi$, $D = D_e + D_h$ is the
sum of the deformation potential constants for electrons and holes, b is
the Burgers vector for the 30°-partial and $f = f(\rho, \theta)$ is a function which
depends on the orientation of b with respect to the dislocation line and
the stacking fault plane. The values of f may vary from - 1 to 1. Exact
calculation of f is difficult.

3.3 Comparison with experiment

In Figure 6 are shown dependences of the experimental values of E_n for the
non-phonon series on inverse stacking fault width $1/r_n$, calculated from (3).
In order to enumerate the lines we used the line E_0 as a starting point, to
which corresponds n = 0. For Si the point n = - 11 is calculated from
TA-phonon replica, and the point n = 12 is the D6-line (Sauer et al 1986).
If the formula (4) is correct, the experimental points have to fit on the
straight line. As can be seen from Figure 6 the points follow a straight
line rather closely. The absence in the experimental spectra of lines with
n = 1 - 11 in Si, and with n = 6 - 10 in Ge can be explained by the small
concentration of dislocations with the corresponding stacking fault widths.
Such a situation has been observed in the microscopic investigations made by
Alexander et al (1980) and Gottschalk (1979). The values of A obtained from
the slope of the straight lines in Figure 6 are A = 4340 meV·Å for Si.
These correspond to values of E_∞ = 598.9 meV for Ge and E_∞ 1037.3 meV for
Si. For Si the value of A is about one half that for Ge, which corresponds
to the deformation potential constant ratio for these semiconductors. It
should be noted that the usual description of the deformation field of a
30°-partial dislocations gives for A values smaller than those observed in
the experiment. We think that this difference is due to the crystal volume
changing near the dislocation core (Friedel 1968), which is neglected in
the ideal elastic approximation. Such changes of the crystal volume can
significantly strengthen the deformation field of a dislocation. This
also enables us to explain the large values of the split-off energies we
have observed for 1d-bands.

4.   RECOMBINATION KINETICS AND HOT LUMINESCENCE

The dislocation energy spectrum is characterized by three main bands: the
half-filled dangling bond band, and the empty and the full 1d-bands split
off from the c- and v-bands by the deformation field of the dislocation.
As was shown by Mergel and Labusch (1977) the 513 meV line does not
correspond to transitions of electrons to the empty part of the half-
filled dangling bond band because this line coincides with the peak in
the photoconductivity. Therefore we think that these transitions occur
between the empty and the full 1d-bands split off by deformation.

4.1 Analysis of recombination processes

In this model the dislocation luminescence intensity I should be proportional to the square of the non-equilibrium carrier concentration n in the 1d-bands. Generally, the recombination probability $R = \alpha_1 n + \alpha_2 n^2 + \alpha_3 n^3 \ldots$. For small excitation power G, the term $\alpha_1 n$ provides the main contribution, and $I \sim n^2 \sim G^2$. In experiments at T > 20 K where the exciton contribution to I may be neglected, I is proportional to G. Therefore at the excitation levels used in our experiments the squared term $\alpha_2 n^2$ in the recombination provides the main contribution. The constant of the quadratic recombination α_2 consists of two terms, $\alpha_2^R + \alpha_2^{NR}$., which correspond to the radiative and non-radiative recombination processes, respectively. We suppose the non-radiative quadratic recombination to be linked with Auger-processes through the half-filled band. The radiative recombination constant for the 1d-bands can be estimated theoretically as $\alpha_2^R \approx 10$ cm/s. For crystals with high dislocation density ($N_D > 10^5 \text{cm}^{-2}$) the recombination through dislocations is the main channel of recombination. Knowing the experimental value of the quantum efficiency for such crystals $q \approx 10^{-5}$ we can obtain $\alpha_2^{NR} = \alpha_2^R/q \approx 10^6$ cm/s. The carrier concentration in the 1d-bands can be estimated as $n = 10^3 - 10^5 \text{ cm}^{-1}$. This gives for the carrier lifetime in the 1d-bands $t_0 = 1/\alpha_2 n = 10^{-9} - 10^{-11}$s. The relaxation of the carrier energy in the 1d-bands is due to the emission of acoustic phonons and the corresponding relaxation time can be calculated as $t_\varepsilon = 10^{-10}$s. Thus at a low excitation level $t_0 > t_\varepsilon$ and PL originates from carriers in thermal equilibrium. At high level of excitation $t_0 < t_\varepsilon$ and the PL is the result of the recombination of the hot carriers because the carriers recombine before reaching thermal equilibrium. This mechanism leads to the saturation of the DE-line (or the line of thermal carriers for T > 20K) at the excitation power where $t_0 < t_\varepsilon$. In fact, in the experiment we observed the saturation of the 513 meV line to occur simultaneously with the appearance of the background, which we ascribe to hot luminescence (Figure 7).

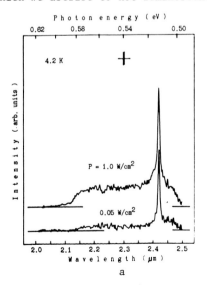 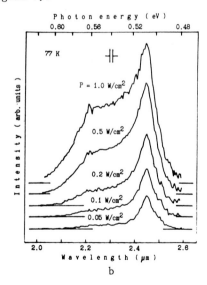

Fig. 7. PL-spectra of Ge with grown-in dislocation density $N_D = 2 \cdot 10^4$ cm^{-2} at T = 4.2K (a) and T = 77 K (b) for various excitation levels.

4.2 Various stacking fault widths and the formation of the background

So far we have considered dislocations as being ideally straight and having
a constant stacking fault width. In reality, the stacking fault width is
not constant. Therefore the edges of the dislocation bands for the carriers
forming excitons change along the dislocation line. This can lead to the
broadening of the DE-line and to the formation of a series of DE-lines.
The relative contribution of these factors to the spectrum depends on the
dislocation morphology. An abrupt change of the stacking fault width
leads to a smooth change of dislocation band edges on a characteristic
scale of $r_0 \approx 50\text{Å}$. Therefore in order to observe the line series in PL it
is necessary that spacing of the segments of dislocations having a non-
equilibrium stacking fault width should exceed $2r_0 \approx 100\text{Å}$. In the other
case the background appears due to strong broadening of the lines, with the
upper edge equal to E_∞ (Figure 7). This background in fact should be
considered as being due to hot luminescence because the lifetime of a
carrier on a segment L/v is smaller than t_0, where v is the velocity of the
carriers and L is the segment length. In some cases, short (L < 100Å) and
long (L > 100Å) segments of the dislocations with the non-equilibrium
stacking fault may exist simultaneously. This leads to the appearance of
a broad band of background with several maxima (quasilines), their positions
being determined by the pecularities of the stacking fault distribution
function (Figure 3). We attribute the luminescence at energies above E_∞
in Figure 8 to the dissociated screw dislocations.

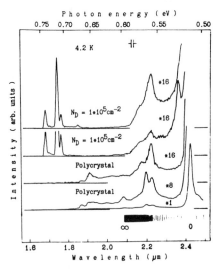

Fig. 8. PL-spectra from different spots of the same germanium ingot with a
complex grown-in dislocation structure. Shallow donor concentration
$N_d = 5 \cdot 10^{13}$ cm^{-3}. Power density P = 0.5 W/cm. | - DE-lines according to (4).

4.3 Recombination kinetics

In order to describe quantitatively the dependences of luminescence
intensity on temperature, on the dislocation density and the excitation
level, the following system of rate equations has been developed:

$$dN/dt = G - NDvN_D - N_{\sigma i} \, vN_i + \delta N_D(n_h + n_t)$$

$$dn_h/dt = NDv - \alpha_2 n_h(n_h + n_t) - \alpha_1 n_h - n_h/t_E - \delta n_h$$

$$dn_t/dt = n_h/t_E - \alpha_2 n_t(n_h + n_t) - \alpha_1 n_t - \beta n_t^2 - \delta n_t - \gamma n_{DE}$$ (5)

$$dn_{DE}/dt = \beta n_t^2 - n_{DE}/t_{DE} - \gamma n_{DE}$$

where N is the carrier concentration in the 3d-band, n_h, n_t are the 1d-hot and thermalised carrier concentrations respectively, n_{DE} is the DE concentration, N_D is the dislocation density, N_i is the concentration of non-radiative recombination centres, t_E is the energy relaxation time of hot carriers, t_{DE} is the lifetime of the DE, γ is the thermal decay probability of the DE, δ is the probability of the carrier thermal escape from the dislocation, α_1 and α_2 are the coefficients of the linear and quadratic recombination in 1d-bands, D is the cross section for the carrier capture by the dislocation, σ_i is the cross section for the recombination through impurities, and v is the mean carrier velocity. The parameters introduced in (5) have been calculated theoretically or estimated from experiment. Their values are:

$N_i = 10^{13}$ cm^{-3}, $D(cm) = 10^{-3}/T^2(K)$, $\alpha_1 = 2 \cdot 10^8$ s^{-1}, $\alpha_2 = 10^6$ cm/s, $\beta = 2 \cdot 10^5$ cm/s, $t_E = 10^{-10}$s, $t_{DE} = 10^{-9}$s, $\delta = (2 \cdot \sqrt{2} \, DvN_3/N_1) \exp(- E_{vi}/kT)$,

$\gamma = \beta \, N_1(m_1^*/\mu)^{1/2} \exp(- E_{DE}/kT)$, $N_1 = (m_1^* kT/2\pi h)^{1/2}$, $N_3 = (m_1^* kT/2\pi h)$,

$m_1^* = 0.07 \, m_o$ and $m_3^* = 0.3 \, m_o$ are mean masses in 1d- and 3d-bands,

$\mu = 0.03 \, m_o$, $E_{DE} = 3$ meV, $E_{vi} = 80$ meV.

To explain the observed PL-intensity increase for the DE and FE-lines in the temperature range of 50K < T < 80K the following model of the recombination centres in the crystal was proposed. The neutral recombination centre has a deep level cross section for one type of carriers. The other type of carriers are captured on the centre, when it is occupied by the first type of carriers and is charged. The capture process for the second type of carriers goes through the shallow Coulomb bond state. Therefore the capture cross section should be equal to that for a shallow centre and the ionization probability should have the corresponding activation energy $E_i \approx 10$ meV. Thus for the σ_i in (5) we obtain

$$\sigma_i = \sigma_1/\{1 + \sigma_1/\sigma_2[\sigma_2 \, t_i \, vN_3 \exp(- E_i/kT) + 1]\}$$ (6)

where σ_1, σ_2 are the cross sections of the capture of the electrons and holes on the recombination centre, t_i is the time of non-radiative recombination of the carriers bound at impurity, E_i is the ionization energy for a shallow centre. The parameters in (6) are $\sigma_1 = 3 \cdot 10^{-14}$ cm^2, $\sigma_2(cm) = 10^{-9}/T^3(K)$, $E_i = 10$ meV, $t_i = 2 \cdot 10^{-7}$ s. In Figures 9-11 the results of the theory and those of the experiments are compared. It can be seen that the theory accords well with the experimental data. Concluding, we enumerate once more the main suppositions that enabled us to explain the basic features of PL of Ge crystals with dislocations. These are the existence of DE, the influence of the stacking fault width on the DE energy, the quadratic recombination in 1d-bands and hot luminescence. Further investigations are desirable to develop the concepts presented in this paper and to obtain additional data substantiating them. We would like to note that the influence of the stacking

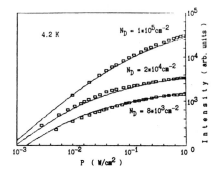

Fig. 9.   DE-line intensity as a function of excitation power for Ge samples
with various dislocation densities.   □ - experiment, ___ - theory.

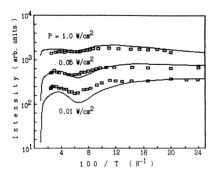

Fig. 10.   DE-line intensity as a function of inverse temperature for
various excitation powers.   □ - experiment, ___ - theory.

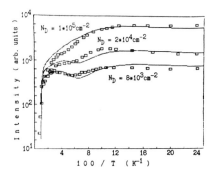

Fig. 11.   DE-line intensity as function of inverse temperature for
various dislocation densities.   □ - experiment, ___ - theory.

fault width on the energy of 1d-dislocation bands described in this paper should also be taken into account when the DLTS, photoconductivity and Hall-effect data are analysed.

ACKNOWLEDGEMENTS

We would like to thanks Professor P Haasen, R Labusch, W Schröter and Dr S A Shevchenko for helpful discussions related to some of the problems considered in this paper.

REFERENCES

Alexander H, Eppenstein H, Gottschalk H and Wendler S 1980 J. of Microscopy 118 13
Bajokin S B, Parschin D A,and Kharchenko V A 1982 Fiz.Tverd.Tela 24 1411
Cockayne D J H and Hons A 1979 J. de Physique 40 C6-11
Demkov Yu N and Ostrovskii V N 1975 The method of zero-range potential in atomic physics (Leningrad University, USSR)
Emtage P R 1967 Phys.Rev. 163 865
Friedel J 1964 Dislocations (Pergamon Press, Oxford, London, Edinburgh, New York, Paris, Frankfurt)
Gippius A A and Vavilov V S 1964 Fiz.Tverd.Tela 6 2361
Gottschalk J J 1979 J. de Physique 40 C6-127
Izotov A N, Kolyubakin A I, Shevchenko S A and Steinman E A 1988 Proc. 8th Int. School of Defects in Crystals (World Scientific, Singapore, New Jersey, London, Hong Kong, 203
Kolyubakin A I, Ossipyan Yu A, Shevchenko S A and Steinman E A 1984 Fiz.Tverd Tela 26 677
Kolyukakin A I, Ossipyan Yu A and Shevchenko S A 1987 Zh. Eksp. Teor, Fiz. 93 248
Kveder V V, Labusch R and Ossipyan Yu A 1985 Phys.Stat.Sol.(a) 92 293
Lelikov Yu S, Rebane Yu T and Shreter Yu A Proc. hI All-Union Conf. on Semiconductor Physics (Kishinev) vol. 3
Mergel D and Labusch R 1977 Phys.Stat.Sol. 41 431, 42 165
Newman R 1957 Phys.Rev. 105 1715
Sauer R, Kisielowski-Kemmerich Ch, and Alexander H 1986 Phys.Rev.Lett. 57 1472
Schröter W and Labusch R 1967 Phys.Stat.Sol. 21 211
Wessel K and Alexander H 1977 Phil.Mag. 35 1523
Winter S 1978 Phys.Stat.Sol.(b) 90 298
Winter S 1977 Electron-Strain Interaction in Crystals with Static Strain Fields (Umea University, Sweden

Inst. Phys. Conf. Ser. No 104: Chapter 2
Paper presented at Int. Symp. on Struct. Prop. Disloc. Semicond., Oxford, 1989

Electron microscopy of process-induced crystal defects

J Heydenreich

Institut für Festkörperphysik und Elektronenmikroskopie der
Akademie der Wissenschaften der DDR, Weinberg 2,
DDR-4050 Halle/Saale, German Democratic Republic

ABSTRACT: For the formation and transformation of process-induced crystal defects in semiconductors the interaction of extended defects with point defects (including
impurities) has to be considered the basis. Practical examples are given related to defects arising from technological processes in microelectronics and optoelectronics,   i.e. patterning (lithography), radiation /
particle beam treatment, heat treatment, and layer
growth.

1. INTRODUCTION

The elucidation of the influence of process-induced extended
crystal defects, especially dislocations, on the achievable
device parameters, is a current task of research in semiconductor physics and semiconductor technology,   (see e.g.
Van Landuyt et al 1986, Heydenreich 1988, Oppolzer 1988). In
micro-electronics this mainly refers to the reduced dimensions in the VLSI/ULSI technique and to related technological
processes. In optoelectronics extremely thin layers and high-
quality layer systems are needed based on epitaxial layers
of compound semiconductors (incl. graded layers) with the
emphasis on superlattices and quantum well structures.

Electron Microscopy with its high spatial resolution is
particularly useful for the characterization of process-
induced crystal defects. Although high-resolution electron
microscopy is the most reliable method of defect characterization in semiconductors (see e.g. Hutchison 1983, Ourmazd
1986, Bourret 1988), still today conventional diffraction contrast electron microscopy is used for many purposes, especially for a quick inspection of defects and their location
within the material and with respect to each other. The electrical activity of geometrically well-defined crystal defects
is investigated by either integral electrical measurements
(in an indirect way) or by using special scanning electron
microscope techniques (EBIC, Cathodoluminescence, SDLTS) for
a direct correlation.

For the division of the technological processes in micro- and
optoelectronics, in the course of which crystal defects are
formed or transformed, a classification shall be used into the
four following categories: 1) Patterning (lithography),
2) Radiation/particle beam treatment, 3) Heat treatment,
4) Epitaxial layer growth. It is obvious that in the majority
of cases the four groups of processes cannot be dealt with independently. Shaping effects by lithographic processes for instance can be turned over into patterned thin film structures
solely by subsequent radiation/particle beam treatment, heat
or growth processes. Epitaxial layer growth presupposes certain processes of heat treatment or of radiation/particle beam
treatment.

In the following sections, examples are given of the electron
microscope detection and characterization of process-induced
crystal defects, classified according to the 4 groups of processes mentioned above. The investigation of extended crystal
defects, especially dislocations, trends from defects in the
bulk of the crystal to defects in interfaces (e.g. in boundary regions between different materials or between differently
doped regions in the same material). In each case the interaction of extended defects with point defects (including e.g.
the impurity decoration of dislocations) and the mutual interaction of extended defects have to be taken into account (for
topical examples see e.g. Lee et al 1988, Hahn et al 1988).

2. DEFECTS IN PATTERNED THIN FILM REGIONS

According to its geometrical position with respect to active
device regions a crystal defect may cause different electronic
effects. Particularly crucial are defects lying in boundary
regions between differently doped areas, or in semiconductor/
conductor interfaces or in semiconductor/dielectric ones. With
respect to crystal defects (especially dislocations) that
cross pn junctions, their break-down voltage may be reduced
due to a certain junction leakage (Ravi 1981). In bipolar
transistors, for instance, local emitter-collector shorts may
appear leading to a decrease in the transistor current gain.
In MOS devices crystal defects may cause leakage currents
between source and drain. Fig. 1 shows the boundary between
the emitter region and the adjoining base region in an FZ

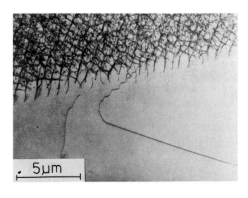

Fig. 1.
Emitter edge dislocation
in processed FZ silicon
(dense dislocation network
in the P$^+$ implanted
emitter region)

5µm

silicon device structure. Characteristic of the emitter region in this case is the presence of a dense dislocation network created by a strong phosphorus implantation carried out for producing the emitter region. This dislocation network lies more or less parallel to the plane of the pn junction. The main feature of the micrograph is the emitter edge disloca-tion starting from the above-mentioned dislocation network. This dislocation having the usual Burgers vector of a/2⟨110⟩ is typical of those defects causing junction leakages.

Besides dislocations in or around pn junctions, also disloca-tions in oxide thin film edges are of importance, which are caused by stresses acting in the transition region between thin gate oxides and thick field oxides of local oxidation structures (LOCOS). The cross section electron micrograph, given in Fig. 2a, clearly shows the so-called bird's beak structure in this transition region. The typical dislocation arrangement parallel to the boundary region is presented in the plan-view micrograph of Fig. 2b, in which the bright area is related to the thick field oxide, which is surrounded by a thin gate oxide area. The dislocations visible have a/2⟨110⟩ Burgers vectors and ⟨110⟩ line directions. Interpreting TEM micrographs and using computer simulations enabled the stresses and strains acting in the film edge to be estimated (cf. e.g. Vanhellemont and Van den Hove 1988, Isomae and

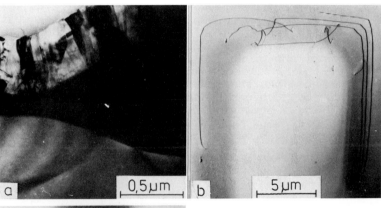

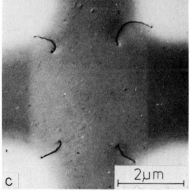

Fig. 2. Dislocations in oxide thin film edges (LOCOS structure)
a) Bird's beak configu-ration (XTEM)
b) Stress-induced dislo-cations parallel to the film edge (bright area: thick field oxide)
c) Dislocations in a LOCOS structure of four-fold symmetry

Aoki 1986). In the practical application of the LOCOS technique often patterned nitride layers are used to optimize the
geometrical configurations. Fig. 2c is related to a special
LOCOS structure consisting of a vertical stripe of gate oxide
crossing the source-drain regions (left, right), which themselves are surrounded by pad oxide areas. Dislocations clearly
appear in the regions of the highest stresses and strains of
the whole system, which shows an almost fourfold symmetry in
viewing direction. The dislocations visible (Burgers vector:
$a/2 \langle 110 \rangle$) are not only stress-induced, but also diffusion-
induced (P diffusion) under the conditions of the injection
of self-interstitials (by the action of the diffusion process).

3. DEFECTS FROM RADIATION/PARTICLE BEAM TREATMENT

For the formation of crystal defects by radiation or particle
beam treatment general effects are to be taken into account,
as also for the thermal treatment (or partly for the mechanical one): Dopants, impurities, interstitials, and vacancies,
arising during this treatment and moving in the material,
cause a distortion of the crystal lattice. Hereby, localized
stresses are created by the action of these point defects,
thus extended crystal defects (dislocations, planar defects,
precipitates) are formed and transformed, respectively. Ion
implantation and subsequent annealing cause secondary defects
(see e.g. Nandedkar and Narayan 1987, Opyd et al 1988,
Pongratz et al 1987), which decisively influence the quality
of the results of technological processes, e.g. the reliable
doping of semiconductors associated with the formation of
shallow pn junctions of well-defined junction depths.

To achieve defect-free shallow pn junctions by ion implantation requires the subsequent annealing process to be optimized. Fig. 3 shows crystal defects generated by As^+ implantation (dose: $5 \cdot 10^{15}$ cm^{-2}, RT, above the dose of amorphi-
zation) and afterwards thermally treated by RTA (halogen lamp

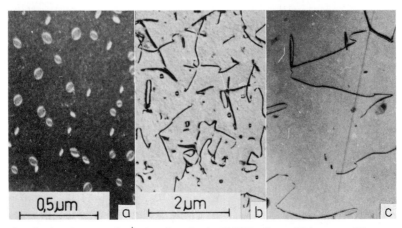

Fig. 3. Defects in As^+ implanted $\langle 100 \rangle$ Cz silicon after
RTA at temperatures of 900°C (a), 1055°C(b) and 1100°C (c)

irradiation at intervals of a few seconds) at different temperatures in ⟨100⟩ Cz Si (D. Baither, R. Kögler, 1989). In
micrograph a), which is related to an annealing temperature of
900°C, Frank faulted loops lying on {111} planes and having the
usual Burgers vector of 1/3⟨111⟩ are formed by interstitial
nucleation (DF micrograph). At the temperature of 1055°C
(micrograph b), these faulted loops have already transformed
into perfect loops having Burgers vectors of the type
a/2⟨110⟩. Loops with their Burgers vectors inclined to the surface glide out of the specimen region, and those with Burgers
vectors parallel to the surface grow along ⟨110⟩ directions to
become dipole-like elongated loops. In the majority of cases
these loops intersect the surfaces leaving behind half-loops
or pairs of parallel dislocations. At still higher temperatures (micrograph c) perfect loops with inclined Burgers
vectors preferentially occur, which leave the specimen region
thus causing a lower density of extended defects. To get
defect-free pn junctions by ion implantation requires still
higher annealing temperatures to be applied. It should, however, be taken into consideration that with increasing temperature the pn junction region moves into the depth of the
material in an unwanted manner. Consequently, a compromise is
aspired between a tolerable defect density and a desired pn
junction depth.

From the more general point of view, ion implantation is used
to form structurally and compositionally well-defined layers,
especially buried layers. The last-mentioned type of layers
is of special importance for SOI structures (silicon on insu-
lator). The main technique here is the formation of a buried
dielectric layer by high-dose implantation of oxygen (sometimes also nitrogen), a method which is called SIMOX (Separation by Implanted Oxygen). Typical defects in the active
superficial silicon layer are silicon oxide precipitates,
dislocation loops and stacking faults (see e.g. Van Ommen and
Viegers 1987). On the other hand, less frequently SOI structures are formed by recrystallizing deposited silicon films
(e.g. by liquid phase processes) on oxidized Si wafers, using
c.w. lasers (strip heaters or lamps). Defects typical of the
recrystallized layer are then low-angle grain boundaries (see
e.g. De Veirman et al 1988).

Although - besides in lithographic processes - electron irradiation is relatively seldom used as a technological process
in microelectronics the formation of crystal defects under
electron irradiation is of interest. This is also true of the
influence of the electron beam of the electron microscope
itself. Fig. 4 shows a defect configuration, which arose
during the EM observation at a beam density of about 5 A/cm^2.
For the investigations, In-doped FZ silicon was used, which
was deformed at 700°C and subsequently annealed at 850°C for
30 minutes. The micrograph shows a complex of extended crystal defects. At the beginning of the observation the Z-shaped
60° dislocation dipole (see arrow) was visible, which obviously acted as the nucleus of the further defect growth,
which was initiated by the electron-beam induced formation of
self-interstitials and the diffusing of the In dopants.

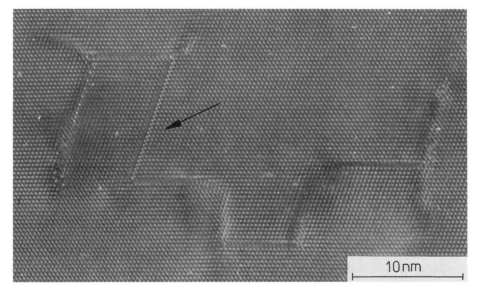

Fig. 4. Extended crystal defects in In-doped FZ silicon:
deformed at 700°C, annealed at 850°C, electron irradiated
(beam density: 5 A cm^{-2})

Starting from the outer partial dislocations of the Z-shaped
dipole, which consisted of the visible stacking fault configuration and was bound by stair-rod dislocations, the growth
of further planar defects (partly stacking faults) was proceeding, being obviously influenced by processes of decoration and precipitation,   of the In doping atoms.

4. DEFECTS FROM HEAT TREATMENT

Of all the processes in semiconductor technology the heat
treatment is most frequently applied. The heating processes
are also important for the above-mentioned patterning procedures and the radiation/particle beam treatment, for the
latter especially as subsequent annealing processes. Furthermore, the technique of doping by diffusion is based on the
thermal treatment. As to the different heating processes,
besides the conventional thermal heating, rapid thermal processes (RTP) carried out in the second range (e.g. by halogen
lamp irradiation), and the topical technique of laser annealing are increasingly used. As mentioned above, the formation, diffusion and agglomeration of point defects, and the
diffusion and agglomeration of impurities and dopants are of
importance for the formation of extended defects.

The gettering of heavy-metal impurities (see e.g. Tan et al
1977, Rozgonyi et al 1976) by purposefully introducing extended crystal defects is one of the examples related to the
effects of heat treatment. Fig. 5 is concerned with a misfit
dislocation network in (100) oriented Cz Si formed by P^{+} implantation and subsequent annealing, shown in a plan-view
micrograph in a), and in a cross section micrograph in b).

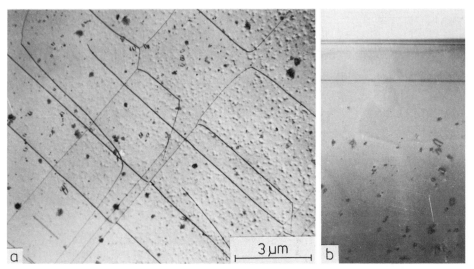

Fig. 5. Plan view (a) and cross section micrograph (b) of crystal defects in P⁺implanted CzSi

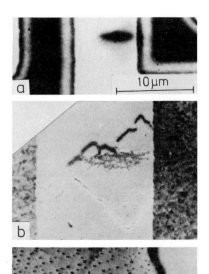

Fig. 6. EBIC(top)/ TEM(middle, bottom)correlation of a colony of typical precipitates (CuSi) in strongly thermally treated silicon

The network consists of 60° dislocations lying in ⟨110⟩ directions. The gettering efficiency of the dislocation network is clearly revealed by a denuded zone of about 3 µm in extension.

Since the first investigations by Ravi et al (1973) and Varker and Ravi (1974) the electrical activity of several crystal defects has directly been evidenced routinely in mány laboratories by combining the SEM (EBIC) technique with transmission electron microscopy. Taking into consideration the position of the defect with respect to the pn junction or the Schottky barrier, the recombination efficiency of a crystal defect for electron beam induced charge carriers can be estimated from the strength of the EBIC contrast present. A relevant example is given in Fig. 6, showing a characteristic group of Cu and CuSi colonies, resp. in EBIC contrast (a) and in the TEM micrograph (b). These colonies arise in strongly heat-treated materials, where the solubility limit is often exceeded and fast diffusing heavy metals, like e.g.

Cu, Fe, and Ni, precipitate. The TEM analysis (diffraction contrast, Moiré contrast, selected area diffraction) points to
planar CuSi particle aggregates on different (110) and (100)
planes with bordering edge dislocations. The EBIC micrograph
shows a homogeneous plane contrast at the surface-parallel
colony without a separate contrast of the bordering edge dislocation itself, which, on the other hand, is clearly visible
in the higher-magnified TEM image (c) of a small part of the
boundary region.

5. DEFECTS OCCURRING DURING LAYER GROWTH

During the growth of layers on bulk substrates, the occurring
defects strongly depend on the parameters of the growth process, on the one hand, and on the quality of the substrate, on
the other, including the transfer of defects from the substrate to the layer of interest.

Of the defects occurring during epitaxial layer growth defects
in interfaces (see e.g. Marée et al 1987) are of special importance. This is true of the interface between a crystalline
substrate and an epitaxial layer as well as of interfaces between epitaxial layers and neighbouring epitaxial layers
(having different structures and/or chemical compositions). An
example is given in Fig. 7 presenting misfit dislocations in a
graded $GaAs_{1-x}P_x$ layer system (Erfurth et al 1988). The analysis of micrograph a) mainly yielded 60° dislocations with
Burgers vectors inclined to the (100) surface (Burgers vector:
$a/2\langle110\rangle$, line directions: $\langle110\rangle$), but also some fraction of
non-glissile edge dislocations (Lomer dislocations). The elec-
trical and optical activities   of these types of dislocations and dislocation bundles   can be estimated e.g.
by applying the SEM cathodoluminescence imaging technique
(Petroff et al 1980, Erfurth et al 1986, Ast et al 1988).

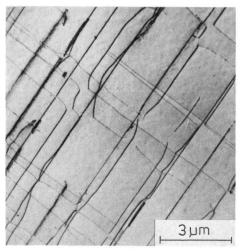

Fig. 7.   Misfit dislocations
in a graded $GaAs_{1-x}P_x$ CVD
grown layer system

High-resolution electron
microscopy has extensively
been used to investigate
hetero-epitaxial layers and
layer systems of different
degrees of lattice mismatch
(see e.g. Humphreys 1986,
Hutchison et al 1986,
Feuillet et al 1988, Gibson
et al 1987). As the main example of a nearly lattice-
matched system the GaAlAs/
GaAs system is usually given.
But many other combinations
of III/V or of II/VI com-
pound semiconductors nearly
fulfil this condition, too.
The main system in the SOS
technique, however, which
is $(100)Si//(01\bar{1}2)$ Al_2O_3
(sapphire), is relatively
strongly mismatched. As far

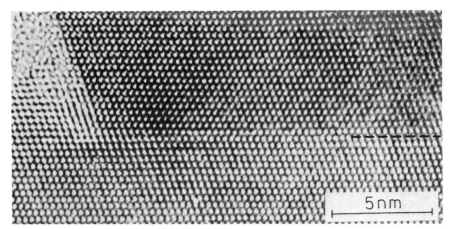

Fig. 8. Silicide/silicon 111 interface
      top: $NiSi_2$, bottom Si (110 orientation)

as it is possible, for all these systems defect-free or nearly
defect-free interfaces are aspired. Fig. 8 gives an example of
the interface of metal silicide on silicon, which has a relatively small lattice mismatch under favourable conditions,
surely depending on the type of the matching lattice planes.
It is a bright-field high-resolution electron micrograph of
the boundary region between $NiSi_2$ (top) and silicon (bottom).
The silicide layer was generated by covering the (111) Si substrate with Ni by electron beam evaporation at a substrate
temperature of 400°C, using an extremely small evaporation
rate (3 pm/s). After evaporation for about 30 minutes a 15 nm
thick $NiSi_2$ layer had formed. The atomically smooth interface, i.e. a (111) plane, is clearly visible.

6. CONCLUSIONS

The paper discussing practical examples, where crystal defects
are formed in technologically processed semiconducting materials, i.e. after patterning processes, during radiation and
particle beam treatments, respectively, during heat treatment,
and in the course of processes of layer growth, points out the
necessity of taking into account interaction processes between
extended crystal defects and point defects. For both types of
defects their geometrical characterization using imaging and
diffraction techniques, resp., and their electronic characterization by applying spectroscopic methods are required,
thus making allowance for the future trend of a uniform consideration of the geometrical and the electronic structure of
crystal defects.

ACKNOWLEDGEMENTS

The electron micrographs presented are of the research work of
the Institute of Solid State Physics and Electron Microscopy
in Halle. The author is indebted to the following colleagues
engaged in this work: Drs. H. Bartsch, H. Blumtritt,W.Erfurth,
R. Gleichmann, M. Reiche, U. Richter, P. Werner.

REFERENCES

Ast D G et al 1988 J. Appl. Phys. 63 4925
Baither D and Kögler R 1989 private communication
Bourret A 1988 Inst. Phys. Conf. Ser. 93 65
De Veirman A et al 1988 Inst. Phys. Conf. Ser. 93 117
Erfurth W, Gleichmann R and Heydenreich J 1986 Proc. 2nd
    Conf. Phys. Techn. GaAs ed E Lendvay (Brookfield: Trans.
    Tech. Publ.) pp 153-8
Erfurth W, Gleichmann R and Heydenreich J 1988 Inst. Phys.
    Conf. Ser. 93 101
Feuillet G et al 1988 Inst. Phys. Conf. Ser. 93 395
Gibson J M et al 1987 Ultramicroscopy 22 35
Hahn S et al 1988 J. Appl. Phys. 64 849
Heydenreich J 1988 J. Micr. Spectr. Electron. 13 479
Humphreys C J 1986 Proc. XIth Int. Congr. Electron Micr. ed
    T Imura, S Maruse and T Suzuki (Tokyo: The Japanese So-
    ciety of Electron Microscopy) pp 105-8
Hutchison J L 1983 J. Phys. (Paris) 44 C4-3
Hutchison J L et al 1986 J. Microsc. (London) 142 153
Isomae S and Aoki S 1986 Proc. 18th Int. Conf. Solid State
    Devices, Tokyo pp 517-20
Lee B T, Gronsky R and Bourret E D 1988 J. Appl. Phys. 64 114
Marée P M J et al 1987 J. Appl. Phys. 62 4413
Nandedkar A S and Narayan J 1987 Philos. Mag. A56 625
Oppolzer H 1988 Inst. Phys. Conf. Ser. 93 73
Opyd W G, Gibbons J F and Madrinly A J 1988 Appl. Phys.
    Lett. 53 1515
Ourmazd A 1986 Mater. Sci. Forum 10-12 735
Petroff P M, Logan R A and Savage A 1980 Phys. Rev. Lett.
    44 287
Pongratz P et al 1987 Inst. Phys. Conf. Ser. 87 497
Ravi K V 1981 Imperfections and Impurities in Semiconductor
    Silicon (New York: Wiley)
Ravi K V, Varker C J and Volk C E 1973 J. Electrochem. Soc.
    120 533
Rozgonyi, G A, Deysher R P and Pearce C W 1976 J. Electro-
    chem. Soc. 123 1910
Tan T Y, Gardner E E and Tice W K 1977 Appl. Phys. Lett.
    30 175
Vanhellemont J and Van den Hove L 1988 Inst. Phys. Conf. Ser.
    93 423
Van Landuyt J et al 1986 Proc. XIth Int. Congr. Electron
    Micr. ed T Imura, S Maruse and T Suzuki (Tokyo: The Japa-
    nese Society of Electron Microscopy) pp 1059-64
Van Ommen A H and Viegers M P A 1987 Inst. Phys. Conf. Ser.
    87, 385
Varker C J and Ravi K V 1974 J. Appl. Phys. 45 272

Inst. Phys. Conf. Ser. No 104: Chapter 2
Paper presented at Int. Symp. on Struct. Prop. Disloc. Semicond., Oxford, 1989

Electrical and optical properties of dislocations in GaAs

J.L. Farvacque, D. Vignaud, E. Depraetère, B. Sieber, A. Lefebvre
Université des Sciences et Techniques de Lille, UA 234 CNRS
Bât. C6   59655 Villeneuve d'Ascq Cedex France

ABSTRACT: The existence of two dislocation bands B1 and B2 associated with α and β dislocations is deduced from the analysis of galvanomagnetic and optical measurements on plastically deformed samples. Photoplastic effects have been experimentally studied in n- and p-type materials. Results show that both the α and β glide directions are photoactivated in n type but that only β glide is activated in p type. These different behaviours are analysed with the help of the dislocation bands and their corresponding configuration coordinate diagram. Degradation phenomena are then explained in terms of such photoplastic properties.

1. INTRODUCTION

GaAs is an interesting semiconductor to study dislocation properties for the two following reasons: 1) GaAs is a direct gap semiconductor so that one may expect to study dislocations by means of optical measurements. These measurements will thus complete the classical conductivity measurements generally used for such kind of investigations. 2) GaAs is a basic semiconductor widely used for optoelectronic devices, the properties of which are strongly dependent on the crystalline quality of the substrates and of the epitaxial layers deposited on them. Not only the device fabrication process can be affected by dislocations or other defects but also operating devices are subject to ageing phenomena frequently connected with point defect and dislocation interactions. Thus, a lot of works have been dedicated to the study of the fundamental properties of dislocations of GaAs in view to answer numerous questions such as: do dislocations introduce extrinsic energy levels in the forbidden band gap? What are the dislocation mechanisms affecting the free carrier mobility? How do dislocations modify the optical properties (absorption or photoluminescence)? How can we understand the role of dislocations in degradation phenomena from their electronic properties? The aim of this paper is double. In the first part, we attempt to make a short point on our knowledge of the electronic properties of dislocations. In the second part, after having described some photoplastic effects and degradation experiments in laser-like GaAs heterostructures, we propose a coherent explanation of these effects in terms of the dislocation electronic properties.

Concerning the general electronic properties of dislocations, most of the experiments have a macroscopic character and thus have been realised on plastically deformed samples in order to introduce large dislocation densities. This is particularly the case for optical absorption, photoluminescence and galvanomagnetic properties. However, plastic deformation not only creates a dislocation substructure but also a large number of unidentified point defects. This makes delicate any direct interpretation. Device degradation has frequently been studied by means of local observations such as transmission electron microscopy, induced current technique (EBIC) or cathodoluminescence. Unfortunately, such investigations are often qualitative. In this paper, we try to show that a confrontation between these two kinds of results may lead to a deeper understanding of dislocation properties and behaviours.

2. ELECTRONIC PROPERTIES OF DISLOCATIONS

2.1 Galvanomagnetic measurements

Numerous galvanomagnetic measurements have been devoted to plastically deformed GaAs samples (Nakata et al. 1977, Gwinner et al. 1979, Suezawa et al. 1986, Gerthsen 1986, Skowronski et al. 1987 and Ferré 1987). They show , in n-type as well as in p-type materials, that the defects introduced by plastic deformation lead to a systematic increase of the Hall coefficient and of the reversal Hall mobility. Thus, plastic deformation introduces extrinsic states in the forbidden band gap. Various classical models (Read 1954, Schröter et al. 1969 and Masut et al. 1982) allow us to calculate the dislocation occupation statistics and have been used in order to fit the experimental results with the presence of only one dislocation band without altering the point defect density. None of the experimental investigations of the Hall constant increase could be so simply interpreted. Thus, there is now some agreement to conclude that a large density of electrically active point defects is also introduced by plastic deformation. This makes any determination of the dislocation characteristics delicate. Identifying the point defects with El2 and evaluating the El2 and dislocation densities led Gerthsen (1986) to describe the dislocations by a single one-dimensional band located at about 0.36 eV above the valence band. Looking at the increase of the reversal Hall mobility, Ferré (1987) concluded that its order of magnitude could not be theoretically obtained by only considering the scattering mechanism associated with the dislocation line charge and thus used the deformation potential mechanism as defined by Farvacque et al. (1977). The dislocation density could then be evaluated and introduced in the theoretical Hall effect calculation. This method led also to the conclusion that the Hall effect of plastically deformed GaAs samples could be described by the simultaneous presence of deep point defect states and of one single dislocation band (called B1 in the following) but located close to the top of the valence band.

However, reversing the well known expression giving the occupation rate of a dislocation gives the expression of the electrostatic interaction energy derivative (the rigid shift):

$$W = F - E_d + KT \log\left(\frac{1-f-2\chi}{f+2\chi}\right) \cong F - E_d \qquad (1)$$

where F is the Fermi level, E_d the dislocation level, f the dislocation occupation rate and χ the occupation rate of the neutral dislocation. As long as galvanomagnetic measurements are done in the extrinsic temperature range the Fermi level is approximately pinned at the impurity levels. Moreover, the logarithmic term of expression (1) is small and can be neglected. Thus, the rigid shift W can be considered as constant. The calculation of W may be done in the depleted region approximation (Read's model, 1954) and is given by:

$$W = f \frac{e^2}{2\pi\varepsilon_o \varepsilon_r b} [\log(R/b) - .866] \qquad (2)$$

where b is the Burgers vector and R the screening radius which only depends on f and the dopant density. Mixing (1) and (2) gives an approximate value of f which occurs to be temperature independent. This show that dislocations associated with a deep band have practically, in the extrinsic temperature range, a constant charge and behave similarly to deep point defects. Thus, the interpretation of Hall effect in terms of additional deep point defect states may be as well done in terms of other dislocation bands deeply located in the band gap.

Finally, let us emphasize that the Hall effect interpretations have been done by assuming that the Hall constant R is equal to 1/ne, the α coefficient depending on the scattering mechanisms being systematically taken equal to one. A new dynamical theory of transport phenomena has been recently proposed by Farvacque (1989) and used for the calculation of the α coefficient associated with the dislocation charge line scattering mechanism. Results indicate that α is strongly temperature dependent and should modify the location of the dislocation band. In conclusion, galvanomagnetic measurements in the extrinsic temperature range show that dislocations introduce at least one band B1 located approximately above the valence band but is totally unable to give a more precise description of the dislocation states which can also eventually be described by more than one band.

2.2 Optical properties

Optical absorption (Böhm et al. 1979, Bazhenov et al. 1982, 1984, Omling et al. 1986, Bray 1986, Skowronski et al. 1987 and Vignaud et al. 1989a, 1989b) studies of dislocations in GaAs are difficult because of the presence of the deep donor El2 which can also be created by plastic deformation. However El2 defects are subject to photoquenching effects when submitted to a photon flux of energy larger than 1.05 eV and at temperatures lower than 140 K for semi-insulating materials and lower than ≅ 50 K for n-doped materials. Optical measurements have thus been done at low temperature after photoquenching of the El2 centres. Comparison between reference and plastically deformed samples do not exhibit any particular new band associated with dislocations but, instead, large exponential-like tails. These features have been interpreted in terms of Franz-Keldysh effects associated with the internal fields of dislocations. Generally speaking, dislocations may be associated with electrical fields issuing either from their charge lines, or from their strain fields (deformation potential and piezoelectric fields in polar crystals). The Franz-Keldysh effect in the case of n type materials can be quantitatively simulated by assuming the existence of the dislocation band B1 as deduced from the galvanomagnetic measurements (Vignaud et al. 1989a). But considering B1 only, the theoretical order of magnitude is found to be too low (or needs a too large dislocation density) to justify any effect in the case of p-type materials (where B1 is practically empty). Electrical fields connected with dislocation strain fields lead also to too weak effects to be evoked in the case of p-type. A reasonable way to explain the Franz-Keldysh effect in the case of p-type materials is to assume that another dislocation band B2 also exists but that it is deeply located in the band gap (Vignaud et al. 1989b). Since no direct optical transitions between this band and the crystal bands can be observed in the absorption spectrum, its configuration coordinate diagram has to be of the form shown on figure 2. This diagram corresponds well to a midgap level in the point of view of thermal equilibrium, but leads to too energetical optical transitions to emerge from the Franz-Keldysh absorption.

Photoluminescence spectra have also been performed in plastically deformed materials (Böhm et al. 1979, Bazhenov et al. 1982, Suezawa et al. 1986 and Depraetère et al. 1987). A very weak band located at 1.13 eV was evidenced in n-type materials by Suezawa et al. (1986). However, we could not reproduce this observation in n-type materials but only in semi-insulating and in p-type samples (Depraetère et al. 1987). Moreover, this band is only visible when plastic deformation is made at relatively high temperature (400°C, i.e. when the dislocation substructure contains a large density of edge-type dislocations), but not when performed at low temperature (200°C, i.e. when the deformed samples mainly contain screw dislocations) (Depraetère 1989). This band has thus been attributed to 60° or edge type dislocations one-dimensional states and identified with the band B1 deduced from galvanomagnetic measurements and optical absorption for the following reasons:

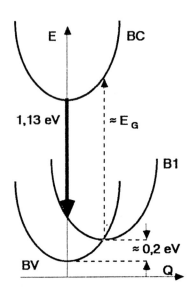

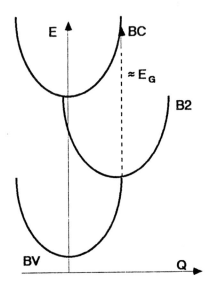

fig. 1: configuration coordinate diagram proposed for dislocation band B1.

fig. 2: configuration coordinate diagram proposed for dislocation band B2.

1) The 1.13 eV light is found to be polarised perpendicularly to the primary glide plane of the deformed sample.

2) Except in plastically deformed materials, no report of this band can be found in literature even when samples have been submitted to various irradiation damages. This band corresponds to some polarised defects which are therefore specifically created by plastic deformation. Why not dislocations themselves?

3) Attributing this band to dislocation levels explains quite well why it is not (or not easily) observable in n-type materials. The Frank-Condon shift of the 1.13 eV band deduced from its half width indicates that B1 corresponds to a distorted configuration (figure 1) whose energy minimum is at about 0.2 eV above the valence band. Such a position is very near the Fermi energy in p-type materials so that the dislocation band is mainly empty in this kind of materials. In contrast, it is strongly filled in n-type materials. In the latter case, dislocations are surrounded by some intense screening fields able to separate the electron-hole pairs and to prevent therefore any radiative recombination at these levels. Instead, in p-type and also partly in semi-insulating materials, the screening electrical field is weak enough so that radiative recombinations can now occur at dislocation levels. Finally, the observation of the very weak 1.13 eV band in n-type materials as described by Suezawa et al. (1986) can be understood by assuming that the incident power light was intense enough to induce some saturation effects tending to decrease the dislocation line charge.

4) The above location of the dislocation band corresponds fairly well to galvanomagnetic measurements. It has finally been used in the Franz-Keldysh effect theoretical simulation and leads to a good quantitative agreement with experimental results in the case of n-type. It has been therefore associated with the previous band B1.

Photoluminescence could not evidence the band B2 as deduced from Franz Keldysh effects in p-type. This is fully explained by the special shape of its configuration coordinate diagram.

In conclusion to this first part, dislocations introduced by plastic deformation seem to be well described by the existence of two one-dimensional bands called B1 and B2. Band B1 stays probably lower than 0.2 eV above the valence band while band B2 is located near midgap. Their configuration coordinate diagrams (figures 1 and 2) explain quite well the set of the experimental galvanomagnetic and optical results. It is very tempting to associate right now these two bands with the two kinds α and β of dislocations which constitute the substructure of the plastically deformed materials.

3. PHOTOPLASTIC EFFECTS

Various experiments show that dislocation mobilities in bulk GaAs depend on the extrinsic electronic characteristics of the material (such as the type of doping, Ninomiya 1979) but are also sensitive to extrinsic excitations such as electron beam interaction (Maeda et al. 1983) or light irradiation (Petroff et al. 1974). Ageing phenomena of optoelectronic devices are a direct illustration of these dislocation properties. Various classes of explanations may be evoked to describe the photoenhancement of dislocation mobility:

1) the double kink formation may be responsible for the introduction of extrinsic electronic states in the forbidden band gap. Thus, the double kink formation energy (and finally the double kink density) would depend on the occupation statistics of these levels or equivalently on the location

of the Fermi level (Hirsch 1979). Optical transition between these states and the crystal bands would modify their statistical occupation and consequently the dislocation mobility.

2) The double kink formation free enthalpy ΔG_o is a given and fixed parameter and the role played by the light is indirect and just leads to a generation of electron-hole pairs. Assuming that dislocations are non radiative recombination centres, the energy E_{opt} released by electron-hole pairs recombination or the carrier capture would either produce a local increase of the temperature or supply part of the ΔG_o energy, in the same way that the applied stresses σ supply a quantity $V\sigma$ (where V is the activation volume) (Depraetère 1989). Such electron-hole pairs are also created by the electron beam and would thus explain the mobility increase of dislocations when observed with an electron microscope (Maeda 1985). A theoretical estimate of the temperature increase (Depraetère 1989) shows that it is too low to play any significant role on the dislocation mobility.

3) Another mechanism has been recently proposed to explain the softening influence of light on deformation experiments of macroscopic samples (Mdivanyan et al. 1988). For these authors, the light does not favour the double kink nucleation like in previous mechanisms, but helps the kink propagation along the dislocation line. Following this model, the kink propagation is controlled by a distribution of some undefined defects located all along the dislocation line. The light flux would then be responsible for transitions between the electronic states associated with these defects and the crystal bands. Thus, the defect charge state and consequently the kink-defect interaction would be modified. Unfortunately, the maximum photoplastic effect is obtained with photon energy of 1.4 eV. Such an energy corresponds to the interband absorption and is absorbed within a few microns.

In order to test these various models, some microhardness tests have been performed on (100) faces at room temperature. Then, they were either followed by annealing treatments at $\cong$ 360°C, or by intense laser irradiation at room temperature. Observations have been made by cathodoluminescence. The α and β directions were identified by chemical etching. Observations show that (Depraetère 1989):

- in n-type materials the rosettes are developed in both the α and β <110> directions after annealing as well as after room temperature light illumination with photon energies larger than the band gap.
- TEM observations on annealed and photoactivated samples show equivalent dislocation substructures, in the case of n-type samples, where the α directions exhibit 30° partials bounding extended stacking faults and where the β directions are constituted by "perfect" (i.e. weakly dissociated) dislocations. Differences can however be noticed between thermal annealing and optical illumination in the β direction: in the former case, dislocations are much more like segments going from one face to the other of the thin TEM lamella while in the latter case, β arms are much more constituted by long screw segments lying parallel to the thin lamella faces. This last remark indicates that the optically induced rosette development can only take place in the crystal region reached by the light flux and therefore limited by the penetration depth of the light. This last conclusion is in desagreement with the explanation of the light softening proposed by Mdivanyan et al. (1988).
- in p-type materials, only one of the rosette arms is developed in the β direction.
- optical illuminations of indentations with photon energies slightly lower than the fondamental absorption edge are totally enable to develop the

rosettes. This clearly indicates that the photoplastic effect at room temperature does not correspond to extrinsic absorption between some kink states and the crystal bands but confirms the conclusions already given by Maeda (1985) that it is indeed connected with electron-hole pair recombinations at dislocation levels. This conclusion is reinforced by the fact that the rosette development does not depend on the polarisation of the incident laser light.

Assuming that photoplastic effects result from non radiative electronic capture at dislocation levels, we can propose a coherent explanation of the different optical sensitivity of α and β arms in terms of the B1 and B2 bands associated with dislocations: the coordinate diagram of band B2 implies that carrier recombination is always non radiative. The mobility of such dislocations is therefore always photoactivated and the corresponding rosette arms develop in every type of materials (n, p and semi-insulating). Thus, band B2 is associated with direction β which may therefore correspond to As(s) cores or to Ga(g) cores. Band B1 leads to non radiative electronic capture in n-type materials only. In p-type materials where dislocations B1 are not charged, a large part (1.13 eV) of the released energy is lost for the dislocation since it is emitted under the form of light. As shown on the configuration coordinate diagram, only a weak part of the energy is given to the dislocation and is not sufficient to help the double kink formation at room temperature. Thus band B1 is associated with the α direction i.e. corresponds either to Ga(s) or to As(g) dislocations.

Theoretical calculations (Jones et al. 1981) indicate that, whatever the actual core structure of dislocations (dissociated, reconstructed or not), the Ga cores would systematically lead to dislocation bands located deeper in the band gap while As atomic cores would correspond much more to dislocation levels located above the valence band. Connecting these theoretical conclusions with our experimental observations indicates that the α direction probably corresponds to As(g) dislocations which are associated with band B1 and that the β direction corresponds to Ga(g) dislocations associated with band B2.

The remaining problem consists in understanding the mechanism which connects the double kink formation with the non radiative capture. Under some applied stresses, the kink formation frequency ν can be expressed in terms of the free enthalpy part $\Delta G = \Delta G_o - \sigma V$ that has to be found in thermal fluctuation by an Arrhenius law:

$$\nu = \nu_o \exp(-\Delta G/KT) \qquad (3)$$

where ν_o is the Debye frequency. Under illumination, part of the double kink formation energy is supplied by the non radiative capture E_{opt}. Thus:

$$\nu' = \nu_o \exp(-\frac{\Delta G - E_{opt}}{KT}) = \nu_o' \exp(-\Delta G/KT) \qquad (4)$$

where:
$$\nu_o' = \nu_o \exp(E_{opt}/KT) \qquad (5)$$

All happens as if locally the Debye frequency were increased due to the non radiative electronic capture. Analysing the physical meaning of a configuration coordinate diagram explains quite well the latter remark. Before any electronic capture the dislocation line stays in a well defined physical configuration (for instance it lies along a Peierls valley).

Before delocalizing along the dislocation line, an electron captured at a given portion of the dislocation line suddenly puts the local states of the dislocation into excited vibronic states which do not correspond to the minimum energy shown in the coordinate diagram. Thus, the dislocation is locally distorted by a coherent phonon emission, the result of which is the formation of a double kink.

4. DEGRADATION PHENOMENA

Laser-like heterostructures constituted by a layer of GaAs between two layers (parallel to the (100) planes) of GaAlAs have been realised by MBE in order to study degradation effects due to intense light illumination. The GaAs layer was not intentionally doped and turned out to be of p-type nature. These samples have been illuminated with 1.92 eV photons, the energy of which is too low to be absorbed by the GaAlAs layers but is totally absorbed by the GaAs layer. They have been then studied by cathodoluminescence.
- Similarly to what happens in real operating systems, but in a shorter time (< 15 hours), such a light illumination is responsible for the appearance of dark line defects (DLD's) in the <110> and <100> directions. However, one notices that in these samples, <110> DLD's only develop along one of the two <110> directions. This last result is coherent with photo-activated glide phenomena as described above since the GaAs layer is p-type.
- A major part of the sample area remained unsensitive to the incident light beam whatever the illumination duration. However we checked that areas where degradation occurs do not exhibit any visible defect such as grown-in dislocations or impurity clusters.
- With the incident power light used, the <110> DLD's systematically appeared before the <100> DLD's. The development of the <100> lines completed until 10 to 15 hours after which the luminescence was stable.
- Use of photons of lower energy than the GaAs band gap was unable to produce any DLD.
- Optical filters were used to select the cathodoluminescence signal issuing from the GaAlAs layers only. DLD's were no more visible. Thus, DLD's are only located in the GaAs layer.

The previous experimental results show that degradation phenomena clearly occur in two steps, the first one being the occurence of some photoactivated glide. Local stresses needed for the dislocation glide are of unknown nature but may correspond to some local variation of the stoichiometry (variation of the Al density at the interfaces, interdiffusion of Al within the GaAs layer ...). Since in p-type materials, only one of the two glide directions (the β one) can be photo activated, local stresses may only be partly released. In order to complete the stress relaxation, one has then to evoke another mechanism which corresponds to the <100> DLD's. These last DLD's have already been studied (Petroff et al. 1974) and correspond to helicoidal dislocations indicating that they result essentially from dislocation climb. This second mechanism constitutes a slower process but is necessary to obtain a more homogeneous relaxation of the local stresses. In order to be efficient, the climb process needs a large quantity of point defects. These last ones probably result from the previously developed dislocations which are ideally located active point defect sources. Our experiments also show that the <100> DLD's only develop under light illumination. This indicates that the point defect motion is also photo activated.

5. CONCLUSION

Tentative explanations of the degradation phenomena have been proposed in terms of intrinsic dislocation properties which appear to be coherently described by two dislocation bands B1 and B2 respectively associated with As(g) dislocations and Ga(g) dislocations. The energy transfer of the captured electron and the resulting double kink is essentially based on the fact that the locally excited part of the dislocation line does correspond to a configuration coordinate diagram exhibiting a strong lattice coupling and the kink formation is viewed to correspond to a configuration change. Obviously our proposal is not totally perfect and particularly raises the following question: an electron captured at the dislocation band must delocalise along the dislocation line. If this delocalisation is instantaneous, it is hard to believe that a small portion of the dislocation line is placed in some excited states. Thus, a possible conclusion is that the dislocation states B1 and B2 do not correspond to classical one-dimensionnal bands expected when the state density is maximum (1/a with a the shortest distance separating two core atoms) but would better correspond to a lower state density associated for instance with structural defects located along the dislocation line (solitons for example). The distance separating these defects has to be sufficiently small in order to explain the electrostatic interaction leading to a self regulation of the occupation statistic and to the existence of the electric field surrounding the dislocation line but sufficiently large to prevent any band formation (no resonance between adjacent defects). This last question remains open.

ACKNOWLEDGMENT

The authors acknowledge the "Pôle microélectronique Nord " (Lille) for supplying the heterostructures.

REFERENCES

Bazhenov A.V. and Krasil'nikova L.L. 1982 Sov. Phys. Semicon. **23,** 2068
Bazhenov A.V. and Krasil'nikova L.L. 1984 Sov. Phys. Semicon. **26,** 356
Böhm K. and Fischer B. 1979 J. Appl. Phys. **50,** 5453
Bray R. 1986 Sol. St. Comm. **60,** 867
Depraetère E., Vignaud D. and Farvacque J.L. 1987 Sol. St. Comm. **64,** 1465
Depraetère E. 1989 Thèse (Lille) - Depraetère E., Vignaud D., Farvacque
        J.L., Sieber B. and Lefebvre A. to be published
Farvacque J.L. and Lenglart P. 1977 Phys. Stat. Sol.(b) **80,**361
Farvacque J.L. 1989 Phys. Rev. **B39,** 1683
Ferré D. 1987 thèse d'état (Lille) - Ferré D. and Farvacque J.L. to be
        published
Gerthsen D. 1986 Phys. Stat. Sol.(a) **97,** 527
Gwinner D. and Labusch R. 1979 J. de Phys. **40,** C6-75
Hirsch P.B. 1979 J. de Phys. **40,** C6-117
Jones R., Oberg A. and Marklund S. 1981 Phil. Mag. B **43,** 839
Maeda K., Sato M., Kubo A. and Takeuchi S. 1983 J. Appl. Phys. **54,** 161
Maeda K. 1985 in "Dislocations in solids" Yamada Science foundation
        University of Tokio press, 425
Masut R., Penchina C.M. and Farvacque J.L. 1982 J.Appl. Phys. **53,** 4964
Mdivanyan B.E. and Shikhsaidov M.S. 1988 Phys. Stat. Sol. (a) **107,** 131
Nakata H. and Ninomiya T. 1977 J. Phys. Soc. of Jap. **42,** 552
Ninomiya T. 1979 J. de Phys. **40,** C6-143
Omling P., Weber E.R. and Samuelson L. 1986 Phys. Rev. **B33,** 5880

Petroff P., Johnson W.D. and Hartman R.L. 1974 Appl. Phys. Let. **25**, 226
Read W.T. 1954 Phil. Mag. **45**, 775 and 1119
Schröter W. and Labusch R. 1969 Phys. Stat. Sol. **36**, 359
Skowronski M., Lagowski J., Milhstein M., Kang C.H., Dabkowski F.P.,
    Hennel A.and Gatos H.C. 1987 J.Appl. Phys. **62**, 379
Suezawa M. and Sumino K. 1986 Jap. J. Appl. Phys. **25**, 533
Vignaud D. and Farvacque J.L. 1989a J. Appl. Phys. **65**, 1261
Vignaud D. and Farvacque J.L. 1989b to be published

Inst. Phys. Conf. Ser. No 104: Chapter 2
Paper presented at Int. Symp. on Struct. Prop. Disloc. Semicond., Oxford, 1989

151

Electronic properties of dislocations and associated point defects in GaAs

T. Wosinski and T. Figielski

Institute of Physics, Polish Academy of Sciences,
Al. Lotnikow 46, 02-668 Warsaw, Poland

ABSTRACT: Effects of plastic deformation on the deep-level spectrum and electronic properties of GaAs single crystals are presented. The deformation creates a high concentration of acceptors. A new level at $E_c - 0.68$ eV has been identified to be associated with dislocation cores. The level exhibits barrier-limited electron capture and governs the recombination of current carriers. The concentration of the EL6 level is enhanced after the deformation and the rate of the normal-to-metastable transformation of the EL2 centre is slowed. The latter results are discussed in terms of the dislocation-related generation of anti-site defects. The role of the conservative climb in the development of dislocation dipoles in strained crystals is emphasized.

1. INTRODUCTION

Many important questions concerning dislocations in semiconducting and semi-insulating GaAs remain still open. These are for instance: what kinds of point defects are created during the dislocation motion and how; what are inherent electron states of dislocations; why dislocation density is suppressed in heavily In-doped GaAs, etc. In this paper we discuss some of the items basing principally on our results elaborated in Warsaw and Goettingen.

Experimental results were obtained with plastically deformed GaAs crystals of commercial origin. The crystals were mainly LEC-grown with the electron concentration, n, of the order 10^{16} cm^{-3} and the etch pit density 10^4 - 10^5 cm^{-2}. The samples were plastically deformed at $400°$C in an argon atmosphere by uniaxial compression along the <123> axis. The deformation was performed at a constant strain rate and terminated when reaching the lower yield point. Different strain rates were used which allowed to produce dislocation densities in the range 10^8 - 10^9 cm^{-2}.

2. POINT DEFECTS IN DEFORMED GaAs

The principal effect of deformation, commonly observed by different authors, is a suppression of the electron concentration in a sample (Gerthsen and Haasen 1986, Suezawa and Sumino 1986, Wosinski and Figielski 1987, Skowronski *et al* 1987). This is due to production of acceptor states with energy levels lying in the lower half of the band gap, but their spectrum remains still unresolved. Figure 1 demonstrates the decrease in n as a function of the deformation, obtained from the capacitance measurements of Schottky diodes prepared on different samples (Wosinski 1988). It is seen in the figure that ca $2 \cdot 10^{16}$ cm^{-3} acceptors are introduced into the sample at the largest deformation, $\varepsilon = 3.5\%$.

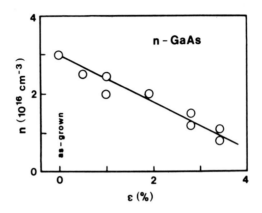

Figure 1. Conduction-electron concentration vs deformation strain $\varepsilon = \Delta l/l$ in plastically deformed n-GaAs samples, measured at the temperature 290 K.

The second pronounced effect of the deformation, observed with the DLTS technique, is an increase in the concentration of the EL6 electron traps having their energy level at $E_c - 0.35$ eV; Figure 2. EL6 are present in as-grown (LEC and HB) GaAs crystals in a concentration which may be comparable, and sometimes even higher than the concentration of the main electron trap EL2. Both EL2 and EL6 are native donors related to excess arsenic. Our DLTS measurements indicate that the sample deformation to about $\varepsilon = 3\%$ produces up to $5 \cdot 10^{15}$ cm^{-3} EL6 traps, while their concentration in original crystals might be several times lower (Wosinski and Figielski 1987).

The controversial problem is whether the plastic deformation has any effect on EL2. In our earlier paper we have reported a remarkable increase in the EL2 concentration after the plastic deformation, observed with the DLTS technique (Wosinski et al 1983). However, in most crystals investigated later on we have not found any remarkable change in the concentration of EL2 up to a strain $\varepsilon = 3.5\%$. These results may be compared with the others reported by different authors. In very early papers (Ishida et al 1980, Kadota and Chino 1983, Hasegawa et al 1984) some increase in the amplitude of the DLTS peak corresponding to the EL2 level was observed after the deformation. Next Omling et al (1986) measured optical absorption of plastically deformed GaAs and concluded that no new EL2 levels had been introduced into the crystal during the deformation. Suezawa et al (1986) inferred from the DLTS measurements that the concentration of EL2 had initially decreased with the strain and then increased at higher strains. In turn, the MIT group did not observe any change in the concentration of EL2 in the deformed samples both with DLTS and optical absorption (Skowronski et al 1987). It must be noticed, however, that their crystals had intentionally been heat-treated prior to the deformation. Recently, Vignaud and Farvacque (1989) investigated

LEC-grown material which had been "EL2-free" before deformation. They concluded, from changes in the quenchable part of the optical absorption, that EL2 had been created during the deformation to a concentration of $6 \cdot 10^{15}$ cm^{-3} at the highest strain, ε = 12%.

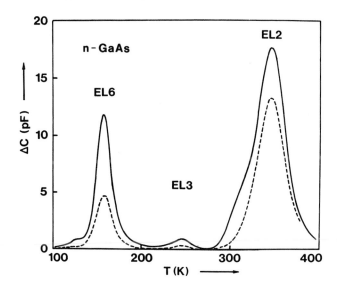

Figure 2. DLTS spectra of undeformed (dashed line) and the deformed to ε = 2.8% (solid line) samples of n-type LEC GaAs, obtained with gold-deposited Schottky diodes at the rate window 5 s^{-1} and the filling-pulse duration time 5 ms.

Those divergent results provide, in our opinion, distinct evidence that the deformation-induced generation of EL2 is a secondary effect (compared e.g. with the generation of EL6) and it may appear or not depending on the original defect structure in a crystal.

Additionally, we have observed a distinct effect of deformation on some properties of EL2. Let us recall that EL2 exhibits a mysterious ability to transform itself from the normal to metastable state under illumination of the crystal with λ = 1.06 μm radiation. This transformation is accompanied by quenching of various properties related to EL2, such as the characteristic optical absorption or photocapacitance of Schottky diodes. We have studied the transient of photocapacitance quenching at 77 K on Schottky diodes made of plastically-deformed samples (Wosinski and Figielski 1987). Prior to the deformation the diodes exhibited a transient which was exactly exponential with time, t (Figure 3), i.e.

$$[C(t) - C_{\infty}]/C_{0} \propto \exp(-t/\tau) \ ,$$

where $1/\tau$ was proportional to the illumination intensity. In the deformed samples the transient became slower and nonexponential, but the effect was reversible, as it ceased after annealing the deformed sample at 490°C for 15 min.

To elucidate the latter results, we invoke the observed correlation between two deformation-induced effects: creation of additional EL6 traps

and appearance of the nonexponential tail in the transient of
photocapacitance quenching. Our explanation assumes that some of
just-created EL6 traps are in the   close vicinity of EL2, so that there
exists a significant probability of electron tunneling between these two

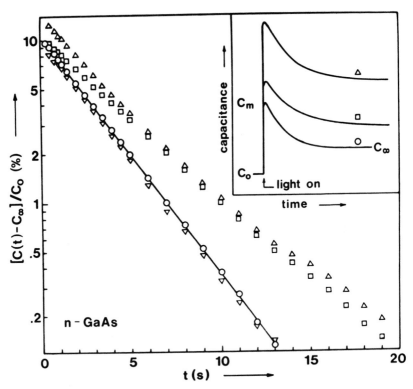

Figure 3. Transients of photocapacitance quenching at 77 K, under
λ = 1.06 μm illumination of the intensity 14 mW/cm^2, for: undeformed
(o), deformed to ε = 0.5% ($\square$), deformed to ε = 2.8% ($\triangle$), and
annealed after the deformation to ε = 2.8% ($\triangledown$) samples. Solid line
represents exponent with the time constant 2.95 s fitted to the
transient in undeformed sample. The overal effect of deformation on
the photocapacitance transient is shown in the inset.

species. Prior to the deformation the EL2 centre, when photoexcited with
1.06 μm radiation, may either undergo autoionization or transform into the
metastable state. After the deformation (which produces new EL6) another
process of deexcitation appears, competing with the previous two ones;
namely, the photoexcited electron may tunnel from EL2 to EL6 with phonon
emission. This new process diminishes the rate of normal-to-metastable
transformation, i.e., of the quenching rate.

3.   GENERATION OF ANTI-SITE-INVOLVED DEFECTS

In order to discuss the mechanisms of generation of EL6, and eventually
EL2, in the process of plastic deformation we should know the physical

nature of these defects. The dispute on the nature of EL2 is still far from being completed, although it seems to be unquestionable now that arsenic anti-site is at least one of the components of EL2 (Bourgoin and Lannoo 1988). There is probably a close relationship between EL2 and EL6 since e.g. a heat treatment at 800°C for 1 h transforms most of EL6 present in a sample into the EL2 traps (Fang *et al* 1987). We have postulated that the simplest defect involving As anti-site is EL6, i.e. the EL6 level corresponds to the (++/+) level of an isolated As anti-site (Wosinski and Figielski 1987). Thus EL2 would contain the As anti-site and something else. For instance, if that "something" were another As anti-site at some distance from the first one along the <110> axis, then the complex defect would be of C_{2v} symmetry and would display a piezo-splitting pattern of the zero-phonon absorption line similar to that observed for EL2 (Figielski and Wosinski 1987).

Anti-site defects can be produced in a process which involves dislocation climb. Dislocations in the sphalerite structure advance by climb in steps requiring interaction of two point defects with a jog. Petroff and Kimerling (1976) and next Weber *et al* (1982) considered some of the possible two-particle steps which would lead to the generation of anti-site defects. In those processes one defect was absorbed at a jog but another was emitted; thus the thermodynamic force driving the process resulted from the difference between chemical potentials of the absorbed and emitted defect. However, under As-rich conditions when arsenic interstitials are mostly available in the crystal, more favourable seems to be the process in which two arsenic interstitials are absorbed at the same jog forming an irregular jog; Figure 4 (Figielski 1985). The driving

Figure 4. Absorption of two arsenic atoms at dislocation jog, followed by dislocation glide, gives rise to the creation of As anti-site defect.

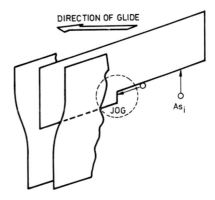

force for this process is determined by the sum of chemical potentials of the two interstitials minus enthalpy of the regular-irregular jog transformation; probably the latter is small for GaAs (Van Vechten 1975). After glide of a dislocation together with its jog, the irregular jog leaves a misplaced atom behind: As in a site of Ga, i.e. the As anti-site defect. We postulate that this mechanism operates during the plastic deformation of GaAs crystals, giving rise to the generation of As anti-sites (see also Alexander and Kisielowski-Kemmerich 1988).

The question arises of the source of arsenic interstitials during the plastic deformation, since isolated interstitials, owing to their high mobility, are very unlikely to be present in GaAs crystals under post-growth conditions. We postulate that the sources of the interstitials are As precipitates. It is now well established by infra-red scattering tomography (Suchet *et al* 1987) and TEM (Cullis *et al* 1980, Lee *et al* 1988) that a great number of small precipitate particles are usually present both in HB- and LEC-grown GaAs crystals. Several electron diffraction analyses identified the particles as elemental hexagonal arsenic precipitates (Cullis *et al* 1980, Lee *et al* 1988). Thus, the As anti-sites can be generated during the plastic deformation of a crystal as a result of the interaction of moving dislocations with arsenic released from the precipitate particles.

To complete this part of the paper we would like to propose the following conclusion. Generation of defects involving anti-sites, like EL2 or EL6, under conditions of plastic deformation is probably not an intrinsic property of an ideal crystal; it may occur only in real nonstoichiometric crystals. Hence, any dispute between different authors whether anti-sites are generated or not in GaAs crystals subjected to plastic deformation is of minor importance, as the effect depends probably on the quality of the material used in the experiment.

4. ELECTRON TRAPS ASSOCIATED WITH DISLOCATIONS

It was noticed already in our early experiments that a shoulder appeared on the low-temperature slope of the EL2 peak in DLTS spectra after the plastic deformation (Wosinski and Figielski 1987). That shoulder was then the subject of systematic studies which confirmed that it corresponded to a new electron trap, labelled ED1, appearing after the deformation. In Figure 5 we demonstrate that the shoulder can be completely separated

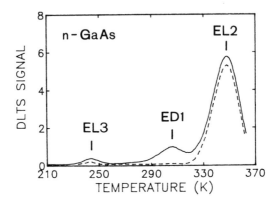

Figure 5.   High-temperature parts of high-resolution DLTS spectra of undeformed (dashed line) and plastically deformed to ε = 2.8% (solid line) samples, recorded at the rate window 4 s^{-1} and the filling-pulse duration time 16 ms.

from the EL2 peak using so-called high-resolution DLTS technique developed by Breitenstein in Halle. Detailed properties of ED1 were studied using standard DLTS and deconvoluting the spectrum into separate peaks (Wosinski 1989); this is exemplified in Figure 6. The ED1 level position determined from an Arrhenius plot is $E_c - 0.68 \pm 0.01$ eV and its cross section for electron capture is $2 \cdot 10^{-14}$ cm^2. The ED1 peak is distinctly broadened with the full-peak half-width FPHW = 80 meV.

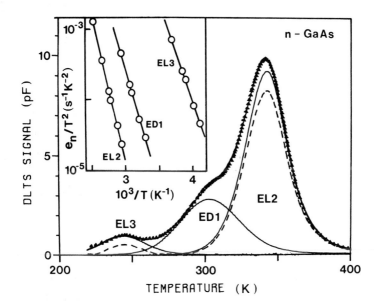

Figure 6. Standard DLTS spectra of undeformed (dashed line) and the deformed to $\varepsilon = 3.4\%$ (triangles) samples, recorded at the rate window 2.5 s^{-1} and filling-pulse duration time 20 ms. Deconvoluted peaks of the latter spectrum are drawn with solid lines. Arrhenius plots are shown in the inset.

The particular property of the ED1 peak is that its amplitude does not saturate but rises linearly with the logarithm of filling time (over six orders of its magnitude!). Such dependence may result from the barrier-limited electron capture and is characteristic of dislocations (Wosinski 1989). The concentration of electrons trapped at the ED1 level, n_T, calculated from the DLTS peak, is shown in Figure 7 as a function of the filling time for various plastically deformed samples. Since there is no saturation of n_T one cannot determine the absolute value of the concentration of ED1 but only its lower limit, which may be assumed to be equal to n_T corresponding to the longest filling pulse used. These lower-limit concentrations in various samples are plotted in Figure 8 as a function of the dislocation density. The concentration of possible core states of dislocations (assuming one state per interatomic distance along the 60°-dislocation) is also indicated on the upper abscissa in Figure 8. The linear dependence demonstrates quantitatively that the ED1 traps might be considered as core states of dislocations with an occupation fraction one-tenth at the longest filling time.

Additionally, global properties of plastically deformed samples were studied using intrinsic photoconductivity; it was hoped to reveal the peculiarities observed previously in deformed germanium and silicon

(Figielski 1978). This study performed by Morawski and Pohoryles (1988) in our laboratory, was successful.   The decay of photocurrent excited with

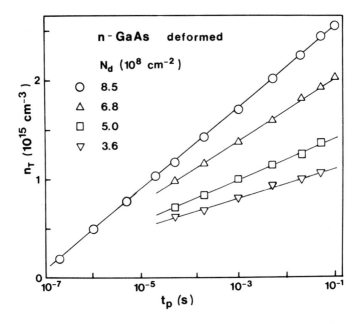

Figure 7. Concentrations of electrons trapped at ED1 traps vs filling-pulse duration time measured in various plastically deformed samples. Dislocation densities are written in the inset.

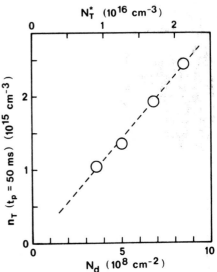

Figure 8. Concentration of ED1 traps occupied with electrons during the filling pulse of 50 ms duration vs dislocation density in various plastically deformed samples. Concentration of possible core states of dislocations is indicated on the upper abscissa.

deeply penetrating inter-band illumination displays, below $\simeq 270$ K, a logarithmic-type process (Figure 9) which obeys the relation:

$$\Delta\sigma/\sigma \propto \ln[(t + t_0)/\tau] \ ,$$

where t_0 is a parameter depending on the carrier-generation rate and τ is the characteristic decay time which depends on temperature but is independent of dislocation density. This relation means that the recombination of excess current carriers is governed by the barrier mechanism that has been discussed in detail for the case of dislocations in Ge and Si. Similar behaviour of plastically deformed GaAs was also observed by Nakata and Ninomiya (1979). So, we have confirmed that the barrier mechanism is common to the two effects occuring in plastically deformed GaAs: the filling of the ED1 traps in DLTS experiments and the recombination of photoexcited carriers. In order to prove whether the ED1 trap controls also the recombination process, the temperature dependence of stationary intrinsic photoconductivity was measured. It turned out to be the case, since the activation energy of the electron capture rate determined from that dependence was 0.67 ± 0.03 eV, which, according to the barrier model by Figielski (1978), should be compared with the value 0.68 eV obtained from DLTS for the position of the ED1 level.

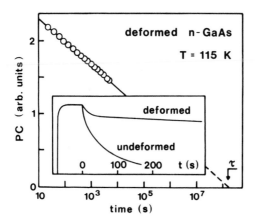

Figure 9. Photoconductivity decay in plastically deformed n-type GaAs after illumination of the sample with a rectangular light pulse. Internal picture - linear time scale, external picture - logarithmic time scale.

Taking into account all the collected data concerning ED1, we postulate this trap to be actually associated with the core states of dislocations. Trying to speculate on the type of dislocation responsible for ED1, we suggest 60°-Ga dislocation (or its partials). The main argument for this suggestion follows from the doping-dependence of the dislocation velocity in GaAs. Only β-dislocation mobility is highly sensitive to the position of the Fermi level within the central part of the energy gap (Maeda 1985), suggesting a midgap level associated with this kind of dislocation; in turn, β-dislocations are Ga-dislocations in the dominating glide set. Perhaps other dislocation-related electron states in plastically deformed GaAs manifest themselves in the form of a broad luminescence band centred at 1.13 eV, as observed by Depraetere *et al* (1987).

5.   CONSERVATIVE CLIMB OF DISLOCATIONS

In the last part of this paper we want to discuss some important features
of the phenomenon of conservative climb of dislocations. The idea has
appeared by considering possible mechanisms of the degradation of junction
lasers.

Let us consider a prismatic dislocation loop which borders two mutually
inclined plane areas inside a crystal. Assume that the Burgers vector of
the dislocation meets each of the plane areas at the same angle but from
opposite sides; Figure 10. In this case each of the two areas has to be of
different type; assume the upper one to represent a deficient atomic layer
and the lower one   an excess atomic layer. Suppose that under favourable

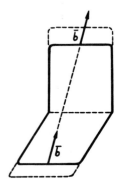

Figure 10. Topological configuration of the
prismatic dislocation loop which expands
owing to conservative climb, as indicated by
the dotted line. If the direction of the
mass transport via pipe diffusion along the
dislocation core were the reverse, the loop
would shrink.

conditions lattice atoms are removed from the upper part of the loop
bordering the deficient atomic layer, transported along the dislocation
via pipe diffusion, and deposited at its lower part bordering the excess
atomic layer. Thus, as long as the process proceeds, the dislocation loop
steadily expands increasing both its area and length. So, this mechanism
may be considered as a counterpart of dislocation multiplication in the
glide process, since they both lead to the growth of the total dislocation
length in a crystal. The driving force for the process considered is the
minimizing of internal stresses in the crystal owing to the mass transfer
between different atomic planes.

It has been shown elsewhere (Figielski 1988) that the conservative climb
of a dislocation threading the active layer of a laser device, driven
by a lattice mismatch between adjacent layers, may lead during device
operation to the development of elongated dislocation dipoles. Their
features resemble very much those of the dipoles actually observed in the
degraded areas of DH GaAs-GaAlAs junction lasers. Moreover, the
conservative climb does not require availability of high concentration of
self-interstitials or vacancies in the active layer and assures a high
speed of the process owing to high diffusivity along the dislocation core.
It has therefore been proposed that the process of conservative climb of
dislocations is responsible for some modes of laser degradation. This
mechanism may operate also in the process of plastic deformation. In
consequence one may notice that the atoms transported by the pipe
diffusion and deposited at new positions may occupy irregular sites in the
lattice. Thus, arsenic may be placed in the site of gallium and vice

versa, which offers an additional intrinsic mechanism for the anti-site defect production.

6. CONCLUSIONS

The effect of plastic deformation on the defects with energy levels lying in the upper half of the band gap of GaAs is rather slight. Only the concentration of EL6 electron traps is significantly enhanced, and (not reported previously) ED1 electron traps are produced which probably are associated with dislocation cores. Probably a richer spectrum of the states created during the plastic deformation appears in the lower half of the band gap but it is still hidden from direct observation. The As anti-site-involved defects seem to be produced during the plastic deformation owing to the presence of internal sources of excess arsenic in the crystal in the form of As precipitates. EL6 is assumed to be the simplest defect involving As anti-site. The often neglected process of conservative climb may contribute to the development of dislocation dipoles in strained crystals.

ACKNOWLEDGMENTS

This work was carried out in part in the IV. Physikalischen Institut der Universitaet Goettingen headed by Professor Schroeter. The authors would like to thank also J. Kronewitz (Goettingen) for his help with the deconvolution of the DLTS spectra and Dr. O. Breitenstein (Halle) for cooperation in the high-resolution DLTS experiments. Part of the work has been supported by the research programme CPBP 01.05.

REFERENCES

Alexander H and Kisielowski-Kemmerich C 1988 *Proc. 8th Int. School on Defects in Crystals* ed E Mizera (Singapore: World Scientific) p 393
Bourgoin J C and Lannoo M 1988 *Revue Phys. Appl.* **23** 863
Cullis A G, Augustus P D and Stirland D J 1980 *J. Appl. Phys.* **51** 2556
Depraetere E, Vignaud D and Farvacque J L 1987 *Solid State Commun.* **64** 1465
Fang Z-Q, Schlesinger T E and Milnes A G 1987 *J. Appl. Phys.* **61** 5047
Figielski T 1978 *Solid-St. Electron.* **21** 1403
Figielski T 1985 *Appl.Phys.A* **36** 217
Figielski T 1988 *Philos. Mag.* **57** 791
Figielski T and Wosinski T 1987 *Phys. Rev. B* **36** 1269
Gerthsen D and Haasen P 1986 *Acta Phys. Polon. A* **69** 415
Hasegawa F, Yamamoto N and Nannichi Y 1984 *Extended Abstracts of 16th Conf. on Solid State Devices and Materials* (Tokyo: Business Center for Academic Societies Japan) p 169
Ishida T, Maeda K and Takeuchi S 1980 *Appl. Phys.* **21** 257
Kadota Y and Chino K 1983 *Jpn. J. Appl. Phys.* **22** 1563
Lee B-T, Gronsky R and Bourret E D 1988 *J. Appl. Phys.* **64** 114
Maeda K 1985 *Dislocations in Solids* ed H Suzuki *et al* (Tokyo: University of Tokyo Press) p 425
Morawski A and Pohoryles B 1988 *Proc. 8th Int. School on Defects in Crystals* ed E Mizera (Singapore: World Scientific) p 454
Nakata H and Ninomiya T 1979 *J. Phys. Soc. Jpn.* **47** 1912
Omling P, Weber E R and Samuelson L 1986 *Phys. Rev. B* **33** 5880
Petroff P M and Kimerling L C 1976 *Appl. Phys. Lett.* **29** 461

Skowronski M, Lagowski J, Milshtein M, Kang C H, Dabkowski F P, Hennel A and Gatos H C 1987 *J. Appl. Phys.* **62** 3791

Suchet P, Duseaux M, Grillardin G, Le Bris J and Martin G M 1987 *J. Appl. Phys.* **62** 3700

Suezawa M and Sumino K 1986 *Jpn. J. Appl. Phys.* **25** 533

Suezawa M, Hara A and Sumino K 1986 *Acta Phys. Polon. A* **69** 423

Van Vechten J A 1975 *J. Electrochem. Soc.* **122** 423

Vignaud D and Farvacque J L 1989 *J. Appl. Phys.* (in the press)

Weber E R, Ennen H, Kaufmann U, Windscheif J, Schneider J and Wosinski T 1982 *J. Appl. Phys.* **53** 6140

Wosinski T 1988 *Proc. 8th Int. School on Defects in Crystals* ed E Mizera (Singapore: World Scientific) p 446

Wosinski T 1989 *J. Appl. Phys.* **65** 1566

Wosinski T and Figielski T 1987 *Solid State Commun.* **63** 855

Wosinski T, Morawski A and Figielski T 1983 *Appl.Phys.A* **30** 233

Inst. Phys. Conf. Ser. No 104: Chapter 2
Paper presented at Int. Symp. on Struct. Prop. Disloc. Semicond., Oxford, 1989

Microwave conductivity in plastically deformed silicon

M. Brohl, H. Alexander

Abt. Metallphysik, Univ. Köln, Zülpicher Str. 77, D-5000 Köln 41

By the cavity perturbation technique we study the mutual influence of dislocation structure, point defects and chemical doping on disclocation related microwave conductivity in plastically deformed FZ-silicon. We show the need for a well defined deformation procedure to gain clear results in n- as well as in p-material. High stress/low temperature deformation minimizes the risk of sample compensation by point defect attributed deep levels; furthermore it allows discrimination between conductivity along screw and 60°-dislocations. We find conductivity under suitable conditions to be much stronger than reported up to now. Various transport mechanisms are discussed.

1. INTRODUCTION

As is now well established by DLTS- and EPR-investigations the electrical and paramagnetic properties of plastically deformed silicon are strongly dependent on the amount of point defect related deep levels introduced into the crystal by moving dislocations (Omling *et al* 1985, Kisielowski-Kemmerich and Alexander 1988). EPR has already been used to identify some of these centers (Weber and Alexander 1977, Brohl *et al* 1987). Therefore one has always to consider that deformation effects, such as the reduction of free carrier concentration (compensation), can have other than directly dislocation related origins. On the other hand, microwave conductivity (σ_{cw}-) investigations (Kveder *et al* 1985) have suffered up to now from the lack of a clear distinction between effects due to dislocation structures and those due to other electrical active levels. Here we expect the explicit balance between dislocation density and structure, chemical doping and point defect concentration to become very important. It is the aim of the following σ_{cw}-analysis to elucidate these interdependences. Extending an earlier study (Brohl *et al* 1989), we show the need for careful sample processing in order to avoid misleading and troublesome results. By means of two-step deformation we produce ordered dislocation geometries; spectrally resolved photoconductivity measurements then help to understand doping and compensation related effects.

2. EXPERIMENTAL

Samples used in this study are FZ-single crystals, always deformed along the [213]-compression axis, first at 800 °C up to different strains. Dislocations introduced into the crystal form irregularly shaped networks with many edge dipoles in the main glide plane ($1\bar{1}1$) perpendicular to the Burgers vector $\underline{b}$=[011]. After a long annealing treatment (up to 16 h at 800 °C) dislocation structures are not greatly altered but all residual deep levels vanish (typically $n_{DEF}<5\times10^{13}$ cm^{-3} from DLTS-data for samples deformed up to $\varepsilon=\Delta l/l=1.6\%$). A *short high stress/low temperature* (HSLT-) deformation now leads, by extending existing dislocation loops, to widely dissociated 60°- and screw dislocations mainly arranged in (1$\bar{1}$1). This step introduces less than 1x10^{14} cm^{-3} DLTS-active centers. In a renewed heat treatment stacking faults and dislocation loops narrow; the number of screw dislocations becomes smaller as we can confirm by TEM and etch pitch density counting. Exact sample parameters are given in the figures. For the σ_{cw}-investigations circular disks (plane normal [1$\bar{1}$1] or [011]) were cut from the central parts of the crystals and rotated with respect to the electrical field $\underline{E}_{cw}$ in the field maximum of a TM_{010} cylinder resonator mode operating at 10 GHz. The evaluation of the measurements followed the method of disturbed resonator in the quasistatic approximation described elsewhere (Altschuler 1963). Dislocation densities N_{EPD} were determined by etch pitch counting.

3. RESULTS

In a first group of experiments crystals with a relatively high chemical doping level ([P]=4.4x10^{15} cm^{-3}) are used to ensure the Fermi level fixed near the conduction band. In figures 1 and 2 we demonstrate the changes in the low temperature (LT-)conductivity ($T_M<25K$) due to the *structural* transition to the HSLT-dislocation arrangement. The strong conductivity increase for $\underline{E}\parallel\underline{b}$ correlates to the high number of screw dislocations generated, whereas for $\underline{E}\parallel[21\bar{1}]$ conductivity can be attributed to the 90/30- and 30/90-dislocations. The activation energies lower from about 1–1.5 meV to 0.2–1.0 meV. The LT-σ_{cw} turns out to be much stronger than reported in the literature up to now. A pronounced anisotropy can be observed when $\underline{E}$ is rotated into the [1$\bar{1}$1]-orientation; this reflects the good single slip in HSLT-deformed crystals. (figure 2)

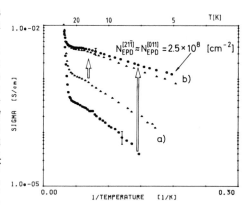

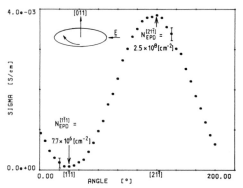

Fig. 1 (upper part): n–Si; [P]=4.4x10^{15} cm^{-3}; ● $\underline{E}\parallel$ [011]; ▲ $\underline{E}\parallel$ [21$\bar{1}$] a) T_D=800°C; ε =1.6 %; 16 h annealing at 800°C; $N_{EPD}^{[011]}\approx2.5\times10^7$ cm^{-2}, $N_{EPD}^{[21\bar{1}]}\approx7\times10^7$ cm^{-2}; b) as a) + second deformation: T_D=420°C; τ= 250 MPa, 30 min

Fig. 2 (lower part): Sample as in fig. 1b) Rotation around $\underline{b}$=[011]; from Brohl *et al* (1989).

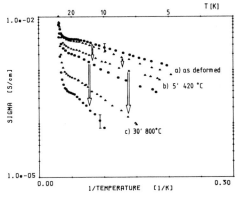

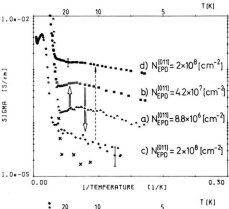

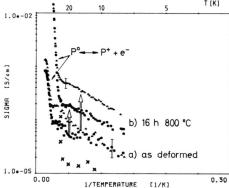

Fig. 3: Sample as in fig. 1b); influence of subsequent heat treatment Values for curve c): $N_{EPD}^{[21\bar{1}]}=1\times10^8$ cm^{-2} $N_{EPD}^{[011]}=5\times10^7$ cm^{-2}

The influence of a subsequent heat treatment is illustrated in figure 3: The drastic breakdown of conductivity can be related to the above described changes in dislocation morphology.

We then investigated the dependence of σ_{cw} on *dislocation densities* N_D at a lower doping level ($[P]=5\times10^{14}$ cm^{-3}) in HSLT–deformed silicon; results are presented in figure 4 (Note there the short intermediate annealing time): An initial increase of LT–σ_{cw} nearly proportional to N_{EPD} is followed by a drastic breakdown at higher strains accompanied by a strong compensation of free phosphorus atoms. The carrier excitation into the conduction band (P° ↔ P$^+$+e$^-$, high temperature (HT–) σ_{cw}, $T_M > 25$K)) is strongly suppressed. The LT–σ_{cw} becomes two orders of magnitude stronger when we prolong the intermediate annealing time between the two deformation steps (fig.4d). Conversely we regain a clear LT–conductivity in strongly compensated crystals also by the changes caused by an annealing treatment *after* deformation, as is illustrated in figure 5. The signal comes up nearly to that of the higher doped annealed sample in fig. 3c. We can say that the Fermi level shift produced by these special deformation conditions is removed by the annealing procedure: Deep levels are annihilated *and* N_{EPD} is decreased.

Fig. 4: $[P]=5\times10^{14}$ cm^{-3}; $\underline{E}$ ∥ [011]; $T_D=800°$C; a) ε =0.11 %; b) ε =0.33 %; c+d) ε =1.6 %; a–c) annealing 30 min at 800°C; d) annealing 16 h at 800°C a) – d): + $T_D=420°$C; τ=250 MPa; 30 min; x control sample.

Fig. 5: $[P]=5\times10^{14}$ cm^{-3}; ● $\underline{E}$ ∥ [011]; ▲ $\underline{E}$ ∥ [21$\bar{1}$]; $T_D=800°$C; ε = 1.6 %; no intermediate annealing; + $T_D=420°$C; τ= 250 MPa; 30 min a) as deformed b) annealing 16 h at 800°C after deformation; x control sample.

Results of photoconductivity measurements at low temperatures in weakly as well as in strongly compensated crystals are shown in figure 6. Parameters are the starting chemical doping level and the duration of annealing between the deformation steps. We observe a steep midgap excitation ($\approx$0.6 [eV]) and a smaller one at $\approx$0.95 [eV], but the latter is only very pronounced in the lower doped samples partly compensated by deformation. Just below the band edge conductivities differ only by a factor of 2. Band edge illumination partly destroys the LT-σ_{cw}.

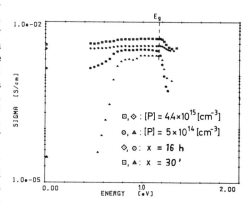

Fig. 6: photoconductivity $\underline{E} \parallel$ [011]; T_M=14.4 K; T_D=800°C; ε =1.6%;. x = annealing time at 800°C; + T_D=420°C; τ=250 MPa; 30 min.

Initial experiments in p-doped silicon showed quite similar phenomena. In figure 7 we compare samples of different N_{EPD}. As long as the Fermi level is close to the valence band the LT-dark conductivity is strong; otherwise only illumination restores the LT-conductivity (figure 8). Anisotropy measurements (not shown) give similar results to n-doped material.

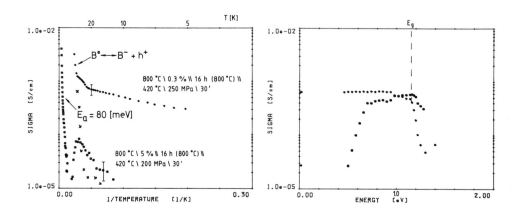

Fig. 7 (left side): p-Si; [B]=2x10^{14} cm^{-3}; $\underline{E} \parallel$ [011]; variation of the extent of predeformation; x control sample; ◆ N_{EPD}[011]=4x10^7 cm^{-2}; ● N_{DIS}≥1x10^9 cm^{-2};

Fig. 8: (right side) photoconductivity, T_M=14.4 K, samples as in fig. 6; ● strong deformation, ◆ weak deformation

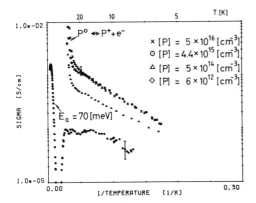

Fig. 9: n–Si; different doping levels; $\underline{E} \parallel [21\bar{1}]$; $T_D=800°C$; $\epsilon =1.6$ %; 16 h annealing at 800°C; $N_{EPD}^{[21\bar{1}]}=(6-8)\times10^7$ cm^{-2}.

The knowledge of the above relations allowed us then to investigate the dependence of σ_{cw} on *chemical doping* in samples carefully annealed at 800°C. The results are shown in figure 9. Down to $[P]=5\times10^{14}$ cm^{-3} the LT-conductivity in those samples which are only predeformed is found to be weakly doping dependent. In the low doped sample conductivity is weak; a strange anomaly in σ_{cw} at 50 K can be observed.

To summarize our results we state the following:
– LT-σ_{cw} in n- and in p-silicon is very strong and parallel to the special dislocation structures chosen, as long as compensation by carrier trapping at other defects is negligible. ($n_{DEF}\ll N_{DOP}$)
– In the case $n_{DEF}\geq N_{DOP}$ the LT-conductivity is always very weak; losses are hardly stronger than those in undeformed control samples that have undergone the same heat treatment. Carriers can be detrapped by photoexcitation and become mobile, being captured by conducting dislocation states.
– The interfering effects of deformation–induced point defects at lower doping levels can only be avoided by a long annealing treatment, at the cost of losing the well-defined dislocation structure. A short HSLT-deformation is the only adequate way to restore the latter, as it minimizes the compensation risk.

4. DISCUSSION

To analyse our results we have generally to take into account different possible dielectric loss mechanisms. First, hopping transport along dislocations via partly compensated donor (acceptor) atoms is discussed. In our opinion two main objections can be made to this view: With regard to the starting chemical doping level, losses in our samples are orders of magnitude higher than the results of cw–investigations on partly compensated doped silicon of Tanaka and Fan (1963). Furthermore, no thermal activation is observed for high frequency hopping conductivity; the whole conductivity at a frequency ν can be expressed in the form $\sigma=\sigma_{dc}+A\nu^s$ with A and s only weakly temperature dependent (Mott and Davis 1971). Nevertheless, the existence of an impurity band formed by doping atoms strongly concentrated along dislocations by some gettering process cannot be excluded.
Polarisable defects on the glide plane also should give dielectric losses, but because of the observed anisotropy and annealing behaviour we believe such an electrical activity in the glide plane (as found by Bondarenko 1986 in EBIC–investigations) to be, for us at least, secondary.
We tentatively explain our LT–results in terms of one acceptor and one donor type shallow dislocation band; they will be partly filled and become conducting in the dark only when the Fermi level is. fixed near the band edges. In the presence of competing midgap acceptors (donors), carriers will be trapped there and become immobile unless they are excited by photogeneration. Dark conductivity in strongly compensated samples is vanishingly weak. Generation

of electron-hole pairs ($\hbar\omega\geq1.16$ eV) destroys the dislocation charge by some recombination process.

This leads us to a schematic energy diagram as proposed in figure 10. Our experiments show LT-dislocation conductivity to be limited either by the number of carriers arising from the chemical dopants or by N_{EPD}. The former can be estimated using results of EPR-spectroscopy on the neutral phosphorus hyperfine signal P° (Kisielowski-Kemmerich and Alexander 1988). For one of our samples (marked by ● symbols in figure 6) we find as an upper limit for the number of conducting electrons $N_e\approx3\times10^{14}$ cm^{-3} and therefore the cw-mobility to be greater than ≈100 cm²/Vsec. Given the dislocation density by $N_{EPD}\approx4\times10^8$ cm^{-2}, the mean distance of conducting electrons along dislocations will be more than 100 A.

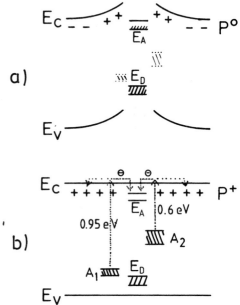

Fig.10: Schematic energy diagram of plastically deformed   n-Si: a) weak, b) strong compensation; A_1, A_2 acceptor levels; E_D, E_A dislocation related bands; arrows indicate possible optical transitions

Further analysis now requires assumptions about the origin and the form of the carrier-attracting potential. Here lattice deformation induced states split off the band edges around dislocations are taken into account by Kveder *et al* (1985) but no direct evidence for conductivity along those states can be given at time. As to the temperature dependence, the origin of the observed thermal activation energies in the LT-dislocation related regime, as well as in the HT-signal of compensated samples (n-Si: 70 meV, p-Si: 80 meV) is still unclear.

The authors thank the Deutsche Forschungsgemeinschaft for financial support.

Altschuler H M 1963 in: *Handbook of Microwave Measurements* Vol II ed. Sucher M and Fox J (New York: Polytechnic Press) p. 495
Bondarenko I E, Blumtritt H, Heydenreich J, Kazmiruk V V Yakimov E B 1986   *phys. stat. sol. (a)* **95** 173
Brohl M, Dressel M, Helberg H W, Alexander H 1989   *Phil. Mag.* in press
Brohl M, Kisielowski-Kemmerich C, Alexander H 1987 *Appl. Phys. lett.* **50** 1733
Kisielowski-Kemmerich C, Alexander H 1988 *Defects in Crystals* ed Mizera E. (New York, Singapore: World Scientific) p 290
Kveder W V, Osip'yan Yu A, Sagdeev I R, Shalynin A I, Zolothukin M N 1985 *phys. stat. sol. (a)* **87** 657
Mott N F, Davis E A 1971 *Electronic Processes in Non-Crystalline Materials* (Oxford: Clarendon Press) p 49 and 184
Omling P, Weber E R, Montelius L, Alexander H, Michel J 1985 *Phys. Rev. B* **10** 6571
Tanaka S, Fan H Y 1963 *Phys. Rev.* **132** 1516
Weber E, Alexander H 1977 *Inst. Phys. Conf. Ser.* **31** 266

Inst. Phys. Conf. Ser. No 104: Chapter 2

169

Paper presented at Int. Symp. on Struct. Prop. Disloc. Semicond., Oxford, 1989

Electrical activity associated with dislocations in silicon

A.Castaldini, A.Cavallini and D.Cavalcoli

University of Bologna, Department of Physics,
CISM-GNSM Via Irnerio 46, I-40126 Bologna, Italy

ABSTRACT: We studied the electrical behavior of dislocations by the Quenched Infra-Red Beam Induced Current (Q-IRBIC) method, which is a development of the LBIC (Light Beam Induced Current) technique of scanning optical microscopy. The Q-IRBIC method combines the LBIC technique with photoconductivity quenching. A Schottky barrier on the sample surface collects the electron-hole pairs generated by a light scanning beam. Simultaneous with the raster sweeping of the charge collection barrier by above-band-gap light, the opposite surface is irradiated by continuous monochromatic below-band gap light. Q-IRBIC contrast at the dislocations versus the quenching light photon energy makes it possible to determine the energy levels associated with the defects, whereas contrast vs. quenching light intensity provides information about the level occupancy degree. Results on Q-IRBIC dependence on specimen temperature are here presented.

1. INTRODUCTION

The electronic properties of dislocations have been subject of study for many years since it is well known that their electrical activity can seriously affect semiconductor materials and device performance. Although the systematic study of defects in semiconductors began about thirty years ago and many questions about the dislocation structure and electrical behavior have been answered from that time on, nevertheless many others are still to be answered. Precisely, many problems arise from the use of bulk techniques of investigation, lacking in spatial resolution, and the difficulty of matching the results so obtained with structural information on individual defects.

In the present paper we report preliminary results of an investigation performed by a recently developed technique called Quenched Infra-Red Beam Induced Current (Q-IRBIC) (Cavallini et al 1986, Castaldini et al 1988a), which is a development of the already well established method LBIC (Light Beam Induced Current) of scanning optical microscopy (SOM).
The Q-IRBIC method consists in combining dislocation imaging by means of the current induced by an above band-gap (ABG) light probe with the photoelectric effect called photoconductivity quenching or also optical quenching by long wavelength secondary illumination (Milnes 1973).

The LBIC technique allows one to obtain spatially resolved information on the electronic properties of dislocations, inferred from the analytical treatment of the defect image formation. In the conventional LBIC method the light source is a laser, whereas we use a halogen lamp, whose wavelength is filtered so as to obtain an infra-red light probe and then to control the beam penetration depth. Thus, we will talk about the IRBIC (Infra-Red Beam Induced Current), instead of the LBIC, method.

To produce the IRBIC signal, a photon probe is used to generate electron -hole pairs in a localized volume in the sample (Wilson et al 1986). The dimensions of this volume are determined by the intensity and wavelength of the light beam, as well as by the absorption of the specimen. The electron-hole pairs reaching the depletion region of a Schottky barrier are divided by the built-in field and produce the IRBIC signal. When a defect lies inside the generation volume, its recombination activity reduces locally the carrier lifetime and, in turn, the induced current. As in the EBIC method, the quantitative parameter from which electrical information about the dislocation recombination activity can be inferred is the contrast $c=(I_b-I_d)/I_b$, where I_b and I_d are the beam induced current far from and at the defect, respectively.

The photoconductivity optical quenching, up to now used as a bulk approach to investigate recombination kinetics at the defects, is used here to force electron transitions from the valence band to energy levels lying in the forbidden gap, or from these levels to the conduction band. Specifically, when the photon energy of the back surface irradiation coincides with the distance E_T of a level within the forbidden gap from one of the allowed energy bands, a level-band transition occurs.

In such a way, the charge at the defect is modified and, in turn, its recombination activity, as well as the IRBIC image and the contrast c, named Q-IRBIC (Quenched Infra-Red Beam Induced Current) contrast in these experimental conditions. From the photon energy and intensity of the quenching, pumping, light the defect energy level E_T in the forbidden gap, as well as its occupancy factor f, can be determined in addition to its spatial position. In particular, Q-IRBIC contrast monitored versus pumping photon energy presents minima corresponding to a defect recombination activity reduction. Electron transitions, therefore, occur at the related pumping energy values, that can be interpreted as the distance between the defect level and allowed bands. By increasing more and more the pumping light intensity with a photon energy corresponding to a defect level, the contrast tends to disappear since the defect recombination efficiency vanishes progressively.

2. EXPERIMENTAL

Silicon single crystals, floating zone grown and doped with phosphorus ($5 \cdot 10^{13} cm^{-3}$) were plastically deformed by creep along $\langle 112 \rangle$ at 650°C. The specimens crystallographic orientation was (111) and their thickness was about 200 μm . After polishing and cleaning, a Schottky barrier was formed by a vacuum deposition of a thin layer (150 Å) of Au. The ohmic contact was made on the opposite surface of the slice by an Hg-In alloy. Since the current signal induced by the infra-red light probe is very low $(10^{-12} - 10^{-14}$ A$)$, a near-ideal behavior of the collecting junction is needed. Thus, it was necessary to check the characteristics of the Schottky diode by measuring its ideality factor n, given by the equation

$$J = A[exp\left(\frac{qV_a}{nkT}\right) - 1]$$
(1)

where J is the current density, V_a the bias and A a constant.

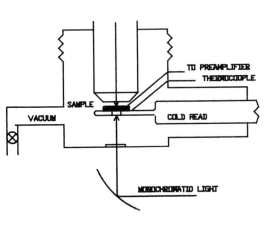

Fig. 1   Schematic diagram of
the Q-IRBIC layout.

For all the specimens the values of n resulting were very close to unity (1.04-1.08) at room temperature. Due to the low light beam irradiance employed $(1.16\cdot 10^{-3}\,W\,cm^{-2})$ and the consequently very small electron-hole generation rate into the specimen, phase-sensitive detection was used, by connecting the back-surface contact to a transimpedance preamplifier and, through this one, to a lock-in amplifier. The electron-hole pair generation rate G was calculated from the expression (Davidson and Dimitriadis 1980):

$$G = P_b\eta(1 - \tau)/(qE_g) \qquad (2)$$

where P_b is the beam power, η is the quantum efficiency, equal to 60 % for carrier generation by a light beam, τ the backscattering coefficient, q the electronic charge, and E_g the band gap. For our experimental conditions G is $1.2\cdot 10^8 s^{-1}$. A sketch of the Q-IRBIC apparatus is shown in Figure 1. Above-band gap light is focused onto the sample surface. The spot diameter ranges from 1.2 to 1.6 μm at the semiconductor surface, depending on the stylus light wavelength. The beam is chopped with a frequency equal to 72 Hz. The opposite surface of the specimen is lit up by monochromatic continuous irradiation.

The wavelength of the secondary quenching light is selected by means of a monochromator so that its photon energy can sweep the whole forbidden energy gap (below band gap (BBG) light). The quenching illumination intensity is adjusted according to the specimen absorption coefficient in such a way that the photon flux is kept constant at the defect examined. In the measurements reported here, the quenching photon flux at the defect was $1.2\cdot 10^{15}$ photons $\cdot cm^{-2}\cdot s^{-1}$. The sample is raster scanned by an x-y holder, whose step size and number are computer driven. A vacuum chamber encloses the microscope objective lens and the specimen holder, which is directly connected to the cold head of a cryogenic system. To minimize the noise, data smoothing by median rank filtering is performed. In such a way the salient features of the dislocations can be imaged. For other experimental details the reader is referred to Castaldini et al (1988a and 1988b).

The expression of the EBIC contrast yielded by Donolato's model (Donolato 1983) is given by

$$c = \gamma F[R(E), L, geometry] \qquad (3)$$

where $\gamma[cm^2s^{-1}]$ is the recombination strength (or line recombination velocity), $R(E)$ is a factor defining the carrier pair generation volume, L is the bulk diffusion length, the term *geometry* refers to the defect-device configuration, and F denotes a functional dependence. Equation (2) also shows that the contrast model is linear, since c and γ are linearly related.

Since the light beam induced current method used in above band gap (ABG) mode is analogous to the SEM EBIC mode (Wilson et al 1979, Donolato 1981), the EBIC mathematical treatment can be exploited to describe light beam induced contrast investigations, as reported by Wilson and McCabe (1987).

Thus, Q-IRBIC investigations on minority carrier recombination at individual defects can be led back to the EBIC theory, bearing in mind, however, that the long wavelength (BBG) illumination affects the charge on the defect, and hence the recombination strength. Therefore, the parameter f can now be changed, in a given specimen, by: (i) a photoelectric excitation by a quenching light, which forces level-band transitions, (ii) a temperature variation that, by moving the Fermi level with respect to the dislocation energy level, changes the charge on the dislocation, and hence its recombination efficiency.

In the present work, Q-IRBIC dislocation contrast was first measured as a function of pumping photon energy to determine the defect energy level position within the forbidden gap. Afterwards, the Q-IRBIC contrast corresponding to two different quenching light wavelengths was monitored as a function of specimen temperature T, with the aim of moving the Fermi level.

3. RESULTS

Q-IRBIC contrast as a function of quenching photon energy is shown in Figure 2 for two specimen temperatures. Contrast minima are shown, whose quenching energy values correspond to dislocation levels found by other authors with different techniques (see, for example, Kimerling et al (1979) and Ono et al 1985)). On the basis of the above description of the Q-IRBIC method, these minima were attributed to transitions from defect levels to allowed bands, since the actual recombination process taking place

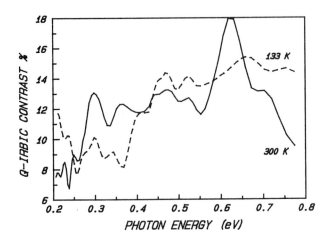

Fig. 2   Plot of Q-IRBIC contrast vs quenching photon energy for two specimen temperatures.

at the dislocation indicates that the efficiency of the defect as a recombination center line decreases. In comparing the contrast measured at room temperature to a low temperature diagram, it was found that the two curves intersect. To investigate the phenomenon, the contrast c was monitored versus the specimen temperature T for different quenching photon energy values. Figures 3 and 4 show the contrast diagrams

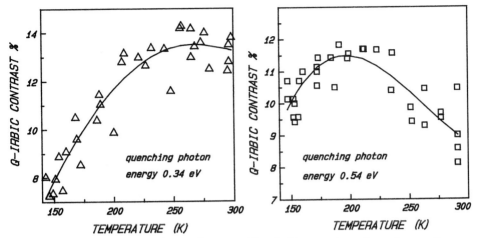

Fig. 3. Temperature dependence of Q-IRBIC contrast for a quenching photon energy equal to 0.34 eV.

Fig. 4. Temperature dependence of Q-IRBIC contrast for a quenching photon energy equal to 0.54 ev.

obtained with quenching photon energy equal to 0.34eV and 0.54eV, respectively. All other experimental parameters (spot irradiance, long wavelength photon flux at the defect, etc.) were kept constant.

The results were surprising since it was always observed (Ourmazd et al 1983), in the EBIC mode, that the contrast increases as T increases. Indeed, the following behaviour was recently observed (Wilshaw et al 1985): (i) at lower temperatures, the trend is linear, (ii) at higher temperatures the contrast variation is sub-linear. Analogous behavior is shown in Figure 3, where a linear variation of the contrast is followed by a sub-linear trend. In that case, however, a marked plateau is evident (the solid lines in Figures 3 and 4 resulted from the elaboration of the experimental data by a multiple non-linear regression to point out the trend). The most unexpected result is, however, reported in Figure 4. Here the contrast behavior is as follows: (i) at low temperatures c increases as T increases, (ii) at intermediate temperatures a fairly constant value occurs, (iii) at high temperatures the contrast decreases as T increases. In LBIC mode, as in the EBIC one, the contrast is related to the minority carrier lifetime τ_p, near the defect, or equivalently to the recombination center N density and capture cross-section σ through

$$\tau_p = (N\sigma v_{th})^{-1} \qquad (4)$$

where v_{th} is the thermal velocity. σ depends on temperature T and occupancy factor f of the defect, which in turn depends on Fermi level position E_f with respect to the defect energy level E_T and, in the present case, on the quenching light photon intensity I_{ph}, too. Thus one may write

$$c = c(E_T, T, E_f, I_{ph}) \qquad (5)$$

In order to predict how the Q-IRBIC contrast will depend on temperature, it will be necessary to introduce this last parameter I_{ph} into the contrast analytical treatment. Although detailed structural information must be obtained so as to yield a functional relation between defect parameters and experimental results, at this time it is interesting to point out the close correspondence between the results of the present work and the theoretical description of the recombination process, given by Wilshaw et al (1988), that defines three regimes of behavior whose trends are similar to those reported in Figures 3 and 4.

4. CONCLUSIONS

The preliminary results described above indicate that:

(i) it is possible to deduce the position of the energy levels associated with an individual defect by monitoring the variation of the contrast as a function of the quenching photon energy,

(ii) the variation of Q-IRBIC contrast with temperature shows different behavior depending on the quenching photon energy, that is on the dislocation level affected by the quenching light,

(iii) the contrast diagrams exhibit a close similarity to the theoretical trend yielded by Wilshaw's model.

Thus, although these results are preliminary, the Q-IRBIC mode promises to provide a means for a detailed study of the recombination processes at defects in semiconducting materials.

5. REFERENCES

Cavallini A, Gondi P and Castaldini A 1986 *Proc. 5th Int.Symp.*
    Structure and Properties of Dislocations in Semiconductors,
    Izv. Akad. Nauk.(Moscow) pp 652-660
Castaldini A and Cavallini A 1988a *Proc.of SPIE Scanning Microscopy*
    Technology and Applications ed E C Teague (Los Angeles) pp 55-62
Castaldini A and Cavallini A 1988b *Workshop: Point, Extended and Surface*
    Defects in Semiconductors ed Benedek G, Cavallini A, Schrter W
    (Erice, Italy) (Plenum Press)in press
Davidson S M and Dimitriadis C A 1980 *J. Micros.*.118 275
Donolato C 1981 *Proc.of Microsc. Semicond. Mater.* (Oxford) **60** pp 215-213
Donolato C 1983 *J. de Physique* **C4** 269
Kimerling L C and Patel J R 1979 *Appl.Phys.Lett.* **34**(1) 73
Milnes A G 1973 *Deep Impurities in Semiconductors* (New York: Wiley & Sons)
Ono H and Sumino K 1985 *Dislocations in Solids* ed Suzuki H, Ninomiya T,
Ourmazd A, Wilshaw P R and Booker G R 1983 *Physica* **116B** 600
Sumino K and Taheuchi S (Univ. of Tokyo) pp 365-368
Wilshaw P R and Booker G R 1985 *Proc.of Microsc. Semicond. Mater.* ed Cullis A G
    and Holt D B (Oxford)**76** pp 329-336
Wilshaw P R, Fell T S and Booker G R 1988 *Workshop: Point, Extended and*
    Surface Defects in Semiconductors ed Benedek G, Cavallini A, Schroter W
    (Erice Italy) (Plenum Press) in press
Wilson T, Osicki W R, Gannaway J N and Booker G R 1979 J. Mater. Sci.
    14 961
Wilson T and McCabe E M 1986 *J.Appl.Phys.***58** 2638
Wilson T and McCabe E M 1987 *J.Appl.Phys.***61**(1) 191

Inst. Phys. Conf. Ser. No 104: Chapter 2
Paper presented at Int. Symp. on Struct. Prop. Disloc. Semicond., Oxford, 1989

175

Effect of He in dislocation pipes on photoconductivity in Ge and Si

B Pohoryles

Instytut Fizki PAN, Al. Lotnikow 32, 02-668 Warsaw, Poland

ABSTRACT: A survey of the photoconductivity (PC) results in plastically deformed Ge is given. The PC dependence on He pressure in plastically deformed Si is then deduced from the model of He-driven reconstruction. Preliminary results for Si coincide with predictions of the model, thus suggesting that dangling bonds exist also within the dislocation cores in Si. These results also indicate that the effect of He pressure on PC is characteristic of the dislocations and not of Ge itself.

1. INTRODUCTION

The problem of the microscopic structures of dislocation cores in Ge and Si remains, despite considerable effort, unresolved. In the past 10 years, discussion was, in particular, focused on that part of this problem which concerns the existence of dangling bonds within the core. The weight of theoretical evidence is now in favour of reconstructed 30° and 90° partials. Reconstruction is possible by bonding along the core for the 30° partials and across the core for the 90° partials.

A number of groups carried out experiments, trying to answer the question of whether dangling bonds exist within the dislocation cores. However interpretation of the results is not straightforward since deformation also generates point defects that are likely to mask the effects searched for.

In order to resolve this problem, an experiment has recently been proposed, in which an attempt is made to change the core structure by He atom diffusion in the dislocation pipes, and to observe this change with the help of photoconductivity (PC) experiments (Pohoryles 1986).

Both DLTS and EPR spectra of plastically deformed Si are generally related to the point defects generated by the deformation procedure (Omling et al 1985, Kisielowski-Kemmerich and Alexander 1988). These spectra are very sensitive to the deformation conditions (Weber and Alexander 1983). In constrastwith the DLTS and EPR results, recent results on the temperature dependence of PC show that the slope of the PC vs T curve does not depend on deformation conditions (Pohoryles, to be published), thereby giving a convincing argument to support the view that the recombination centers examined by the PC experiments are related to the dislocation core.

Results of the PC experiments carried out under high He pressure conditions suggest that dangling bonds exist within dislocation cores in Ge and Si.

2. THE RELATIONSHIP OF THE PC TO DISLOCATION ELECTRON STATES

Theoretical calculations show that the one-dimensional band or the two
overlapping bands, created by the dangling bonds of a non-reconstructed
30° or 90° partial dislocation in the band gap of Ge and Si, split, after
reconstruction, into two bands which lie outside the band gap (Veth and
Teichler 1984). Depending on the distance s between the core atoms that
couple upon reconstruction, the energy position of these bands should
vary in the range given in the above-mentioned calculations, and crudely
outlined in Figure 1. Discrepancies concerning the width and precise
energy position

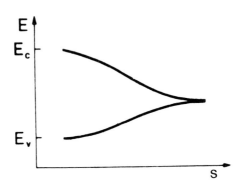

Figure 1. Energy positions of the dislocation bands vs the distance **s**
between the core atoms which couple upon reconstruction.

of the dislocation bands may exist, but there is general agreement that
for both partials the dependence of the dislocation bands' position on s
looks like that shown in Figure 1.

For sufficient concentration of free carriers, these fill one dislocation
band completely and partly fill the other one, thus pinning the Fermi level
to the latter band. Under these circumstances, Figure 1 may be regarded as
a schematic diagram of the dependence of the Fermi level E_F on **s**, with the
upper branch representing this dependence in an n-type material and the
lower branch in a p-type material.

An exponential dependence of the PC on $E_{CF} = E_C - E_F$ (or on $E_{FV} = E_F - E_V$,
for a p-type material) at a dislocation can be expected from the model of
recombination via dislocations given by Figielski (1978).

3. PC vs He PRESSURE MEASUREMENTS

Details of the experimental procedure, sample preparation and the results
of measurements of the intrinsic PC vs the pressure p exerted by He, Ar
and N on both n- and p- Ge, are given elsewhere (Pohoryles and Jung 1984).

One can qualitatively outline all the results obtained in one figure. This
is done in Figure 2, in which the pressure axis has been rescaled for each
set of data to allow easy comparison. The dashed line represents the

characteristics obtained under Ar or N_2 pressure. The solid line stands for the results obtained under He pressure. It is worth noting that, apart from differences in slopes, one finds that the results for undeformed Ge are, regardless of the gas used as a pressure transmitting medium, qualitatively similar to those obtained on deformed Ge under Ar or N_2 pressure.

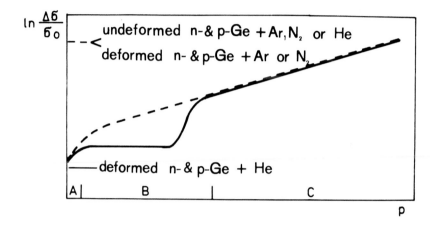

Figure 2. Results of the PC vs pressure (p) measurements on Ge. $\Delta\sigma$ – change of conductivity under illumination; σ_0 – dark conductivity of the sample.

The experimental results show unambiguously that, unlike the other two gases, He penetrates into Ge via dislocations.

On the basis of this experimental conclusion a phenomenological model of He-driven reconstruction was proposed (Pohoryles 1988) in order to explain the peculiarities observed. In this model, it was assumed that the dislocation cores are unreconstructed and that the He atoms interact with core atoms in such a fashion as to promote reconstruction. In other words, subsequent He atoms are built in those sites of a dislocation pipe from which the distance to the nearest Ge atoms is the greatest, causing further confinement of the already squeezed core atoms. Thus, subsequent He atoms decrease the distance s and thereby E_{CF} (E_{FV}) at the dislocation (see Figure 1). A decrease of the E_{CF} (E_{FV}) caused by the He atoms is compensated for by an increase of the E_{CF} (E_{FV}), caused by the external pressure – both E_{CF} and E_{FV} pressure coefficients are positive (Pohoryles et al 1980). In consequence, a distinct cut appears in the PC(p) curve in the pressure range B in Figure 2. At the beginning of pressure range C in Figure 2, the distance between some neighbouring dangling bonds becomes sufficiently small for them to bind with each other. The reconstruction proceeds along a dislocation line over a distance that depends on the He concentration in a dislocation pipe. Finally, the dislocation breaks up into two subsystems:

1. Reconstructed segments containing He atoms;
2. He-free, non-reconstructed segments, with interatomic distances in their cores similar to those they would have under the same pressure of a gas which does not penetrate into Ge.

The second subsystem determines the magnitude of the PC signal, which thus comes back to the value it would have had under appropriate Ar or N_2 pressure (Pohoryles 1986).

If we assume that reconstructed dislocation segments impede the diffusion of He atoms into Ge, then, in the pressure range C in Figure 2, only external pressure affects the E_{CF} (E_{FV}), resulting, due to the linear pressure dependence of E_{CF} (E_{FV}), in an exponential increase of the PC signal with pressure.

As far as Si is concerned, most EPR experiments on deformed Si support the hypothesis that the 30° and 90° partials are reconstructed.

In view of the results obtained for Ge, the problem of dangling bonds in plastically deformed Si again becomes of interest.

We now proceed to use the model of He-driven reconstruction to examine the character of the PC dependence on the He pressure applied to Si. We would expect to find negative pressure coefficients of both E_{CF} and E_{FV}. The actual findings are presented in Figure 3. The dashed straight lines represent the PC(p) dependence resulting just from the dependence of E_{CF} (E_{FV}) on external pressure. The different lines correspond to the different values of the E_{CF} (E_{FV}) resulting from the different atomic configurations of a dislocation core, i.e. different values of **s**. It seems reasonable to assume that the effect of He atoms on **s** is similar in Ge and Si. If so, the different lines correspond to the different He contents in the dislocation pipes – the higher the He concentration N_i, the smaller the **s**, and thus, according to Figure 1, the smaller the E_{CF} (E_{FV}) and the PC signal.

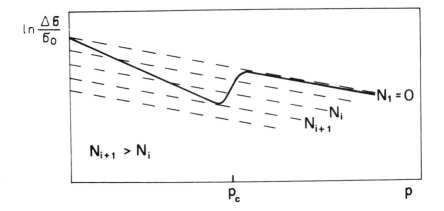

Figure 3. PC vs He pressure dependence expected for Si on the basis of the model of He-driven reconstruction.

As the He content in the dislocation pipes increases with He pressure, the PC signal drops to the value determined by the lower straight line, corresponding to a new value of N_i, until a critical pressure p_c is reached at which reconstruction occurs. After reconstruction, the PC(p) curve follows the straight line for $N_i = 0$, as was the case for Ge. The dependence of the PC on He pressure, taking into account both the effect of external pressure and the effect of He atoms in the dislocation pipes on the PC, is drawn with the solid line.

Preliminary results, obtained on plastically deformed n-Si, coincide with the predictions of the model of He-driven reconstruction. The PC vs He pressure measurements were performed at room temperature. For the samples measured, $p_c \simeq$ 5kbar. Measurements of the PC vs the pressure exerted on Si samples by various gases are under way.

4. FINAL REMARKS

The phenomenon of the influence of He on the PC of plastically deformed Ge turned out to be more general. It could also be observed in plastically deformed n-Si, indicating that the influence of He pressure on the PC is a characteristic of dislocations in Si or Ge, and not that of Ge itself. The model of He-driven reconstruction describes properly the PC vs He pressure characteristics in both materials. This suggests that dangling bonds also exist within the dislocation cores in Si.

Bearing in mind the results of the EPR experiments performed by the group of Prof. Alexander, we must reflect on the possible reason, intrinsic to the dislocation core, whereby the dangling bonds in Si do not generate an EPR signal. If negative U is the reason for this, it might be possible to reveal an EPR signal using background illumination with hν=0.2eV. According to Ossipyan (1983), this is the energy gap between the donor and acceptor dislocation bands.

ACKNOWLEDGEMENT

I am indebted to Professor T Figielski for valuable discussions.

REFERENCES

Figielski T 1978 Sol. St. Electron. 21 1403
Kisielowski-Kemmerich C and Alexander H 1988 Proc. 8th Int. School on
    Defects in Crystals, Szczyrk, Poland, 22-29 May, 1988 ed E Mizera
    (World Scientific) pp 290-305
Omling P, Weber E R, Montelius L, Alexander H and Michel J 1985 Phys.
    Rev. B 32 6571
Ossipyan Yu A 1983 J. de Physique C4-103
Pohoryles B 1986 Mat. Sci. Forum 10-12 815
Pohoryles B 1988 Phil. Mag. Lett. 58 1
Pohoryles B and Jung J 1984 J. Phys. Chem. Solids 45 671
Pohoryles B, Jung J and Figielski T 1980 phys. stat. sol.(b) 100 K87
Veth H and Teichler J 1984 Phil. Mag. B 49 371
Weber E R and Alexander H 1983 J. de Physique C4-319

Inst. Phys. Conf. Ser. No 104: Chapter 2
Paper presented at Int. Symp. on Struct. Prop. Disloc. Semicond., Oxford, 1989

181

Influence of non-stoichiometric melts on the defect structure of n-type bulk GaAs crystals

R Gleichmann, A Höpner, R Fornari[+] and C Frigeri[+]

Academy of Sciences of GDR, IFE, Weinberg 2, Halle 4050, GDR
[+]CNR MASPEC, Via Chiavari 18/A, 43100 Parma, Italy

ABSTRACT: LEC GaAs:Si single crystals were grown from non-stoichiometric melts with Ga/As ratios varying between 0.93 and 1.35. Considerable changes in properties (e.g. dislocation density, mobility, EL 2 content) were detected by TEM, selective etching and electrical measurements. The extended defects preferentially studied (dislocations polycrystalline particles, dislocation loops) were shown to reduce their density and/or size under Ga-rich conditions, with As_i being the dominating point defect. The loops were found to generate dislocations via point defect condensation by dipole formation.

1. INTRODUCTION

With the growing demands to the substrate materials for various semiconductor applications there is a permanent interest in improving the growth conditions. This also holds for GaAs, where a thorough understanding of the complex point defect mechanisms is necessary to enable a suppression of disturbing defects. Recently undoped semi-insulating GaAs crystals suitable for the IC production and ion implantation were grown from As-rich melts (Holmes et al 1982), stimulating further investigations of crystals grown from non-stoichiometric melts (see e.g. Miyazawa et al 1982, Katsumata et al 1986, Giling et al 1986, Brozel et al 1987, Fornari et al 1989). In addition to the possibility of influencing important properties in such a way, the non-stoichiometric growth allows a deeper insight into the point defect mechanisms. The aim of the present paper is the study of the role of melt stoichiometry for the structural properties of LEC GaAs, doped with silicon (amphoteric), which is expected to be seriously affected by changes in the overall point defect distribution.

2. EXPERIMENTAL

The <111> oriented crystals were all grown under the same conditions from silica crucibles (water content of the boron oxide < 250 ppm). The ingots of 35 mm in diameter had usually a weight of about 300 g. To realise Ga- or As-rich conditions an excess of known weight was added to the starting material, causing a continuously increasing deviation from the stoichiometry as the weight of the melt was reduced. The Ga/As ratio influencing the corresponding part of the crystal was calculated subsequently taking into account the almost unchanged stoichiometry of the crystal. The realised Ga/As ratio R varied between 0.93 and 1.35. By adding pure silicon to the starting charge the doping level was adjusted to be between 1.10^{16} and 2.10^{18} cm^{-3}, well below the Si solubility limit.

The crystals were sliced, polished and characterised structurally (KOH etching, TEM, SEM/EBIC) as well as electrically (Hall effect, DLTS). TEM samples were prepared by cleaving rectangular sections (up to 6x6 mm^2), which were then chemically etched with aqua regia at 60°C. A high-voltage electron microscope operating at 1000 kV and equipped with a 45° double-tilt goniometer stage was used for the diffraction contrast analysis. In order to obtain reliable results, up to eight specimens of each wafer were investigated.

3. RESULTS

The principal type of defects were grown-in dislocations with Burgers vectors 1/2<110> (mainly 60° type). In the present paper, mainly referring to results obtained by TEM, precise systematic studies were made of three different Ga-rich crystals, whereas various other crystals were investigated using random samples. The dislocation density was determined by TEM to be 2 to 4 times higher than the results obtained by selective etching. This might be ascribed to the problems inherent in such TEM measurements around 10^4 cm^{-2}, but also to the fact that solely areas closer to the rim of the wafers were left for the TEM analysis. Furthermore, each individual dislocation was counted, including those in dislocation dipoles, representing a considerable fraction of all dislocations observed. However, there is no discrepancy as to the relative density changes along the crystal axis.

The most interesting type of defects - being very sensitive to the melt stoichiometry - turned out to be the well-known precipitates decorating the grown-in dislocations. In undoped stoichiometric material the density of the particles was around 5.10^8 cm^{-3}, agreeing well with data in the literature (see e.g. Cornier et al 1984). The size of these defects is around 120 nm. Diffraction contrast analysis as well as selected area diffraction (SAD) proved several particles to be poly-crystalline with few grains per particle. A typical example is shown in

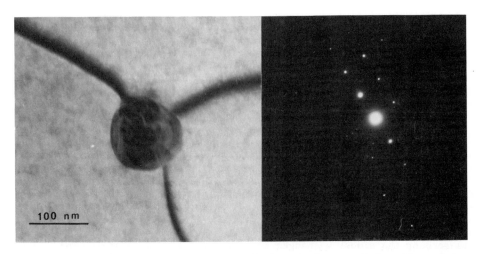

Fig. 1: As-rich particle with two grain boundaries at dislocations in GaAs; typical SAD pattern (right)

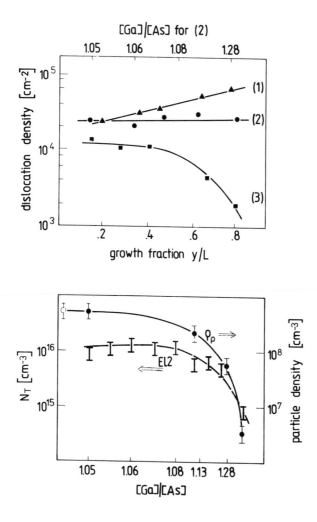

Fig. 2: Variation of the dislocation density along the crystal axis for three crystals (see text), and change of the precipitate density and EL 2 concentration for crystal (2).

fig. 1 with two grain boundaries being distinctly imaged. Applying SAD to several particles revealed more than 20 different lattice plane distances. Some of them might be spurious (polycrystal) and only few reflections coincided with those of the presently accepted analysis as hexagonal arsenic (Cullis et al 1980, Booker 1985). This suggests the presence of other arsenic phases of low-atomic-number elements (e.g. oxygen or boron, not detectable by EDX analysis). Nevertheless, arsenic represents the main constituent. While for undoped material and R lying between 0.98 and 1.1 no noticeable changes were detectable by TEM a further shift to Ga-rich melts drastically reduced both the density and

size of the precipitates.   Some quantitative data are shown in fig. 2
(bottom) for the particle density of crystal "2" (Ga-rich, n < 5.10^{16} cm^{-3}),
which are compared with the EL2 concentration determined by DLTS (Fornari
et al 1986).   Examples of the size change of the precipitates are given in
fig. 3.   The dislocation density determined by selective etching is shown
in fig. 2 (top) as a function of the growth fraction y/L related to R
(L - crystal length) for three selected crystals; (1) As-rich,
n $\approx$ 5.10^{17} cm^{-3}; (3) Ga-rich, n $\approx$ 10^{18} cm^{-3}.   Considerable effects of the
silicon content and the melt stoichiometry are observable.   Under Ga-rich
conditions the crystals reacted most sensitively to an increase of the
silicon content: for R = 1.004 and n = 2.10^{17} cm^{-3} the particle density
is reduced by one order of magnitude, and with R = 1.014 (n = 3.10^{17} cm^{-3})
no precipitates can be detected any longer, unlike As-rich conditions.
The varied growth conditions strongly affect also the electrical parameters

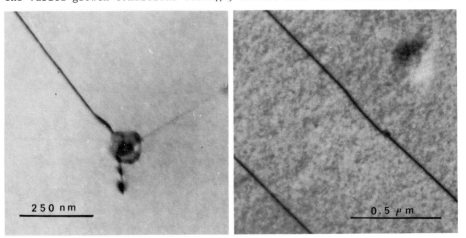

Fig. 3:   Decrease in the size of precipitates under Ga-rich conditions;
stoichiometric crystal (left), R = 1.35 (right)

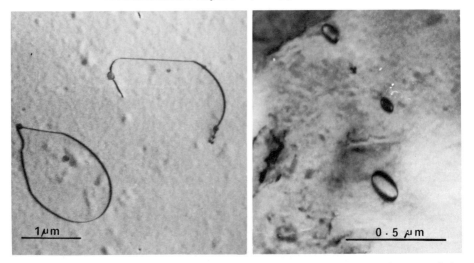

Fig. 4:   Reduction in the size of extrinsic loops under Ga-rich conditions
(left: R $\approx$ 1, right: R = 1.35)

such as the room temperature mobility [≈ 3200 (1), ≈ 2800 (2) and
≈ 1800 cm^2/Vs (3)] or the compensation ratio N_a/N_d [≈ 0.3 (1), > 0.7 (2)
and ≈ 0.5 (3)].

The third type of extended defects observed were perfect extrinsic loops
having a density in the order of 10^8 cm^{-3}. Their size ranges between
about 1 and 3 μm for nearly stoichiometric crystals and reduces to 0.1 –
0.7 μm under highly Ga-rich conditions as demonstrated in fig. 4. In
crystals with a higher Si content (> 10^{17} cm^{-3}) no individual loops were
observed. The loops are necessarily formed by interstitials of As and Ga,
where only one species should be the driving one due to its supersaturation.
The suppression of loops for R > 1 shows As$_i$ to be the initiating species,
where the process should be restricted to the crystal part close to the
melt. At lower temperatures the supersaturation is supposed to reduce by
another mechanism via gettering dislocations resulting in the described
As-rich particles (see e.g. Gleichmann et al 1987). The occurrence of
particles at some of the loops supports this view (fig. 4).

A considerable fraction of the dislocations in the crystals studied were
dipoles. In some cases it was possible to observe the origin of the
dipole formation, this being the dislocation loops described above. Two
examples are shown in fig. 5. Close to the seed (y/L < 0.2) the fraction
of dipoles in crystal (3) was about 15%, while at the tail of the crystal
this fraction increased to nearly 60% though their density was decreasing.
Thus, the dipole mechanism seems to be a limiting factor in the growth of
dislocation-free "pure" GaAs.

The dipoles are obviously generated once the climb of the loops is faster
than the crystal growth. As soon as they reach the liquid-solid interface
they can grow further without the necessity of the point defect absorption.

4. CONCLUSIONS

The melt stoichiometry R was shown to strongly influence many important
parameters of n-doped LEC GaAs crystals, which should be taken into

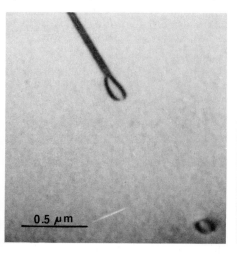
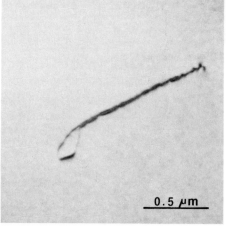

0.5 μm   0.5 μm

Fig. 5: Examples of dislocations generating at loops

account in any tailoring of materials. All the changes observed seem to be related to one dominating point defect, i.e. arsenic interstitials. As_i are present in all crystals (either Ga-rich or As-rich) in a high concentration, which is reduced, but not completely suppressed, even by a considerable Ga excess of the melt. The precipitates, EL 2 and the extrinsic loops reduce their size and/or density with an increasing ratio R, which clearly proves As_i to be one of the main constituents of all these defects. This is rather obvious for the As-rich particles; it supports the assumed concept of EL 2 intensely discussed at present (see e.g. von Bardeleben and Stievenard 1988) and was shown here for the dislocation loops. The latter are responsible for the dislocation density once the generation due to thermal stresses makes a small contribution. It explains the constant (not increasing) dislocation density along the crystal axis in the undoped crystal (2) and the considerable reduction in the density of Si-doped crystals such as (3). Under Ga-rich conditions the changes considered in the material essentially increase with silicon ($n > 10^{17}$ cm^{-3}) being added. This suggests the formation of $Si-As_i$ complexes or silicon/antisite complexes, which will strongly affect electrical parameters such as mobility or compensation ratio.

REFERENCES

von Bardeleben H J and Stievenard D 1988 MRS Symp. Proc. vol. 104 351
Booker G R 1985 Inst. Phys. Conf. Ser. 76 201
Brozel M R, Foulkes E J, Grant I R and Hurle D T J 1987 J. Cryst. Growth
    80 323
Cornier J P, Duseaux M and Chevalier J P 1984 Appl. Phys. Letters 45 1105
Cullis A G, Augustus P D and Stirland D J 1980 J. Appl. Phys. 51 2556
Fornari R, Gombia E and Mosca R 1986 Ital. Conf. Solid State Phys., Genoa
    1986, to be publ. J. Electr. Mat. 18 (1989)
Fornari R, Frigeri C and Gleichmann R 1989 J. Electr. Mat. 18 185
Giling L J, Weyher J L, Montree A, Fornari R and Zanotti L 1986 J. Cryst.
    Growth 79 271
Gleichmann R, Menniger H and Raidt H 1987 Cryst. Prop. Prep. (Switzerland)
    12 17
Holmes D E, Chen R T, Elliott K R and KirkpatrickC G 1982 Appl. Phys.
    Letters 30 46
Katsumata T, Okada H, Kimura T and Fukuda T 1986 J. Appl. Phys. 60 3105
Miyazawa S, Mizutani T and Yamazaki H 1982 Jpn. J. Appl. Phys. 21 L542

Inst. Phys. Conf. Ser. No 104: Chapter 2
Paper presented at Int. Symp. on Struct. Prop. Disloc. Semicond., Oxford, 1989

187

LCAO analysis of dislocation-related EPR spectra in deformed silicon

C. Kisielowski-Kemmerich

Abt. f. Metallphysik, Universität Köln, Zülpicher Str. 77,
D-5000 Köln 41

ABSTRACT: A LCAO analysis of reconstruction defects, threefold coordinated vacancies (V_{3c}) and their complexes in terms of one electron
"defect molecules", leads to microscopic models for the dislocation
related EPR-centers Si-K1, Si-K2 and Si-Y. It is proposed to identify
Si-Y with V_{3c}, Si-K1 with V_{3c} trapped at a reconstruction defect and
Si-K2 with multi-vacancy complexes ($n*V_{3c}$). The results suggest that
impurities are involved in the defect formation and that they are
preferably formed in the core of those 30^0 partial dislocations which
are part of screws.

1. INTRODUCTION

Si-K1, -K2 and -Y are different paramagnetic defects detectable by
Electron Paramagnetic Resonance (EPR) in high resistivity FZ-silicon
crystals which were subject to a deformation below $\approx 700^0$C. Their
properties have recently been reviewed and it has been stressed that all
three defects are influenced by a particular crystal field of orthorhombic I symmetry (Kisielowski-Kemmerich and Alexander 1988). Since one of
the symmetry axes coincides with the line direction of screw dislocations,
the results also suggest that the defects are preferably formed on screw
dislocations. In this paper topological models of defects in the core of
30^0 partial dislocations (PD) are presented which can explain the EPR-data
of these defects. The approach is stimulated by the experience that the
electronic structure of vacancies and vacancy related defects in irradiated Si often can be approximated by a linear combination of the atomic
orbitals (LCAO) of the ligand atoms (Watkins 1975).

2. MODEL

Figures 1, 2 show topological models of a reconstruction defect (RD) and a
V_{3c} vacancy. The main slip system for a deformation along [213] is
$(1\bar{1}1),[011]$ and this leads to the geometry of the "defect molecules" shown
in fig.3. The structures will probably relax to some unknown extent and -
in this first approach - a relaxation of the direction AD towards [100]
with equal bond angles is assumed, as shown in fig. 3b. Both defects are
close to trigonal symmetry with their threefold rotation axes parallel to
$\approx[011]$! This uncommon result (in Si the C_3 axes are parallel to $\langle 111\rangle$)
comes from the fact that atom D does not occupy an ordinary lattice site.
It is part of the stacking fault (SF) in contrast to atoms C,E.

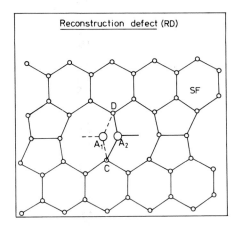

Fig.1: Atomic positions of a reconstructed 30°PD in the glide plane
(1$\bar{1}$1). A₁ , A₂= two possible configurations of a RD.

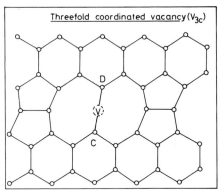

Fig.2:   Vacancy substituting atom A
in the core of the 30°PD.   (1$\bar{1}$1)
plane.

Application of elementary group
theory to the RD in C_{3v} symmetry
gives the molecular orbitals shown
in fig.4. In the picture it is
assumed that the energy of the $2a_1$
orbital $E(2a_1)$ is larger than $E(e)$.
This depends on the hybridisation
energies of the atoms A,C,D,E ($\Phi_i =$
$a_i s + \beta_i p$; i=A,C,D,E; $a_i^2, \beta_i^2 =$
hyperfine parameters of the s,p
wavefunctions on atom i). $\Phi_{Aj} = \Phi_A +$
Φ_J are the orbitals which bind atom
A to atoms J=C,D,E. The defect is
paramagnetic if one electron occupies the $2a_1$ level. It exists in the
two orientations A₁,A₂ (fig.1) which
are not distinguishable in an EPR
experiment.

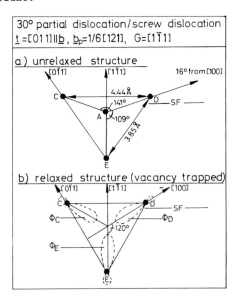

Fig.3: "Defect molecules": a)RD,
b) V_{3c} vacancy in the "relaxed"
configuration. ≈(011) plane.

Treating the V_{3c} vacancy in its neutral and paramagnetic charge state
in C_{3v} symmetry (fig.5), involves
the occupation of the electronically
degenerate e-level with one electron. In this case a Jahn-Teller
distortion is expected which lowers the symmetry of the defect ($C_{3v}-$ C_{2v})
as shown in fig.5. It might also be expected that strain in the core of
the PD itself leads to an arrangement of the atoms with lower than C_{3v}
symmetry. From the resulting geometry of the defect the following
assumptions are resonable for a LCAO estimation:
1) $\theta < 120°$: $S_{CD}= S_{ED}= 0$, $S_{CE}= S$; $E_{CD}= E_{ED}= 0$, $E_{CE}= E_1$
2) $\theta > 120°$: $S_{CD}= S_{ED}= S$, $S_{CE}= 0$; $E_{CD}= E_{ED}= E_1$, $E_{CE}= 0$
with the bond- and Coulomb integrals: $S_{ij} = \langle \Phi_i | \Phi_j \rangle = S$, $S_{ii}=1$; $E_{ij}=$
$\langle \Phi_i | H | \Phi_j \rangle = E_1$, $E_{ii}= E_0$ (i,j = labeling atoms C,D,E). It gives the order of
the energy levels and the molecular orbitals shown in fig.5. In case of
$\theta< 120°$ the resonance electron occupies a pure dangling bond orbital $\Phi_2 \approx \Phi_D$

located at atom D, in contrast to
$\Theta > 120°$ where it shares atoms C,E in
an antibonding orbital $\Phi_2' \approx \Phi_C - \Phi_E$.
There is an additional threefold
<u>orientational</u> degeneracy (I,II,III)
for the two <u>energetical</u> different
configurations (1,2). It is remov-
able in an EPR experiment and shown
in fig.6. As long as atoms C,D,E are
equivalent, the defects exist in
three orientations (1I, 1II, 1III or
2I, 2II, 2III). Distinguishing atom
D (SF) from atoms C and E, predicts
that the defects should either occur
in single orientations (1I or 2I) or
in pairs (1II, 1III or 2II, 2III).

A fourfold coordinated vacancy (V_{4c})
can be modeled by attaching a RD of
type A_1 or A_2 to the V_{3c} vacancy .

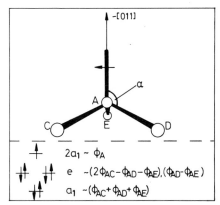

Fig.4: Term scheme of a RD in C_{3v}
symmetry. Paramagnetism requires
occupation of the $2a_1$-level with
one electron.

Fig.5: Term schemes of a
V_{3c} vacancy. For $\Theta=120°$
the upper (e) level is
occupied with one elec-
tron. A symmetry lowering
Jahn-Teller distortion is
expected, generating the
configurations:
1) $\Theta<120°$, 2) $\Theta>120°$.

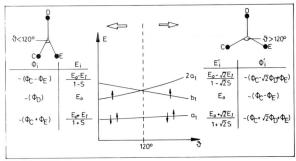

Both possibilities are equivalent
and fig.7 shows the $V_{3c} + A_1$ constel-
lation. Energetically, it requires
the addition of the Φ_A orbital of
fig.4 to the molecular orbitals
listed in fig.5. It is shown in
fig.8 that, at least, it is necessary to distinguish case a) from
cases b) and b') which are the
extreme possibilities. Symmetry
arguments cannot longer be used for
the construction of the molecular
orbitals because of the low symmetry
of the defect (triclinic). But compared with irradiation defects in
Si, case a) is typical for vacancy
and vacancy-oxygen related defects
while cases b) and b') involve impurities (B,P,As,Sb) which substitute
one Si-atom in the nearest- or next
nearest neighbour position (Sieverts
1983 and references therein). The
shift of one energy level associated
with the impurity (Φ_A) with respect

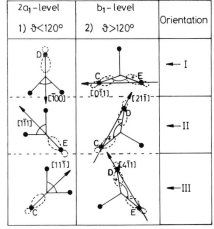

Fig.6: Three possible orientations
(I,II,III) for the two energeti-
cally different configurations 1,2.
Directions are indicated where the
EPR g-value is supposed to be: $g \approx$
g_{free}

to the vacancy levels originates from the different hybridisation energies and from the charge states of the impurities resulting in lone pair- and dangling bond orbital formation. In fig.8 the V_{3c} vacancy levels associated with $\theta < 120^\circ$ have been chosen with a view to the experimental results. For $\theta > 120^\circ$ there are always bonding and antibonding orbitals involved in the binding of atoms C,D and E (fig.5) which excludes the existence of a dangling bond orbital on one of the three Si-atoms.

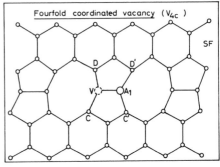

Fig.7: V_{4c} vacancy modeled by: V_{3c}+ RD (A_1 type). Exchange of V and A_1 generates V_{3c}+ A_2. They are equivalent.

Fig.8: Extreme cases for addition of the Φ_A orbital (RD) to the V_{3c} vacancy levels. Only cases b,b' allow for paramagnetism in the dangling bond orbital Φ_D as shown in case b', only.

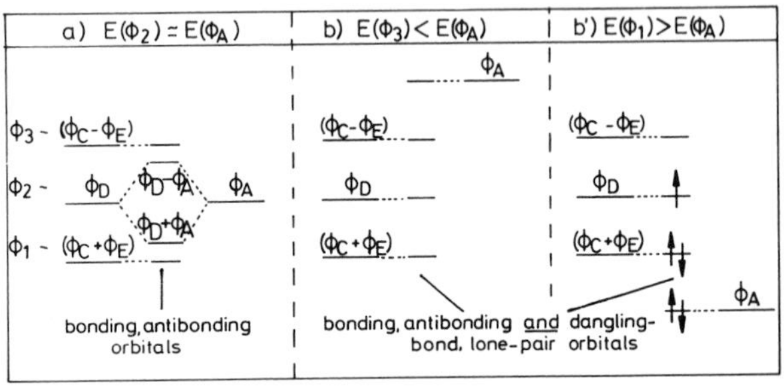

a) $E(\Phi_2) = E(\Phi_A)$	b) $E(\Phi_3) < E(\Phi_A)$	b') $E(\Phi_1) > E(\Phi_A)$
		ϕ_A
$\phi_3 \sim (\phi_C-\phi_E)$	$(\phi_C-\phi_E)$	$(\phi_C - \phi_E)$
$\phi_2 \quad \phi_D \quad \phi_A$	ϕ_D	ϕ_D
$\phi_1 \sim (\phi_C+\phi_E)$	$(\phi_C+\phi_E)$	$(\phi_C+\phi_E)$
		ϕ_A
bonding,antibonding orbitals	bonding, antibonding and dangling-bond, lone-pair orbitals	

$n*V_{3c}$ vacancy-complexes (fig.9 ,n=3) are of orthorhombic I symmetry, too. They show the same two mirror planes and the same C_2 rotation axis as a single V_{3c} vacancy does. Irrespective of the two posible energetic constellations of a single V_{3c} vacancy (1,2), the dangling bonds or the antibonding orbitals Φ_2 or Φ_2' (fig.5) are spaced parallel to each other. Thus, there is little overlap among them. In the most simple picture this causes a multiplet ground state with spin $S > 1/2$ (fig.9) where $\theta > 120^\circ$ is chosen, which accounts for the experiments.

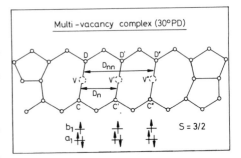

Fig.9: Topological model and term scheme of a $n*V_{3c}$ complex with n=3. Neglect of overlap of the b_1 orbitals leads to a multiplet ground state with spin S= n * 1/2.

3. DISCUSSION

Modeling of defects within the frame of a simple one electron LCAO "defect molecule" picture opens up the possibility for direct comparison of the EPR-data (g-tensor, fine- and hyperfine structure) with the model (Watkins 1975). A detailed description of the spectroscopic parameters with respect to the model which includes the bistable transition Si-K1 - Si-K2 (Erdmann and Alexander 1979) and stress experiments on Si-K1 (Jung 1981) goes beyond the space of this paper and will be given elsewhere (Kisielowski-

Kemmerich 1989b). But the results can be summarized as follows:

Fig.10: g-values as a function of rotation in the (011) plane. Angle (α): magnetic field $\underline{H}$ and [1̄1̄1̄]. K1/L1, K1/L2 correspond to the two orientations of the defect with dangling bonds on atoms C,E, respectively. Y,Y' are two possible fits to the data point within exp. errors. Symmetry/spin: K1 (triclinic/ 1/2), Y (monoclinic I/ 1/2), Y'(orthorhombic I/ 1/2), K2 (orthorhombic I/ n*1/2, 1<n<6).

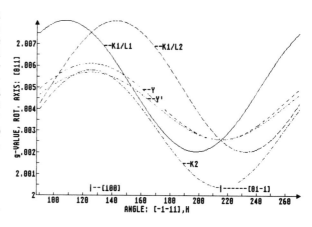

Fig. 10 shows the g-value variation of the considered defects , rotating the crystal around [011] (Burgers vector). The prominence of the orthorhombic I axes [100], [01̄1̄] and [011] is obvious. Both Si-K2 and Si-Y (or Y') show their minimum of g ($\approx$ g_{free}= 2.0023) along [01̄1̄]. Both defects exist in only one orientation. From the model these properties are expected for the V_{3c}vacancy in the configuration 2I (fig.6). It implies that atom D must be distinguished from atoms C,E. Si-Y is a spin S=1/2 defect and, therefore, is identified with a single V_{3c}vacancy. Si-K2 is a spin n*S center with 1< n <6 and is identified with a multi-V_{3c}vacancy-complex. At least, the Si-K2 20K-lines (Bartelsen 1977) show fine structure values (D_n and D_{nn} in fig.9) which indicate a separation of the interacting spins (dipol-dipol interaction) of one or two times the Burgers vector, respectively. These values are in resonable agreement with what is observed in case of multi-vacancy complexes in irradiated Si (Lee and Corbett 1976). But there are several series of S > 1/2 centers (Bartelsen 1977, Zolotukhin 1986). This is as expected because multi-vacancy complexes must have more variances than the single one mentioned here.

Si-K1 is triclinic. It exists in two orientations with paramagnetic electrons in dangling bond orbitals on Si-atoms C, E and, therefore, is identified with a V_{4c}vacancy. Its two defect orientations (dangling bond ‖ [1̄11] and ‖ [11̄1] , configuration 1II and 1III in fig.6) supports the point that atom D (SF) must be distinguished from atoms C and E. (It is noted that some build-in tensile strain along [01̄1̄] is an equivalent assumption to the distinction of atom D from atoms C, E. It also causes the three configurations 2I, 1II and 1III on the V_{3c}vacancy site.) The identification involves a shift of the Φ_A level with respect to the V_{3c}vacancy levels indicating the presence of an impurity atom (case b or b' in fig.8) which replaces atom A in the most simple case.

The similarity of Si-K1 and the impurity-vacancy complex is shown in table 1. It suggests the presence of an impurity with a spin-orbit coupling constant smaller or equal to that of silicon. From the concentration of Si-K1 ($\approx$ 2 10^{14}cm^{-3}) and the purity of the crystals, only the concentrations of C, O and perhaps also of H (Kisielowski-

Kemmerich and Beyer 1989a) are large enough ($\approx 5\ 10^{15}\mathrm{cm}^{-3}$) to create Si-K1. But it cannot be excluded that impurity complexes involving N and B are hidden below the line width of Si-K1. In this connection it is remarkable that the concentration of Si-Y $\approx 2\ 10^{15}\mathrm{cm}^{-3}$ is most likely too large to assume that impurities are <u>always</u> involved in the formation of this defect. It is, therefore, modeled as a pure V_{3c} vacancy. But it is known (Kisielowski-Kemmerich and Alexander 1988) that there are different contributions to that resonance.

Table 1:
Spectroscopical parameters of vacancies trapped at impurity atoms compared with Si-K1

DEFECT	g_1, g_2, g_3	HYPERFINE PARAMETERS α^2 β^2 n^2	SYMMETRY	REMARK
(PV)*	2.005, 2.0112, 2.0096	0.14, 0.86, 0.59	MONOCLINIC I	P replaces Si-atom
(AsV)*	1.9991, 2.0117, 2.0081	0.14, 0.86, 0.59	MONOCLINIC I	As rep. Si-atom
(SbV)*	1.9958, 2.0163, 2.0058	0.15, 0.85, 0.57	MONOCLINIC I	Sb rep. Si-atom
(BV)**	2.0011, 2.0052, 2.0090	0.13, 0.87, 0.55	TRICLINIC	B rep. Si-atom in second shell
Si-K1	2.0018, 2.0092, 2.0074	0.12, 0.88, 0.54	TRICLINIC	RD rep. Si-atom

*= Elkin and Watkins 1968, **= Watkins 1976, $g_{max}-g_{min}$ depends on spin-orbit coupling, n^2= fractional amount of wavefunction on Si-atom.

In this paper no explanation can be given concerning the mechanism which leads to the preferential formation of the defects on screw PD's.

In conclusion, the defects are attributed to vacancies and vacancy-impurity complexes which are formed by dislocation motion and impurity gettering during the deformation procedure.

ACKNOWLEDGEMENT

The author acknowledges detailed discussions with Prof. H. Alexander.

4. REFERENCES

Bartelsen L. 1977 phys. stat. sol. (b) <u>81</u> 471
Elkin E.L., Watkins G.D., 1968 Phys. Rev. <u>174</u> 881
Erdmann R., Alexander H. 1979 phys. stat. sol. (a) <u>55</u> 251
Jung C., Köln 1981 unpublished
Kisielowski-Kemmerich C., Alexander H. 1988 Defects in Crystals (Mizera E. ed., Singapore: World Scientific) p.290
Kisielowski-Kemmerich C., Beyer W. 1989a, J. Appl. Phys, in press
Kisielowski-Kemmerich C. 1989b, to be published
Lee Y.-H., Corbett J.W. 1976 Phys. Rev.B <u>13</u> 2653
Sieverts E.G. 1983 phys. stat. sol. (b) <u>120</u> 11
Watkins G.D. 1975 Point Defects in Semiconductors (Crawford J.H., Slifkin L.M. eds., New-York: Plenum) <u>2</u> p.333
Watkins G.D. 1976 Phys. Rev.B <u>13</u> 2511
Zolotukhin M.N., 1986 Sov. Phys. Sol. State <u>28</u> 1862

Inst. Phys. Conf. Ser. No 104: Chapter 2
Paper presented at Int. Symp. on Struct. Prop. Disloc. Semicond., Oxford, 1989

Lineshape of combined resonance on dislocations in semiconductors

A E Koshelev, V Ya Kravchenko and D E Khmel'nitskii

Institute of Solid State Physics, Academy of Sciences of the USSR, Chernogolovka, Moskow district 142432

ABSTRACT: The nature of the resonance phenomena at the Zeeman frequency under the action of rf electric field, recently discovered in plastically deformed Si, is studied theoretically. Observed peculiarities of absorption and dispersion signals are explained as the manifestation of polarization effects at conducting dislocation segments.

1. INTRODUCTION

The symmetry of arrangement of dislocation core atoms is lower than the ideal crystal local symmetry. In particular, the inversion centre is absent, therefore a crystalline electric field $\mathbf{F}$ of significant magnitude directed across the dislocation line $\mathbf{l}$ may arise and act on electrons in the core region. We suppose that a one-dimensional electronic band is associated with the dislocation. The motion of electrons along $\mathbf{l}$ at a velocity $\mathbf{v}$ gives rise, through a relativistic effect, to a magnetic field $\mathbf{h} = - [\mathbf{v} \times \mathbf{F}]/c$, which directly affects the spins. In an external field $\varepsilon(\omega)$, the velocity $\mathbf{v}$ acquires an rf component, proportional to $\varepsilon(\omega)$, so an rf part of $\mathbf{h}$ arises, which leads to transitions between Zeeman sublevels. This picture qualitatively describes the nature of so called combined resonance (CR, predicted by Rashba (1960)). The amplitude A of the CR possesses a specific orientational dependence in the one-dimensional case: $\Lambda \sim [\cos\phi \sin \theta]^2$ where ϕ is the angle between the vectors ε and $\mathbf{l}$, θ is the angle between external magnetic field $\mathbf{H}$ and $\mathbf{h}(\omega)$, i.e. between $\mathbf{H}$ and $\mathbf{l} \times \mathbf{F}$. The resonance effect at the paramagnetic frequency under the action of $\varepsilon(\omega)$ has been recently observed in plastically deformed Si and interpreted, by its angular dependence, as the CR at the dislocation electronic band (Kveder et al 1986). Some unusual peculiarities of the CR lineshape have been established: in the absorption case the line is similar to the usual dispersion curve and vice versa. In the present paper we examine the problem of the CR lineshape.

2. KINETIC EQUATION AND TRANSPORT EQUATIONS

Let us take the most simple form of the one-dimensional electronic Hamiltonian in the presence of crystalline field normal to the dislocation line:

$$\hat{\varepsilon}(p) = p^2/2m - V p \hat{\sigma}_x/2 - g \mu_b H \hat{\sigma}/2 \qquad (1)$$

where **p** is the electronic momentum parallel to **1**. V is a constant proportional to F. Let the z-axis be parallel to the line direction **1** and the y-axis to **F**. As can be shown, the spin density matrix $f_{\sigma\sigma}(p)$ satisfies the equation

$$\frac{\partial \hat{f}}{\partial t} + i[\hat{\epsilon},\hat{f}]_- + \frac{1}{2}[\frac{\partial \hat{\epsilon}}{\partial p},\frac{\partial \hat{f}}{\partial z}]_+ + eE_z\frac{\partial \hat{f}}{\partial p} = \hat{I}_p(\hat{f}) \qquad (2)$$

Here $[\ ,\ ]_{+\ -}$ is the commutator or anticommutator respectively. $\hat{I}$ is the collision integral. We consider the case when the collision length is much smaller than other characteristic lengths. Under such conditions, instead of kinetic equation, transport equations for densities and currents may be used. Presenting $\hat{f}$ in the form

$$\hat{f}(p) \approx \frac{1}{2}(f_0(p) + f(p).\sigma)$$

where f_0 coincides with a scalar distribution function and **f** corresponds spin polarization, we obtain for densities and currents respectively:

$$\rho_0 = \int (dp)\, f_0, \qquad\qquad M = \int (dp)\, f$$

$$J_0 = \int (dp)\, (\frac{p}{m} f_0 - \frac{V_f}{2} f_x) \qquad J = \int (dp)\, (\frac{p}{m} f - \frac{V}{2}\rho_0 n_x) \qquad (3)$$

Here n_x is the unit vector in x-direction. The used electron–phonon collision integrals acquire rather simple forms at $V(mT)^{1/2} \ll \Omega_0 \ll T$, where T is the temperature, $\Omega_0 = g\mu_B H$. As a result eq. (2) yields (Koshelev et al 1988):

$$\frac{\partial \rho_0}{\partial t} + \frac{\partial J_0}{\partial z} = 0, \qquad \frac{\partial M}{\partial t} + \frac{\partial J}{\partial z} + [\Omega_0 \times M] + mV[n_x \times J] + \frac{M}{T_1} = 0$$

$$(4)$$

$$\left(\frac{\partial}{\partial t} + \frac{1}{\tau}\right)J_0 + \frac{D}{\tau}\frac{\partial \rho_0}{\partial z} - \frac{V}{2}\frac{\partial J_x}{\partial t} + \frac{V}{2}\frac{\partial M_x}{\partial t} = \frac{eE}{m}\rho_0$$

$$\left(\frac{\partial}{\partial t} + \frac{1}{\tau}\right)J + [\Omega_0 \times J] + \frac{V}{2}\rho_0 [\Omega_0 \times n_x] + VT[n_x \times M] + \frac{D}{\tau}\frac{\partial M}{\partial z} - Vn_x\frac{\partial J_0}{\partial z} = \frac{eE}{m}M$$

Here τ is the electron–phonon transport relaxation time, T_1 is the spin-lattice relaxation time introduced phenomenogically (for simplicity it is taken that $T_1 = T_2$). $D = T\tau/m$ is the diffusion coefficient. Eqs. (4) contain the acting electric field E. Starting from the experimental situation, we will take into account the case when the alternation of conducting and non-conducting sections is realized along the dislocation line. At the segment ends the currents J_0 and **J** must vanish, therefore a redistribution of the charge density ρ_0 along the segment is obtained. As a result, screening of external field ϵ arises so that ϵ may differ from the acting field **E**. The solution of the Poisson equation, with the charge density $\rho_0(z)$ distributed along the segment $-L \le z \le L$ and

localized in cross directions over the length b << L, gives a relationship
between E, ε and ρ_0:

$$E = \mathcal{E}_z + \frac{e}{\epsilon_0} \int_{-L}^{L} \frac{dz' \, (z-z') \, \rho_0(z')}{[\, b^2 + (z-z')^2 \,]^{1/2}} \simeq \mathcal{E}_z - \frac{e}{\epsilon_0} \Lambda \frac{\partial \rho_0}{\partial z} \tag{5}$$

where $\Lambda = \ln(L/b)^2$, ε_0 is the static dielectric constant. Eqs. (4) and (5)
yield the expression for the conduction current J_0, which presented in the
usual form $J_0 = \sigma(\omega)\varepsilon_z/e$ gives the effective conductivity $\sigma(\omega)$, including
spin resonance parts. The effective dielectric susceptibility $\hat{\varepsilon}$,
responsible for absorption (Re$\hat{\varepsilon}$) and dispersion (Im$\hat{\varepsilon}$) has the form:

$$\epsilon_{zz} = \epsilon_0 + i \, n_d \frac{4\pi\sigma(\omega)}{\omega} \tag{6}$$

where n_d is the total length of conducting dislocation segments per unit
volume.

3. LINESHAPE OF THE CR

The solution of (4) and (5) shows that the z-dependence of densities and
currents manifests itsef at the characteristic lengths:

$$L_D = [(D + e\mu\Lambda N\epsilon_0)/\omega]^{1/2}, \qquad L_S = (D/w)^{1/2},$$

$$L_R = |D / (\omega - \Omega_0 + \frac{iD(mV)^2}{2} + \frac{i}{T_1})|^{1/2} \tag{7}$$

Here $\mu = e\tau/m$, N is the concentration of electrons per unit length. Under
resonance conditions ($\omega = \Omega_0$) $L_R \gg L_D \cdot L_S$. If $L \gg L_R$, we are dealing
with the uniform situation and obtain an effective conductivity

$$\sigma(\omega) = e\,\mu(\omega)\,N - \frac{iN}{8T} \frac{[\, mV\Omega_0\mu(\omega)\sin\theta \,]^2}{\omega - \Omega_0 + iD(mV)^2 \,[\cos^2\theta + \frac{\sin^2\theta}{2(1-i\omega\tau)}] + \frac{i}{T_1}} \tag{8}$$

$\mu(\omega) = \mu'(1 - i\omega\tau)$. As can be seen from (6) and (8), at $\omega\tau \ll 1$ the CR
leads to absorption diminishing (antiresonance). If $\omega\tau \gg 1$, the ordinary
resonance lineshapes occur with the usual Lorentz curves for absorption
and dispersion.

Consider the situation of short segments, i.e. $L \ll L_R$. If conductivity
is averaged over the segment length, then we obtain

$$\sigma(\omega) = e\,\mu(\omega)\,N\,S\,(L/L_D) - \frac{iN}{8T} \frac{[\, mV\Omega_0\mu(\omega)S(L/L_D) \,]^2}{\omega - \Omega_0 + iD(mV)^2 \, \frac{S(L/L_D)}{2(1-i\omega\tau)} + \frac{i}{T_1}} \tag{9}$$

(therefore shortening has taken the angle θ = π/2).   Eq. (9) differs from
eq. (8) by the presence of structure factors S:

$$S(x) = 1 - x^1 e^{i\pi/4} th[x.e^{-i\pi/4}] \qquad (10)$$

If x << 1 (very short segment lengths), $S(x) = - (ix^2)/3$ and CR curves
have the usual resonant shape.   The effect of depolarization is most
noticeable when x ∿ 1 (L ∿ L_D << L_R).   In this case ReS ∿ ImS and the
resonant lineshape is strongly deformed.   In particular, at L = 1.592 L_D
the factor S^2(L/LD) in (9) acquires a purely imaginary value and
consequently the absorption and dispersion curves interchange.   The
inequality x >> 1 leads to (8).   The dependence of dimensionless con-
ductivity $\tilde{g} = 4T^2\Delta\sigma(\omega)/\sigma(0)\Omega_o^2$ (Δσ is the second term in (9)) on the
dimensionless magnetic field h - $\tilde{h}_r$ = 2(gμ_BH - ω)/D(mV)^2 is depicted in
Fig. 1 a, b.   The average of Δσ over a Poisson distribution with lengths
P(L) = exp(- 1/L_o)/L_o is used.

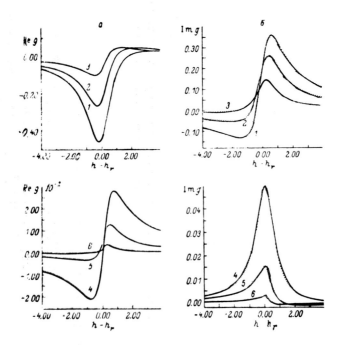

Fig. 1 a, b.   The averaged resonance conductivity versus magnetic field
curves (in dimensionless units) at L_D = 3L_S and various values of L_o/L_D :
1. - 4; 2. - 2; 3. - 1; 4. - 0.5; 5. - 0.25; 6. - 0.17.   a - the real
part of conductivity, b - its imaginary part.

4. CONCLUSION

Experimentally observed   CR lineshapes show that effective dielectric susceptibility must contain an additional phase factor which is responsible for mixing up the usual Lorentz resonance curves. We have described the physical picture which leads to the CR phenomenon and explains the origin of this phase factor. It is necessary to emphasise that such a situation is possible only if dislocation conductivity occurs.

REFERENCES

Koshelev A E, Kravchenko V Ya and Khmel'nitskii D E 1988 Fiz.Tverd.Tela
    30 433 [Sov. Phys. Solid State]
Kveder V V, Kravchenko V Ya, Mchedlidze T R, Ossipyan Yu A, Khmel'nitskii
    D E and Shalynin A I 1986 Pis'ma Zh. Eksp.Teor.Fiz. 43, 202 (JETP Lett.
    43, 255)
Rashba E I 1960 Sov. Phys. Solid State 2 1109

Inst. Phys. Conf. Ser. No 104: Chapter 2
Paper presented at Int. Symp. on Struct. Prop. Disloc. Semicond., Oxford, 1989

199

High spatial resolution cathodoluminescence from dislocations in semiconductors studied in a TEM

J W Steeds

Physics Department, University of Bristol, Bristol BS8 1TL

ABSTRACT: Results are presented of details of cathodoluminescence spectra obtained from individual dislocations or groups of dislocations directly resolved and characterised by transmission electron microscopy. The general conclusion of this work is that the luminescence effects we have observed are the result of impurity decoration of the dislocations.

1. INTRODUCTION

In seeking an understanding of the relationship between the structure of dislocations and their luminescence properties it is clearly beneficial to be able to study individual dislocations which can be characterised directly. As transmission electron microscopy (TEM) is the only really satisfactory method of performing the characterisation, I will restrict my comments to those examples which examine the luminescence of dislocations directly characterised by TEM. This selection necessarily excludes the case of silicon when the carrier diffusion lengths and non-radiative surface recombination preclude the collection of measurable luminescence signals from foils thin enough to be transparent to high energy electrons. Research work in this laboratory has concentrated on three examples where rather thorough analysis has been possible. These are the Y_0 emission associated with dislocations in ZnSe grown as epitaxial layers to GaAs, the luminescence of misfit dislocations in AlGaAs/GaAs quantum well structures, and the loss of exciton luminescence from dislocations in LEC InP. In each of these cases sufficiently strong cathodoluminescence signals and short effective carrier diffusion lengths made possible a careful study of the characteristics of the associated emission spectra.

2. ZnSe ON GaAs

So-called Y_0 emission (see Dean 1984) at approximately 2.60eV was found in both undoped and doped samples. However, the dislocation arrangement in the two samples took a very different form. In the undoped material the dislocations were approximately uniformly distributed and frequently associated with widely extended stacking faults while the doped (Al or In) samples contained much lower dislocation densities, no stacking faults but occasional extensive clusters of dense dislocation tangles. In addition, both sets of samples contained misfit dislocations at the GaAs interface which gave no detectable luminescence effects. As the isolated dislocations in the undoped samples gave rather weak effects of a similar character to those in the doped samples, the most detailed analysis was made of the doped material. The stacking faults themselves had only weak luminescence effects associated with them. For the best spectral detail rather thick regions of the epitaxial layers were studied by CL at 120kV and later in transmission by 300kV electrons. Y_0 emission is quite unusual in that, for a relatively deep level, it has rather weak phonon coupling (low Huang Rhys factor) and a short lifetime ($\sim$4ns). Also, unlike donor-acceptor-pair (DAP) emission the intensity of the Y_0 emission is strongly excitation-dependent and the temperature dependence of the emission energy follows that of excitonic emission rather than that of DAP emission (Batstone & Steeds, 1986). Previous photoluminescence spectroscopy had indicated that Y_0 emission was not very sharp, with some evidence of substructure, a conclusion positively confirmed by mono-chromatic images formed in the Y_0 band at a variety of temperatures below 100K (Myhajlenko et al, 1984). Dean (1984) took this evidence to indicate that the line was the result of free exciton emission in the vicinity of defects which were simultaneously excited. He assumed that defects were extended on the scale of the exciton size ($\sim$ 5nm) with strong delocalisation of the binding potential, and that there was a large number of nearly equivalent binding sites associated with what was probably a dislocation. Our results gave direct evidence in support of Dean's conclusions except that the dislocations concerned were in tangles rather than in the form of small loops. The dislocations were associated with strong quenching of the excitonic emission. We were unable to find any difference by TEM between the various regions of a dislocation tangle giving rise to the different temperature-dependent behaviour in monochromatic CL maps, and freshly formed dislocations behaved in a similar fashion to those grown in tangles (Myhajlenko et al, 1984).

3. MISFIT DISLOCATIONS IN AlGaAs/GaAs QUANTUM WELL STRUCTURES

The samples studied were nominally lattice-matched and gave excellent luminescence emission. The defects studied were widely separated and might not have been noticed but for the large field of view available in selectively-etched plan-view samples (Steeds et al, 1989). A good reason for the existence of the misfit dislocations has not been found, but it has been suggested that their existence can be correlated with a strong arsenic excess during growth (private communication, Mme Hubert).

The misfit dislocations were present at different levels in the structure. The majority lay at the substrate/AlGaAs buffer layer interface. Some shorter segments ran along the interfaces of the quantum wells themselves. Although the experiments were performed in plan-view samples, it was possible to distinguish between these two cases without difficulty. I shall concentrate on the dislocations at the level of the quantum wells in a particular five quantum well sample. The existence of the quantum wells parallel to the surface prevented loss of excitation to the surface so that particularly strong CL was obtained.

The form of the quantum well luminescence was that of a single narrow peak at 772nm (half width $\sim$1nm) with a very weak and slightly structured long wavelength tail (Steeds & Bailey, 1987). In the vicinity of the dislocation the CL spectrum showed some enhancement of the tail emission and accompanying changes of structure. Each dislocation gave a slightly different form of this tail emission and there was no systematic dependence on dislocation character. The dislocations ran parallel to the two <110> directions in the (001) growth plane and both 60° and Lomer Cottrell dislocations were studied.

Monochromatic CL images were formed by selecting wavelength bands in different positions in the tail emission. For emission close to the exciton peak the monochromatic images were of rather poor resolution and the dislocation image was broad and unstructured. As the selected wavelength moved progressively farther towards the red end of the spectrum the images became systematically sharper. Structure was observed in the emission pattern along the dislocation and the regions of highest intensity of emission became strongly concentrated along the core of the dislocation line. A tentative explanation for these results will be given in the discussion.

4. GROWN-IN DISLOCATIONS IN LEC GROWN InP

Single isolated dislocations were studied by TEM and CL in thick regions of chemically polished samples. For samples with a high temperature heat treatment (at 1023K in a hydrogen gas-flow with 1% phosphine) remarkable and characteristic results were obtained. The CL spectrum, which normally contained lines from excitons bound to neutral donors in $10^{16}cm^{-3}$ n type material, donor-acceptor pair emission, and C band deep level emission, showed complete local quenching of the excitonic emission in the vicinity of the dislocation without any detectable change of the other emission lines (Myhajlenko et al, 1984). These results were found to be independent of dislocation character but other samples not heat treated in this way failed to show any effect. It is perhaps important to emphasize the difference between this result and the well-known dark spots often observed at dislocations by CL in SEMs. First, the scale of the observations is quite different, the exciton quenching being localized in this case to a region of about $1\mu m$ diameter around the dislocation. Second, the quenching was specific to a particular (and rather weak) emission process and would not have been detected by integrated CL imaging. Finally, the result was only obtained at low temperatures of about 30K.

5. DISCUSSION

Many possible explanations exist in the literature for luminescence effects associated with dislocations. One whole category of explanations has to do with intrinsic properties of dislocations such as core charge, jog charge or piezo-electricity. Systematic differences would then be expected between screw and edge dislocations, and between curved and straight dislocations. Our observations do not support explanations of this type. The results were found to depend on sample history, heat treatment and impurities rather than dislocation character or curvature. AlGaAs/GaAs samples grown with AlAs barrier layers to absorb impurities were found to have less intense tail emission. Samples known to be less pure or found to have regions of enhanced impurity were those giving the strongest tail emission. It is also well-known that dislocation strain fields attract Cottrell atmospheres of impurities with high core concentrations and decreasing concentration with distance. The AlGaAs/GaAs quantum well results are apparently consistent with this picture. High impurity concentrations at the dislocation cores will give rise to large red-shifts in the local luminescence spectrum (Yoshimura et al, 1988). Monochromatic images formed at long wavelengths are therefore expected to be concentrated on the dislocation cores, as observed. For wavelengths close to the excitonic emission, the low concentration regions distant from the core would play a major part, hence giving rise to wide dislocation images. The structure visible along the dislocation lines could well be the consequence of transfer of the dislocation line from one interface to another of the quantum wells.

In the case of the Y_o emission from ZnSe it is believed that free excitonic emission with simultaneous dislocation impurity excitation is responsible. For the InP samples the excitons either fail to form near the dislocations or else they are attracted to the dislocation core where they recombine non-radiatively.

In conclusion, although these studies are in a state of comparative infancy and much work remains to be done, it seems likely that the luminescence effects associated with individual dislocations could play an important part in understanding the properties of impurity segregation associated with dislocations in semiconductors. It remains to be established whether intrinsic dislocation properties can also be studied in this way.

ACKNOWLEDGEMENTS

I wish to thank Joanna Batstone, Stefan Myhajlenko, Simon Bailey and Harry Hutchinson for many fruitful discussions and for freedom of access to their results, reviewed in this paper. This work was partly supported by the Science & Engineering Research Council.

REFERENCES

Batstone J L & Steeds J W 1986, Proceedings 44th Annual Meeting of EMSA, San Francisco Press p 818
Dean P J 1984 Phys Stat Sol 81(a), 625
Myhajlenko S, Batstone J L, Hutchinson H J & Steeds J W 1984 J Phys C. Solid State Phys 17, 6477
Steeds J W & Bailey S J 1987, Atti de XVI Congresso di M E Bologna, p 107
Steeds J W, Bailey S J, Wang J N & Tu C W 1989, Proceedings of Bristol NATO Workshop, to be published
Yoshimura, Bauer G E W & Sakaki H 1988, Phys Rev B38, 10, 791

Inst. Phys. Conf. Ser. No 104: Chapter 2
Paper presented at Int. Symp. on Struct. Prop. Disloc. Semicond., Oxford, 1989

203

Dislocation-related deep levels in GaAs: an optical absorption study

D Vignaud and J L Farvacque

University of Lille Flandres Artois, U.A. 234 CRNS-Bât C6, 59655
Villeneuve d'Ascq Cedex, France

Optical absorption studies of plastically deformed semiconductors are
generally disappointing in the sense that direct absorption involving
dislocation related levels is never observed, at least in GaAs. One
systematically obtains a shift of the band-to-band transitions (red shift)
towards shorter photon energies. In a previous paper (Vignaud and
Farvacque 1989), we have shown that this effect cannot result from an
absorption involving point defects (either intrinsic or impurities).
Instead, we propose a mechanism in which electric fields induced by
dislocations are responsible for the red shift, by some internal
equivalent of the Franz-Keldysh effect.

Quantitative calculations of this absorption show that it results mainly
from the electrostatic field induced by the carriers trapped by dislocation
levels (the charged line effect). In some specific cases, the piezoelectric field also induces an optical absorption which cannot be
neglected. The absorption resulting from the deformation field is
always very weak in GaAs, when compared to the two other already mentioned
components of the electric fields.

Comparing the relative experimental absorption in three kinds of plastically deformed GaAs materials (n^+, p^+ and semi-insulating), these
calculations imply the existence of two dislocation-related deep levels.
The first one is an acceptor level (efficient mainly in n^+ materials),
which may be the same as the one already observed by Hall effect and
photoluminescence measurements. Following these last results, its
location should be about 0.22eV above the valence band edge. The
observation of a red shift in p^+ material, slightly weaker than in n^+
material but stronger than in semi-insulating GaAs, requires that
dislocations also introduce some donor levels in the upper half of the
band gap. The calculations are not sensitive enough to allow precise
location of this donor level. Since the optical absorption involving
this donor level is never observed, this may imply that the reconstruction
at the core of the corresponding dislocation is variable.

Vignaud D and Farvacque J L 1989 J. Appl. Phys. 65 1261

Extended Abstract only

Inst. Phys. Conf. Ser. No 104: Chapter 2
Paper presented at Int. Symp. on Struct. Prop. Disloc. Semicond., Oxford, 1989

205

Calculations for defect strength determinations

D.B. Holt, E. Napchan and C.E. Norman

Department of Materials, Imperial College of Science, Technology and Medicine, London SW7 2BP.

ABSTRACT: Monte Carlo electron trajectory simulations or semi—empirical expressions can be used to compute electron hole pair distributions induced by the SEM beam. These were used in programs based on the Donolato phenomenological theory of dark EBIC defect contrast, to determine the values of a number of parameters including defect strengths. Determinations of the variations of the strengths of clean and decorated defects of many types as functions of beam operating parameters, temperature etc. are needed to test the applicability of the non—interacting point recombination centre model and the electrostatic barrier model for dark contrast. The principles of the computations and the experience of their applications to dislocations in photodetector diodes are reported.

1. INTRODUCTION.

EBIC and CL have the spatial resolution to measure the electronic properties of individual defects but few studies have been carried out. The necessary evaluations are greatly simplified by the use of microcomputer programs which will be outlined, and the implications of the results discussed.

1.1 Donolato's Theory and the Definition of Defect Strength

The phenomenological theory of EBIC dark contrast, due to Donolato (1978/79, 1988) with, where necessary, the second order corrections due to Pasemann (1981, Pasemann et al 1982) defines and makes it possible to determine the recombination strength of individual defects from observations of local reductions in EBIC current.

Figure 1. The Donolato model of the EBIC contrast of a defect beneath a Schottky barrier (after Donolato (1985)).

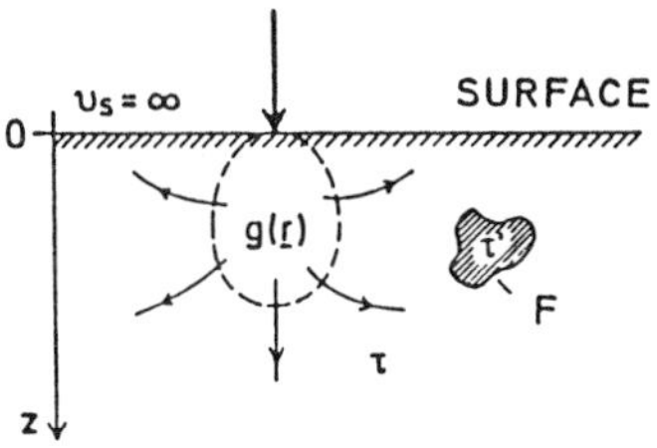

A defect is modelled phenomenologically (Figure 1) as a volume F in which the minority lifetime is reduced from τ to τ' so the charge collection current is reduced from the perfect Schottky diode value Io to

$$Icc = Io - I^* \qquad (1)$$

where I*, the reduction in EBIC due to the defect, is

$$I* = \gamma \int p_o(\underset{\sim}{r}) \, dl \qquad (2)$$

where $p_o(\underset{\sim}{r})$ is the minority carrier distribution. γ the dislocation strength (line recombination velocity) can be written (Kittler and Seifert 1981):

$$\gamma = \sigma(1/L_d^2 - 1/L^2) = \pi r_d^2/D(1/\tau' - 1/\tau) \qquad (3)$$

where σ is the cross-sectional area of the cylindrical defect volume (F) around the dislocation line and r_d is the radius of this cylinder.

EBIC contrast is defined as:

$$C = I*_{max}/Io \qquad (4)$$

and the Donolato treatment leads to the relation

$$C = \gamma F(R, L, \text{ geometry of the specimen and defect}) \qquad (5)$$

where R is the Gruen range, L the minority carrier diffusion length and F a complicated function, different for each specimen geometry.

From measured dislocation EBIC contrast, γ can thus be obtained but an explicit expression for F is required for each arrangement such as a dislocation parallel to the surface and lying above a p-n junction (Pasemann et al 1982). The parameters involved, such as L and the dislocation depth must also be measured for each specimen. Consequently few dislocation strength determinations have been made.

1.2 The Joy Calculation and the Dislocation Active Radius

Microcomputer calculations of EBIC defect contrast profiles (Joy 1986) use Monte Carlo electron trajectory simulation (MCS) (Joy 1988) to obtain the carrier generation distribution under the beam. The method of images is used (Figure 2) and the

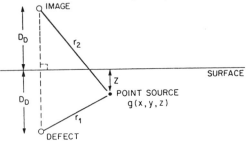

Figure 2. The Joy calculation of the EBIC contrast of a dislocation beneath a Schottky barrier (after Joy 1986).

essential steps in Joy's method are expressed in the following lines from the Turbo Pascal program.

$$CC = CC + EH*EF \qquad (6)$$

i.e. the CC current is that calculated previously plus the contribution of the number of electron-hole pairs (EH) at the next point of the distribution, computed earlier by MCS, times the charge collection probability ('efficiency' EF). This is:

$$EF = 1 - Ro*EF = 1 - Ro \left[\frac{\exp \, (-R_1/L)}{R_1} - \frac{\exp(-R_2/L)}{R^2} \right] \qquad (7)$$

That is, the unity CC 'efficiency' in perfect material in the depletion region is reduced by the dislocation 'effective radius' Ro times its charge collection 'efficiency'. The first exponential term is the no. of carriers from the point source diffusing to and recombining via the dislocation i.e. contributing to I*, while the second is the number collected by the Schottky barrier (modelled as diffusing to the image) and thus contributing to Io. (Ro is the cut-off radius arising in the calculation of the electrostatic field of a charged dislocation line by Mil'stein and Senmderichin (1982) and Joy adopted their sugested value of Ro = 0.1μm.) Neither Ro nor the dislocation strength k in Joy's program are readily related to the Donolato–Kittler–Seifert dislocation recombination velocity expression, however.

One of us (E.N.) has, therefore, written an MCS generation distribution based program for the geometry of Figure 3, using the Donolato model for the dislocation.

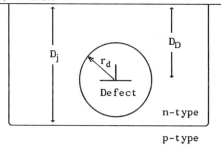

Figure 3. The geometry of the observations on dislocations in Si photodetectors and the image used in the MCS programs.

1.3 EBIC Contrast of Dislocations in Si Photodiodes

Sets of diffusion-induced dislocations at particular depths, were studied in two types of silicon photodiodes in the geometry of Figure 3: quadrant photodetectors (QPDs) with p–n junctions at a depth of about 3.4 μm and avalanche photodetectors (APDs) in which they were at a depth of 6.5 μm, that were kindly supplied by Centronic Ltd. Further details of the experiments and the computer program (written by C.N.) to compute the EBIC contrast C for a beam incident verticaly above a dislocation, from a simple analytical carrier generation distribution, are given by Norman and Holt (1989).

2. RESULTS

In determining dislocation strength parameters, it is first necessary to find the current collected in the absence of the defect, Io. This is speeded up by fitting computed curves to experimental data. By doing this for values of Io measured for many beam voltages and hence penetration ranges, values of L above and below the junction can be found. In the photodiodes, it was found necessary to use L varying with depth i.e. P concentration in the heavily doped, P-diffused n-region. This was similar to the experience of Davidson et al (1982).

The EBIC contrast profiles in the QPDs were recorded across the dislocations shown in Figure 4.

Figure 4. EBIC micrograph of dislocations in a Si QPD (quadrant photodetector) with superimposed EBIC linescans./

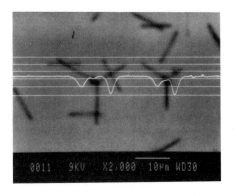

Figure 5.   Schematic illustration
of the effects of the dislocation
parameters used in fitting computed
curves to experimental data.

(a) The effect of D_D the defect depth
is to shift the peak along the beam
voltage axis.

(b) Variation of the dislocation
cylinder radius r_d alters the width of
the peak and

(c) changes in the percentage
recombination in the cylinder
affect the height of the peak.

Figure 6.   Computed curve of
EBIC contrast versus beam
voltage and experimental points
for one dislocation like
those in Figure 4.

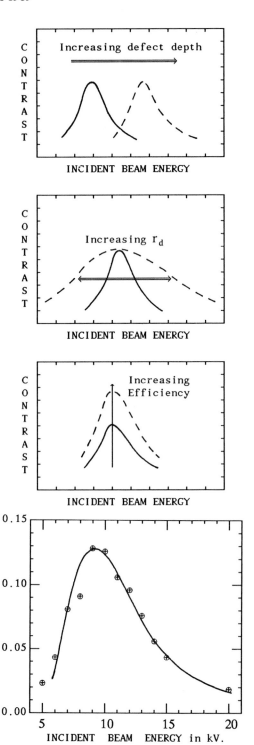

It was found that the three dislocation parameters depth D_D, cylinder radius r_d, and percentage of carriers recombining in that volume could be found since they affected the fit differently as indicated in Figure 5. For the dislocations of Fig. 4, as found e.g. in Figure 6, the depths were $0.8\mu m$, $r_d = 0.5$ μm and 18% of the carriers in the cylinder were lost by recombination. The depth was also determined by ion beam milling a hollow into the top of a QPD and observing the depth at which dislocation dark contrast disappeared and good agreement was obtained.

In the APDs the dislocations and the junction were both much deeper and the values found by curve fitting were $D_D = 2.45\mu m$, $r_d = 0.93$ μm and recombination loss was 14%. Determining the dislocation depth by scans over a cleaved cross section through the device gave agreement with the curve fitting value and with previous determinations (Lesniak and Holt 1983). The value of C obtained when the beam was incident on the point of emergence of misfit dislocations through the cross-sectional surface also agreed with the percentage recombination in the dislocation cylinder found by curve fitting.

A few MCS-derived EBIC contrast profiles were generated, using the E.N. program, and gave results in broad agreement with the earlier, simpler, program, and a value for the Donolato dislocation strength for the QPD dislocations of 0.025.

3. DISCUSSION

The result of fitting computed curves to the EBIC contrast measurements on dislocations in the QPDs and APDs was that r_d always lay between 0.5 and 1 μm and the percentage recombination was 10 to 20%. The finding of relatively large values of r_d and small values of recombination effectiveness is in agreement with the experience of P. Wilshaw (private communication) and the values found in early work on the electrical effects of deformation of Ge (Read 1954a,b, 1955).

The present work demonstrated the speed and convenience of computer EBIC profile fitting to experimental curves which makes it possible to systematically extract dislocation data as a function of a number of experimental variables. In previous studies of EBIC profiles across Si p-n junctions (Hungerford and Holt 1987), by curve fitting, the changes in surface recombination velocity due to the electron beam became experimentally determinable. In the present work, it is variations of not merely dislocation strength, but separately of the dislocation recombination radius r_d and % recombination effectiveness that can be determined as functions of dislocation depth i.e. local doping density, etc. Such information is needed to elucidate the physics of the recombination processes involved.

Two treatments have been put forward for the recombination strength of dislocations: the non-interacting recombination centre picture and the electrostatic barrier model. Kittler and Seifert (1981) considered a model of dislocations decorated by non-interacting point recombination centres. This led to the suggestion that the values of γ for the impurity-decorated dislocations observed in devices (a few tens of percent) could readily be accounted for by modest numbers of non-interacting impurity atoms per unit length while no possible density of dangling bonds would suffice. Closely spaced centres, however, should interact. Their wave functions will overlap, leading to energy band formation and the carriers they trap will strongly repel one another, limiting the line charge.

The electronic effects of dislocations in semiconductors have been modelled by a line charge and surrounding space charge cylinder (Read 1954, 1955b; Figielski 1978) and Wilshaw and Booker (1985) applied this model to EBIC dark contrast.

The published values of γ, and those measured here, lie in the range 0.01 to a few times 0.1.   The electrostatic potential at a charged dislocation line cannot rise above a modest level due to band bending and the possibility of carrier inversion.   Hence, whether the dislocation is clean, so dangling bonds are involved, or decorated, so impurities are responsible, the electronic effects would seem likely to be of the same order of magnitude.   Non-interacting impurity recombination centre decoration would seem to impose no such limits, however.   On the other hand, the larger values of γ but smaller values of recombination effectiveness were found for the APD dislocations subject to more severe decoration. Thus it may be that both point centre densities and the electrostatic barrier play a role in EBIC contrast.   To test these models and study the variation of EBIC contrast with temperature, decoration, dissociation, defect polarity (in the compounds) etc.   numerous measurements are needed and computer curve fitting makes this practical.   Care in interpretation will be required, however, as the measures of defect recombination effectiveness in the different computer programs are not the same.

4.   REFERENCES

Davidson S M, Innes R M and Lindsay S M 1982   Sol. State Electron. 25 261
Donolato C   1978/79 Optik 52 19
Donolato C   1988 Scanning Microscopy 2 801
Figielski T   1978 Sol. State Electron. 21 1403
Hungerford G A and Holt D B 1987   Microscopy of Semiconducting Materials 1987.
     Conf. Series No. 87 (Inst. Phys.: Bristol)   721
Joy D C 1986 J. Microscopy 143 233
Joy D C 1988   EUREM 88. Conf. Series No. 93 (Inst. Phys.:Bristol) 23
Kittler M and Seifert W 1981 Phys. Stat. Sol. A66 573
Lesniak M and Holt D B 1983   Microscopy of Semiconducting Materials 1983.   Conf.
     Series No. 67.   (Inst. Phys.:Bristol)   439
Mil'stein S and Senderichin A 1982 Phys. Stat. Sol. B109 429
Norman C E and Holt D B 1989   Microscopy of Semiconducting Materials 1989.
     Conf. Series (Inst. Phys.:Bristol) to be published.
Pasemann L 1981 Ultramicroscopy 6 237
Pasemann L, Blumentritt H and Gleichmann R 1982 Phys. Stat. Sol. A70 197
Read W T 1954a Phil. Mag. 45 775-796;   1954b Phil. Mag. 45 1119-1128;   1955
     Phil. Mag. 46 111
Sze S M 1981 Physics of Semiconductor Devices.   2nd Ed. (Wiley: New York).
Wilshaw P and Booker G R 1987 Izv. Akad. Nauk Ser. Fiz.51 1582

Inst. Phys. Conf. Ser. No 104: Chapter 2
Paper presented at Int. Symp. on Struct. Prop. Disloc. Semicond., Oxford, 1989

211

Electronic properties of dislocation segments in plastically deformed silicon

Yu A Ossipyan, V V Kveder and E A Steinman

Institute of Solid State Physics, USSR Academy of Sciences, Chernogolovka, Moscow district, 142432, USSR

ABSTRACT: Presented are the experimental data on photoluminescence and electrodipole resonance in plastically deformed Si and Ge, yielding information about the electronic properties of straight dislocation segments.

1. INTRODUCTION

The electronic properties of dislocations are appreciably affected by various defects residing in dislocation cores and in their vicinity. Meanwhile one of the most interesting fundamental questions is the question concerned with electronic properties of ideal segments of dislocation cores possessing the translational symmetry. In particular, there is a question about the existence of deep one-dimensional Bloch bands. Two groups of the experimental data have been presented in this work, which may be interpreted in terms of electronic properties of dislocation segments.

2. ELECTRODIPOLE RESONANCE

The question about the existence of deep dislocation bands may be answered by the experiments dealing with registration of electrodipole spin resonance in dislocated semiconductors. Let us imagine an electron in a deep one-dimensional dislocation band. A local symmetry of the crystal in the dislocation core is appreciably lower than that of the ideal crystal, and for many dislocation types the inversion centre is missing even in Si and Ge. In these conditions the spin-orbital interaction leads to the formation of the coupling between the electron impulse directed along the dislocation and its spin. If a high-frequency electric field E_ω is applied to the sample, then for $E_\omega \parallel L$, where L is the dislocation direction, one will observe the energy absorption caused by the conductivity along dislocation segments $\sigma_d = eN\mu$, N being the concentration of electrons per unit of the dislocation length and μ their mobility. If a constant magnetic field H_0 is applied to the sample, then with the resonance condition $h\omega = g\mu_B H_0$ one will observe a decrease of the conductivity σ_d due to the resonance energy transfer from the translational electron motion to its spinal degree of freedom. This effect, called electrodipole spin resonance, was observed by us in 1986 in plastically deformed silicon. Si samples deformed at 700°C by compression along [110] and annealed at 800°-1000°C in order to eliminate dangling bonds, were used. A resonance change of the microwave conductivity of the sample $\sigma(H_o)$ was observed at a frequency of 9300 mHz. We called it the Ch-line. The g-factor of the Ch-line was anisotropic and close to 2. An analysis of the dependence of the amplitude (A) of the Ch-line on the E_ω orientation showed that the Ch-line corresponds to quasi-one-dimensional defects lying in the direction L = [1$\bar{1}$0] (corresponding to the intersection of glide planes in the samples). Without illumination the Ch-line was observed only in the samples doped by shallow donors (P), but it appeared in p-type samples as well under illumination. It was found from the Ch-line photoexcitation spectra (Kveder et al 1987), that it corresponds to the electrons trapped on deep electronic states $E_{ch} \simeq E_v + 0.8$ eV which are empty in the neutral state. These states are likely to correspond to one of the dislocation types (or other one-dimensional defects) and they are stable up to annealing temperature of 1000°C. Now comes the question of whether these quasi-one-dimensional states are localized or they form a sufficiently wide Bloch band. In the latter case one has to observe a decrease of σ_d with growing temperature, characteristic of metals. (Naturally, in the case of the Bloch band the states must be localized along the dislocation too because of the dislocation-core defects. However, the localization length L in this case may be very large (hundreds of the lattice constants) and the conductivity $\sigma_d(T)$ of finite conductive segments will have metallic character down to very low temperatures). The exact equation for $\sigma(\omega)$ in this case at the electrodipole resonance was obtained by Koshelev et al (1988):

(1)

$$\frac{\sigma(\omega)}{\mathcal{L}_D} = eN\mu(\omega) \left\langle S\left[\frac{L}{L_D}\right] \right\rangle_L - i\frac{N}{8kT} \left\langle \frac{[\mu(\omega)S(L/L_D) m v]^2 \Omega^2}{\omega - \Omega + i/\tau_2 + iD(mv)^2 S(L/L_s)/2(1-i\omega\tau)} \right\rangle_L$$

where: $S(x) = 1 - \exp(i\pi/4)\text{th}(\exp(i\pi/4)x)/x$ is the structural factor, <lll> stands for averaging with respect to lengths L of the conducting segments, N is the electron concentration per unit of the dislocation length, $\mu = e\tau/m(1-i\omega\tau)$ is the high-frequency mobility, $D = T\tau/m$ is the coefficient of electron diffusion along the dislocation $L_D = ((D + e\mu N\ln(L^2/b^2)/\epsilon)/\omega)^{1/2}$ where b is the localization radius perpendicular to the dislocation (b is of the order of the lattice parameter), $L_s = (D/\omega)^{1/2}$, $\Omega = g\mu_B H/h$. As seen from (1), $\sigma_d = eN\mu \neq \sigma(\omega)$ and the Ch-line amplitude

(2) $A^{1/2} = C\frac{\sigma_d}{\sigma_0} S_\sigma\left[\frac{\sigma_d}{\sigma_0}\right]$ where $\sigma_0 = \omega L_0 \epsilon_0 / \ln(L_0^2/b^2)$

Due to a complex character of the function S(x), the shape of the absorption line $\sigma_d(H_o)$ is absorption-dispersion mixture. Therefore, the temperature dependence of $\sigma_d = \mu(\omega)eN$ may be obtained both from an analysis of the temperature dependence of the Ch-line amplitude (A) and from an analysis of the temperature dependence of its shape. As an example, fig. 1 presents on the left the experimental shape of the Ch-line (d σ/d H_o)(H_o) and on the right the calculated one.

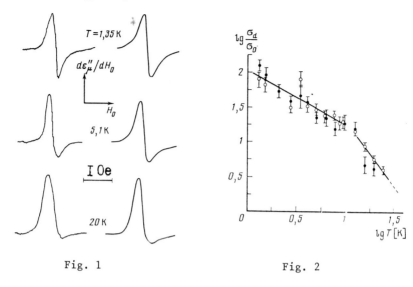

Fig. 1   Fig. 2

Fig. 2 demonstrates the temperature dependence of σ_d, obtained from
experimental data by these two procedures. Coincidence of σ_d, obtained
from the amplitude and from the Ch-line shape indicates the applicability
of the eq. (1), and a decrease of σ_d with growing T indicates the existence
of one-dimensional Bloch band. So, regular straight segments of the cores
of one of the dislocation types in silicon are connected with deep one-
dimensional Block band $E_{ch} \simeq E_v + 0.8$ eV. One cannot say unequivocally
what type of dislocation it conforms to. There are grounds to believe
that these are Lomer dislocations. The absence of the electrodipole
resonance from dislocations of other types implies either the absence
of a deep and sufficiently wide band for them or the presence of a greater
number of defects in their cores (smaller localization length L).

3. FINE STRUCTURE OF PHOTOLUMINESCENCE SPECTRA

The plastic deformation of covalent crystals at lowered temperatures and
high stresses (LTHS) involves a considerable change of the dislocation
structure which manifests itself as a straightening of dislocation lines
along the <110> direction and a deviation of the value of dislocation
splitting from the equilibrium state (Wessel et al 1977; Gottschalk et al
1982). The possibility to obtain relatively long defectless segments of
dislocations enables one to expect the observation of specific phenomena
associated with the geometric one-dimensionality of dislocation lines.
The appearance of fine structure lines in the spectra of dislocation
germanium and silicon samples subjected to LTHS deformation seems to be
just this type of phenomena (Ossipyan et al 1983; Sauer et al 1985:
Izotov et al 1988). The employment of extra LTHS stage of deformation
changes the distribution of intensity in the DPL spectrum and brings
about the formation of a fine structure. The main changes occur with
line D4, which due to the second stage of deformation splits into two
bands lying at the short-wave and long-wave sides from the initial line
and consisting of unresolved lines of fine structure. The splitting grows

Table 1. Line maxima positions calculated from eqn. (3)-(E*) and obtained experimentally-(E). The lines marked with letter H and digit are predicted by eq. (3) in the long-wave region of the DPL spectrum. These digits have nothing in common with the exponent in formula 3. E_∞ = 1013 meV, α = 264 meV, β = 1.131 meV. *-the values are borrowed from the paper by Sauer et al (1985); **-the values are borrowed from the paper by Gwinner et al (1981).

NO	E*	E	Notation
H7	525.4	540**	
H6	581.8		
H5	631.8		
H4	675.9		
H3	715	709**	
H2	749.5		
H1	780	780	
1	807	807	
2	831		
3	852		
4	871	870	D2
5	887	887*	
6	902	904*	
7	915	910*	
8	926	925*	
9	936	935	D3
10	945	947*	
11	953	953*	
12	960	960*	
13	966	965*	
14	971	971*	
15	976	976*	
16	980	980*	
17	984	984*	
18	988		
19	991		
24	1002	1000	D4
∞	1013	1013	E_∞

with increasing load at the second stage of deformation. Table 1 illustrates the positions of the fine structure lines. The long-wave lines, D1 and D2, do not undergo any shift under the action of the second stage of deformation and only decrease in intensity. The equilibrium value of DPL spectrum in silicon is recovered after annealing at T = 265°C.

Today it may be considered established that the DPL spectra of both germanium and silicon are due to the recombination of non-equilibrium carriers via the energy states connected with dislocations. The absence of the effects of thermalization and the similar dependence of the lines on the excitation level suggest that each DPL line is connected with its own type of dislocation optical centres (DOC) pertaining to a certain structure configuration of dislocations. The intensity transfer, resultant from the second stage of deformation, from some DPL lines to the others

Fig. 3. Intensity distribution in the DPL spectrum of silicon: 1-after the first deformation stage 1000°C, 2-after the second deformation stage at 400°C and as a result of subsequent 265°C izothermal annealing, 3-15 min, 4-30 min, 5-45 min, 6-135 min, 7-255 min, 8-920 min.

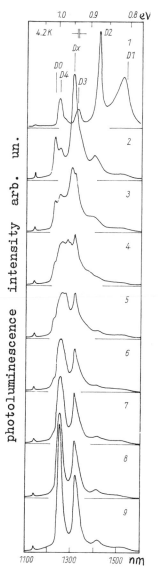

thus means that some DOC are transformed to the others. The common property of the DPL lines in germanium and silicon is their crowding to a certain threshold energy in the short-wave spectrum region. This regularity is well described by the empirical formula

$$(3) \qquad E_n = E_\infty - \alpha/\beta^n$$

The values of the line maxima positions obtained experimentally and calculated using eq.(3) are listed in Table 1 together with the values of α and β. A fascinating agreement between the experimental and the calculated values shows that all the fine structure lines are connected with one and the same type of centres differing in a certain discrete parameter. The latter is numbered by integers starting from unity. One may also introduce the notion of an equilibrium parameter corresponding to an equilibrium dislocation structure. In silicon this parameter is determined by the D4 line. The evaluation of possible dislocation splittings satisfactorily agrees with the total number of the observed fine structure lines. One could assume that the exponent n in eq. (3) is connected with the value of the perfect dislocation splitting. However, the appearance of narrow lines most likely is connected with the formation of rather long, regular, defectless segments of dislocations. In particular this follows from a rapid rise of the intensity of line D4 in silicon at relatively small shear stresses at the second stage of deformation. At this low level of shear stresses it is hard to imagine any change in the dislocation splitting So, the length of defectless dislocation segments and the value of their splittings both seem to be responsible for the appearance of a fine structure in the DPL spectra.

REFERENCES

Gottschalk H Proc. 10th Intern.Conf. on Electron Microscopy Hamburg 1982 11 527
Gwinner D and Labusch R 1981 Phys.St.Sol.(a) 65 K99
Izotov A N, Kolyubakin A I, Shevchenko S A and Steinman E A Proc. 8th Intern School on Defects in Crystals, Szczyrk, Poland 1988 203
Izotov A N and Steinman E A 1988 Fiz.tverd.Tela 30 3177
Kolyubakin A I, Ossipyan Yu A, Shevchenko S A and Steinman E A 1984 Fiz.tver Tela 26 677
Koshelev A I, Kravchenko V Ya and Khmelnitsky D E 1988 FTT 30 433
Kveder V V, Kravchenko V Ya, Mchedlidze T R, Ossipyan Yu A, Khmelnitsky D E and Shalynin A I 1986 JEP Le, 43 255
Kveder V V, Mchedlidze T R, Ossipyan Yu A and Shalynin A I 1987 Sov.Phys. JETP 66 838
Ossipyan Yu A, Rtishchev A M and Steinman E A 1983 J.Physique 44 255
Sauer R, Weber J, Stolz J, Weber E R, Kusters K H and Alexander H 1985 Appl.Phys. A36 1
Wessel K and Alexander H 1977 Phil.Mag. 35 1523

Inst. Phys. Conf. Ser. No 104: Chapter 2
Paper presented at Int. Symp. on Struct. Prop. Disloc. Semicond., Oxford, 1989

217

Optical studies of cadmium sulphide crystals plastically deformed at low temperatures

Yu A Ossipyan and V D Negriy

Institute of Solid State Physics, USSR Academy of Sciences, Chernogolovka, Moscow district, 142432, USSR

ABSTRACT: Piezospectroscopy and space distribution of radiation of defects in the bulk and on the surface have been studied in plastically deformed CdS crystals. The character motion of edge and screw dislocations has been determined under deformation in basal and prismatic systems of glide.

$A^{II}B^{VI}$ crystals are classic objects of semiconductor physics due to strong sensitivity of their physical properties to illumination. Quantitative parameters, which characterize the band gap magnitude and the arrangement in it of local centres responsible for photoconductivity and recombination, enable one to control the electrical properties of these crystals by means of visible light and to investigate easily their spectral sensitivity. The discovery of the photoplastic effect and related phenomena has shown that dislocations and other defects of the crystalline structure in these crystals interact strongly with the electronic subsystem that, in turn, changes its state appreciably under illumination. This has made it possible to employ effectively modern optical methods for investigation of dislocation structure and dynamics as well as of the motion and evolution of other defects in $A^{II}B^{VI}$ crystals. Further investigations were also facilitated by the discovery that sufficiently pure CdS, ZnS, ZnSe crystals could be plastically deformed at low temperatures (4.2 - 77K), where we observed the motion of both individual dislocations and their groups. In this case we succeeded in observing the difference in the behaviour of "fresh" dislocations, formed at low temperatures, and those kept at room temperature ("aged" dislocations).

We have already reported that the photoluminescence spectrum of plastically deformed CdS crystals exhibits the appearance of a group of lines in a narrow spectral range (λ = 505-510 nm). We termed this emission "dislocation photoemission" (Negriy and Ossipyan 1978). Subsequently we carried out detailed studies of spectral, time and polarizational characteristics of this emission, as well as its excitation spectrum. The observation of specific features of this emission from different faces of the crystal plastically deformed in different geometry and crystallography suggested the conclusion that the emission sources are not only the moving dislocations but, also, traces on the crystal surfaces left behind by the moving dislocations (Negriy and Ossipyan 1979, 1982a, 1982b). A nonstationary character of the emission and presence of polarizational domains in the traces of moving dislocations served as the basis for the model of

configurational defect chains ("bimodular dipoles") arising from low
temperature dislocation motion (Ossipyan and Negriy 1987). Piezospectroscopic studies, using spectral splitting under the action of
elastic deformation gave information about symmetry and possible
geometry of the emissive defects. All the studies in question were
conducted in conditions when the energy of the laser-induced excitation
of photoluminescence was not lower than the interband transition energy.
In these conditions the luminescence under study was emitted from a thin
near-surface crystal layer. Now comes a natural question of what is the
volume distribution of dislocations and defects, arising from the process
of low temperature deformation. In order to elucidate this question we
used argon laser radiation with   energy somewhat lower than that of the
interband transition in CdS but, however, lying in the region of the
dislocation photoemission excitation spectrum (Negriy and Ossipyan 1979).
In fact, we had an argon laser working at two wavelengths, λ_1 = 4579Å and
λ_2 = 4965Å, one of which corresponded to   energy of a larger ($h\nu_1 > E_g$)
the other to a smaller ($h\nu_2 < E_g$) value than the band gap of CdS. So, with
$\lambda = \lambda_1$ the photoluminescence was excited due to the interband electronic
transitions and the luminescence was recorded from the crystal surface and
visually observed in different spectral regions. With $\lambda = \lambda_2$ the defect-free
fraction of the crystal appeared transparent, and only the defects emitted
light as λ_2 fell into the region of their excitation spectrum. So one could
get information about defects lying in the crystal bulk.

The plastic deformation by uniaxial compression was carried out in a low
temperature cryostat so that glide should occur either on basal (0001) or
prismatic {1$\bar{1}$00} planes. Observations were conducted, respectively, on
prismatic and basal surfaces of the crystal. The results of the observation
may be summed up as follows.

During glide in prismatic planes the traces left by moving dislocations on
the surface parallel to the basal plane form lines, oriented along <11$\bar{2}$0>.
The emission from these lines has a stationary character both in polarized
and nonpolarized light.

When observing the surface parallel to planes (11$\bar{2}$0) or (1$\bar{1}$00), one can see
individual bright points, forming rows, and short (several tens of microns)
bright linear segments directed perpendicular to the c axis. On external
loading one can observe their multiplication and the motion of the segments
in the direction parallel to the c axis. No contrast traces of the moving
segments are formed on the (11$\bar{2}$0) or (1$\bar{1}$00) surfaces. When observed in
polarized light, said segments are divided into domains with different
polarizations and the emission becomes nonstationary. In fixed polarization one can observe scintillations of the point and segment intensities.

The employment of long-wave excitation λ_2 = 4965Å made it possible to
establish that in prismatic planes there occurs generation and elongation
of specific defects in the form of narrow rays oriented in the <11$\bar{2}$0>
directions. The sources of their generation are the outer crystal faces
or small-angle block boundaries. In the prismatic plane these defects
move as a whole in the c axis direction. If the crystal is deformed so
that the glide occurs in the basal plane then there one can also observe
analogous defects in the form of narrow rays directed along <11$\bar{2}$0>, but
no glide of such rays were observed in the basal plane. The analogously

Fig. 1. Photograph of
the dislocation rosette
formed by moving
dislocations (in the
direction of the type
<11$\bar{2}$0>) in the basal plane
(0001) at the (1120) surface
indentation.

arranged rays were also observed in the case of indentation of a lateral
crystal face of the type {11$\bar{2}$0} (Fig. 1).

The luminescence spectra upon excitation by both wavelengths used (λ_1 and
λ_2) appeared identical.

At the volume excitation (λ_2) in polarized light of the deformation-induced
defects no nonstationary behaviour of the luminescence was observed either
in time or in space.

Piezospectroscopic studies of "fresh" and room-temperature "aged" defects
revealed that on splitting of the spectra the displacement for equal elastic
stresses is somewhat larger for the lines obtained from "fresh" defects
(Figs. 2 and 3).

Note, in conclusion, that in order to construct authentic models of the
defect   structure of the CdS crystals, plastically deformed at low
temperature, one has to carry out complementary studies using high
resolution structural methods.

As a working hypothesis one can assume that the extended defects being
formed are narrow dislocation loops with extended screw components. As
for the phenomena of nonstationary luminescence, they may be associated
with presence of local dipole defects forming on extension of the loops due
to elongation of their screw components.

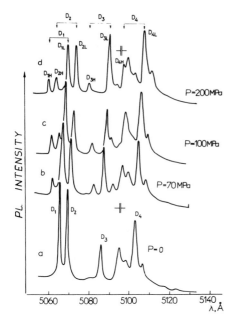

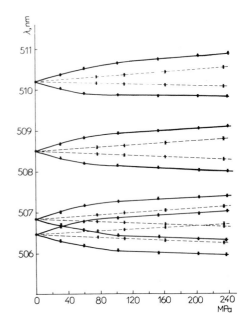

Fig. 2. Luminescence spectrum of defects as a function of a uniaxial stressing along the <11$\bar{2}$0> axis

Fig. 3. Luminescence spectrum line position as a function of stressing
a) solid line - "fresh" defects;
b) dashed line - "aged" defects

REFERENCES

Negriy V D and Ossipyan Yu A 1978 Fizika tverd. Tela 20 744
Negriy V D and Ossipyan Yu A 1979 Phys.Stat.Sol.(a) 55 583
Negriy V D and Ossipyan Yu A 1982a Pis'ma Zh. eksper. teor. Fiz. 35 484
    (JETP Lett. 35 598)
Negriy V D and Ossipyan Yu A 1982b Fizika tverd. Tela 24 344
Ossipyan Yu A and Negriy V D 1987 Inst.Phys.Conf.Ser.N87 333

Inst. Phys. Conf. Ser. No 104: Chapter 2
Paper presented at Int. Symp. on Struct. Prop. Disloc. Semicond., Oxford, 1989

221

Theoretical investigations of combined CL and EBIC measurements on crystal defects

S Hildebrandt, W Hergert, and J Schreiber

Sektion Physik der Martin-Luther-Universität Halle-Wittenberg, Postfach, Halle, DDR-4010, GDR

ABSTRACT: The fundamentals for the unified theoretical description of the cathodoluminescence (CL) and EBIC contrast of crystal defects in semiconductors are reviewed. Especially the CL and EBIC contrast from an individual surface-parallel dislocation is considered in detail. Introducing the contrast profile area, a realistic depth distribution of minority carrier generation is applied in the calculations. The behaviour of the contrast profile area in dependence on various parameters is discussed. From beam-voltage dependent measurements of the CL contrast profile area the depth position of the dislocation is determinable.

1. INTRODUCTION

It is well known that crystal defects strongly influence the electrical and optical properties of optoelectronic semiconductor materials. For the study of individual defects with high lateral resolution the scanning electron microscope (SEM) has been proved to be a suitable instrument. By means of investigations of the CL (cathodoluminescence) and EBIC (electron beam induced current) signal from homogeneous semiconductor regions the great ability of this technique for the determination of local electrical and optical parameters of the defect-free region like minority carrier diffusion length L, absorption coefficient $\alpha(h\nu)$, and surface recombination velocity v_s has been demonstrated (Hergert et al 1987a, Koch et al 1988, Hildebrandt et al 1988).
In the same way it is possible not only to image defects due to their signal contrast in SEM micrographs but also to determine quantities which characterize the carrier recombination activity as well as geometrical parameters of a special defect from its contrast. Due to similar mechanisms of contrast formation a **unified** theoretical description of the CL and EBIC contrast has been developed (see Section 2). Especially, in the theory of the CL contrast some important progress has been achieved (Pasemann and Hergert 1987, Pasemann 1987, Hergert 1988). In this paper we will present the application of a realistic electron beam generation function to the theory by introduction of the contrast profile area (see Section 3) which is a substantial improvement compared with the unified sphere generation model used so far.

2. FUNDAMENTALS OF THE THEORETICAL DESCRIPTION OF THE CL AND EBIC CONTRAST

Only the basic formulas for the general theoretical description of the CL and EBIC contrast can be reviewed here. For a more detailed representation see e. g. Pasemann 1982, Pasemann and Hergert 1986, Pasemann 1987, Hergert 1988.

The contrast at a certain distance ξ of the electron beam incidence point from the defect is defined as

$$c(\xi) = \frac{I(\xi) - I_o}{I_o} . \qquad (1)$$

The function $c(\xi)$ is the so-called contrast profile, $I(\xi)$ the signal at the defect, and I_o the signal in the defect-free region, that means far

away from the defect. Further on, as a special example, an individual surface-parallel dislocation will be considered (Figure 1). The dislocation is characterized by two regions Ω_d and $\Omega_d^{(r)}$ where the total and radiative minority carrier lifetimes have values τ' and τ_r' , respectively, different from those of the defect-free matrix region (τ, τ_r). In the theoretical model both regions are of cylindrical form.

The CL and EBIC contrast is caused by the distribution of the excess minority carrier density $q(\underline{r})$ in the sample containing the dislocation. The minority carrier density $q_o(\underline{r})$ in absence of the defect can be calculated by the method of Green's function $G(\underline{r}, \underline{r}')$:

$$q_o(\underline{r}) = \frac{1}{4\pi D} \int_{\Omega_s} d^3\underline{r}' \ g(\underline{r}') \ G(\underline{r}, \underline{r}') . \qquad (2)$$

Here, D is the diffusion coefficient of the minority carriers and $g(\underline{r})$ the excess carrier generation rate. Ω_s denotes the volume of the

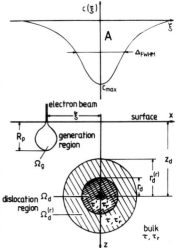

Fig.1. Schematic illustration of the contrast profile and the sample containing a surface-parallel dislocation

sample. Starting from the basic formulas for the calculation of EBIC and CL signals (cf. Hergert et al 1987b, e. g.), we obtain for the EBIC contrast

$$c_{EBIC} = - \frac{\tau/\tau' - 1}{\tau \ I_o^{EBIC}} \int_{\Omega_d} d^3\underline{r} \ q(\underline{r}) \ j_{EBIC}(z) \qquad (3a)$$

where I_o^{EBIC} is the EBIC signal from the defect-free region, and $j_{EBIC}(z)$ describes the EBIC signal for a unit point source at the depth z. In the case of a Schottky contact at the surface we get

$$j_{EBIC}(z) = e^{-z/L} \qquad (3b)$$

In a similar way the CL contrast is given by

$$c_{CL} = \frac{1}{\tau_r \ I_o^{CL}} \int_0^{\Theta_c} d\Theta \ \sin \Theta \ \overline{c_{CL}} \qquad (4a)$$

(I_o^{CL} - CL signal from the defect-free region) with

$$\overline{c_{CL}} = - \left[\frac{\tau}{\tau'} - 1 \right] \int_{\Omega_d} d^3\underline{r} \ q(\underline{r}) \ j_{CL}(z) + \left[\frac{\tau}{\tau'_r} - 1 \right] \int_{\Omega_d^{(r)}} d^3\underline{r} \ q(\underline{r}) \ e^{-\hat{\alpha}z} \quad (4b)$$

where Θ_c denotes the critical angle of total reflection at the sample surface ($\hat{\alpha} = \alpha/\cos\Theta$). c_{CL} (4b) consists of two parts of which the first one is analogous to EBIC since

$$j_{CL}(z) = \frac{1}{1 - \hat{\alpha}^2 L^2} \left[e^{-\hat{\alpha}z} - \frac{\hat{\alpha}L+S}{1+S} e^{-z/L} \right] \qquad (4c)$$

is the CL signal for a unit point source at z ($S = v_s\tau/L$). The second term is a CL specific one which occurs due to a local variation of the **radiative** recombination lifetime.

In the following the **first-order approximation** often used in the contrast theory will be applied that means $q(\underline{r})$ is replaced by $q_0(\underline{r})$ in (3a) and (4b), respectively.

3.   THE CL AND EBIC CONTRAST PROFILE AREA

It is a serious handicap of the theoretical description of the contrast that even the undisturbed minority carrier distribution $q_0(\underline{r})$ can only be evaluated analytically if the uniform sphere generation model is applied. In many experimental situations, especially if dislocation and generation region are located close to each other, one has to expect serious deviations of the calculated contrast from the measured value. These difficulties can be overcomed by introduction of the **contrast profile area**

$$A = \int_{-\infty}^{\infty} d\xi \; c(\xi) \qquad (5)$$

Donolato and Bianconi (1987) made a first successful attempt to use the EBIC contrast profile area for the determination of the recombination strength $\gamma = \tau/\tau' - 1$ of dislocations in Si basing on experimental material from Pasemann et al (1982). According to the given geometrical situation the authors were able to use the simple point source generation model. Here we will extent the theoretical description to the case of an arbitrary generation function, and the CL contrast profile area will be included, too.

3.1   Calculations

In order to calculate the EBIC contrast profile area we start from the EBIC contrast formula (3) in the first-order approximation ($q(\underline{r}) \approx q_0(\underline{r})$). Then, in (3) we insert equation (2) which contains Green's function. Performing now the integration (5) of the contrast profile, due to the translation invariance of

$$G(\underline{r},\underline{r}') = G(x-x',y-y',z,z') \qquad (6)$$

the evaluation of A reduces to a one-dimensional problem (Donolato and Bianconi 1987). For the EBIC contrast profile area we obtain

$$A_{EBIC} = -\frac{1}{\tau \, I_0^{EBIC}} \left(\frac{\tau}{\tau'} - 1 \right) \iint_{f_d} dx \; dz \; j_{EBIC}(z) \; Q_z(z) \qquad (7)$$

and for the CL contrast profile area

$$A_{CL} = \frac{1}{\tau_r \, I_0^{CL}} \int_0^{\Theta_c} d\Theta \; \sin\Theta \; \overline{A_{CL}} \qquad (8a)$$

with

$$\overline{A_{CL}} = -\left(\frac{\tau}{\tau'} - 1\right) \iint_{f_d} dx \; dz \; \alpha_z z) j_{CL}(z) + \left(\frac{\tau_r}{\tau'_r} - 1\right) \iint_{f_d^{(r)}} dx \; dz \; \alpha_z z) e^{-\hat{\alpha}z} \tag{8b}$$

f_d and $f_d^{(r)}$ are the cross-section areas of Ω_d and $\Omega_d^{(r)}$, respectively. The function

$$Q_z(z) = \frac{1}{4\pi D} \int_0^\infty dz' \; g_z(z') \; G_1(z,z') \tag{9}$$

is the depth distribution of the minority carrier density where $g_z(z')$ denotes the depth distribution of the generation rate, and $G_1(z,z')$ is the one-dimensional Green's function. In contrast to (2) $Q_z(z)$ can be calculated using a realistic depth distribution of the generation rate and can be written in terms of a universal function containing an arbitrary $g_z(z)$ which was introduced for the description of CL and EBIC signals from a defect-free region (Hergert et al 1987b) .

If $r_d \ll L$ and $r_d^{(r)} \ll L$, resp., the cross-section integration in (7) and (8b) is approximated as follows

$$\iint_{f_d} dx \; dz \; F(z) \sim \pi \; r_d^2 \; F(z_d) \tag{10}$$

so that a final expression for the contrast profile area is obtained immediately. Otherwise, if (10) is not fulfilled, the cross-section integration may be performed if for a realistic generation rate (for instance Wu and Wittry 1978) an exponential approximation is used (Hergert and Hildebrandt 1988, cf. Hildebrandt and Hergert 1989).

3.2 Discussion of the theoretical results

Applying the approximation (10) the EBIC and CL contrast profile area were calculated in dependence on the electron beam-voltage U_b for three different depth positions z_d of the surface-parallel dislocation. The generation models used were point source, uniform sphere generation and the realistic generation rate by Wu and Wittry (1978).

Figure 2 shows the calculated EBIC contrast profile area. It is seen that in principle with increasing U_b the contrast profile area A_{EBIC} increases where the shape of the curves depends first of all on the depth position z_d of the disloca- tion. As it is expected, the point source and the generation sphere, however, are adequate models for the real carrier generation dis- tribution only if the primary electron range R_p (U_b) is substan- tially less than z_d . In other cases drastic deviations of the different models are visible so

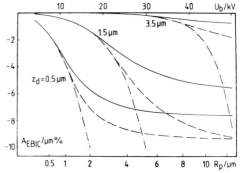

Fig.2. EBIC contrast profile area ($L=1\mu m$, defect strength $\lambda = \gamma (r_d/L)^2 \pi$ = 0,1π). $g_z(z)$: —— Wu and Wittry,

that only the Wu and Wittry function is applicable.

Considering the CL contrast profile area, the agreement of the results obtained from the three generation models is still poorer already at low beam-voltages (Figure 3). In all examples the second, "radiative" part of the CL contrast (8b) is neglected ($\tau_r = \tau_r'$).

$S \rightarrow \infty$ is assumed. Compared with EBIC, the beam-voltage dependence of the absolute value $|A_{CL}(U_b)|$ of the CL contrast profile area is mainly characterized by a maximum at the position U_b^{max} which depends on the dislocation depth z_d. The position of the maximum as well as the shape of the curves is also strongly affected by the generation model $g_z(z)$ used in the calculations. This is very important since the experimentally abtained maximum position could be used for the determination of z_d. Thus, for these purposes the application of a realistic generation function should be absolutely necessary. In Figure 4 the dependence of A_{CL} (U_b) on the dislocation depth z_d is illustrated more in detail. The maximum position does not only depend on z_d but also on the parameters of the defect-free region L

--- sphere generation, -.-. point source.

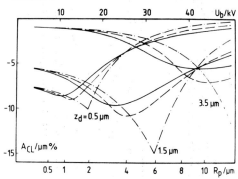

Fig.3. CL contrast profile area (L=1μm, α=0, λ=0,1π). g_z (z) see Fig.2.

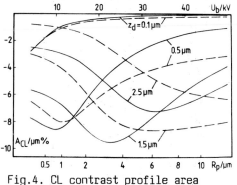

Fig.4. CL contrast profile area (L=1μm, λ=0,1π). —— α = 0, --- α=0,75 μm^{-1}. g_z(z) by Wu and Wittry

and α. For the determination of the dislocation depth z_d from beam-voltage dependent CL contrast measurements the function $U_b^{max} = U_b^{max}(z_d; L, \alpha)$ has to be known. In Figure 5 this function is shown in its general form $R_p^{max}/L = f(z_d/L, \alpha L)$ given for an arbitrary semiconductor material.

With the analysis of the beam-voltage dependent Cl contrast profile area basing on the relation shown in Figure 5 we propose a new, alternative method for the <u>determination of the depth position of a surface-parallel dislocation</u>. Whereas in the work of Pasemann and Hergert (1986) a combined CL and EBIC contrast experiment is needed for the evaluation of z_d from the ratio c_{EBIC}/c_{CL}, here one can restrict to CL contrast measurements. Moreover, Pasemann and Hergert used the generation sphere model only.

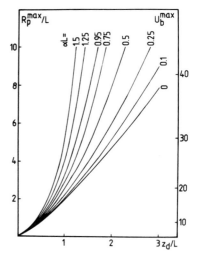

Fig.5. Maximum position of the CL
contrast profile area

REFERENCES

Donolato C and Bianconi M 1987 phys. stat. sol. (a) **102** K7
Hergert W, Reck P, Pasemann L, and Schreiber J 1987 phys. stat. sol.
  (a) **101** 611
Hergert W, Hildebrandt S, and Pasemann L 1987b phys. stat. sol. (a)
  102 819
Hergert W 1988 Thesis (Halle)
Hergert W and Hildebrandt S 1988 phys. stat. sol. (a) **109** 625
Hildebrandt S, Schreiber J, Hergert W and Petrov V I 1988 phys. stat.
  sol. (a) **110** 283
Hildebrandt S and Hergert W 1989 submitted for publication in phys. stat.
  sol. (a)
Koch F, Hergert W, Oelgart G, and Puhlmann N 1988 phys. stat. sol. (a)
  110 283
Pasemann L 1981 Ultramicroscopy **6** 237
Pasemann L, Blumtritt H, and Gleichmann R 1982 phys. stat. sol. (a)
  70 197
Pasemann L and Hergert W 1986 Ultramicroscopy **19** 15
Pasemann L 1987 Thesis (Leipzig)
Pasemann L and Hergert W 1987 Izv. Akademii Nauk SSSR, Ser. Fiz. **51** 1528
Wu C J and Wittry D B 1978 J. Appl. Phys. **49** 2827

Inst. Phys. Conf. Ser. No 104: Chapter 2
Paper presented at Int. Symp. on Struct. Prop. Disloc. Semicond., Oxford, 1989

227

Recombination at dislocations in the depletion region in silicon

T S Fell and P R Wilshaw

Department of Metallurgy & Science of Materials, University of Oxford, Parks road, Oxford OX1 3PH

ABSTRACT: An analysis of recombination at dislocations in the depletion region in silicon is presented which is in good agreement with all the experimental results. The prospect of using the analysis to determine dislocation energy level positions is discussed.

1. INTRODUCTION

Recent developments in the theory of recombination at dislocations, (Wilshaw et al 1986, 1989), have enabled the EBIC technique to be used to obtain the concentration of electrically active states at dislocations in bulk silicon,(Wilshaw et al 1989). However, the analysis of dislocations in the bulk does not allow the position of the energy levels within the band gap to be determined. In order to extend the usefulness of the EBIC technique and enable the results to be compared with those of techniques such as DLTS (Omling 1985), a method of establishing the energy level position from simple EBIC contrast measurements is required. It is the purpose of this paper to extend the recombination theory to dislocations in the depletion region, present experimental results which verify this treatment and hence lend further support for the treatment of dislocations in the bulk, and finally to discuss how EBIC measurements of dislocations in the depletion region may enable the position of the dislocation energy level to be found.

2. THEORY

Wilshaw's theory of recombination at dislocations may be applied to a dislocation in n-type silicon which introduces acceptor states into the band gap normally below the position of the Fermi level. Figure 1a. shows the band structure of such a dislocation where some of the states are occupied and so the dislocation becomes negatively charged. This negative charge, Q per unit length, increases the energy of neighbouring electrons and is represented on a band diagram by the conduction and valence bands bending up. As it is normally assumed that the addition of electrons to the dislocation level does not alter its position relative to the band edges, the entire energy level can be considered to be rigidly shifted by the electrostatic potential of its charge. In this model the recombination of an electron-hole pair via such a dislocation level can be considered to be controlled by 2 steps. 1) The capture of a hole into the bound hole states at the top of the valence band. In this process hole capture is determined by the volume of the region surrounding the dislocation in which there is an attractive field for holes towards the dislocation line. Detailed analysis shows the hole capture rate J_h, to be directly proportional to the dislocation line charge for dislocations in the bulk, ie.

$$J_h = A Q \qquad (1) \qquad \text{A is a constant}$$

Therefore as the dislocation line charge is increased so the region with an attractive field, (the space charge region in the bulk), is increased and hence so is J_h. 2) The capture of an electron to the dislocation level. This step proceeds as an electron is thermally excited over the repulsive barrier Ø, of the space charge region.

$$J_e = B \exp[-q\emptyset/kT] \qquad (2) \qquad \text{B is a constant}$$

The important features are the exponential dependence of electron capture J_e, on the electrostatic barrier surrounding the dislocation and the dependence of hole capture J_h on the dislocation line charge. For steady state recombination the overall recombination rate J, and the rate of capture of electrons and holes must be equal, ie.

$$J = J_e = J_h \qquad (3)$$

Further detailed analysis shows Ø is also a function of Q and thus both hole and electron capture rates can be described solely in terms of Q the dislocation line charge. As EBIC contrast C, can be shown to be proportional to the recombination rate it also is proportional to Q, ie.

$$C \approx F Q \approx G \emptyset \qquad (4) \qquad \text{F and G are constants}$$

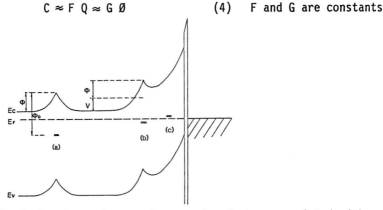

Figure 1.   Band structures in an n-type semiconductor associated with,
a) a charged dislocation in the bulk
b) a charged dislocation in the depletion region
c) an uncharged dislocation in the depletion region.

Expanding this model to include recombination at dislocations in the depletion region requires an adjustment to the electron capture rate in order to account for the superposition of the depletion region potential on the band bending associated with the dislocation energy level, see figure 1b. As the potential V, at the dislocation due to the depletion region field is an additional barrier to thermal excitation of an electron to the dislocation energy level it must be added to Ø, ie.

$$J_e' = B' \exp[-q(\emptyset+V)/kT] \qquad (5) \qquad \text{B' is a constant}$$

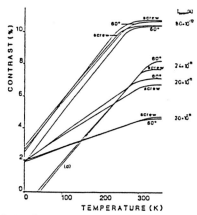

Figure 2. Plot of contrast vs temperature for dislocations at different depths. Curves marked (d) are from dislocations in the depletion region

Figure 2 shows contrast vs temperature curves for two dislocations in the depletion region which are in agreement with the model described above. They show that C varies linearly with temperature at low temperatures and constant beam current and also when extrapolated give a negative contrast at zero temperature. This is in agreement with the theoretical treatment and is due to the band bending V as described in equation (9). Dislocations in the bulk, also shown in figure 2, show similar behaviour to those in the depletion region except that when extrapolated they show positive contrast at zero temperature. Although equation (9) predicts that for the bulk the intercept should be zero, a more detailed analysis, not shown here,(which uses a more accurate expression for the relation between Ø and Q) fully accounts for this observed effect. In addition, an analysis can be performed at constant temperature as a function of beam current which describes the contrast vs beam current measurements shown in figure 3.

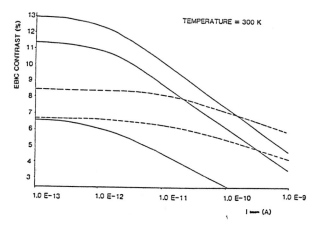

Figure 3. Plots of contrast vs ln(I beam) for screw dislocations in the bulk, (dashed lines), and 60° dislocations in the depletion region, (solid lines).

Considering hole capture at a dislocation in the depletion region, there is still a volume of space surrounding the defect in which there exists an attractive field for holes but this volume is reduced in size from that in the bulk due to the superposition of the depletion region field. The superposition of this field will also disturb the shape of the volume compared to the cylinderical space charge region present for dislocations in the bulk. An approximate estimate of the reduction of this volume due to the depletion region field may be calculated by noting that the field E, at the edge of the volume r_0, closest to the dislocation is equal to zero. The field in the depletion region may be expressed as,

$$E_1 = (d-x)n_0 q/\epsilon \epsilon_0 \qquad (6)$$

where d is the depletion region width, x is distance measured from the Schottky barrier and n_0 is the doping concentration. The field surrounding a negatively charged dislocation can be written as,

$$E_2 = Q/2\Pi r \epsilon \epsilon_0 \qquad (7)$$

and directed towards the dislocation. Therefore at r_0, E=0 and hence,

$$r_0 = Q/2\Pi(d-x)n_0 q \qquad (8)$$

Thus it will be assumed that in the depletion region as in the bulk, hole capture and hence EBIC contrast are proportional to Q, but are reduced relative to the bulk because of the depletion region field. This allows a detailed analysis resembling that performed by Wilshaw et al (1989) to be performed for dislocations in the depletion region in order to obtain the dislocation line charge and ultimately the concentration of recombination centres at the defect. However, there is one further effect to consider which occurs when the band bending due to the depletion region is sufficient to raise the dislocation states above the Fermi level. In this situation the dislocation charge Q, is now zero and so, in this model, the recombination at the dislocation will also be zero since the attractive potential for minority carriers is removed, see figure 1c. Thus, the effect of the depletion region field is to reduce the recombination rate and hence the EBIC contrast of dislocations lying in the depletion region relative to a dislocation of the same charge lying in the bulk, ie. from equations (2),(4) and (5),

$$C \propto [kT/q \ \ln(B'/J_e')] - V \qquad \qquad C \propto [kT/q \ \ln(B/J_e)] \qquad (9)$$
$$\text{depletion region} \qquad \qquad \qquad \qquad \text{bulk}$$

3. EXPERIMENTAL

In this work contrast measurements of dislocations were made using an EBIC system similar to the one previously described by Wilshaw et al (1983), which is now based around a Philips 505 SEM with a LaB_6 gun. High purity, swirl free, float zone, n-type, 10^{15} cm^{-3} silicon that has been deformed under clean conditions by a two stage compression at 850°C and 420°C has been studied. The specimens contain hexagonal loops and straight segments of dislocation up to several hundred microns in size. Contrast measurements of individual straight, dissociated, screw and 60° dislocations in both the depletion region and the bulk were made using an accelerating voltage of 15kV. The EBIC signal was collected at an evaporated Au/Pd Schottky contact.

4. DISCUSSION

This analysis points to the possibility of exploiting EBIC measurements of dislocations in the depletion region to obtain values for the position of the dislocation energy level. It can be seen that the dislocation charge and hence recombination efficiency tends to zero when $V = \emptyset_0$ in figure 1c. Since V is known as a function of depth through the depletion region, by noting the depth at which a dislocation inclined to the specimen surface disappears a value for $\emptyset_0$ could be obtained. Alternatively, the value of V at any point in the depletion region may be increased by applying a reverse bias to the Schottky barrier. Thus a dislocation at such a depth that its dislocation level lies below the Fermi level and hence is charged may be rendered neutral by applying a sufficiently large reverse bias so that V now exceeds $\emptyset_0$. Thus the value of V at which the dislocation disappears will yield the value of $\emptyset_0$. V may be obtained knowing the dislocation depth x, from,

$$V = V_s - x[2n_0 qV_s / \epsilon \, \epsilon_0]^{\frac{1}{2}} + n_0 qx^2 / 2\epsilon\epsilon_0 \qquad (10)$$

where

$$V_s = V_{bi} + V_{RB} \qquad (11)$$

V_s is the voltage at the surface, V_{bi} is the built in voltage due to the Schottky barrier and V_{RB} is the applied reverse bias voltage.

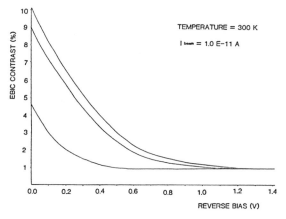

Figure 4. Plots of contrast vs reverse bias for the three 60° dislocations in figure 3.

Figures 4 and 5 show EBIC contrast does indeed decrease with increasing reverse bias as predicted, but approaches a finite contrast contrary to zero contrast as predicted from the above analysis. This residual recombination is not described by this analysis since the model used assumes a depletion region depleted of majority carriers and only considers thermally excited electrons from the conduction band. However, under experimental conditions the injected carriers produced within the generation volume of the electron beam creates an electron quasi Fermi level within the depletion region. Hence under these conditions it may still be possible to charge a dislocation even if the dislocation states are above the position of the Fermi level at equilibrium.

(a) 0.0 V Reverse bias

(b)   0.0 V Reverse bias

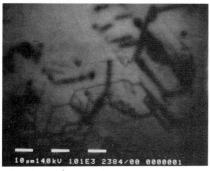

(c) -0.6 V Reverse bias

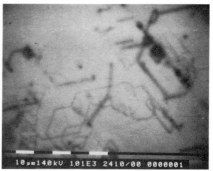

(d) -1.2 V Reverse bias

Figure 5. (a) An EBIC micrograph taken at a low accelerating voltage of 10 kV. Only those dislocations which lie in the depletion region are visible. (b),(c) and (d) are EBIC micrographs taken at 14 kV with different reverse bias voltages. These clearly demonstrate the reduction in contrast of dislocations lying in the depletion region field as increasing reverse bias is applied. Those present in the bulk show little change.

In summary we believe this technique has great potential to determine the position of dislocation energy levels within the band gap from simple EBIC measurements once the depth of the dislocations has been measured.

5. ACKNOWLEDGEMENTS

The authors would like to thank SERC for financial support.

6. REFERENCES

Omling P, Weber E R, Montelius L, Alexander H and Michel J 1985 Phys. Rev. **B32** 6571
Wilshaw P R, Ourmazd A and Booker G R 1983 J. Physique **44** C4 445
Wilshaw P R, Booker G R 1986 5th Int.Sym. Structure and properties of dislocations. Moscow
Wilshaw P R, Fell T S and Booker G R 1989 Point, Surface & Extended Defects. ed. Benedek G, Schröter W, Cavallini A. (NATO ASI series,Plenum)

Inst. Phys. Conf. Ser. No 104: Chapter 2
Paper presented at Int. Symp. on Struct. Prop. Disloc. Semicond., Oxford, 1989

233

On the formation of bright EBIC contrasts at crystal defects

H. Blumtritt[+], M. Kittler[++] and W. Seifert[++]

Academy of Sciences of the GDR,
[+] Institute of Solid State Physics and Electron Microscopy, PF 250
  Halle(S), DDR-4010
[++] Institute of Semiconductor Physics, PF 509, Frankfurt(O), DDR-1200

ABSTRACT: The paper reviews different formation mechanisms of bright EBIC contrasts at crystal defects. Contrasts due to enhanced lifetime in getter zones, variation of doping, charge collection by defect-own depletion layers, and repulsive action of charged defects are illustrated by examples.

1. INTRODUCTION

As a rule crystal defects are imaged by EBIC in the well-known dark contrast, which results from an enhanced carrier recombination at defect-related centres and has been widely studied both experimentally and theoretically in recent years. The explanation of occasionally observed bright defect contrasts, however, is more complicated, since different physical effects may be involved, sometimes even acting simultaneously, and a unique theoretical modelling has not yet been performed. In general, one has to distinguish between indirect defect actions, connected with a defect-induced redistribution of dopant and/or deep-centre impurities, and the direct electronic effects of charges in defect-related centres on the minority carriers. The paper will give a review on the different mechanisms of bright contrast formation, illustrated by examples of dislocations, precipitates, and of planar defects in silicon.

2. LIFETIME CONTRAST OF GETTER ZONES

In the neighbourhood of decorated crystal defects zones of bright contrast may be observed due to a slightly enhanced diffusion length (DL), which results from a denudation of recombination-active impurities within these so-called getter zones. Under optimum imaging conditions (material of a short diffusion length, p-n junction EBIC with a surface-near generation, or Schottky barrier EBIC using a large generation depth) variations of DL by only some per cent are easily detectable. The common appearance of strong recombination contrast and bright getter zones directly proves the dominant role of the defect decoration and allows comparative studies to be made of the gettering capability of defects of different types (Blumtritt et al. 1979).
Fig.1a shows high-active, strongly decorated dislocations (60°-type, dissociated) within the p region of a diffused p[+]-n diode with recombination contrasts up to 12 % and getter zones, and contrary to those also low-

active, non dissociated dislocations without a getter zone contrast.
Fig. 1b presents getter zones of gold atoms near deformation-induced
dislocations in an intentionally contaminated n-type Fz-crystal after
annealing at 800°C for 1 h, imaged by a surface Schottky barrier (sample
prepared by I.E.Bondarenko, Chernogolovka).

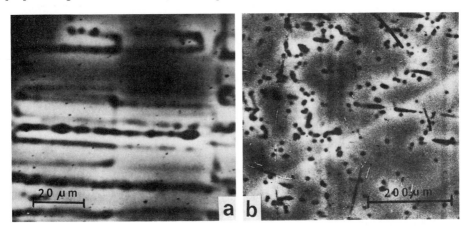

Fig. 1. Getter zones around dislocations in a process-contaminated
p⁺-n diode (a: E_o=10keV) and in gold-doped, deformed and annealed n-type
Fz-Si (b: E_o=20keV)

The contrast values of the extended getter zones and the material parameters in defect-free regions can be used to estimate the corresponding
increase of DL by applying expressions of the charge collection efficiency
of planar barrier structures (Pasemann et al. 1982) assuming, in a rough
approximation, a homogeneous DL within the getter zone.
In the case presented in Fig. 1a (L=1.5/um, S=55, X_J=2/um, R= 1/um) an
increase by about 0.05/um is obtained; for Fig.1b (L=5/um, W_{scr}=1/um,
R=4/um) an increase by up to 10 % results.

3. CHARGE COLLECTION IN DEFECT-OWN DEPLETION LAYERS

A major reason for enhanced defect signals is an additional charge collection within the space charge region (SCR) around defects charged due to an
occupation of centres by majority carriers. The effect has occasionally
been observed both in EBIC and OBIC studies (Castaldini et al. 1987,
Turner et al. 1982). Formation and detection of this signal component,
however, presuppose a conductivity of the defect plane itself (which is
contacted by the metal layer) occuring solely under special structural
conditions.
The Schottky barrier micrographs in Fig.2 show small-angle grain boundaries (GB) in p-type silicon, produced by hot-pressing of well-oriented
slices (prepared by T. Sullivan at Cornell University). While most GB
occur in the usual, strong recombination contrast, at a high beam energy
part of the segments have a remarkably bright contrast due to the additional charge collection by the GB SCR. The origin of the bright signals from
the barriers in vertical direction to the surface can be proved by their
energy dependence being different from that of the signal of the surface
Schottky barrier. TEM investigations showed the bright contrast to be
associated with a very high density of (oxygen?) precipitates on these
boundaries obviously being the origin of a high density of interface

states necessary for the formation of a conducting inversion layer at the GB plane (in analogy to a model of Henry et al. 1986).

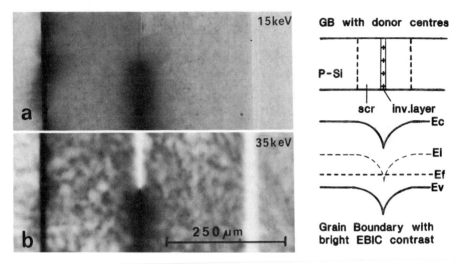

Fig. 2. Small-angle GB in p-type hot-pressed Si, segments partly with bright contrast due to an additional charge collection in defect-own depletion barriers

4. CONTRAST DUE TO A REPULSIVE DEFECT POTENTIAL

In principle, a repulsion of minority carriers by defects charged by carriers of the same sign should influence the charge collection process and, depending on the geometry, should result in enhanced or decreased signals. While for line defects such an effect should probably not suffice to be detected, for extended planar defects and precipitates under special

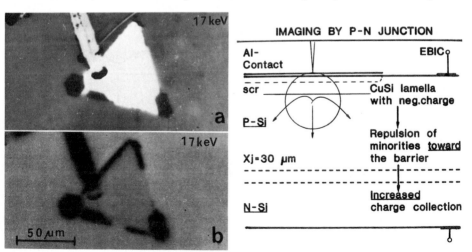

Fig. 3. Negatively charged CuSi planar defect in p-type Si, EBIC imaging by a deep p-n junction (a) and by a surface Schottky barrier (b), schematic illustration (c)

imaging and detection conditions bright contrasts well above the detection limit have been found (Gleichmann et al. 1983, Jakubowicz and Habermeier 1985).

Fig. 3a shows a complex defect consisting of different morphological types of copper silicide precipitates near the surface of a p^+-n structure. Between recombination-active compact precipitates a monolamellar CuSi-planar defect appears with a bright contrast up to 30% provided the majority of carriers is generated below the defect plane. This is explained by a shielding of the minorities (electrons) against the surface influence by the negatively charged defect plane, reducing the carrier loss due to surface recombination. Large values of the junction depth, the beam energy, the surface recombination velocity, and a small defect depth increase the bright EBIC contrast. The essential influence of the charge collection geometry on the detection of such a repulsive defect action is demonstrated by Fig. 3b, where the same defect is simultaneously imaged by an additional surface Schottky barrier, now showing - contrary to Fig. 3a- only a weak dark contrast as a result of the repulsion of minority carriers down to the bulk material. This unusual contrast behaviour was regarded as a strong indication of the defect not to be a normal stacking fault as might be concluded from a diffraction contrast analysis alone.

5. CONTRAST DUE TO DOPING INHOMOGENEITIES

As the properties of the collecting junction, particularly SCR width and electrical field strength, depend on doping and may influence charge collection significantly, contrast effects can be also due to redistribution of dopants. Depending on the experimental conditions both enhanced and decreased doping may lead to bright contrasts.

Fig. 4 shows results on oxygen-induced defects in initially p-type Cz-Si annealed 20h at 1100°C. After thermal donor (TD) formation at 450°C the sample became n-type and bright contrasts were found for $E_0 > 15 keV$ (Fig.4a). These contrasts could be explained both by a 20% increase of DL and by a 50% decrease of doping (i.e. increase of SCR width) in the defect neighbourhood, without possibility to decide about the acting mechanism. Additional information was provided by investigating the sample after TD

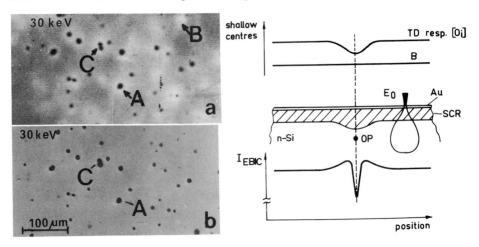

Fig. 4. Contrast at oxygen induced defects: after TD formation, Au Schottky contact (a); after TD destruction, Al Schottky contact (b); schematic illustration (c)

destruction (compare Fig. 4b). The experimental data gave rise to the following interpretation (Fig. 4c) of the contrast, supported by a good correspondence between dimensions of bright zones and oxygen diffusion length: The contrast is due to a reduced net doping (increased width of the SCR) caused by a reduced TD concentration. The TD distribution on its turn reflects the denudation of interstitial oxygen around the defect (Seifert and Kittler 1987).
A similar effect might be responsible also for bright contrasts after annealing in the new-donor temperatures range as observed e.g. by Batistella and Rocher (1987).
The second case of doping contrast may be observed at low beam energies and high injection when the generation volume is inside the SCR and a relatively stable electron-hole plasma is formed. Under these conditions SCR regions of stronger electrical field (higher doping) give a larger EBIC signal (Leamy 1982). Fig. 5 demonstrates this behaviour for GB in a thick phosphorus-doped SOI layer (for details see Kittler et al. 1988). The dramatic contrast change at the outer GB is a clear indication of this contrast mechanism and shows that the near surrounding of the bright GB (Fig. 5b) is enriched with phosphorus. This explanation agrees nicely with results of detailed consideration of the recrystallization process which indicate that the GB regarded are sites where two solidification fronts have met and where segregation effects could be strong.

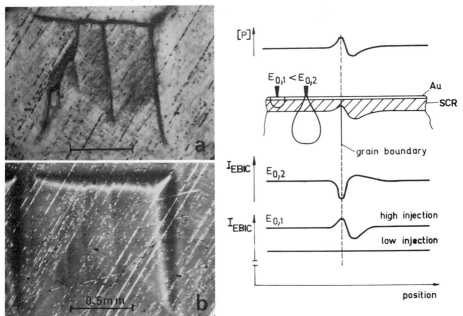

Fig. 5. Bright contrast due to enhanced doping at GB in a SOI layer: 30keV, low injection (a); 5keV, high injection (b); schematic illustration (c); contrasts passing the micrograph diagonally are due to surface scratches

6. CONCLUDING REMARKS

The examples given above illustrate main mechanisms leading to bright EBIC contrasts. Avalanching of carriers is a possible contrast cause too, however, this process is unlikely for unbiased junctions and is therefore not considered here. Simultaneous charge collection by more than one junction may also result in unexpected bright contrast effects. Because of the variety of existing contrast mechanisms the interpretation of contrast features in unknown samples is difficult and should be carried out with care. For this purpose as much as possible information should be collected, either by the EBIC method itself (e.g. by variation of beam energy, beam current and collection geometry) or by other experimental techniques. Then, one has to check whether a supposed mechanism is likely to be responsible for the behaviour observed. Although there are no adequate theoretical models available to verify the bright contrast origin, some insight can often be obtained on the ground of the already existing charge collection models. So, as shown above, corresponding calculations are able to give an idea on the probable candidates for the contrast origin and to provide an estimate of the material characteristics necessary to produce the observed contrast.

To conclude, more systematic experimental and numerical investigations of bright contrast effects are desired in order to draw conlusions about properties of defects and their interaction with the surrounding material.

The authors acknowledge fruitful collaboration and discusion with I. E. Bondarenko, R. Gleichmann and T. Sullivan.

REFERENCES

Batistella F and Rocher A 1987 Semicond. Scie. Technol. 2 226

Blumtritt H, Gleichmann R, Heydenreich J and Johansen H 1979 phys. stat. sol.(a) 55 611

Castaldini A, Cavallini A, Poggi A and Susi E 1987 Proc. 2nd Int. Autumn Meeting 'GADEST' ed H Richter (Acad. Sci. GDR) pp. 248-51

Gleichmann R, Blumtritt H and Heydenreich J 1983 phys. stat. sol. (a) 78 527

Henry A, Pautrat J L and Saminadayar K 1986 J. Appl. Phys. 60 3192

Jakubowicz A and Habermeier H-U 1985 J. Appl. Phys. 58 1407

Kittler M, Tillack B, Hoppe W, Seifert W, Banisch R, Richter H H and Rocher A 1988 Revue Phys. Appl. 23 281

Leamy H J 1982 J. Appl. Phys. 53 R51

Pasemann L, Blumtritt H and Gleichmann R 1982 phys. stat. sol.(a) 70 197

Seifert W and Kittler M 1987 phys. stat. sol. (a) 99 K11

Turner G B, Tarrant D, Aldrich D, Pressley R and Press R 1982 Grain boundaries in Semiconductors ed H J Leamy, G E Pike and C H Seager (New York, Amsterdam, Oxford: North Holland) pp. 241-6

Inst. Phys. Conf. Ser. No 104: Chapter 2
Paper presented at Int. Symp. on Struct. Prop. Disloc. Semicond., Oxford, 1989

239

Deep states associated with platinum decorated stacking faults in silicon

G R Lahiji, A R Peaker, and B Hamilton

Centre for Electronic Materials and Department of Electrical Engineering and Electronics
University of Manchester Institute of Science and Technology
P O Box 88, Manchester M60 1QD, U K

ABSTRACT: We have studied the electron emission from deep states associated with oxygen induced stacking faults in silicon with and without platinum decoration. The activation energy for electron emission increases systematically with the concentration of platinum incorporated in the lattice by 700 °C diffusions. At higher diffusion temperatures the activation energy decreases and it is postulated that this is due to a change in the platinum siting.

1. INTRODUCTION

A very considerable amount of work has been expended in trying to elucidate the electronic nature of extended defects in silicon. Most of the published results have used Hall measurements or DLTS to determine electron binding energies associated with various types of dislocation (Kimmerling & Patel 1979, Kveder et al 1982, Whitwer et al 1986, Szkielko et al 1981). In some cases these studies have been supplemented by Electron Paramagnetic Resonance (EPR) and Photoluminescence (PL) (e.g. Weber et al 1987, Sauer et al 1984). Although there is by no means a consensus, there is a sufficient degree of similarity in the results to be able to draw a number of generalisations.

The electron emission characteristics associated with dislocations in silicon seem to fall into three main groups, one with an activation energy of $\sim$ 190 meV, one with $\sim$ 280 meV and a much broader band of results with an activation range from 370 – 600 meV. Some reviews subdivide this last group into three sections making a total of five groups of deep states associated with dislocations observable in n–type material. In all cases the dislocations have been produced by plastic deformation, often at very high densities ($\sim 10^9$ cm^{-2}). In all the publications referred to, no attempt has been made to decorate the dislocations intentionally.

In our work we have examined the deep states associated with oxygen–induced stacking faults and have selected material and processing conditions which we believe to be 'clean', i.e. in which the level of any unintentional decoration should be slight. We have then introduced impurities into the silicon. In this paper we report results on platinum decoration.

2. EXPERIMENTAL

The work was performed on 3 inch (100) slices of epitaxial silicon grown on n$^+$ Czochralski substrates. The layers were n–type (phosphorus doped) with n = 2 x 10^{15}cm^{-3}. Each slice was cleaved into two. One half was lightly damaged by polishing with 1 μm diamond slurry, while the other half was left as received. Both parts were then cleaned and oxidised in dry oxygen at atmospheric pressure at 1050 °C for 120 minutes. The half of the slice which had been mechanically damaged had a high concentration of stacking faults ($\sim 10^7$ cm^{-2}), as revealed by etching (Yan 1984), while the undamaged half showed no stacking faults. Transmission electron microscopy showed a distribution of oxygen–induced stacking faults (OISF) between 2 and 4 μm long. The concentration of dislocations was very low in the undamaged material, and in the damaged layers the dislocation density was less than the stacking fault density.

Slices with and without stacking faults were then subjected to further heat treatments in order to diffuse platinum into the silicon. In addition, a group of slices was subjected to identical heat treatments but without platinum. The experiments were undertaken at low temperatures in order to minimise any further unwanted contamination and to avoid the formation of new stacking faults. To contaminate the wafers with platinum the back side oxide was removed while protecting the front side oxide. The slice was then dipped in a diluted solution of a proprietary platinum source (one part DNS solution/five parts DI water).* Samples were heated to 700°C or 750°C for 15, 30 and 5 minutes in a double tubed furnace in a nitrogen atmosphere.

After the heat treatment, low temperature electroless nickel plating (< 100°C) was used to form the back side ohmic contact. Gold Schottky diodes were fabricated on the front surface after removing the oxide. Capacitance–voltage measurements were used to determine the carrier concentration and Deep Level Transient Spectrometry (DLTS) to observe the electron emission characteristics from the deep states.

Consequently, the sample set contained the following groups:

    i)   no stacking faults, no contamination
    ii) 10^7 cm^{-2} stacking faults, no contamination
    iii)   no stacking faults, platinum contamination
    iv) 10^7 cm^{-2} stacking faults, platinum contamination

Each group had been subjected to the range of heat treatments referred to earlier.

3.   RESULTS AND DISCUSSION

Figure 1 shows DLTS spectra from representative slices. Group (i) consisting of as–received slices exhibited no significant DLTS signal on the concentration scale shown. Group (ii) which was made up of slices with stacking faults which were not deliberately contaminated gave the form of trace shown in (a). This did not change perceptably with the 700 or 750°C heat treatments but its magnitude did depend linearly on the concentration of stacking faults over the range measured (2 x 10^6 to 3 x 10^7 cm^{-2}). The DLTS trace shown is for a sample with 10^7 cm^{-2} stacking faults. Spectrum (b) is typical of group (iii), the dominant feature being the well–established platinum acceptor (E_a = 227 meV, e_n = 100 s^{-1} at 122.1 K). This is widely believed to be due to platinum located as a simple substitutional impurity. The concentration of the centre as derived from the DLTS trace was in the range 5 x 10^{12} to 5 x 10^{13} cm^{-3} dependent on both the diffusion temperature and the diffusion time. There was also some variation in concentration between slices which had undergone nominally the same heat treatments. The solid solubility of platinum in silicon is reported as 1.9 x 10^{14} cm^{-3} at 700°C and 3.5 x 10^{14} cm^{-3} at 750°C (Mantovani et al 1986). This is about an order of magnitude higher than the values of the electrically active species measured in our slices. This is expected as the diffusion must proceed via a 'kick out' mechanism (Goesele et al 1980) and equilibrium conditions will not be reached in our experiments. The trace shown as (c) in Figure 1 is typical of the spectra obtained from samples in group (iv), i.e. those containing stacking faults and contaminated with platinum. The peak associated with the isolated substitutional platinum can be seen at $\sim$ 120 K but a larger peak with emission characteristics similar to that of the undecorated stacking fault can also be seen.

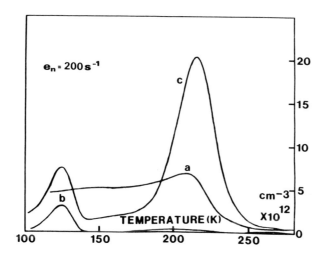

Figure 1 DLTS spectra of electron emission from deep
states associated with a) stacking faults without
platinum; b) platinum without stacking faults;
c) stacking faults decorated with platinum.

A number of points are worthy of note in relation to the concentration of the electrically detected defect states.   Invariably, the height of the DLTS peak associated with the decorated stacking fault is greater than either that associated with the undecorated stacking fault or the substitutional platinum.   The detailed interpretation of this requires rather more knowledge than we currently have regarding the interaction of the platinum with the stacking fault.   The DLTS plot merely tells us that the amount of trapped charge on the decorated stacking faults is greater than on the undecorated ones. This is likely to be due to an increase in the number of trapping sites.   What is also interesting is that the increase in peak size associated with the stacking fault is not accompanied by a decrease in the size of the substitutional platinum peak.   Indeed the very opposite appears to be true.   In the samples with stacking faults the overall concentration of the 227 meV state (substitutional platinum) is generally higher than in the silicon samples without extended defects which have been diffused under the same conditions.   It seems that the presence of the stacking faults has resulted in a substantial degree of diffusion enhancement.

Figure 2 shows an Arrhenius plot of the electron emission giving the peak related to the stacking faults.   The line with an activation energy of 415 meV is that associated with the undecorated stacking fault while the other lines are from slices with increasing levels of platinum in the lattice.   All the diffusions of platinum for the results shown in this diagram have been at $700\,^{\circ}C$ and the highest activation energy corresponds to the highest concentration of platinum.

In all our results we have taken the DLTS peak associated with the platinum acceptor (substitutional platinum) as a measure of the platinum concentration.   We know from previous work which compared neutron activation analysis and electrical measurements that this is an accurate measure of the platinum concentration in material with a low level of extended defects (Lisiak and Milnes 1975).   However, in the present case the proximity of a high concentration of stacking faults may make this assumption somewhat dubious and this should be borne in mind when interpreting the results, especially as mentioned earlier, the DLTS peak associated with the substitution platinum increases in magnitude as the stacking fault density increases.   We are assuming this is a real increase in the concentration of substitutional platinum due to enhanced diffusion.

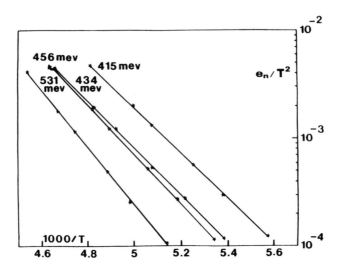

Figure 2   Arrhenius plot of electron emission from the deep states associated with stacking faults.   The line on the right (lowest energy) corresponds to the undecorated defect.   The activation energy increases and the line shifts to the left with increasing platinum decoration.

It is interesting to look more closely at the dependence of the activation energy for electron emission on the amount of platinum present.   The best quantitative measure we have of this is the substitutional platinum concentration, as observed from the 227 meV DLTS peak, although as mentioned above, this is not independent of the presence of stacking faults.   Figure 3 shows the variation of activation energy of the stacking fault associated peak with platinum concentration.   Two groups of samples have been measured, one group diffused with platinum at 700°C and the second group at 750°C. The points linked by the line on the left represent data associated with the 700°C diffusion.

As indicated by Figure 2 there is a systematic dependence of activation energy on platinum concentration at 700°C.   The activation energy increases to mid gap as the platinum concentration increases.   This occurs until the substitutional platinum reaches a concentration of $10^{13} \mathrm{cm}^{-3}$.   This is the maximum concentration observed for a 700°C diffusion in our samples.

Higher concentrations of platinum were obtained in the samples diffused at 750°C.   The range of substitutional platinum in this case was $2.7 \times 10^{13} - 5.2 \times 10^{13} \mathrm{cm}^{-3}$.   The activation energy tended to decrease with increasing platinum concentration in this region although there is very considerable scatter.

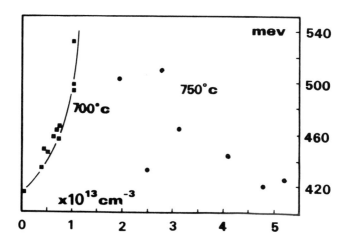

Figure 3 *The dependence of the activation energy for electron emission from the deep states associated with stacking faults as a function of the concentration of platinum in the lattice as measured by the amplitude of the DLTS peak associated with substitutional platinum. Data are shown for platinum diffusion at 700°C (■) and 750°C (●).*

We have not as yet examined the electron capture properties of the stacking fault related defects in detail. However, there are marked differences between the capture behaviour of the deep states associated with the 'clean' stacking faults and those associated with decorated ones.

4.   COMPARISON WITH PREVIOUS WORK

As outlined in the introduction, there has been a substantial amount of work on the electrical activity of defects produced by the plastic deformation of silicon. The collated results of the published work in the field (Peaker and Sidebotham 1988) show three main groups of deep states associated with dislocations. Our results on stacking faults fall into the highest energy group which display quite a wide range of emission characteristics (E_a = 370 – 540 meV; T = 205 – 262K for e_n = 100s^{-1}). However, at the Aussois meeting in this conference series, a state with very similar emission characteristics to the undecorated stacking faults was described by Weber and Alexander (1983) by Schröter and Seibt (1983) and more recently by Omling et al (1985) and by Schröter (1989). The emission has been tentatively ascribed to the core states associated with incompletely reconstructed 60° partial dislocations. A salient feature of this work was the observation of a quasi–saturation of filling of the defect which could be approximated by a logarithmic relationship. In many cases we do not observe such behaviour. Although the filling behaviour is not the same as that normally observed for an isolated point defect it does not approach logarithmic behaviour. It seems possible that the difference in electron emission dependence on the concentration of platinum after diffusion at 700°C and at 750°C is due to a change in the siting of the platinum on the dislocation. A similar transition in the behaviour of tantalum in MBE silicon layers containing dislocations has been described previously (Sidebotham et al 1988).

We have previously published a brief report on the decoration of stacking faults with gold (Lahiji et al 1988) and in that case also observed a systematic dependence of activation energy on the degree of decoration.   However, a very significant observation is that although the electron emission characteristics of substitutional gold are very different to those of platinum (Pt: 227meV, e_n = 100s^{-1} @ 122.1K, Au: 560meV, e_n = 100s^{-1} @ 272.5K), the effect of decorating the stacking fault associated defect with either gold or platinum is very similar.   If we compare our results on stacking faults with those previously reported on dislocations there is a marked similarity between the behaviour of our decorated defects and the previous work on as–deformed material.   Our undecorated stacking faults are distinctly different to those reported previously and although we are comparing stacking faults with dislocations, the possibility that previous work has described unintentionally decorated dislocations must be considered.

5.   CONCLUSIONS

We have observed that the electron emission behaviour of a deep state associated with oxygen–induced stacking faults is dependent on the amount of platinum we diffuse into the lattice. The activation energy systematically increases with increasing platinum decoration for diffusion at 700°C.   The capture behaviour of the defects associated with undecorated stacking faults is not logarithmic unlike states with similar emission characteristics previously reported as being associated with dislocations.

ACKNOWLEDGEMENTS

This work was conducted in part under a UK government research contract (ALVEY VLSI 022) and has been supported by the Science and Engineering Research Council. We would like to thank our collaborators in this contract for their help and advice.   One of us (GRL) would like to thank the University of Tehran for leave of absence.

REFERENCES

Goesele U, Frank W and Seeger A, 1980 *Appl. Phys.* **23** 361
Kimerling L C and Patel J R, 1979 *Appl. Phys. Lett.* **34** 1 73
Kveder V V, Osyipan Yu A and Schroter W, 1982 *Phys. Status Solidi a* **72** 701
Lisiak K P and Milnes A G 1975 *Sol. St. Electron* **18** 53
Lahiji G R, Hamilton B and Peaker A R, 1988 *Electronics Letters* **24** 1341
Mantovani S, Nava F, Nobili C and Ottaviani G, 1986 *Phys. Rev B* **33** 5536
Omling P, Weber E R, Montelius L, Alexander H M and Michel J 1985 *J. Phys. Rev. B* **32** 6571
Peaker A R and Sidebotham E C, Dislocation related deep levels in Si: *EMIS data review RN17836* in 'Properties of Silicon' IEE London, 1988, 221.
Sauer R, Weber J, Stoltz J, Weber E R, Kusten K H and Alexander H, 1984 *Appl. Phys. A* **36** 1.
Schröter W and Seibt M 1983 *J. Phys. Colloq.* **44** no.C-4 329–37
Schröter W 1989 (this volume)
Sidebotham E C, Peaker A R, Hamilton B, Hopkinson M, Houghton R, Patel G, Whall T E and Parker E H C, *Proc. 2nd International MBE Symposium Hawaii, 1987 Electrochem. Soc.* **Vol 88-8** 355.
Szkielko W, Breitenstein O and Pickenhein R 1981 *Cryst. Res. & Technol.* **16** 197
Weber E R and Alexander H 1983 *J. Phys. Colloq.* **44** no.C-4 319
Weber E R , Omling P, Kisielowski–Kemmerich C and Alexander H, 1987 *Izvestiya Akademii Nauk SSR. Seriya Fizicheskaya* **51** No. 4 644
Whitwer F D, Haddad H and Forbes L 1986 *Mater. Res. Soc. Symp. Proc.* **71** 53
Yan K H 1984 *J. Electrochem. Soc.* **131** 1140

* DNS Platinum solution contains 19g Pt/100cc of solution and is supplied by Johnson Matthey Metals, Hatton Garden, London, U.K.

Inst. Phys. Conf. Ser. No 104: Chapter 3
Paper presented at Int. Symp. on Struct. Prop. Disloc. Semicond., Oxford, 1989

Impurity effects on dynamic behaviour of dislocations in semiconductors

Koji SUMINO

Institute for Materials Research, Tohoku University, Sendai 980, Japan

ABSTRACT :   This paper gives a review of the dynamic behaviour of dislocations in semiconductors, especially in silicon and GaAs, with a special emphasis on the influence of impurities on the basis of the work of the author's group. Electrically inactive impurities do not affect the mobility of dislocations when the impurity atoms are dispersed individually within the crystal. They usually show strong immobilizing effects on dislocations through the occurrence of some reaction at the dislocation core. Electrically active impurities give rise to quite a rich variety of effects concerning both the mobility and immobilization of dislocations.

1. INTRODUCTION

Semiconductor crystals are commonly observed to be very brittle at room temperature and become ductile at high temperatures. This observation leads to the idea that a covalent crystal is the typical material in which the intrinsic barrier (the Peierls potential barrier) against dislocation motion is extremely high. The development in semiconductor device technology in the last few decades has brought about magnificent progress in the techniques of growing highly perfect semiconductor crystals which are substantially free from dislocations or are of low dislocation densities. This has facilitated the study of the dynamic behaviour of individual dislocations in semiconductor crystals. Thus, in comparison with other kind of materials, much more reliable results of measurements on the velocities of individual dislocations are available in semiconductor crystals.

However, in spite of such advantages in conducting experiments, the theoretical interpretation of the observed facts is usually not very easy. A considerable part of the difficulties seems to be related to direct and/or indirect effects of impurities on the nature of a dislocation in semiconductor. The core of a dislocation or a kink on it may have local electronic levels within the bandgap. Impurities which are electrically active as donors or acceptors control the Fermi level and, in turn, the electron occupation of such levels as a function of the temperature. The character of a dislocation or a kink may be influenced by the occupation state. A semiconductor crystal, even very pure in the starting material, is easily contaminated with heavy metals especially at elevated temperatures. Such metal impurities usually have very high diffusivities and low solubilities at low temperatures. This may result in a strong interaction between dislocations and impurity clusters.

This paper reviews impurity-related problems in the dynamic behaviour of dislocations in semiconductors on the basis of the work of the author's group.

2. PROBLEMS ON DISLOCATION MOBILITY RELATED TO IMPURITIES

A number of reports have so far been published on dislocation velocities in various kinds of semiconductors (see reviews, Alexander and Haasen 1968, Alexander 1986). It is now well established that the dislocation velocity in any of these materials under a given stress depends very sensitively on the temperature and, at a given temperature, rather insensitively on the stress. The velocity v of a dislocation in a semiconductor is generally expressed empirically by a simple equation of the type :

$$v \propto \tau^m \exp \ (-Q/kT) \tag{1}$$

as a function of the stress τ and the temperature T, where m and Q are material constants and k the Boltzmann constant. The magnitudes of m and Q are measured to be $1-2$ and $1.5-2.5$ eV, respectively. Equation (1) seems to hold in the stress range which is much lower than the Peierls stress of the material.

To interpret experimental observations on the dislocation velocity in semiconductors, theories have been put forward on the basis of the model that the elementary process of dislocation motion in a crystal is the thermally activated nucleation of a double kink on a straight dislocation line lying along the Peierls valley, followed by the expansion of the generated kink pair. Usually, any theory of double kink nucleation leads to a stress-sensitive activation energy for kink pair formation (Celli et al 1963, Dorn and Rajnak 1964). On the other hand, experimentally determined activation energies for dislocation motion in various kinds of semiconductors are insensitve to the stress, being almost constant with respect to the stress under which measurements are practically possible.

Two modifications of the theory have thus emerged. One assumes that some localized obstacles residing discretely along the dislocation line impede the kink motion and that the process of overcoming such obstacles plays an important role in determining the dislocation velocity (Celli et al 1963, Rybin and Orlov 1970). If we fit the experimental results on the dependence of the dislocation velocity on the stress and temperature in silicon to the results of the theory, the activation energy for overcoming the local defects turns out to be considerably higher than that for the double kink nucleation. This leads to a somewhat strange picture in the sense that the temperature-sensitive mobility of dislocations that is common in many kinds of semiconductors originates, not directly from the intrinsic nature of atomic bonding in semiconductors, but rather from some accidental or extrinsic defects in the crystal.

The magnitudes of the stress exponent m close to unity and the practically stress-independent activation energies Q of Eq. 1 in a variety of semiconductors remind us of the characteristics of the Einstein relation for the motion of a particle drifting under the action of a low force. Thus, the second model assumes that the intrinsic barrier for a kink motion (the second Peierls potential) is high in a semiconductor. The drift velocity of a kink is assumed to obey the Einstein relation while the nucleation rate of double kinks or the concentration of kinks is taken to be approximately independent of the stress. The latter assumption might be justified by the fact that the applied stress in usual velocity measurements is thought to be much lower than the Peierls stress.

On such a model Hirth and Lothe (1982) derived the dislocation velocity given by

$$v = (2 b_p b h^2 v_D/kT) \ \tau \ \exp \ [- (E_k + W_m)/kT] \tag{2}$$

where b_p and b are the magnitudes of the Burgers vectors of a partial and a total dislocation, respectively, h is the period of the Peierls potential, v_D the Debye frequency, E_k the energy of a single kink, and W_m the activation energy of kink motion along the dislocation line. Equation (2) has a form similar to the empirical one of Eq. (1) with respect to the dependence of the dislocation velocity on the stress and the temperature. Q in Eq. (1) may be interpreted to be the sum of E_k and W_m. The problem was also treated by Kawata and Ishioka (1983) as the Brownian motion of a one-dimensional object, on the basis of the idea that the motion of a smallest element of a dislocation obeys the Einstein relation. They succeeded in showing that the dislocation velocity depends linearly on the stress at low stresses.

From the experimental point of view, the magnitude of the stress exponent m has so far been determined to be unity only in the case of silicon when the motion of dislocations is traced by means of *in situ* X-ray topography (Sumino and Harada 1981, Imai and Sumino 1983). Those reported in other work on both silicon and other kind of semiconductors are larger than unity but smaller than 2. Such measurements have commonly been made by the intermittent technique, in which the position of a dislocation is determined at room temperature by the etch pit technique, or by X-ray topography, while they are displaced by stress at elevated temperature. The intermittent technique involves several sources of error in determining the accurate velocity of dislocations (Sumino 1987a). Among such sources the most important and unavoidable one is that related to the pinning of dislocations by impurities. As demonstrated in a later section, most impurities in a semiconcuctor crystal are very effectively gettered by dislocations and, as a consequence, the dislocations are immobilized. In some cases the effect is observed as the existence of a critical stress to start a pinned dislocation moving. In other cases a pinned dislocation spends some incubation period before starting after application of the stress. The latter case is probably related to local pinning of the dislocation. In such a case the stress exponent tends to be measured as apparently larger than unity even if the true value is unity. The pinning of a dislocation takes place by impurities, not only those introduced at the time of crystal growth, but also those introduced by contamination during heating of crystals at the time of velocity measurements. Under usual experimental conditions for dislocation mobility measurements it is practically impossible to avoid contamination by heavy metals such as Fe, Cu, etc. at elevated temperature. They become supersaturated when the specimen is cooled down to room temperature to determine the locations of dislocations and segregate on the latter, resulting in the immobilization of the dislocations.

A dislocation loop generated from some surface source in a highly pure silicon crystal usually assumes the shape of a half-hexagon, each segment of which is straight along ⟨110⟩ during motion under stress. However, *in situ* X-ray topographic observations sometimes reveal that the shape of a moving dislocation is disturbed from the half-hexagon (Harada 1980, Imai and Sumino 1983). The character of the shape disturbances is such that they seem to be caused by the retardation of dislocation motion in the region close to the specimen surface due to some dragging. Shape disturbances of this kind are observed to take place rather at random along the specimen and are not affected by the oxygen content of the crystal. Dislocations in motion at some places are observed to keep the regular half-hexagonal shape and those at other places to be disturbed within the same specimen. One of the plausible causes for this shape disturbance is thought to be the local pinning of dislocations in the surface region by impurities which are taken into the specimen from the surface when the specimen is kept at elevated temperature to observe the dislocation motion. A break on the dislocation line related to the shape disturbance sometimes extends over a depth of more than 100 µm from the surface. Thus, one should be careful in analysing the velocity data of dislocations that are obtained by the etch pit technique.

Now, a question arises as to whether the true magnitude of m is unity in all kinds of semiconductors as the theories based on the Einstein relation in kink motion require. Is the deviation from unity in the magnitude of *m* reported in a number of papers just related to experimental error? This is quite important a question to be answered. Even apart from achieving reliability in the measurements of dislocation velocity, the exact mechanism of impurity gettering at dislocations is by itself a very important problem to be clarified. The simplest answer for this effect is that dislocations are preferential nucleation sites for supersaturated impurities. However, as will be shown in a later section, the situation is much more complicated.

3. PINNING OF DISLOCATIONS BY IMPURITIES

Detailed studies have been conducted on how originally fresh dislocations are immobilized by gettering of various kinds of impurities in silicon and GaAs when they are aged at elevated temperatures. An extra stress is needed to start an aged dislocation moving. Such stress is termed here *the release stress*. The magnitude of the release stress is nearly equal to zero for dislocations aged in highly pure silicon or undoped GaAs.

The typical magnitude of the energy of interaction between an impurity atom and a dislocation related to the size-misfit of the impurity is evaluated to be lower than 0.5 eV. It can readily be shown that, with such a low interaction energy, an individual impurity atom can never pin a dislocation at temperatures where the measurements of dislocation velocities are possible in semiconductors. Dislocations are immobilized by impurities when the latter develop into clusters or complexes on the former, the magnitude of the interaction energy being higher than $2 - 3$ eV between them.

Figure 1 shows the release stress at 647°C for 60° dislocations measured against the duration of aging at the same temperature in silicon doped with O impurity at concentrations of 1.5×10^{17} and 7.5×10^{17} atoms/cm³, P impurity at a concentration of 1.2×10^{19} atoms/cm³, N impurity at a concentration of 5.5×10^{15} atoms/cm³ (Sumino and Imai 1983). The experiments were conducted by means of in-situ X-ray topographic observations. The release stress increases with increasing duration of aging, namely, with the increase in the number of impurity atoms gettered by the dislocations. The number of impurity atoms accumulated on a unit length of the dislocation core can be estimated from the concentration and diffusion coefficient of the impurities as a function of the temperature and the duration of aging (Sumino and Imai 1983, Yonenaga and Sumino 1985).

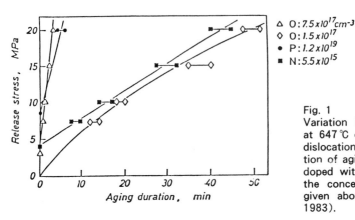

Δ O: 7.5×10^{17} cm⁻³
◇ O: 1.5×10^{17}
● P: 1.2×10^{19}
■ N: 5.5×10^{15}

Fig. 1
Variation in the release stress at 647°C of initially fresh 60° dislocations against the duration of aging at 647°C in silicon doped with O, P or N impurity the concentration of which is given above (Sumino and Imai 1983).

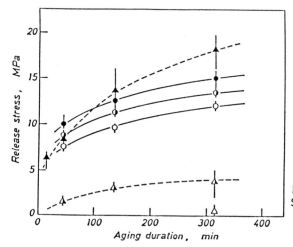

Fig. 2
Variation in the release stress
for initially fresh α dislocations
at 350 ℃ against the duration
of aging at 350, 450 and 550℃
in GaAs doped with In or Si
impurity of which concentration
is given below.

$$In:2x10^{20}cm^{-3}$$ 350 450 550°C
                          o   o   ●
$$Si:4x10^{18}cm^{-3}$$ △   △   ▲

It is shown that the strength of immobilization of a dislocation per gettered atom increases in the order O, P, and N. However, because of a high diffusion rate and a rather high solubility of O impurity at high temperature, immobilization of dislocations due to impurity gettering proceeds most efficiently in O-doped silicon.

The stress to release a dislocation from the impurities gettered due to a given thermal treatment has been measured on Czochralski silicon as a function of the temperature at which the stress is applied (Sato and Sumino 1985). The release stress decreases with an increase in the stressing temperature, showing that the release process of an aged dislocation from the gettered impurities is a thermally activated one . The energy of interaction of a pinning agent with a dislocation and the mean separation of the pinning agents along the dislocation line are deduced from the release stress versus temperature relation experimentally measured by the analysis based on a well accepted theoretical model. They have been determined to be of magnitudes around 3 eV and $3 \times 10^{-7} - 1 \times 10^{-6}$ m, respectively.

Figure 2 shows the release stress at 350℃ for α dislocations measured against the duration of aging at 350, 450 and 550℃ in GaAs doped with In impurity at a concentration of 2×10^{20} atoms/cm³ or with Si impurity at a concentration of 4×10^{18} atoms/cm³ (Yonenaga and Sumino, unpublished). The release stress increases very slowly in the Si-doped GaAs with increasing duration of aging at temperatures lower than 450℃, in comparison to the case of the In-doped GaAs. α dislocations are, however, very much more effectively immobilized by aging at higher temperatures in Si-doped GaAs than in In-doped GaAs. This implies that Si impurity immobilizes α dislocations very strongly if it is gettered by the dislocations. The low diffusivity of Si impurity, however, manifests the effect only at high temperatures.

One of the most striking facts concerning the dislocation immobilization due to impurity gettering in GaAs is that the efficiency of the immobilization depends not only on the species of impurities gettered but also very sensitively on the type of dislocations by which the impurities are gettered. Figures 3 (a) and (b) demonstrate this fact. Fresh dislocations are introduced into GaAs crystals doped with various kinds of impurities and subjected to a certain prescribed aging, in the present case at 550℃ for 315 min. Figure 3 (a) shows the release stress for α dislocations measured against the stressing temperature and (b) for β dislocations (Yonenaga and Sumino 1987, 1989). Except for undoped GaAs, the

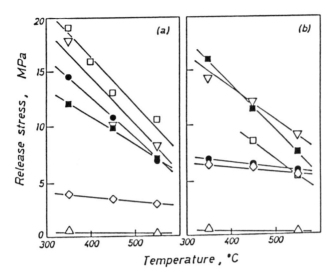

Fig. 3
Dependence of the release stress on the temperature in various GaAs aged at 550°C for 315 min.
(a) α dislocations
(b) β dislocations
(Yonenaga and Sumino 1987, 1989)

△ undoped
■ Zn:2x10^{19}cm^{-3}
▽ Si:4x10^{18}
□ Te:6x10^{18}
◇ Al:3x10^{18}
● In:2x10^{20}

immobilization takes place effectively for both α and β dislocations due to the above aging. It is seen that an isovalent impurity of In immobilizes α dislocations more strongly than β dislocations while Al, also an isovalent impurity, immobilizes β dislocations more strongly than α dislocations. It is to be noted, however, that the above kind of preferential immobilization of α dislocations by In impurity occurs only in the case when the In concentation is of the order of 10^{20} atoms /cm^3 or higher. In impurity at lower concentrations immobilizes α and β dislocations with approximately equal strength, the release stress being much lower. Contrarily, preferential immobilization of β dislocations by Al impurity takes place at an Al concentration as low as 3×10^{18} atoms/cm^3. A donor impurity of Te immobilizes α dislocations more strongly than β dislocations while Si, acting dominantly also as a donor, immobilizes α and β dislocations at approximately the same strengths. An acceptor impurity of Zn immobilizes β dislocations more strongly than α dislocations.

The magnitudes of the energy of interaction between a pinning agent and a dislocation and of the density of the pinning agents along the dislocation line are deduced from the data in Figs. 3 (a) and (b) and are determined to be $2.8 - 3.8$ eV and $2 \times 10^6 - 2 \times 10^7$ m^{-1}, respectively, depending upon the species of the impurity (Yonenaga and Sumino 1989).

At present, only the characteristics of the preferential immobilizing effect of In impurity on α dislocations in GaAs have successfully been interpreted in terms of the idea which takes into account of the size-misfit of an In atom occupying the Ga site and the high concentration of In at which the effect is observable (Yonenaga and Sumino 1987). However, no unified model is available to give an explanation for a variety of the observed facts mentioned above. It is to be noted that significant immobilizing effects are observed at rather low concentrations of the impurities, which are well below their solubility limits in the matrix crystal. A theoretical calculation shows that no Cottrell atmosphere is developed around a dislocation in such cases (Sumino 1987b). The author believes that some chemical reaction takes place at the dislocation core which incorporates the particular impurity atoms concerned together with other residual impurities. It is probable that the occurrence of such a reaction depends on the species of atoms located at the center of the dislocation core. The reaction products formed at the core of an α dislocation may generally differ from those formed at the core of a β dislocation. The reaction products developed at the core stabilize the system and lead to the immobilization of the dislocation.

An oxygen-related precipitate developed on a dislocation in silicon has been found to have atomic structure which is different from that in the matrix region developed at the same temperature. This gives evidence for the fact that some special reaction, which is absent in the bulk crystal region, can take place at the core region of a dislocation to incorporate an O impurity (Koguchi *et al* 1982, Yonenaga and Sumino 1985).

4. IMPURITY EFFECTS ON THE DISLOCATION VELOCITY

4.1. Silicon

When dislocations in motion are followed in highly pure silicon by means of *in situ* X-ray topography, the dislocation velocity v is measured to be linear with respect to the stress τ, and the activation energy Q is independent of τ as long as the shape of the moving dislocation is the regular half-hexagon. This has been confirmed to hold in the stress range $1 - 40$ MN$/$m^2 and the temperature range $600 - 800\,°C$. v is expressed as a function of τ and T by the following equation :

$$v = v_0 \ \tau \ \exp \left(- Q / k T \right) \tag{3}$$

where the magnitudes of v_0 are 1.0×10^4 and 3.5×10^4 m$^3/$MN $\cdot$ s and those of Q are 2.20 and 2.35 eV for $60°$ and screw dislocations, respectively (Imai and Sumino 1983).

The experimental magnitudes of v_0 in Eq. (3) are higher than the pre-stress factor in Eq. (2) by three to four orders of magnitude in the relevant temperature range. Thus, we may conclude that the microscopic model of Hirth and Lothe (1982) does not describe the elementary process of dislocation motion in a semiconductor crystal correctly.

Most dislocations in motion in highly pure silicon assume the shape of a regular half-hexagon irrespective of the magnitude of the applied stress in the above range, each segment of the half-loop being straight along $\langle 110 \rangle$. This is true also in impure silicon when dislocations move under high stresses. However, segments of a moving half-loop in impure silicon, originally straight under a high stress, are observed to be perturbed from $\langle 110 \rangle$ straight lines when the applied stress is reduced. Further reduction of the applied stress leads to the halting of dislocation motion. The above type of perturbation in the shape of moving dislocations is reversible. When the specimen in which dislocations are moving with the perturbed shape is subjected to high stresses, the segments quickly regain their straightness along $\langle 110 \rangle$. On reducing the stress again, the shape is perturbed again (Imai and Sumino 1983).

The above observations lead to the following picture. In impure silicon the motion of a kink is strongly retarded by some impurity-related obstacle formed on the dislocation line. Overcoming of such an obstacle by the kink is less frequent than the nucleation of double kinks in the obstacle-free region, at least under low stresses. Pile up of kinks takes place against the obstacle and, consequently, a depression is developed on the dislocation line. Once the piled-up kinks overcome the obstacle, the dislocation has a strong tendency to become straight along $\langle 110 \rangle$. This phenomenon is best interpreted in terms of the idea that the intrinsic barrier against the kink motion is small enough to allow kinks to have a high mobility along the dislocation line. The impurity-related obstacles are not individual impurity-atoms since the energy of interaction between a kink and an impurity atom is so small that they do not affect the kink motion in the relevant temperature range.   The obstacles are thought to be clusters or complexes of impurities which are developed at the core of a slowly moving dislocation. Probably, the pipe diffusion of impurities along the dislocation core plays an essential role in developing such pinning agents. An extra stress is needed to restart fully stopped dislocations. This is the release stress for dislocations described in the preceding section.

In situ observations of dislocations in motion with a high voltage electron microscope equipped with a high temperature tensile stage have also revealed interesting features in the dislocation motion in silicon on a microscopic scale (Sato and Sumino 1977, Sumino and Sato 1979). In accord with the results of *in situ* X-ray topography, fast moving dislocations are observed to consist of straight segments parallel to $\langle 110 \rangle$. In impure silicon the shape of dislocations becomes irregular when they move at low velocities. Segments of the dislocation are observed to move by repeated local depression and quick recovery of it. The dislocation in such a case is observed to be very flexible and behaves in such a way that the concept of line tension of a dislocation works quite well. In terms of the concept of kink motion this again suggests a high mobility of kinks along the dislocation line.

Light element impurities dissolved in silicon are found not to affect the velocity of dislocations which are moving under high stresses. Dislocations in motion are observed to keep the shape of the regular half-hexagon in such cases.   Under low stresses the shape of a moving dislocation is perturbed and the velocities are measured to be lower than those in highly pure silicon. In such a stress range the stress exponent m is measured to be larger than unity. This situation is shown in Fig. 4 (Imai and Sumino 1983). These phenomena are attributed to the effect of the impurity-related obstacles developed on the dislocations.

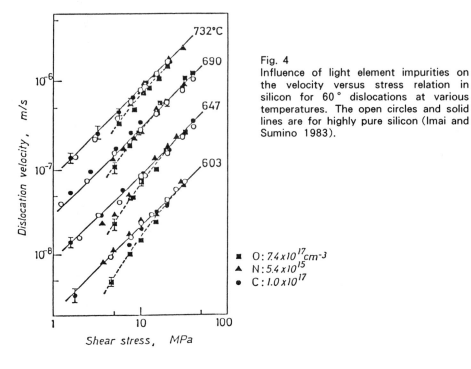

Fig. 4
Influence of light element impurities on the velocity versus stress relation in silicon for 60 ° dislocations at various temperatures. The open circles and solid lines are for highly pure silicon (Imai and Sumino 1983).

■ O : $7.4 \times 10^{17} cm^{-3}$
▲ N : 5.4×10^{15}
● C : 1.0×10^{17}

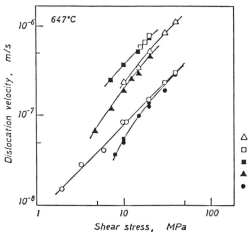

Fig. 5
Effect of electrically active impurities on the velocity versus stress relation of 60° dislocations at 647 °C in silicon. The open circles are for highly pure silicon (Imai and Sumino 1983).

△ P : $6.2 \times 10^{18} cm^{-3}$
□ P : 1.5×10^{19}
■ As : 1.5×10^{19}
▲ Sb : 6.5×10^{18}
● B : 1.4×10^{19}

Figure 5 shows the effects of electrically active impurities on the velocity versus stress relation for 60° dislocations at 647 °C measured by means of *in situ* X-ray topographic observations (Imai and Sumino 1983). The breakdown of the linear relation in the low stress range is evident in all the doped silicon and is attributed to the local pinning of slowly moving dislocations already discussed. The velocity in silicon doped with donor impurities is higher than that in highly pure silicon and increases with increasing concentration of donor impurities. The same effect has been reported earlier (Erofeev *et al* 1969, Kulkarni and Williams 1976, Patel *et al* 1976, George and Champier 1979). The stress exponent *m* in the high-stress range is unity and dislocations in motion are observed to retain their regular half-hexagonal shape there. The increase in the mobility is related to decreases in the magnitudes of both v_0 and Q in Eq. (3). Acceptor impurities seem to have no appreciable influence on the dislocation velocity in the high-stress range.

The above effects of electrically active impurities may be interpreted in terms of the idea that the core of a dislocation or a kink in the saddle point configuration for motion accompanies a deep acceptor level within the bandgap. The reduction in the formation energy of a kink or in the activation energy of kink motion may result from the doping of donor impurities in such a case (Hirsch 1979, Jones 1980).

4.2. GaAs

The experimental investigation on dynamic characteristics of dislocations in GaAs is much more difficult in comparison with the case of silicon for several reasons. First of all, except for GaAs doped with In impurity, crystals free from dislocations or of low dislocation densities are difficult to grow. Usually, the crystals contain dislocations at densities higher than 10^4 cm^{-2}. To follow the motion of freshly introduced dislocations by means of the X-ray topography or the etch pit technique is quite difficult in such a crystal . Only a limited number of specimens suitable for investigation are available by chance. Secondly, the absorption coefficient for X-rays is much higher in GaAs than in silicon. Thus, one must utilize the Borrmann effect to observe dislocations. The effect, however, does not work well at elevated temperatures because of thermal vibration of the lattice. So, one has to adopt more or less the intermittent technique that is inaccurate in measuring the dislocation velocity. Since most impurities immobilize dislocations very effectively, some error is unavoidable in the measurements with this

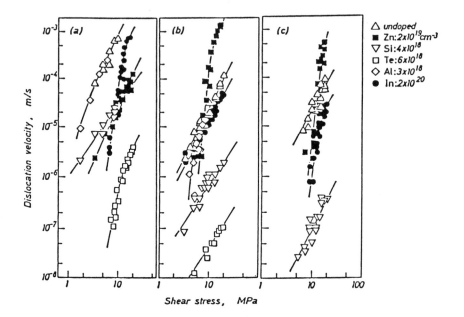

Fig. 6 Velocity versus stress relations at 550 °C for (a) α dislocations, (b) β dislocations and (c) screw dislocations in GaAs doped with different impurities (Yonenaga and Sumino 1989)

technique. Thirdly, α dislocations usually have a much higher mobility than β or screw dislocations in GaAs. This makes it difficult to measure the velocities of all kinds of dislocations simultaneously with the same specimen.

Figures 6 (a), (b), and (c) show the velocities of α, β, and screw dislocations, respectively, at 550 °C measured against stress in GaAs doped with different kinds of impurities (Yonenaga and Sumino 1989). The velocities were measured by means of the intermittent technique. The velocity of α dislocations is higher than those of β and screw dislocations by more than one order of magnitude in undoped GaAs. Dislocations are not moved under low stresses in GaAs doped with the impurities that are effective in immobilizing the dislocations. The velocity of dislocations is determined to be zero in the low-stress range in such GaAs. This is the essential deficiency of the intermittent technique. The velocity of dislocations in GaAs increases rapidly with stress once the applied stress exceeds the release stress and the apparent magnitude of the stress exponent m is determined to be high in the low-stress range. In such a case the velocity versus stress relation shows a break at some high stress, from where the velocity increases rather slowly with increasing stress, at approximately the same rate as that in undoped GaAs. This behaviour is similar to that seen in impurity-doped silicon shown in Fig. 4.

The dislocation velocity in the high-stress range after the break is thought to reflect the effect of impurities dispersed within the crystal. Expressing the dislocation velocity in the high-stress range by Eq. (1), we obtain the magnitudes of m and Q given in Table I. The magnitude of m ranges between 1.4 and 2.1, showing no systematic dependence on the immobilizing efficiency of the impurities shown in the preceding section. However, because of the errors which

Table I   Magnitudes of *m* and *Q* for α, β and screw dislocations in GaAs

	α dislocation		β dislocation		screw dislocation	
	m	*Q* (eV)	*m*	*Q* (eV)	*m*	*Q* (eV)
undope	1.7	1.30	1.6	1.30	1.8	1.4
GaAs : In		1.40	1.8	1.40		1.3
GaAs : Al	1.7	1.3		1.3		
GaAs : Si	1.4	1.10	1.4	1.60	1.9	1.7
GaAs : Te		1.10	1.5	1.60		
GaAs : Zn	2.1	1.40		1.15		1.6

are thought to be inherently involved in the intermittent technique to determine the dislocation velocity, we do not dare to conclude here that the exact magnitude of *m* in GaAs is not unity.

In GaAs doped with an isovalent impurity of In the velocities of α and screw dislocations after the breaks are a little lower than those in undoped GaAs, while the velocity of β dislocations is also slightly lower than that in undoped GaAs over the whole stress range investigated. The velocity of α dislocations in GaAs doped with the isovalent impurity of Al is approximately equal to that in undoped GaAs over a wide stress range, while the same holds for the velocity of β dislocations after the break. We conclude here that, as in the case of silicon, electrically inactive impurity atoms dispersed within a GaAs crystal do not effectively impede the motion of any type of dislocations, in agreement with the theoretical expectation.

As to the effects of electrically active impurities, donor impurities such as Si and Te reduce significantly the velocities of all the types of dislocations. The acceptor impurity Zn reduces the velocity of α dislocations. On the other hand, Zn impurity enhances strongly the velocities of β and screw dislocations under high stresses after the breaks. As a result, contrary to the case of undoped GaAs or that doped with other kind of impurities, the velocities of β and screw dislocations are higher than that of α dislocations under high stresses. This fact has been reported also by other workers (Erofeeva and Ossipyan 1973, Osvenskii *et al* 1973, Choi *et al* 1977, Steinhardt and Haasen 1978).

It is interesting to note that the impurities that are effective in immobilizing dislocations at rest are not necessarily effective in reducing the velocity of dislocations in motion. Zn impurity, which is very effective in immobilizing β and screw dislocations, enhances the motion of these types of dislocations in the high-stress range. The enhancement of the velocities of β and screw dislocations due to the doping of acceptor impurities may be interpreted with the idea that the core or the kink on a 30° partial dislocation of β type accompanies some donor level within the bandgap. Donor impurities retard the motion of all types of dislocations. The retarding effect is observed at a temperature as high as 550°C. If we interpret this fact in terms of the interaction of a dislocation in motion with some obstacles dispersed within the crystal, the magnitude of the interaction energy turns out to be higher than 3 eV (Yonenaga and Sumino 1987). Such a high interaction energy is available only when donor impurities develop clusters or complexes with a large misfit strain. A dislocation in motion in such a case is thought to assume an irregular shape which originates from local retardation by the discretely distributed obstacles. The stress exponent *m* is also expected to be high in such a case. Contrary to such expectations, X-ray topographic observations show that dislocation half loops moving in Si-doped GaAs at the reduced velocities assume a regular half-hexagonal shape and the magnitude of *m* is measured to be close to that in undoped GaAs over a wide range of stress, showing no break. These seem to show that the reduction in the dislocation velocity due to the doping of donor impurities is associated with the increase in the inherent resistance of the GaAs lattice to the motion of any type of dislocations due to the elevation of the Fermi level.

5. CONCLUSION

Many problems remain to be clarified with respect to the dynamic characteristics of dislocations in semiconductors, which may be regarded to be simple materials having extremely high Peierls potential barriers. Especially, the effects related to impurities are rich in their variety and are quite interesting from the fundamental viewpoint and, at the same time, are very important with respect to the practical device production technology.

REFERENCES

Alexander H 1986 *Dislocations in Solids* **7** ed F R N Nabarro (Amsterdam : North-Holland) pp 113 – 234
Alexander H and Haasen P 1968 *Solid State Physics* **22** ed F Seitz, D Turnbull and H Ehrenreich (New York : Academic Press) pp 27 – 158
Celli V, Kabler M, Ninomiya T and Thomson R 1963 *Phys. Rev.* **131** 58
Choi S K, Mihara M and Ninomiya T 1977 *Jpn. J. Appl. Phys.* **16** 737
Dorn J E and Rajnak S 1964 *Trans. Met. Soc. AIME* **230** 1052
Erofeev V N, Nikitenko V I and Osvenskii V B 1969 *Phys. Stat. Sol.* **35** 79
Erofeeva S A and Ossipyan Yu A 1973 *Soviet. Phys. Solid State* **15** 538
George A and Champier G 1979 *Phys. Stat. Sol.*(a) **53** 529
Harada H 1980 *PhD Thesis* Tohoku University
Hirsch P B 1979 *J. Phys. Paris* **40** C6-117
Hirth J P and Lothe J 1982 *Theory of Dislocations* (New York : John Wiley & Sons) pp 531 – 545
Imai M and Sumino K 1983 *Phil. Mag. A* **47** 599
Jones R 1980 *Phil. Mag. B* **42** 213
Kawata Y and Ishioka S 1983 *Phil. Mag. A* **48** 921
Koguchi M, Yonenaga I and Sumino K 1982 *Jpn. J. Appl. Phys.* **21** L411
Kulkarni S B and Williams W S 1976 *J. Appl. Phys.* **47** 4318
Osvenskii V B, Kholodnyi L P and Mil'vidskii M G 1973 *Soviet. Phys. Solid State* **15** 661
Patel J R, Testardi L R and Freeland P E 1976 *Phys. Rev. B* **13** 3548
Rybin V V and Orlov A N 1970 *Soviet. Phys. Solid State* **11** 2635
Sato M and Sumino K 1977 *Proc. 5th Int. Conf. on High Voltage Electron Microscopy* ed T Imura and H Hashimoto (Kyoto : Jpn. Soc. of Electron Microscopy) pp 459 – 462
Sato M and Sumino K 1985 *Proc. Yamada Conf. IX on Dislocations in Solids* ed H Suzuki, T Ninomiya, K Sumino and S Takeuchi (Tokyo : Univ. of Tokyo Press) pp 391 – 394
Steinhardt H and Haasen P 1978 *Phys. Stat. Sol. A* **49** 93
Sumino K 1987a *Proc. 7th Int. School on Defects in Crystals* ed E Mizera (Singapore : World Scientific) pp 495 – 513
Sumino K 1987b *Proc. 2nd Int. Autumn School on Gettering and Defect Engineering in Semiconductor Technology* ed H Richter (Frankfurt GDR : Academy of Sciences GDR) pp 218 – 231
Sumino K and Harada H 1981 *Phil. Mag. A* **44** 1319
Sumino K and Imai M 1983 *Phil. Mag. A* **47** 753
Sumino K and Sato M 1979 *Kristall und Technik* **14** 1343
Yonenaga I and Sumino K 1985 *Proc. Yamada Conf. IX on Dislocations in Solids* ed H Suzuki, T Ninomiya, K Sumino and S Takeuchi (Tokyo : Univ. of Tokyo Press) pp 385 – 390
Yonenaga I and Sumino K 1987 *J. Appl. Phys.* **62** 1212
Yonenaga I and Sumino K 1989 *J. Appl. Phys.* **65** 85

Inst. Phys. Conf. Ser. No 104: Chapter 3
Paper presented at Int. Symp. on Struct. Prop. Disloc. Semicond., Oxford, 1989

Experimental investigation of kink motion in a field of random forces

V I Nikitenko and B Ya Farber

Institute of Solid State Physics Academy of Sciences of the USSR, 142432
Chernogolovka, Moscow district, USSR

ABSTRACT: Dislocation displacements were measured in Si single
crystals under the action of loading pulse sequences and static
loading using selective etching to reveal the dislocations.
Investigations were made of the dependence of the individual 60°
dislocation glide distances on the pulse duration, and pause
durations as well as on the value of stress acting during the
pulse separations, with the maximum amplitude of the stress being
constant. Starting stresses for the dislocation motion were
measured as well. The results obtained are discussed in the
framework of theories taking into account the influence of point
defects on the processes of kink pair formation and spreading
along the dislocation line.

1. INTRODUCTION

As is known the first observations of dislocations in semiconductors
revealed that dislocations tend to lie along crystallographic directions.
It seems to be evident that dislocation glide in such materials is
determined by the Peierls relief. The theory of dislocation mobility in
this relief (Hirth and Lothe 1968) takes into consideration that applied
stress σ lowers the dislocation energy in an adjacent valley of the
potential relief (Fig. 1a, b). At stresses lower than the Peierls stress
$\tau_p = [1/b(\partial W_p(y)/\partial y)]_{max}$ kink pairs are formed on the dislocation line due
to thermal fluctuations, where b is the Burgers vector and $W_p(y)$ is the
line energy of the dislocation. They diffusively overcome the secondary
Peierls relief and grow up to the critical size l_c (Fig. 1c), determined
by the stability of a kink pair against collapse. After that, kink pairs
spread in a drift regime till the end of the dislocation segment or till
annihilation with kinks of opposite sign from adjacent kink pairs.

However, it has been shown (Celli et al 1963, Nikitenko 1975, Nikitenko and
Farber 1985, Alexander 1986) that the theory of dislocation glide in a
perfect crystal fails to explain the experimental data. It was established
that point defects affect strongly the electrical activity of dislocations
as well as dislocation mobility, producing large changes in dislocation
velocity and its apparent activation energy. In the first theory proposed
by Celli et al for the explanation of the influence of point defects on the
Peierls mechanism of dislocation glide it was supposed that they create
additional barriers for kink motion (Fig. 2). At small stresses this leads
to an essential decrease of the velocity of the kinks and of the dislocation
as a whole, increasing the apparent activation energy for its glide. Later

Petukhov (1971, 1988), Vinokur (1986) and Vinokur and Sagdeev (1987)
treated the case when the effect of point defects was attributed to a
change of the straight dislocation segment energy due to the dislocation
point defect interaction.

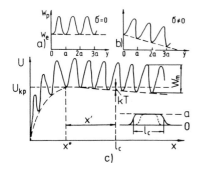

Fig. 1. Energy of dislocation core in
a Peierls relief at $\sigma = 0$ (a) and $\sigma \neq 0$
(b). Theoretical dependence of the kink
pair free energy on its width (c). W_m
is the secondary Peierls barrier height
for the migration of the kink along the
dislocation line.

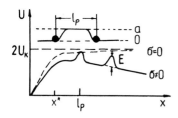

Fig. 2. Kink pair energy vs its width
in a random potential created by point
defects (Celli et al 1963). $W_m = 0$.
E is the energy of dislocation point
defect interaction.

Absorption (or desorption) of point defects on the segment during the kink
motion results in step like changes of the kink pair energy by a value of u
(Fig. 3). As a result, during doping, the apparent activation energy for the
dislocation motion may be decreased, increasing the dislocation velocity.
In the latter case kink motion proceeds in a field of random forces rather
than in the random potential treated by Celli et al, and has distinctly
different characteristics. In particular the drift velocity of the kinks
may be nonlinearly time dependent $x \approx t^{\delta}$, where $\delta < 1$.

Recently, due to the work of Louchet (1981, 1988), Hirsch et al (1981),
Gottschalk et al (1987), Jendrich and Haasen (1988), Nikitenko et al (1985),
substantial progress has been made in the experimental investigation of
dislocation mobility in conditions when the apparent activation energies
for kink pair formation and spreading can be measured separately. It has
been shown that the kink migration energy is significant, making up an
essential part of the apparent activation energy for the dislocation motion.
However the question concerning the relation between the contributions of a
secondary Peierls relief and point defects to the process of the dislocation
glide is as yet unsolved. To study the kink-point defect interaction and
reveal distinctions in the types of the potential created by point defects
for kink motion, it is necessary to obtain detailed information about the
kink mobility along the dislocation line. To solve this problem the method
of periodic pulse loading proposed by Nikitenko et al (1985, 1986, 1987)
has been used in this work.

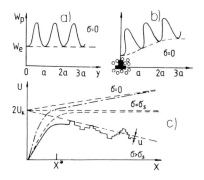

Fig. 3. Peierls relief for (a) dislocation motion in a perfect crystal and (b) in a crystal with point defects in the dislocation atmosphere. (c) Kink pair energy vs its width in a field of random forces, u is the energy of the dislocation point defect interaction, σ_s is the starting stress for the dislocation motion.

2. EXPERIMENTAL

The samples with individual dislocations were deformed by a sequence of load pulses with the amplitude stress σ_i, the duration of a single pulse t_i being comparable with the mean time of the dislocation displacement over one lattice parameter (a) $t_a = a/v_d$, where v_d is the steady state dislocation velocity. The pulses were divided by pauses with duration t_p when the applied stress $\sigma = \sigma_p$ (σ_p was zero in some experiments). During the pulse action stable kink pairs can be formed and will spread along the dislocation line. During the pulse separation the kink pairs may become unstable and begin to shrink to the formation centre. If the pulse separation is long enough the kink pairs formed will collapse. If the pulse separation is short, any kink pairs preserved will be broadened by the next load pulse, giving rise to a macroscopic displacement of the dislocation. So varying the durations of the pulses and pauses and using the dislocation displacement as an indicator one can obtain information about the characteristic times of stable kink pair formation and kink motion till annihilation. It allows one to estimate in the framework of rather general assumptions the main quantitative parameters of the processes of the kink pair formation and spreading.

The investigated samples were rectangular prisms with edge orientation $[1\bar{1}0]$, $[11\bar{2}]$, $[111]$ and with dimensions 35x4x1.5 mm^3. They were cut from dislocation-free single crystals of n-type Si, grown by the floating zone method and doped with P up to a conductivity of 150 Ohm cm. Individual dislocations were introduced into the samples by a scratch and revealed using selective etching. The displacements measured were of individual 60° dislocations which had no bends beneath the surface. The samples were deformed by four point bending around the $[11\bar{2}]$ axis with the loads being a sequence of square pulses driven from a function generator with the required pulse ratio through an electromagnetic-force transducer. It was possible to maintain the desired stress level during the pulse separation due to the magnetic hanging of the loading system elements. The active loading time (t_{st}), i.e. the sum of the stress pulse durations was maintained constant and equal to the static loading time over which the dislocations covered a distance $\bar{1}_{st} \approx$ 20-30 µm. The deformation temperature (T) was measured by a thermocouple near the samples and maintained constant with an accuracy of 1 K.

3. RESULTS AND DISCUSSION

Fig. 4 shows the dependences of the normalized mean 60° dislocation displacements on the relative pulse duration (in the units t_i/t_a) at $t_i = t_p$
(curve a) and on the relative pause duration (in the units t_p/t_i) at $t_i =$
Const. = 94 ms (curve b). Mean dislocation displacements measured in a
pulse experiment were normalized to the average dislocation displacements
measured during static loading. It is seen that dislocation displacements
drop from $\bar{l}_{st}$ down to zero as the pulse duration decreases (at $t_i = t_p$) or
during increase of the pulse separation (at t_i = Const.). There are
inflection points on the curves $\bar{l}(t_i)$ and $\bar{l}(t_p)$ at $t_i^* \approx 38$ ms and $t_p^* \approx$
169 ms respectively.

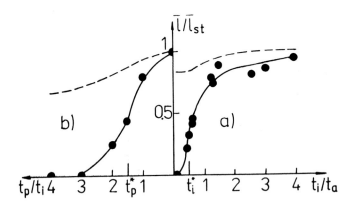

Fig. 4. Average normalized dislocation displacements vs pulse duration (a)
at $t_i = t_p$ and pause duration (b) at $t_i = 94$ ms = Const. T = 600°C, $\sigma_i =$
7 MPa, $\Sigma t_i = t_{st} = 7200$ s. Dotted lines correspond to the theoretical
dependences $\bar{l}(t_i)$ and $\bar{l}(t_p)$ calculated from (10-12).

Let us analyze the results obtained in the framework of the diffusion theory
of the kink pair formation (Hirth and Lothe 1968). It is assumed that there
is an equilibrium kink density on the dislocation line at T = 0 even at
$\sigma = 0$

$$c_k = (2/b) \exp (-U_k/kT) \qquad (1)$$

where U_k is the energy of formation of a single kink, and k is the Boltzmann
constant. Application of stresses results in a directional drift of the
kinks with the velocity:

$$v_k = (D_k/kT)\sigma ab, \quad \text{where} \quad D_k = \nu_D b^2 \exp -W_m(\sigma)/kT \qquad (2)$$

Here D_k is the kink diffusivity, W_m is the apparent activation energy for
its glide and ν_D is the Debye frequency. In the range of very small
stresses the kink density is practically independent of σ and dislocation
velocity is determined by c_k

$$v_d = a \, c_k \, v_k \tag{3}$$

When the applied stress is increased the kink density on the dislocation line increases due to the formation of additional kink pairs. Only the stable ones can contribute to the dislocation velocity, having reached the size $l_c = x' + x^*$ (Fig. 1) where

$$x^* = (\alpha/\sigma ab)^{1/2} \quad \text{and} \quad x' = kT/\sigma ab \tag{4}$$

Here α is a constant of the elastic interaction between the kinks forming a pair. At the critical distance x^* the mutual attraction between kinks is balanced by the external force. At this point the energy of the kink pair would be

$$U_{kp} = 2U_k - 2(\sigma\alpha ab)^{1/2} \tag{5}$$

x' is the distance from the saddle point over which the energy U_{kp} decreases by kT (Fig. 1). The process of the kink pair expansion up to the stable configuration (from x^* to l_c) is governed by a random walk of the kinks in the potential relief shown in Fig. 1c, since the work done by the applied forces in the course of the pair expansion to a distance $x < x'$ is less than kT. Kink pairs with the size $x > l_c$ expand in the drift regime. A static load establishes a steady state kink flux (J) from the saddle point, with J being equal to the rate of stable kink pair formation. The steady state dislocation velocity is given by

$$v_d = aJL, \quad J = (v_k/b^2)\exp(-U_{kp}/kT) \quad \text{for} \quad L < L_k = (2v_k/J)^{1/2} \tag{6}$$

$$v_d = aJL_k = a \, (2v_kJ)^{1/2} \quad \text{for} \quad L > L_k \tag{7}$$

L_k is determined by the steady state condition $(L_k/2v_k) = (1/JL_k)$.

Nikitenko et al (1986, 1987) proposed a scheme for the processing of experimental data obtained from periodic pulse loading experiments. It is based on the following assumptions: the inflection point on the $1(t_i)$ dependence corresponds to the condition when a stable kink pair is formed on the dislocation line during single load pulse action. The inflection point on the curve $1(t_p)$ corresponds to the condition when the displacement of the kinks in a pair during the pulse action is equal to the back displacement during the pulse separation. This allows one to obtain equations setting up correlations between the critical parameters (t_i^* and t_p^*) and the characteristics of the kink mobility:

$$x' = (2D_p \, t_i^*)^{1/2}, \quad \text{where} \quad D_p = 2D_k \tag{8}$$

$$x' + 2v_k(t_i - t_i^*) = (2D_p \, t_p^*)^{1/2} \tag{9}$$

Substituting the values of $t_i^* = 38$ ms and $t_p^* = 169$ ms measured from Fig. 1 in (8,9) the values $D_k = 1.2 \cdot 10^{-11}$ cm^2/s and $v_k = 1.33 \, 10^{-5}$ cm/s were estimated. This allows us to estimate from (2) the apparent activation energy for the kink motion $W_m = 1.58$ eV. The measurement of the temperature dependence of the kink diffusivity (Nikitenko et al 1987) confirmed the thermally activated nature of the process of kink migration along the dislocation line and gave values of W_m correlating with the above mentioned estimation. So the scheme for the treatment of the results of the pulse experiments allows one to explain qualitatively the results of the pulse experiments and to obtain self consistent estimations of the kinks' dynamical properties.

However, as was mentioned by Nikitenko et al (1987), taking into consideration the dislocation motion in a perfect crystal, it is impossible to describe the whole set of the experimental data obtained. In particular, carrying out experiments with periodic pulse loading one can compare the equilibrium density existing on the dislocation at $\sigma = 0$ with the density of stable kink pairs determining the steady state dislocation velocity at a given stress level. According to (1-7) their ratio should not exceed the value of $\exp{(\alpha\sigma ab)^{1/2}/kT}$, i.e. about 40% for these particular experimental conditions, which essentially contradicts the experimental results shown in Fig. 4, where the $\bar{1}$ value drops by two orders of magnitude at least (at $t_i < t^*$ and $t_p > t^*$). This discrepancy would be even more pronounced if one took into account that during diffusive motion kink pairs may expand or shrink with equal probability and there is no reason for the existence of the relaxation process. Some decrease of the dislocation displacements with increasing pulse separation may be described in principle by taking into consideration the process of kink random walks with absorbing walls (Petukhov 1989). If during the random walk the kink pair shrinks to the size x* it will collapse. The probability for a kink pair with size x_m to reach the saddle point would be

$$P_c(x_m, t) = (x_m/ (4\pi D_k)^{1/2} t^{3/2}) \ \exp - \ x_m^2/(4 \ D_k \ t) \tag{10}$$

Then the probability for the kink pair to be preserved on the dislocation line till the end of the pause duration would be

$$P(x_m, t) = 1 - \int_0^{t_p} P_c(x_m, t)dt = (2/\pi) \int_0^{x_m/(4Dt_p)^{1/2}} \exp-(z^2)dz \tag{11}$$

This would result in the renormalization of the stationary rate of stable kink pair formation and accordingly it leads to a decrease in the dislocation velocity and mean dislocation displacements

$$\bar{1}/\bar{1}_{st} \approx P(x_m, t_p) \tag{12}$$

Here x_m is the relative displacement of the kinks in a pair, which is determined from the left hand side of (9) at a given pulse duration. The dotted lines in Fig. 4 show the theoretical dependence $\bar{1}(t_i)$ and $\bar{1}(t_p)$ calculated from (10-12). It is seen that these curves differ considerably from the experimental ones.

Due to these discrepancies, Nikitenko et al (1987) concluded that the effects observed may be produced by the influence of point defects on the processes of kink pair formation and spreading along the dislocation line. It should be noted that the method of periodic pulse loading was used by K Maeda (1988) for the investigation of dislocation mobility in GaAs.

The results obtained were similar to those shown in Fig. 4. Maeda also failed to explain his results in the framework of Hirth-Lothe theory and proposed that the effect observed might be explained by taking into account the influence of strong obstacles on the dislocation line, impeding the kink-antikink annihilation.

Let us analyze the results shown in Fig. 4 in the framework of theories taking into account the influence of point defects on dislocation mobility in semiconductors. It is known that due to the dislocation point defect interaction the accumulation of point defects takes place in the dislocation stress field and in the dislocation core, giving rise to a decrease of the dislocation elastic energy (Fig. 3) and the appearance of a starting stress (σ_s) for dislocation motion.

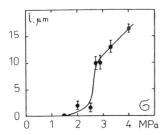

Fig. 5. Mean dislocation displacements vs static stresses at loading duration t_{st} = 9000 s and T = 600°C.

They create an additional force which, similar to the attraction between kinks forming a kink pair, would impede the kink pair expansion. In this case the apparent force acting on the kinks would be $F_{ef} = \sigma_{ef}ab = (\sigma_i - \sigma_s)ab$ (Petukhov 1982). So the contradictions mentioned above seem to be qualitatively explained by taking into account the existence of the starting stresses. During the pulse unstable kink pairs may be formed on the dislocation line, which will shrink under the action of the internal force $(-\sigma_s)$ in the drift regime rather than due to a random walk with absorbing walls as was discussed above. It may explain the sharp decrease of the average dislocation displacements with rising t_p or decreasing t_i. If one compensates for the action of internal stresses by applying an external stress during the pulse separation the dependences $\bar{l}(t_i)$ and $\bar{l}(t_p)$ should be changed. To verify this assumption we carried out such experiments when a small stress (σ_p) was maintained during the pulse separation.

To determine the starting stresses we measured the dependence of the average dislocation displacements vs applied stresses in a constant duration experiment: t = 9000 s at T = 600°C (Fig. 5). It is seen that dislocation displacements decrease with decreasing σ and sharply drop to zero at $\sigma < 3$ MPa. The starting stress is assumed to be the smallest stress $(\sigma_s = 1.5$ MPa) which gives rise to a displacement of the dislocations. Fig. 6 shows the dependences of the average normalized dislocation displacements on the pause duration at t_i/t_a = 0.5 obtained in the conditions with the constant stress σ_p being applied during the pulse separation $(\sigma_p = 0$, curve 1, $\sigma_p = 1$ MPa (2), 1.5 MPa (3)). It is seen that the application of σ_p shows down the decrease of l with rising t_p and at $\sigma_p = \sigma_s$ the results obtained are in qualitative agreement with the predictions of the theory based on a random walk of the kinks with absorbing walls. The dotted line in Fig. 6 corresponds to the $l(t_p)$ dependence calculated from (10-12), where $\sigma_{ef} = (\sigma_i - \sigma_s)$ was used instead of σ_i. Measurements of $\bar{l}(t_i)$

dependence at $t_i = t_p$ and $\sigma_p = 1.5$ MPa gave results which also qualitatively agree with the predictions of the diffusion theory, with the decrease of l from the steady state value not exceeding 40%. So the results obtained in the experiments with a constant stress acting during the pulse separation provide evidence that the most probable reason for the collapse of the kink pairs should be attributed to the action of point defect atmospheres, which result in the appearance of a force returning the dislocation to its initial valley of the potential relief.

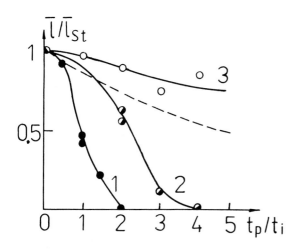

Fig. 6. Average normalized dislocation displacements vs pause duration $(t_i/t_a = 0.5 =$ Const.) at different stresses (σ_p) applied during the pulse separation: (1) $\sigma_p = 0$, (2) $\sigma_p = 1$ MPa, (3) $\sigma_p = 1.5$ MPa. $T = 600°C$. $\sigma_i = 7$ MPa, $\Sigma t_i = t_{st} = 7200$ s. The dotted line is the theoretical curve $l(t_p)$ calculated from (10-12).

It should be noted that point defects are distributed inhomogeneously on the dislocation line. As was mentioned above, two distinct cases are possible. Kink motion may proceed in a random potential (Fig. 2) as was treated by Celli et al (1963). In this case the expansion of the kink pair may be characterized by the drift velocity, but its value will be significantly smaller than that for the perfect crystal. Petukhov (1971, 1988) and Vinokur (1986) treated the other case when the kinks move in a field of random forces (Fig. 3). Here some specific 'memory' effects are possible due to the superposition of the influences from different point defects. It was shown that in a field of random forces two regimes of kink motion may exist. At high stresses kink pair expansion may be characterized by ordinary drift with the kink displacements being linearly time dependent. But at small stresses $\sigma < \sigma_o = (c_1 + c_2)u^2/2kTab$ a regime of nonlinear drift was predicted, when the kink displacements would be

$$x = x_o(t/t_o)^\delta, \text{ where } \delta = 2\sigma abkT/[(c_1 + c_2)u^2] \qquad (13)$$

Here $t_o = (x_o)^2/(2D_p)$ is the characteristic time of the kink displacements over the diffusion length x_o during the random walk and c_1 and c_2 are the concentration of point defects in the neighbouring valleys of the potential relief.

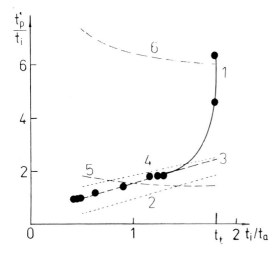

Fig. 7. Dependence of the critical pause duration t_p^* (in units of t_i) on the relative pulse duration (t_i/t_a). 1 - experimental data (2-5) - calculations using (15) at $\delta = 0.4$ (2), 0.5 (3), 0.6 (4), 1 (5). 6 corresponds to the $t_p^*(t_i)$ dependence obtained from (14). T = 600°C, σ_i = 7 MPa. σ_p = 0, $\Sigma t_i = t_{st}$ = 7200 s.

In conditions of periodic pulse loading it is possible to investigate the $x(t)$ dependence measuring the times of relaxation of kink pairs of different sizes in the field of internal stress $[\sigma_s = (c_1 - c_2)u/ab]$ where the size of the pair is determined by the durations of the external stress pulses. Fig. 7 shows the dependence of the critical pause duration t_p^* (in the units of t_i) on the relative pulse duration (t_i/t_a). This dependence was plotted by determining the inflection points on curves similar to that shown in Fig. 4b for different relative pulse durations. We will now try to describe the $t_p^*(t_i)$ dependence taking into account the kink motion in a random potential and in a field of random forces. During the pulse action with $\sigma_i > \sigma_0$ the kink pair expands by ordinary diffusion and drift. In this case the maximum relative displacements of the kinks may be estimated from the left hand side of (9), with the σ_i value being replaced by σ_{ef}. (It should be noted that this only slightly changes the values of W_m, D_k and v_k, estimated above.) Significant distinctions arose when we analyzed the dependence $x(t_p)$ during the relaxation of the kink pair under the action of internal stress (σ_s). In a random potential we would expect $x(t_p) = 2v_k' t_p$, where $v_k' = D_k\sigma_s ab/kT$. In this case the $t_p^*(t_i)$ dependence would be:

$$x' + 2v_k (t_i - t_i^*) = 2 v_k' t_p^* \qquad (14)$$

In a field of random forces $x(t_p)$ is described by (13), where $x_0 = kT/\sigma_s ab$, and instead of (14) we obtain

$$x' + 2 v_k(t_i - t_i^*) = x_0(t/t_0)^\delta \qquad (15)$$

Theoretical dependences of $t_p^*(t_i)$ calculated from (14,15) are shown in Fig. 7 by dotted lines. As is seen $t_p^*(t_i)$ calculated from (14) (curve 6) differs even qualitatively from the experimental data. The best fit was obtained at short pulse durations $(t_i/t_a < 1.4)$ for the case of kink motion in a field of random forces with $\delta = 0.5$ (curve 3).

The deviation of the experimental data from the calculated ones at large

t_i may be caused by the fact that in this case the kink pair may expand up to a large size when the annihilation of kinks in adjacent pairs is more likely than the collapse of the kink pair to its nucleation centre. In this case the dislocation may pass completely to a neighbouring valley and will not return to its initial one no matter how long the pause. This is indeed observed in Fig. 7 as the sharp increase of t^* with rising t_i. The vertical asymptote to the experimental curve $(t_t \approx 1.8 \, t_a)$ should then correspond to the time of the kink free path to annihilation. It should be noted that $t_t \approx t_a$. This seems to show that at high stresses $(\sigma_i \gg \sigma_s)$ the dislocation moves apparently as a straight line even on the micro scale.

The theory of kink motion in a field of random forces predicts the possibility for the regime of kink self localization to be realised at small stresses, with the mean kink velocity being equal to zero. This would result in qualitative changes in the mechanism of the dislocation glide. The sticking of the kinks after motion suggests the possibility for the spreading of dislocation glide over several Peierls valleys resulting in superkink formation. The superkinks formed stimulate the escape of the kinks from the traps of the field of random forces. At smaller stresses, when superkinks cannot be formed, the linearity of the dislocation form would be restored, and in this case a sharp drop in dislocation mobility would be observed due to the fact that the kinks can leave the traps by thermal fluctuations only. We succeeded in revealing the manifestation of these processes when investigating the dependence $\bar{l}(t_i)$ at $t_i = t_p$ in the range of small stresses. Fig. 8 shows the experimental data obtained at T = 600°C and σ_i = 7 MPa (curve 1), 4 MPa (2) and 3 MPa (3). It is seen that at high stresses the dislocation displacements measured in the pulse experiment reach the static values at $t_i \approx t_a$, corresponding to a characteristic kink height $h_c \approx a$. As the stresses decrease down to 4 MPa, mean dislocation displacements reach the l_{st} value at $t_i \approx 10^3 \, t_a$. In this case superkinks are formed with a characteristic height $h_c \approx 10^3 a$. During the latter decrease of σ_i down to 3 MPa l reaches l_{st} at much smaller t_i, but as is seen from Fig. 5 in this stress range a sharp drop of the dislocation mobility was observed. It should be noted that deviations of the dislocation form in Si from the straight line have been observed earlier using X-ray topography (Erofeev et al 1971, Imai and Sumino 1983) and were attributed to the influence of strong obstacles on the dislocation motion, producing the formation of superkinks.

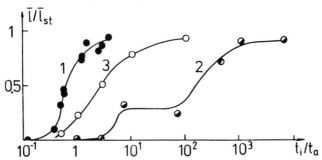

Fig. 8. Normalized mean dislocation displacements vs relative pulse duration in the units t_i/t_a ($t_i = t_p$) at different applied stresses: (1) σ_i = 7 MPa (t_a = 76 ms), (2) σ_i = 4 MPa (t_a = 129 ms), (3) σ_i = 3 MPa (t_a = 228 ms).

4. CONCLUSIONS

The discussion presented shows that the significant discrepancies revealed
between the predictions of the theory of the dislocation motion in a
perfect crystal and experimental data obtained using periodic pulse
loading may be explained to some extent taking into account the influence
of point defects creating a field of random forces for kink motion. It
produces the nonlinear regimes of kink drift in the course of kink pair
relaxation during the pulse separation and it may be the reason for the
sharp decrease of the dependences $l(t_i)$ and $l(t_p)$ at short pulse durations
and long pulse separations. To obtain quantitative information, additional
experiments are needed for the investigation of the kink pair relaxation
process at different stresses acting during the pulse separation.

5. ACKNOWLEDGEMENTS

The authors are very grateful to Prof V Ya Kravchenko, Prof B V Petukhov
and Dr V M Vinokur for their interest in this work and stimulating
discussions.

6. REFERENCES

Alexander H 1986 Dislocations in Solids ed F R N Nabarro (Amsterdam: North
   Holland) pp 115-234
Celli V, Kabler M, Ninomiya T and Thomson R 1963 Phys Rev 131 58
Erofeev V,N, Nikitenko V I, Polovinkina V I and Suvorov E V 1971
   Kristallografia 16 190
Farber B Ya, Iunin Yu L and Nikitenko V I 1986 Phys.Stat.Sol.(a) 97 469
Gottschalk H, Alexander H and Dietz 1987 Inst.Phys.Conf.Ser. 87 339
Hirsch P B, Ourmazd A and Pirouz P 1981 Inst.Phys.Conf.Ser. 60 29
Hirth J P and Lothe J 1968 Theory of Dislocations (New York: McGraw Hill)
Imai M and Sumino K 1983 Phil.Mag. 47 599
Jendrich U and Haasen P 1988 Phys.Stat.Sol.(a) 104 553
Louchet F 1981 Inst.Phys.Conf.Ser. 60 35
Louchet F 1988 Phil. Mag. A 57 327
Maeda K 1988 8 Int. School on Defects in Crystals ed E Mizera (Singapore:
   World Scientific) pp 153-68
Nikitenko V I 1975 Dinamica Dislokatsii (Kiev: Naukova Dumka) pp 7-26
Nikitenko V I and Farber B Ya 1985 Proc. Yamada Conf IX on Dislocations
   in Solids ed H Suzuki (Tokyo:University of Tokyo Press) pp 417-20
Nikitenko V I, Farber B Ya and Iunin Yu L 1985 JETP Letters 41 124
Nikitenko V I, Farber B Ya and Iunin Yu L 1987 Sov. Phys. JETP 66 738
Petukhov B V 1971 Fiz.Tverd.Tela 13 1445
Petukhov B V 1982 Fiz.Tverd.Tela 24 439
Petukhov B V 1988 Fiz.Tverd.Tela 30 2893
Petukhov B V 1989 Private communication
Vinokur V M 1986 J. Physique 47 1425
Vinokur V M and Sagdeev I R 1987 J. Physique 48 1395

Inst. Phys. Conf. Ser. No 104: Chapter 3
Paper presented at Int. Symp. on Struct. Prop. Disloc. Semicond., Oxford, 1989

269

Kink formation and migration in covalent crystals

K Maeda and Y Yamashita

The Department of Applied Physics, The Faculty of Engineering,
The University of Tokyo, Hongo, Bunkyo-ku, Tokyo 113, Japan

ABSTRACT: Critical remarks are made on the kink diffusion model often used to interpret experiments on dislocation velocity in tetrahedrally bonded semiconductors. Theoretical analyses shows that the intermittent loading effect observed under the usual conditions is not a direct manifestation of the kink diffusion mechanism. Dislocation pinning is proposed as an alternative model.

1. INTRODUCTION

It is believed that dislocation glide in tetrahedrally bonded semiconductors is controlled by the Peierls mechanism since it requires breaking of covalent bonds which should form a high energy barrier against dislocation motion. For the same reason it is sometimes inferred that the secondary Peierls potential, a potential undulation that a kink experiences during its motion for every atomic distance along a dislocation line, is also quite large. The kink width in tetrahedrally bonded crystals is estimated to be of the order of atomic distances and therefore one expects that kink migration will experience the crystalline periodicity more strongly than smooth kinks. Thus, the widely accepted picture is that the dislocation mobility in semiconductors is governed not only by the double kink formation but also by the kink diffusion process (kink diffusion control). This is in contrast with metals in which the secondary Peierls potential is considered to be so small even in dislocations associated with high Peierls potentials (e.g., screw dislocations in bcc metals) that the double kink formation solely forms a bottle-neck of dislocation motion (kink formation control).

Straight dislocations in elemental semiconductors are considered to be reconstructed to eliminate dangling bonds. If kinks in the stable configuration are also reconstructed with no broken bonds, the formation energy of a reconstructed kink $F_k^{rec.}$ will probably be small due to elimination of dangling bonds. Concomitantly in that case, however, the secondary Peierls potential or the energy barrier for kink migration E_m will be rather high (the solid curve in Fig. 1). If the kink is unreconstructed, the corresponding kink formation energy $F_k^{unrec.} \approx F_k^{rec.} + E_m$ will be large and accordingly E_m may be diminishingly small (the dash curved in Fig. 1). Although a theoretical calculation assuming reconstructed kinks (Jones 1980) showed that E_m is comparable with $F_k^{rec.}$, no calculation based on unreconstructed kinks has been made. Thus, theoretically it is not quite obvious which is the real case.

On the experimental side, there have been some reports which claimed the secondary Peierls potentials are really high. From *in-situ* TEM observations of dislocation motion, the kink migration energy was evaluated to be as large as 1.2 eV in Si (Hirsch et al. 1981; Louchet 1981) and 0.8–0.9 eV in Ge (Louchet et al. 1988) based on the assumption that kinks, whose motion is limited by the secondary Peierls potential, collide with each other when the dislocation segment is longer than a critical length. The internal friction (IF) peaks observed in plastically deformed Ge (Jendrich and Haasen 1988) were attributed to migration of built-in kinks and double kink formation. The kink migration energy was evaluated to be 0.7–1.1 eV, variable depending on the doping condition, substantially large in comparison with the activation energy of dislocation motion in Ge (1.1–1.6eV). The identification of the low temperature IF peak as kink migration relies on the fact that the magnitude of prefactors of observed relaxation time is considerably large, and it decreases with deformation, which are in agreement with predictions of theories for internal friction due to kink migration (Southgate and Attard 1963). However, physically different models, such as the solute dragging model by Schoeck (1982) and the dispersed obstacle model by Ninomiya et al. (1964), also could explain the above observations, both qualitatively and quantitatively. Thus, although all of these results so far obtained appear consistent within the framework of the kink diffusion model, the evidences are more or less indirect and the interpretation of these experimental observations is not completely unique.

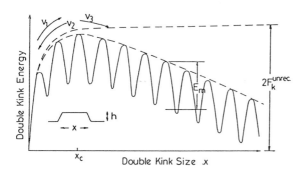

Fig. 1. Double kink energy as a function of double kink size expected for reconstructed kinks (solid curve) and unreconstructed kinks (broken curves, upper without stress and lower under stress). x_c is the critical size of unreconstructed double kink.

Recently Nikitenko and his collaborators (1985) performed dislocation velocity measurements on Si which showed that dislocations do not move when the crystal is subject to intermittent loading with pulse durations less than a time required for a dislocation to advance to the next trough of Peierls potential. Qualitatively the same result was obtained also for α-dislocations in n-GaAs more recently by one of the present authors (Maeda 1988). Although Nikitenko et al. (1985) tried to explain this intermittent loading effect in terms of the kink diffusion model, our theoretical analysis and computer simulations of kink formation and migration based on the same model indicate that, as far as we assume an acceptable value for kink formation energy, such quenching of motion should occur only at pulse durations much shorter than that observed experimentally. This serious conflict of the kink diffusion picture with experimental facts prompted us to develop a new model which takes into account strong obstacles against kink propagation (Maeda 1988).

However, in a more recent paper (Nikitenko et al. 1987) it was reported that the intermittent loading effect becomes very weak as the pulse is shaped to more ideal rectangles. Since the practical absence of intermittent loading effect in the studied range of pulse lengths is exactly what is expected from the kink diffusion model, there arose a doubt that the previous observations might have been brought about by some experimental artefact. Thus the situation is very critical and the necessity for refinements of experimental conditions is realized. The present paper first discusses the quantitative features of the intermittent loading effect to be expected theoretically from the kink diffusion model, then presents some new findings recently added by the present authors, and finally explains the experimental results in terms of a newly proposed model.

2. KINK DIFFUSION UNDER INTERMITTENT LOADING

The intermittent loading experiment measures glide distances of dislocations in crystals subjected to a periodic pulsed loading, each pulse lasting in a duration of t_a and being separated by an unloading interval t_p (Fig. 2). The intermittent loading effect is referred to the phenomenon that the glide distance traveled by dislocations in a total loading time Σt_a which is fixed constant depends on the pulse lengths t_a or t_p.

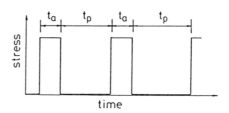

Fig. 2. Intermittent loading

In the kink diffusion model or the model assuming a high secondary Peierls potential, kinks can exist thermally in metastable states. At sufficiently low stresses, the kink density per unit length of dislocation does not differ much from its thermal equilibrium value $c_k^{eq.}$ and hence the dislocation velocity is given by

$$V_d = 2c_k^{eq} h\, V_k \tag{1}$$

where h is the kink height and V_k the drift velocity of a kink which proportionally depends on the applied stress σ_a. Since the dislocation velocity in this case varies simultaneously with the applied load in a linear manner, no effect of intermittent loading is expected.

At high stresses, a condition usually met in experiments, the kink density can vary with the applied stress because the thermal equilibrium balance will be lost between generation and annihilation of kinks. For the steady state, the dislocation velocity in the high stress regime is given by

$$V_d = \frac{2\sigma_a bah^2 v_D}{kT} \exp\left(-\frac{F_k + E_m}{kT}\right) \tag{2}$$

in the case that kinks collide with each other before they sweep up the whole segment of a dislocation, or

$$V_d = \frac{\sigma_a bh^2 \Lambda v_D}{a^2 kT} \exp\left(-\frac{2F_k + E_m}{kT}\right) \tag{3}$$

in the case that the dislocation segment is too short for kink collision to occur (Hirth and Lothe 1968). Here b is the Burgers vector, a the crystal periodicity along the dislocation, Λ the length of dislocation, v_D a frequency factor of the order of the Debye frequency, k the Boltzman constant, T the temperature, and F_k the formation energy of a single kink. The *stationary* velocity depends linearly on σ_a as in the low stress regime; however, it does *not* mean that the intermittent loading effect would not be expected.

Theoretical analysis of kink generation and migration under *time-varying* load was attempted in a previous paper (Maeda 1988) and in the present paper, after a minor modification, has been applied for combinations of parameters that have been claimed to explain experimental observations in terms of the kink diffusion model. Two approaches are employed: The first approach is computer simulations by which the problem can be treated as exactly as possible and the second is analytical formulation of the model which serves to provide us with insight to the role of involved parameters.

In the computer simulations, we suppose a dislocation line divided into n sites at every atomic distance a. Each site i is characterized by an index $S(i)$ expressing whether the site is occupied with a positive kink ($S=+1$), a negative kink ($S=-1$) or no kink ($S=0$). The effective stress on the i-th site including the internal stress exerted by other kinks on the same dislocation line is calculated by

$$\sigma_{eff} = \sigma_a - \sum_{j \neq i} \frac{K S(j)}{sign(j - i)\{(j - i)a\}^2} \tag{4}$$

where K is the kink-kink interaction coefficient (with the shear modulus G and the Poisson ratio v, $K = Gbh (1+v)/ 8\pi(1-v)$ for screw dislocations and $K = Gbh (1-2v)/ 8\pi(1-v)$ for edge dislocations). For every site and for every time step Δt, either jump of a kink (at such sites that $S \neq 0$) to a neighboring site or generation of a kink pair (at such sites that $S(i)=0$ and $S(i+1)=0$) separated by two atomic distances (smallest double kink generation) was allowed to take place in a random manner according to the probability

$$P_{\pm}^{X} = \frac{f_{\pm}^{X}}{f_{+}^{X} + f_{-}^{X}} f^{X} \Delta t \tag{5}$$

with

$$f_{\pm}^{X} = f_{0}^{X} \exp\left(\frac{E^X \mp \Omega^X \sigma_{eff}}{kT} \right)$$
$$= f^{X} \exp\left(\pm \frac{\Omega^X \sigma_{eff}}{kT} \right) \tag{6}$$

$$f^{X} = f_{0}^{X} \exp\left(-\frac{E^X}{kT} \right) \tag{7}$$

Here the superscript X stands for kink migration (KM) and smallest double kink generation ($SDKG$), and the sign $+$ ($-$) indicates that the event is assisted (resisted) by the applied stress. The activation energy E^{KM} represents the kink migration energy E_m for kink jump, and E^{SDKG} the activation energy for generation of a smallest double kink E_{dk} (not equal to the double kink formation enthalpy F_{dk}). In the present study, the activation volume Ω^X was, for both kink migration and generation, taken to be b^3. The glide distance in the duration of a time step Δt is calculated by

$$\Delta L = \frac{h}{n}\left\{ \sum_i n^{KM}(i) + \sum_i n^{SDKG}(i) \right\} \tag{8}$$

where $n^X(i)$ is the number of events actually activated at the i-th site during the time step, according to random numbers generated in a computer; $n^X(i)$ is set to +1 (-1) if a stress-assisted (-resisted) event takes place or otherwise to 0. The glide distance in each step is averaged over all the steps to obtain the mean glide velocity

$$\bar{V}_d = \frac{\sum \Delta L}{\sum \Delta t} \cdot \frac{t_a + t_p}{t_a} \tag{9}$$

For dimensionless analyses, units $1/f^{KM} = \exp(E_m/kT)/f_0^{KM}$ for time, K/a^2 for stress, b^3K/a^2 for energy, E_m/k for temperature and b for distance were used.   Bearing Si in mind, we examined a case that E_m=12.5 ($\sim$1.5eV (Nikitenko et al. 1987)), E_{dk}=20.0 ($\sim$2.4eV, corresponding to $F_k\sim$0.5eV). A model dislocation of length n=1000 (or 2000) was set under a periodic boundary condition, and the time step was chosen to be Δt=0.1. To reduce statistical dispersions in the simulations, it was necessary to use a temperature of T=0.1 ($\sim$1700K) and a stress of σ_a =0.1 ($\sim$230MPa), both considerably higher than normally used in experiments.

Results obtained as a function of t_a=t_p (for the sake of brevity, we consider mainly the case of t_a=t_p in the following unless otherwise stated) are shown in Fig. 3, in which V_∞ is the glide velocity achieved under continuous loading and t_f (=290/f^{KM}) the time required for the model dislocation to advance by h, a period of the Peierls potential. One should notice an evident reduction of the mean glide velocity at short pulses. However, it is more important to note that the critical pulse length t_c , below which the reduction becomes appreciable, is much shorter than t_f.   This is in serious conflict with experimental results shown later in Figs. 5 and 6.

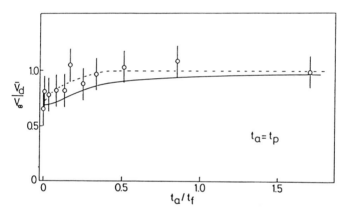

Fig. 3. Variation of the normalized velocity with the pulse length obtained by computer simulation based on the kink diffusion model. The solid line presents the theoretical curve calculated by eq.(14).

Analytical treatment , even if not exact, can be complimentary to computer simulations because it has the merit of allowing one to see how the model behavior depends on the parameters involved. The most simple approximation for the kink diffusion model is to consider only two states with regard to dislocation configuration, one being state 1 in which there is no kink in a unit length of dislocation and the other being state 2 in which there is a double kink in the critical size x_c (Fig. 1). The two states are interchangeable via double kink formation (1→2) at a rate v_1, via double kink contraction (2→1) at a rate v_2, and via double kink expansion and annihilation with neighboring kinks of opposite sign (2→1) at a rate v_3. Only the last process produces net glide of dislocations to the next Peierls valley and at the same time resumes state 1 in a statistical sense. Therefore, with C_2 , the probability of finding the dislocation in the transition state 2, the mean glide velocity is expressed as

$$\bar{V}_d = \frac{1}{t_a + t_p}\int_0^{t_a+t_p} C_2(t)\, h\, v_3(t)\; dt \qquad (10)$$

The probabilities C_1 to find the dislocation in state 1 and C_2 obey the following rate equation

$$\frac{dC_1}{dt} = -\frac{dC_2}{dt} = -v_1 C_1 + v_2 C_2 + v_3 C_3 \qquad (11)$$

which can be solved to give:

$$C_2(t) = \frac{1}{a}\left\{\frac{v_1}{v} + \left(C_2(0)a - \frac{v_1}{v}\right)\exp(-v t)\right\} \qquad (12)$$

where $v \equiv v_1 + v_2 + v_3$. Since $v_2 >> v_1$ and v_3, $v \approx v_2$. When a load is applied periodically, the probability $C_2(t)$ varies in a cyclic manner. Since $v_2^p/v_2^a \sim \exp(\sigma_a bh x_c/kT) \equiv \alpha \geq 1$ (the superscript a (p) refers to the value maintained during t_a (t_p)), the rate of increase of C_2 under stress is slower than that of decrease when the stress is removed. The transition rate $v_3(t)$ is approximately given by

$$v_3(t) = \frac{2V_k(t)}{\lambda} \qquad (13)$$

with λ being the mean free path of a kink. As the drift velocity of kinks V_k varies proportionally with the applied stress, v_3 switches between zero and a finite value following the load. Assuming that $\lambda \sim a \exp(F_k/kT)$ is insensitive to the loading condition, and taking account of $v_1^p/v_2^p \sim \exp(-2F_k/kT) \equiv \beta$, and $v_1^a/v_2^a \sim \exp\{-(2F_k - \sigma_a bh x_c)/kT\} = \alpha\beta$, one obtains for the mean glide velocity as a function of $v_2^a t_a \equiv x$

$$\frac{\overline{V}_d}{V_\infty} = 1 - \frac{(\alpha - 1)\{1 - \exp(-x)\}\{1 - \exp(-\alpha x)\}}{\alpha\, x[1 - \exp\{-(1 + \alpha) x\}]} \qquad (14)$$

where

$$V_\infty = \alpha\sqrt{\beta}\,\frac{h}{a}V_k^a \qquad (15)$$

is the dislocation velocity under continuous loading. For various values of α, the right hand side of eq. (14) is drawn in Fig. 4 to show how the average velocity depends on the pulse length. For large values of x or long pulses, the average velocity approaches, as expected, to V_∞, and for short pulses it decreases asymptotically to $2/(1+\alpha)$. Using $t_f = h/V_\infty$, $V_k^a = D_k \sigma_a bh/kT$ with the kink diffusion constant D_k, where $v_2^a \sim D_k/x_c^2$, and $x_c \cong (Gbh/8\pi\sigma_a)^{1/2}$, we have

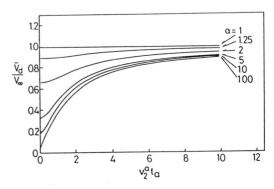

Fig. 4. $\overline{V}_d$ vs t_a relations calculated from eq.(14) for various values of α

$$v_2^a t_f \approx \frac{8\pi akT}{Gb^2h^2}\exp\left(\frac{F_k - \sigma_a bh x_c}{kT}\right) \qquad (16)$$

The solid line in Fig. 3 is a theoretical curve calculated with eq. (14) for values corresponding to the computer simulation, $\alpha \sim 1.9$ and $v_2{}^a t_f \sim 9.2$. The good agreement of the calculated curve with the result of computer simulation demonstrates the validity of the analytical treatment discussed above.

3. INTERMITTENT LOADING EXPERIMENTS IN Si AND GaAs

The intermittent loading effect first found by Nikitenko et al. (1985) attracted much attention of those who are greatly concerned with the magnitude of the secondary Peierls potential in semiconductors. In experiments performed on $60°$ and screw dislocations in Si, they observed that, when t_a is shorter than a critical pulse length t_c, the mean glide velocity decreases sharply from the value obtained in continuous loading.

Quite similar effects were also observed by one of the present authors from α-dislocations in n-GaAs (Maeda 1988). The present paper adds some more recent findings obtained by lowering test temperatures compared to the previous set of experiments, so that the pulse widths set around the characteristic time t_f are sufficiently long in comparison with the rise time of pulses ($\sim$2ms or less). Nevertheless, a peculiar situation in n-GaAs is that the mobility of α-dislocations is, particularly at low temperatures, much higher than the other dislocations; as a result, dislocation loops tend to be so elongated that the α-segments are likely subject to large back stresses due to trailing of screw segments. Since this back stress, if present, would bring about an effect similar to the intermittent loading effect, special measures were taken to avoid such a complication: Instead of the single etching method previously employed, a double etching method was used to trace dislocation movements; in addition, initial dislocation loops were injected at a relatively high temperature, so that the loops are deep enough for the back stress to be negligible.

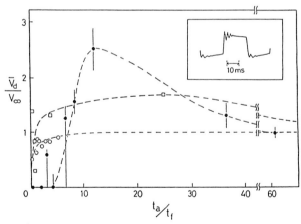

Fig. 5. The intermittent loading effect in α-dislocations in n-GaAs. Filled circles: L crystals (t_f=82ms, V_∞=4.2×10^{-9}m/s). Open circles: H1 crystals (t_f=19ms, V_∞=1.8×10^{-8}m/s). Open squares: H2 crystals (t_f=40ms, V_∞=8.6×10^{-9}m/s). The inset shows the shape of the *shortest* load pulse used for the L crystals.

Two kinds of Si-doped GaAs crystals were used: One is highly doped (n=1.4×10^{18}cm^{-3}) and the other is lightly doped (n=8.6×10^{16}cm^{-3}). Figure 5 presents results obtained with the pulse duty of $t_a = t_p$ for highly doped crystals (called H1 crystals prepared with loop injection at 230°C for 150 s, loaded at 24 MPa for Σt_a=9000 s at 167°C) and for lightly doped crystals (called L crystals prepared with loop injection at 400°C for 30 s followed by restressing at 200°C for 60 s, loaded at 16MPa for Σt_a=5400 s at 135°C) . Previous data for highly doped crystals (called H2 crystals loaded at 24 MPa for Σt_a=9090 s at 183°C) obtained by a single etching method are included for comparison.   The glide distance scatters significantly: Figure 6 shows typical histograms of glide distances obtained for L crystals of GaAs.   One   should note that the scatter is prominent particularly around t_c. Figure 7 shows also for comparison the results on Si obtained by Nikitenko et al. (1987). The most common feature in these crystals, except for data indicated with filled circles, is that $t_c \approx t_f$. However, this is not necessarily the case with the L crystals of GaAs and crystals of Si measured using pulses with short rising time.

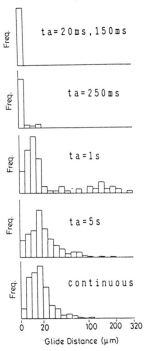

Fig. 6. Histogram of glide distances for experiments for H crystals of GaAs. The horizontal axis for glide distance is scaled by 5 μm from 0 to 20 μm, by 10 μm from 20 to 100 μm, by 20 μm from 100 to 200 μm, and by 40 μm from 200 to 320 μm.

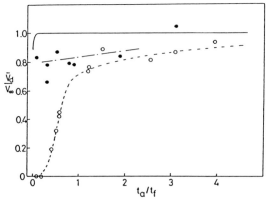

Fig. 7. The intermittent loading effect for 60°-dislocations in Si (Nikitenko et al. 1987) for rising time of loading pulse = 4 ms (open circles) and 1ms (filled circles). The solid line indicates the theoretical curve predicted by the kink diffusion model (eq.(14)).

Nikitenko et al. (1985, 1987) tried to explain the phenomenon on the basis of the kink diffusion model and claimed that the value of kink migration energy can be deduced from this type of experiment.   However, as shown in the following, great caution should be exercised in using this approach.

The value of α in eq. (14) which determines the variation of $\bar{V}_d/V_\infty$ with t_a is estimated to be ~ 1.25 for $\sigma \sim 7\text{MPa}$ and $T \sim 873\text{K}$, experimental conditions employed in the measurements of Si (Nikitenko et al. 1987). The parameter $v_2{}^a t_f \sim 720$ if one uses $F_k \sim 0.6\text{eV}$, a value previously proposed (Farber et al. 1986). The theoretical curve calculated with these parameters is shown in Fig. 7 for comparison with corresponding experiments. Evidently the typical experimental results shown with open circles are completely in disagreement with the theoretical prediction based on the kink diffusion model. The disagreement is found in two respects: In the magnitude of the asymptotic value of $\bar{V}_d/V_\infty$ at short pulses which is much smaller in the experiment (~ 0) than in the theory (~ 0.44), and in the critical pulse length which is much longer in the experiment ($\sim t_f$) than in the theory ($\sim t_f/720$). If we persist with the kink diffusion model to interpret the intermittent loading effect, we have to assume $F_k \sim 0.1\text{eV}$, which is much smaller than the elastic energy ($\sim 0.1Gb^2 \approx 0.7\text{eV}$ in Si) associated with a kink (Takeuchi and Suzuki 1988), and therefore too small to be acceptable physically. The case of GaAs confronts the same difficulty.

Such discrepancy between theory and experiment may have arisen from deviation of actual conditions in experiments from the ideal ones which are envisaged in theoretical analyses. As a matter of fact, Nikitenko et al. (1987) found that, if they performed intermittent loading experiments using pulses with short rising time, in other words, in shape of ideal rectangles in stead of blunted shapes used in their early studies, the intermittent loading effect was very weak even at pulses much shorter than t_f as shown with filled circles in Fig. 7. Thus the idealized experiment is rather in better agreement with the theoretical predictions and therefore one may conclude that the kink diffusion model itself is valid.

However, our data for L crystals of GaAs (Fig. 5), although the experiments were performed with the use of fast rising pulses, as the shortest pulses shown in an inset of Fig. 5, still exhibit an intermittent loading effect. Moreover, the critical pulse length t_c is considerably *longer* than t_f. A similar result of long t_c was reported also for Si in a specific loading condition (Nikitenko et al. 1987).

4. DISLOCATION PINNING MODEL FOR INTERMITTENT LOADING EFFECT

As discussed in our previous paper (Maeda 1988), the dispersed obstacle model (Celli et al. 1963) which presumes point obstacles for kink motion dispersed along a dislocation line, although not excluded as a possible mechanism of internal friction, is ruled out as a mechanism to explain the above fact because it predicts $t_c \ll t_f$. Previously Maeda (1988) proposed a new model named strong obstacle model, which presumes, similarly to the dispersed obstacle model, point obstacles against kink motion that are strong enough to block kink propagation but weak enough to be overcome on arrival of another kink with an opposite sign generated in the adjacent dislocation segment (Fig. 8). This model could naturally explain the experimental observations that $t_c \approx t_f$; however, the original form of the model can not explain the result that $t_c \gg t_f$ either.

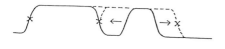

Fig. 8. Strong obstacle model.

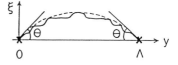

Fig. 9. Dislocation bowing-out between pinning points

The result that $t_c \gg t_f$ is hardly explicable by the Peierls mechanism alone because the long t_c corresponds to dislocation movement over much more than one atomic distance which is a characteristic dimension in the lattice friction mechanism. The most plausible explanation for the long $t_c \geq t_f$ is to consider dislocation pinning by point obstacles: If the stress is removed before depinning occurs, segments bowing-out under stress will shrink back due to their own line tension (equivalently due to kink-kink interaction). The equation of motion of a dislocation segment with length Λ pinned at both ends as shown in Fig. 9 takes the form (Granato and Lücke 1956)

$$A\frac{\partial^2 \xi}{\partial t^2} + B\frac{\partial \xi}{\partial t} + C\frac{\partial^2 \xi}{\partial y^2} = \sigma b_p \tag{17}$$

where $\xi(y)$ expresses the loop shape of dislocation segment, $A = \rho b^2$ the dislocation mass, $C = 2Gb_p^2/\pi(1-v)$ the dislocation line tension, and b_p the Burgers vector of the perfect dislocation. The parameter B represents the viscosity for dislocation motion the expression of which will be given later. Neglecting the inertia term as a good approximation, integrating eq. (17) from y=0 to Λ, and assuming that the dislocation loop expands in the form of an arc, one obtains for the expansion angle $\theta(t)$

$$\frac{B\Lambda^2}{6}\frac{\partial \theta}{\partial t} + 2C\theta = \sigma b_p \Lambda \tag{18}$$

which can be solved to give:

$$\theta(t) = \frac{1}{2C}\left[\sigma b\Lambda - \{\sigma b\Lambda - 2C\theta(0)\}\exp\left(-\frac{t}{\tau}\right)\right] \tag{19}$$

where

$$\tau = \frac{B\Lambda^2}{12C} \tag{20}$$

The viscosity parameter B arising from the lattice friction connects the stress σ with the dislocation velocity as

$$V_d = \frac{b}{B}\sigma = V_0\left(\frac{\sigma}{\sigma_0}\right)^m \exp\left(-\frac{U}{kT}\right) \tag{21}$$

where U is the activation energy for dislocation motion. Thus, when $m=1$,

$$B = \frac{\sigma_0 b}{V_0}\exp\left(\frac{U}{kT}\right) \tag{22}$$

On removal of the stress, if kink diffusion controls the dislocation mobility, the shrinkage of a loop is resisted by the same frictional stress as in its expansion; however, if double kink formation controls the mobility, the loop shrinks simply via back motion of kinks without lattice friction ($B=0$) because no double kink formation is necessary for the shrinkage. The time-variation of θ is schematically illustrated in Fig. 10. The maximum angle θ^a is written as

$$\theta^a = \frac{\sigma b_p \Lambda\left(1 - \exp\left(-\frac{t_a}{\tau}\right)\right)}{2C\left\{1 - \exp\left(-\frac{t_a + t_p}{\tau}\right)\right\}} \qquad \text{for kink diffusion control} \tag{23}$$

or

$$\theta^a = \frac{\sigma b_p \Lambda}{2C}\left\{1 - \exp\left(-\frac{t_a}{\tau}\right)\right\} \qquad \text{for kink formation control} \qquad (24)$$

If depinning of the dislocation from point obstacles is to take place at a critical angle θ_c, the depinning can actually occur only if t_a is sufficiently long so that θ_a exceeds θ_c (Fig. 11). Therefore, whichever the case, kink diffusion control or kink formation control, the dislocation pinning will bring about an intermittent loading effect similar to that observed experimentally. Since the relaxation time τ depends on Λ, and Λ should have a distribution around a mean value, the transition will be gradual as also observed. Since the critical pulse length t_c varies with τ and θ_c, it can differ from t_f. In any case t_c will be of the order of τ. Substituting appropriate values for the lightly doped GaAs tested at 135°C into eq. (20), one obtains $t_c \sim \tau \sim 4\times10^{-6}(\Lambda/b_p)^2$ s, which reproduces experimental t_c when $\Lambda \sim 400 b_p$, a possible value for pinning lengths.

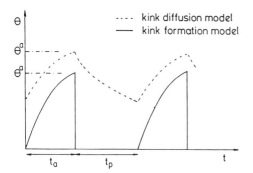

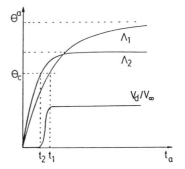

Fig. 10. Variation of expansion angle with time under intermittent loading.

Fig. 11. Dependence of the maximum angle on pulse length for two pinning lengths Λ_1 and Λ_2 ($\Lambda_1 > \Lambda_2$). Depinning occurs at $\theta^a = \theta_c$.

The origin of the pinning obstacles is not clear at present. Nevertheless, a possible candidate is the same as those obstacles assumed in the strong obstacle model. If one takes into account the distribution of the segment length which was ignored in the original model, there may be a segment narrower than x_c, in which formation of expandable double kinks is not possible; as a result, strong obstacles at both ends of the short segment effectively act in pair as a much *stronger obstacle* and depinning of dislocation from it cannot occur until the neighboring segments bow out sufficiently for some other depinning mechanism (e.g., thermal depinning assisted by line tension of the bowing-out dislocation) to operate.

As far as the above argument is concerned, the obstacles may be dispersed in the crystal. However, the scatter in the glide distance prominent especially around $t_a = t_c$ suggests that the obstacles are such defects that *lock* the dislocations at their initial positions. In the pinning model, the distribution of Λ will lead to a distribution of τ. When the loading pulse is long enough for all the loops to reach the depinning configuration and subsequently glide for considerable distances, the distribution of Λ will have no significant effect on the resultant glide distance. In contrast, when the pulse length is around some critical value, we will have a state that some loops can escape from locking while others remain locked, which leads to a considerable scatter in the glide distance. The similar scatter of glide distance around $t_a = t_c$ is also observed in Si (Nikitenko et al. 1987), which may indicate that the intermittent loading effect in Si also is brought about by dislocation locking effect.

One may notice in Fig. 5 that the mean glide velocity in L and H1 crystals at pulse lengths just above t_c exceeds considerably the velocity under continuous loading. This anomalous increase of velocity, also evident in Fig. 6, suggests some dynamical effect of pulse loading which might be the cause of the increase of velocity in fast rising pulses observed in Si (filled circles in Fig. 7); however, the mechanism is not known at present.

The conclusions of the present study may be summarized as follows:
(1) It is difficult to consider that the intermittent loading effect so far observed is a direct manifestation of the kink diffusion mechanism.
(2) The observed intermittent loading effect is explained by considering dislocation pinning probably of locking type.

ACKNOWLEDGMENTS

The authors are grateful to Sumitomo Electric Industries, Ltd. for supplying GaAs single crystals. They would like to thank Prof. Takeuchi and his colleagues for their permission to use facilities for sample preparation.

REFERENCES

Celli V, Kabler M, Ninomiya T and Thomson R 1963 Phys. Rev. **131** 58
Farber B Ya, Iunin Yu L and Nikitenko V I 1986 phys. stat. sol. **97** 469
Granato A and Lücke K 1956 J. Appl. Phys. **27** 583
Hirsch P B, Ourmazd A and Pirouz P 1981 Inst. Phys. Conf. Ser. **60** 29
Hirth J P and Lothe J 1968 *Theory of Dislocations* (New York: McGraw-Hill)
Jendrich U and Haasen P 1988 phys. stat. sol. (a) **108** 553
Jones R 1980 Philos. Mag. B **42** 213
Louchet F 1981 Inst. Phys. Conf. Ser. **60** 35
Louchet F, Cochet Muchy D, Brechet Y and Pelissier J 1988 Philos. Mag. A **57** 327
Maeda K 1988 *Proc. Int. Summer School on Defects in Crystals* ed E Mizera (World Scientific: Singapore) pp 153–168.
Nikitenko V I, Farber B Ya and Iunin Yu L 1985 JETP Lett. **41** 124
Nikitenko V I, Farber B Ya and Iunin Yu L 1987 Sov. Phys. JETP **66** 738
Ninomiya T, Thomson R and Garcia-Moliner F 1964 J. Appl. Phys. **35** 3607
Schoeck G 1982 Scripta Met. **16** 233
Southgate P D and Attard A E 1963 J. Appl. Phys. **34** 855
Takeuchi S and Suzuki T 1988 Proc. 8th Int. Conf. on Strength of Metals and Alloys ed P O Kettunen, Lepisto and M E Lehtonen (Oxford: Pergamon) pp 161–6

Inst. Phys. Conf. Ser. No 104: Chapter 3
Paper presented at Int. Symp. on Struct. Prop. Disloc. Semicond., Oxford, 1989

281

Mobility of dislocations in gallium arsenide

H Alexander and H Gottschalk

Abteilung für Metallphysik, II. Physikalisches Institut, Universitaet zu Koeln, Zuelpicher Str. 77, D-5000 Koeln, FRG

ABSTRACT: Measurements of the stress and temperature dependence of the velocity v of 60° and screw dislocations in GaAs, both doped and undoped , are reviewed. Dislocation motion is thermally activated, the activation energy Q being determined by the formation energy of kink pairs and this in turn by the Peierls relief. It is the reduced Peierls potential which makes the dislocations more mobile in III-V-compounds than in Si and Ge. 30° partials are less mobile than 90° partials; the least mobile is the 30°β partial in undoped SI GaAs, in p-type GaAs the 30°α partial.

1. INTRODUCTION

The dynamics of the motion of dislocations in semiconductor crystals has been under investigation since Patel's pioneering paper (Chaudhuri, Patel and Rubin, 1962). What is measured more recently by different authors does not perfectly fit together, especially in case of III-V-compounds, but there is agreement about the principal character of the influence of various experimental parameters on the dislocation velocity. In contrast there is little understanding of the phenomena on an atomic level. In this paper we are going to collect the experimental facts and we will try to draw some conclusions from the overview over the numerous effects which are observed when dislocations move in GaAs under well defined conditions.

2. BACKGROUND

Comparing III-V-compounds with elemental semiconductors, like Ge and Si, there are several complications: Firstly, the two fcc-sublattices are occupied by chemically different atoms. This doubles the number of dislocation types. Secondly, exact stoichiometry of the crystals in general cannot be presumed; since 1981 a large portion of GaAs crystals is grown with As excess. So we have to take into account large numbers (up to 10^{18} cm^{-3} (Hurle, 1979)) of structural point defects, which may be in various positions: so an excess of As atoms may be adjusted by As interstitials, Ga vacancies, antisite defects (ADs) As$_{Ga}$ and by As-precipitates. Most of these point defects may exist in two or three different charge states so that the actual spectrum of defects in a given crystal strongly depends on its thermal history and on electronic doping. Additionally the purity of III-V-compounds is much inferior to that of Si or Ge. It is plausible that any measurement is the more valuable the better the material in consideration is characterized. Unfortunately, in most papers this information is very limited.

Considering dislocations nearly all of them are found dissociated into two Shockley partials (Gomez and Hirsch 1978, Feuillet 1982, Jimenez-Melendo et al. 1986, Küsters et al. 1986). Only in studies by high resolution electron microscopy (HREM) undissociated dislocations are observed, too (Ponce et al. 1986), but this seems to indicate metastability of the dissociation in very thin areas (Tanaka and Jouffrey (1984)).

The stacking faults always were found to be of intrinsic character. In all III-V-compounds the dislocations are dissociated (Gottschalk et al. 1978). If the stacking fault energy (SFE) is related to the area of a unit cell, it decreases with increasing ionicity because of an extra Coulomb attraction across the plane of the stacking fault (Gottschalk et al. 1978). By weak beam electron microscopy the SFE was measured by several groups (Table 1).

Author		SFE	
Gomez and Hirsch	(1978)	48 ± 6	
Gottschalk et al.	(1978)	55 ± 5	
Feuillet	(1982)	50 ± 7	(α)
		41 ± 7	(β)
Nakada and Imura	(1987a)	42 ± 7	
Snigireva et al.	(1988)	48 ± 3	
Boivin	(1988)	47 ± 9	
Jimenez-M. et al.	(1988)	49	

Table 1 SFE of GaAs (mJ/m²)

The search for a dependence of the SFE on doping failed (Nakada and Imura 1987a, Boivin 1988). Only with the highest concentration of In atoms (1 to $4*10^{20}$ cm^{-3}) Jimenez-Melendo et al. (1988) found surprisingly wide dissociations corresponding to a SFE of 27 mJ/m². It is worth noting that also grown-in dislocations were found dissociated with no significantly different SFE (Nakada and Imura 1987b). Nakada and Imura conclude from this insensitivity on doping (i.e. Debye screening length) and Cottrell clouds around grown-in dislocations that the partials do not carry strong electrical charge in any case.

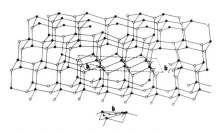

Fig. 1 Dissociated 60° dislocation
    30° partial (left),
    60° partial (right)

Since the stacking fault ribbon has to be placed between two closely spaced {111}-planes it is natural to assume that also the bounding partials move in the same plane (so called "glide set", Hirth and Lothe (1968)). Fig. 1 shows a sketch of a dissociated 60° dislocation (glide set). One notices in the core of both partials rows of atoms (ions) of the same species which have lost one of their four ligands. There is little doubt that in Si these broken bonds pairwise are saturated. In III-V-compounds (and the more in II-VI-compounds) this is not so obvious because reconstruction needs bonding between two ions of the same sign. But the existence of numerous ADs shows that this is energetically not as expensive as one could believe. Jones (1981) proposes some periodic displacements of the atoms in the core of the partials analogous to the superstructure of the {110} crystal surface. Referring to the long standing discussion whether partial dislocations in diamond-like crystals in fact have the simple glide set structure shown in Fig. 1, we see no reason to doubt that for most of their length. However, in Si there are shuffle points (i.e. core vacancies) distributed along the 30° partials the density of which increases during dislocation motion. There are indications that these core vacancies interact with reconstruction defects and impurity atoms (Kisielowski-Kemmerich, this conference). Probably in GaAs the conditions are similar: plastic deformation does in fact produce point defects (about 10^{17} cm^{-3} for 10% strain), closely related to the antisite defect As$_{Ga}$, but not metastable (Wattenbach et al. 1989).

Electron microscopy of deformed GaAs reveals that dislocations prefer ⟨110⟩ orientations, indicating a relatively strong Peierls potential. Particularly after deformation at lower temperature ($\leq$ 300°C) 60° and screw dislocations are prevalent. This means: Only two types of partials dominate: 30° and 90° partials which are shown in Fig. 1. In 60° dislocations both partials are of the same character (i.e. containing As or Ga atoms in the core). Screws consist of one 30° partial with As atoms and the other with Ga atoms. Since we do not discuss shuffle dislocations we choose the simple nomenclature now widely used: dislocations with anions (arsenic) in the core are called α dislocations (extended: As(g)), while dislocations with cations (gallium) are called β dislocations (Ga(g)). Compared with the older literature, an α dislocation stays as an α dislocation, though it was believed in former times to be a shuffle dislocation and to contain Ga atoms. Paufler (1989) reports on frequent 30° dislocations which should consist of one 0° and one 60° partial, both being more complicated than those shown in Fig. 1. Moreover, after deformation at higher temperatures many edge dislocations are observed, like in Ge and Si. They consist of two 60° partials.

3. DIRECT MEASUREMENTS OF DISLOCATION VELOCITIES

3.1 Methods

Most measurements of the dislocation velocity in GaAs have been carried out by the double etching method. However, first observations of moving dislocations by X-ray topography (350-500°C) and in an HVEM (450-800°C, Sato et al. 1985) revealed that α dislocations move in an extremely jerky manner and also screws did not move uniformly. So double etching yields only mean values of the velocity, composed of periods of much faster "free flight" and waiting periods at unidentified local obstacles. Moreover, screw dislocations become very slow in regions swept by many dislocations. All these findings prove that there are local obstacles distributed in the glide plane, which are at least partly produced by other moving disloca-tions. (By the way: also de Cooman and Carter (1987) have clear indications that it is 30° partials which are most frequently pinned, and screws seem particularly often to interact with point defects). Although being concerned with "mean" velocities macroscopic measurements bring to light a lot of interesting tendencies concerning the long-range mobility and its role for plastic deformation. Theories to be compared with these measurements should be, however, adapted to the two-stage motion.

3.2 Undoped GaAs

Most authors represent their results using the well known empirical relation (Chaudhuri et al. 1962)

(1) $$v = v_0 \, (\tau/\text{MPa})^m \, \exp(-Q/kT)$$

which proved to be useful for semiconductors in all cases investigated so far. However, the stress exponent m turns out to decrease with increasing temperature and the activation energy Q depends on the resolved shear

Author		Material	α			β		
			v_0 (m/s)	m	Q (eV)	v_0 (m/s)	m	Q (eV)
Erofeeva et al.	1973	?			1.2			1.4
Osvenskii et al.	1973a	n = 1..3*10^{16}			.93			1.57
Choi et al.	1977	n = 10^{17}/HB	2.35	1.4	1.0	13	1.6	1.35
Steinhardt et al.	1978	n = 10^{17}			1.2			
Matsui et al.	1986	n = 1.4*10^{17}/LEC	3.7	1.7	.89	38	1.5	1.24
Yonenaga et al.	1989	Si: n = 10^{13}/Boat	1900	1.7	1.3	59	1.6	1.30

Table 2. Parameters of eq. (1) for undoped GaAs
Q measured at 10 or 20 MPa; HB, LEC, Boat: growth method
n: carrier density (cm^{-3}) at room temperature

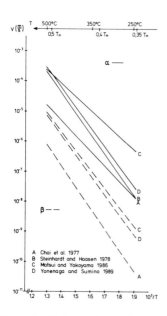

Fig. 2 Velocities of α and β dislocations in undoped GaAs (calc. using data from Table 2, τ = 10 MPa))

stress τ. Fig. 2 gives the velocities of α and β dislocations in undoped GaAs measured by various groups. The velocity is calculated for τ = 10 MPa (0.35 ≤ T/Tm ≤ 0.5) from values of the parameters v_0, m and Q given by the authors (Table 2). Undoped means not intentionally doped; carrier densities (at room temperature) of the order 10^{17}cm^{-3} show that those crystals contain donors (Si?).

With the exception of Yonenaga and Sumino (1989) the activation energy of the dislocation motion is found to be smaller for α than for β dislocations (Table 2).

Data for screw dislocations are scarce. Osvenskii et al. (1973a) give the most extensive information in this respect. For undoped GaAs they fit their data to

(2) $v = v_0 \exp(-\tau_1/\tau) \exp(-Q_s/kT)$

with τ_1 = 2.3 MPa and Q_s(20 MPa) = 1.11 eV. For 350°C and τ = 10 MPa one obtains the velocity of screws as $v_s \approx 10^{-7}$ m/s. Q_s decreases from 1.4 eV (1 MPa) to 1 eV (50 MPa). In their material the ratio of velocities of the three basic dislocation types at 300°C is $v_\alpha : v_s : v_\beta = 300 : 9 : 1$.

Distinguishing the leading 30° partial from the trailing one, there are two types of moving screw dislocations: αβ and βα which, surprisingly, have different mobilities.

Concluding this part, we state that dislocations move in undoped GaAs much faster than in Si (2 to 3 orders of magnitude at 0.5 Tm); the activation energy Q is roughly half that of Si. This difference decreases at higher temperatures (if we assume that the physical nature of the process of motion does not change; measurements above 0.5 Tm are difficult because the crystal decomposes by evaporation of As). In each of the investigated materials α dislocations are more mobile than β dislocations. About the ratio there is no agreement (40 to 400 at 300°C). Screw dislocations apparently have behaviour in between that of the two 60° dislocations.

3.3 Doped GaAs

Impurity atoms can influence dislocation mobilities if they interrupt the lattice periodicity by either different size (misfit) or different bonding (change of local elastic properties). This solution hardening needs typically impurity concentrations in the percentage region. In semiconductors there is a second interaction between dislocations and electrically active impurities (donors or acceptors) which works only at temperatures where the intrinsic carrier density is below the density of ionized dopants. This "Patel effect" is non-local, depending on a change of the charge state of dislocations and possibly of point defects. The Patel effect due to the small number of intrinsic carriers in pure GaAs could be expected even at low doping levels in this (fictitious) material: 10^{13} cm^{-3} at 300°C. Actually one has e.g. in LEC GaAs high concentrations (10^{16} cm^{-3}) of shallow acceptors (mostly carbon) and more or less the same concentration of deep donors (EL2). GaAs crystals grown in a quartz boat usually contain rather high concentrations of Si, strongly interacting with dislocations (see below). LEC grown crystals incorporate boron from the encapsulant. So the influence of a given doping has to be considered on the background of the position of Fermi level without this doping.

To begin with the group of electrically inactive elements (B, Al, In, N, P and Sb), an enormous effort is directed towards understanding the fact that alloying GaAs with In starting at $10^{19} cm^{-3}$ In (i.e. $5*10^{-4}$ of the cations) strongly reduces the density of dislocations in crystals grown by the LEC technique (similar data exist on Sb, Blom and Woodall 1988). The first idea, of course, is reduction of the dislocation mobility impeding multiplication of the first dislocation nuclei in the growing crystal. In fact the misfit of In instead of Ga (7%) is large enough to cause solution hardening (the effective solute being $InAs_4$ tetrahedra) (Ehrenreich and Hirth 1985).

Dislocation velocities have been measured in GaAs:In by Yonenaga et al. (1987, 1989), Matsui and Yokoyama (1986) and Burle-Durbec et al. (1987). While Matsui et al. ([In]: 10^{20} cm^{-3}) did not find any effect on α or β dislocations within $4 \leq \tau \leq 40$ MPa, Yonenaga et al. ([In]: $3*10^{19}$ to $2*10^{20}$ cm^{-3}) report all three dislocation types to be only slightly slower than in undoped GaAs (350-720°C); however, for α and screw dislocations this holds only above a critical stress τ_c where these particular dislocations were unlocked from In clusters. The critical stress decreases with increasing temperature and should not be responsible for the reduction of grown-in dislocations as these dislocations are generated at higher temperatures.

Summarizing so far we follow Yonenaga et al (1989) (who compare the effects of many different impurities on dislocations in GaAs and stress), that one has carefully to distinguish between pinning (locking) effects by impurities gettered by the dislocation (at rest or slowly moving) and changes of mobility caused by impurity atoms dispersed in the lattice. The latter is a solution hardening and/or Patel effect which may be negative or positive due to its electrical nature. A striking example for the separate character of pinning and Patel effect is Zn, which effectively pins β dislocations and screws but makes the same dislocations <u>more</u> mobile than in undoped material when unlocked (Yonenaga et al. 1989). Indium causes

only weak retardation of moving dislocations;   locking - particularly of α
and screw dislocations - does exist but is weakened at higher temperature

What remains to explain the spectacular effect of In on the grown-in
dislocations? Several groups of authors take the view that alloying with
In may broaden the existence region of the phase GaAs or decrease the
"nonstoichiometry of the crystal" (Lagowski et al. 1984, Blom and Woodall
1988, Bourret et al. 1987). In effect this means that fewer vacancies are
at hand for dislocation nucleation and climb.

Remembering that GaAs crystals grown by the LEC technique contain two
families of grown-in dislocations   differing in their arrangement
(cell walls and slip lines),   and by the presence or   absence of
EBIC contrast (Takebe et al. 1986) the present authors believe that those
two kinds of dislocations are generated at different temperatures: those
which form cell walls from vacancy agglomeration followed by climb at high
temperature, and the slip dislocations at lower temperature by multiplication of dislocations from surface sources (Alexander 1989). The first
named are suppressed by In via its influence on point defects; the second
named are somewhat retarded by the Patel effect. Moreover, the rate of
widening of slip lines which happens by double cross slip seems to be
strongly reduced (Sato et al. 1986) - possibly confirming the reduction of
the SFE claimed by Jimenez-Melendo et al. (1988).

Aluminum is a special case because it fits perfectly into the GaAs lattice
and should not be electrically active, either. In fact it does not change
the mobility of free dislocations but it locks β dislocations (Yonenaga
and Sumino 1989). This is not understood at all.

Turning now to the electrically active dopants we expect superposition of
mechanical hardening and electronic Patel effects; the latter in principle
could be separated by counterdoping experiments which have been successfully applied by Lagowski et al (1984) to prove that the Fermi level is
the decisive parameter for suppression of grown-in dislocations by Si.

In Fig. 3 and Fig. 4 activation energies of the motion of α and β dislocations and the dislocation velocities in n- and p-doped GaAs are collected.
Overlooking the whole of the data one gets the impression that there is
qualitative agreement, with the definite exception of the region of high p-
type doping (p ≥ 10^{18} cm^{-3}) with respect to β dislocations. This latter
can be understood as an influence of the particularly strong pinning
effect of Zn on β dislocations (Yonenaga and Sumino 1989). Admittedly
α dislocations should be locked even stronger by Si, but Si is a slow
diffusor, so that it seems to be ineffective at 300°C. Without claiming
perfect separation of pinning, solution hardening and Patel effect at the
present time we state the following tendencies:

The activation energy Q_α of α dislocations increases for both   n- and p-
type doping. Q_β on the other hand also increases with n-type doping, but
p-type doping decreases Q_β (Ninomiya 1979) till p ≈ 10^{18} cm^{-3}. For higher
p-doping, the results are not systematic, probably due to other unknown
effects.

The dislocation velocity (Fig. 4) also behaves differently for α and β
dislocations . α dislocations are slowed somewhat (by one and a half orders
of magnitude) by both doping types;   β dislocations in contrast are
very strongly retarded by n-doping (by 3 orders) and they are <u>mobilized</u>

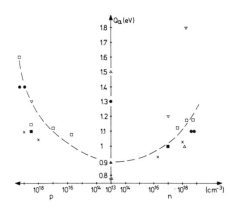
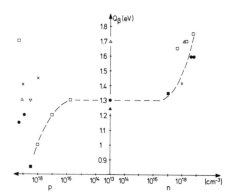

Fig. 3 Activation energies of the α and β dislocation motion in doped GaAs vs doping level (p-type: Zn, n-type: Si, Te) (See Fig. 4 for symbols).

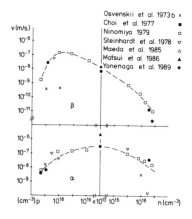

Osvenskii et al. 1973b x
Choi et al. 1977   ■
Ninomiya 1979   □
Steinhardt et al. 1978 ▽
Maeda et al. 1985   △
Matsui et al. 1986   ▲
Yonenaga et al. 1989   ●

Fig. 4 Velocities of α and β dislocations vs doping level (p-type: Zn, n-type: Si, Te) T = 300°C, τ = 10 MPa; some data extrapol. from meas. at higher temp. and higher stress.
- - - - - - - - - - - - - - - - - - -
(1.5 orders of magnitude) by p-doping. (The decrease of v beyond p = 10^{18} cm^{-3} may be a consequence of pinning effects, as mentioned above). Qualitatively the same statement was reached by Yonenaga and Sumino (1989) for T = 550°C and τ=20 MPa on the only basis of their own measurements. Discussion of these results is being postponed until some other related effects are described.

3.4 Recombination enhanced dislocation glide (REDG)

In 1981 Maeda and Takeuchi discovered strongly enhanced mobility of dislocations in n-type GaAs (HB grown) as soon as a beam of 30 keV electrons hit the sample. The temperature dependence of the effect is characterized by a break at a certain temperature T_c: while for $T \geq T_c$ the velocity v is independent of irradiation, the activation energy Q at temperatures below T_c is decreased under irradiation (Table 3). Variation of the beam intensity I influences only the prefactor v_0. Maeda and Takeuchi describe their results by superposition of two contributions to the dislocation velocity

(3) $$v = v_{0d} \exp(-Q/kT) + v_{0i}(I) \exp(-(Q-\Delta Q)/kT).$$

The authors did not investigate whether the irradiation had an influence on the stress dependence of v. Mdivanyan and Shiksaidov (1988) report on a strong change of the stress exponent m (from 3 in the dark to 1.75 under the light of xenon lamp). It is not clear what material is concerned, but from the activation energies (Q = 0.9 eV, ΔQ = 0.6 eV) one may conclude that the data belong to n-type GaAs ($2*10^{18}$ cm^{-3} Te).

Material	Q_α	ΔQ_α	Q_β	ΔQ_β	Q_s	ΔQ_s
n $(1.4*10^{18}\,\mathrm{cm^{-3}}$ Si)	1.00	0.71	1.7	1.1	1.7	1.1
intr. $(2*10^{17}\,\mathrm{cm^{-3}}$ Cr)	1.5	1.5 (!)	1.7	1.3	-	-
p $(1.2*10^{19}\,\mathrm{cm^{-3}}$ Zn)	-	-	1.3	0.6	-	-

Table 3. Change of the activation energy Q by REDG (cf. eq.(3)) in eV
        (From Maeda and Takeuchi (1985))

It was of course checked that the effect is not due to heating the
specimen. It also is different from the Patel effect; this is evident from
the simultaneous occurence of both effects in Si and also from their
different influence on different types of 60° dislocation in this element
(Küsters and Alexander 1983). The REDG effect can be produced by light as
well as by electrons (for a quantitative comparison see Fujita et al.
1988).

Maeda and Takeuchi called this effect "Radiation Enhanced Dislocation
Glide"; they came to the same conclusion as did Küsters and Alexander
(1983), i.e. it belongs to the class of Recombination Enhanced Defect
Reactions and it influences the kink motion along the (partial) dislocations which is enhanced by the energy released when injected minority
carriers recombine. The critical temperature T_c marks the transition from
the intrinsic range to the range where the injected carriers dominate.

Later on, Maeda and Takeuchi (1985) showed the REDG effect to exist (with
different ΔQ) for intrinsic (Cr doped) and p-type GaAs , too. Particularly
striking is evidence for REDG at 80K presented by Schreiber and Leipner
(1988): dislocations may move at 80K with $v \approx 4*10^{-8}$ m/s in the stress
field of surface scratches under an electron beam density of 300 A/m². The
authors point to the small difference between the velocity at 300K and
80K. In fact Maeda and Takeuchi (1985) also found under the electron beam
perfectly athermic motion of α dislocations in GaAs doped with Cr (Table
3). The importance of the REDG for all applications of semiconductors with
injection of minority carriers (light emitting diodes, lasers, devices
under ionizing radiation etc.) is selfevident; moreover: not only dislocation glide but also dislocation climb is strongly enhanced (Petroff and
Hartmann 1974). Therefore, research on dislocations using electron microscopy of any type has to be done considering this drastic change of
activation energies.

4. OTHER SOURCES OF INFORMATION

4.1 Dislocation motion under high stress

It is of great interest to investigate the mobility of "isolated" partial
dislocations and to draw conclusions from the analysis of dislocation
arrangements formed after having applied high stress by one of the
following techniques:
a) uniaxial deformation under high stress ($\geq$ 100 MPa) at low temperature
   ($<$ 0.5 T_m) (Küsters et al. (1986), de Cooman and Carter (1987))
b) uniaxial deformation under confinement (i.e. hydrostatic pressure) at
   very low temperature (down to room temperature) (Lefebvre et al.(1985),
   Rabier et al. (1985), Androussi et al. (1987), Lefebvre et al. (1987a))
c) microindentation (Höche und Schreiber (1984), Hirsch et al. (1985),
   Lefebvre et al. (1987b)).

Using these techniques important properties of GaAs; e.g. the anisotropy of hardness of {111} surfaces and the consequences of misfit in multilayer structures are explained on the basis of different mobilities of the dislocation types which are needed to adapt the strain forced by external conditions. Because of limited space we can only refer to information on the different mobilities of single partials which are drawn from those experiments.

Since dislocations are more straight and parallel to <110> directions (parallel to Peierls valleys) after high stress deformation, one observes many pure 60° and screw dislocations, which are all composed of 30° and 90° partials. Evaluating those dislocations taking into account α- and β-modifications one is left with four basic types of partial dislocations: 30°α, 30°β, 90°α and 90°β. In crystals deformed with very high stress (under confining pressure) additionally to glide of perfect dislocations microtwinning as a secondary deformation mode is observed. This can be understood if one realizes that the forces working on the two partials bounding a stacking fault can be of very different size. This may be either on account of the deformation axis, but in most cases it results from different mobilities of the two partials: If for instance the second partial, trailing behind the stacking fault ribbon is less mobile one can treat this by a friction stress diminishing the effective driving force. If the difference of effective driving forces exceeds the stacking fault tension the two partials are separated and a macroscopically wide stacking fault ribbon is produced. Bundles of such ribbons are equivalent to microtwins. (A microtwin of 5nm thickness is a very effective obstacle to dislocation motion on other glide planes, Gottschalk et al., 1987). By analysis of extended stacking faults, microtwins and the spectrum of dislocation types in high stress deformed crystals the following order of mobilities for partials has been stated (Androussi et al. 1987):

undoped SI GaAs: $\mu(30°\beta) < \mu(30°\alpha) < \mu(90°\alpha/\beta)$
p-type GaAs (10^{18}cm^{-3} Zn + 10^{20}cm^{-3} In): $\mu(30°\alpha) < \mu(30°\beta) < \mu(90°\alpha/\beta)$

As in Si 90° partials generally are more mobile than 30° partials. But more interesting is the change of the order of the 30° partials with doping: it indicates that it is in fact the 30° partial which is influenced by p-doping and causes the ratio v_β/v_α to increase going from intrinsic to p-type GaAs (Fig. 4).

The most striking result with respect to partial mobilities was obtained by Sato et al. (1985) by X-ray topography and by Küsters et al. (1986) by TEM: out of the two screw dislocations, distinguished by the order of their 30° partials, the 30°α/30°β is about twice as mobile as its opposite. (The partials are quoted in the order they move). De Cooman and Carter (1987), in an informative paper on high stress deformed GaAs; report on 60°β dislocations: the 30°β/90°β type is more mobile than the 90°β/30°β type, in contrast to what has been found in Si and Ge. The authors stress the significance of the mobility (and density?) of the <u>kinks</u> for the mobility of a certain type of dislocation. This is so at least at lower temperature where without any doubt surmounting of the Peierls potential is the rate determining process of dislocation motion in semiconductors.

4.2 Macroscopic plastic deformation

Dynamic deformation of semiconductor single crystals starts with dislocation multiplication resulting in a yield point phenomenon. At the lower yield point a state of optimal plasticity (SOP) is reached where the density of (mobile) dislocations depends only on an elastic constant A (taking into account dislocation interaction, Haasen 1962) and the stress exponent m of the dislocation velocity

$$(4) \qquad N^{SOP} = (2 \ \tau_{ly} \ / \ (A \ (m + 2)))^2 \qquad \text{(Alexander and Haasen, 1968).}$$

This means that the dislocation velocity can be calculated from the lower yield stress τ_{ly} and the plastic strain rate $\dot{\varepsilon}$ (as far as m and A are known or constants with respect to the investigated dependence):

$$(5) \qquad v^{SOP} = \dot{\varepsilon}/bN^{SOP} = (\dot{\varepsilon}/b) \ (A \ (m + 2) \ / \ 2 \ \tau_{ly})^2$$

(Analogously v^{SOP} can be calculated from the maximum creep rate). It has to be stressed that the effective shear stress acting on a dislocation at SOP is $\tau_{ly}m/(m + 2)$ (Alexander and Haasen 1968).

First attempts to use this method for determining a characteristic dislocation mobility from deformation tests in GaAs were made in 1978 (Gottschalk et al., Steinhardt and Haasen). Benefits are obvious: the measurements are less time consuming and are not prone to surface effects so that they can be extended to much higher temperature. But there are complications when dislocation sources are not homogeneously distributed (as in case of III-V-compounds where all dislocations are generated at the surface (Tohno et al. 1986)). Also interpretation is less straightforward when dislocation types with different mobilities are contributing to $\dot{\varepsilon}$. Careful analysis of experimental results should have the power to clear up the cooperation of the various dislocation types producing macroscopic strain. Out of regard of space we have to postpone this to a forthcoming paper.

5. DICUSSION

Due to the high and variable concentration of structural point defects and impurities in crystals of III-V-compounds it is difficult to make a guess for the activation energy Q of dislocation motion in "pure" GaAs. From the collection of the data in Fig. 3 we propose: $Q_\alpha = (1 \pm 0.1)$ eV and $Q_\beta = (1.3 \pm 0.1)$ eV. For screw dislocations Yonenaga et al. (1989) give $Q_s = 1.4$ eV. Where the stress exponent m has been measured it was found to be between 1.4 and 1.8; but m is not a true physical parameter.

The stress dependence of Q as measured for 60°α and 60°β dislocations in undoped GaAs by Erofeeva et al. (1973) can perfectly be fitted to the socalled abrupt-kink-model (Ninomiya et al. 1964) which proposes

$$(6) \qquad Q = Q_0 + A \ \tau^{-\frac{1}{3}} - B \ \tau^{\frac{1}{2}}.$$

From A and B the Peierls energy E_P per unit length can be calculated.

	Q_0 (eV)	E_P (10^7 eV/m)
60° α	0.98	4.7
60° β	1.28	6.15

Table 4 (left) contains the values we obtained. It is interesting to note that both energies are greater by 31% for β dislocations indicating that Q_0 is determined by E_P. If kinks annihilate by collision Q_0 is the energy of one kink, otherwise Q_0 means the energy of two separate kinks.

It is to be assumed that the two coupled partial dislocations move uncorrelatedly and the values given in Table 4 concern the (less mobile) 30° partials. The Peierls energy of GaAs here turns out rather small; it corresponds to a Peierls stress $\tau_P = \pi E_P / ab$ = 226 MPa (α disl.) and 294 MPa (β disl.). Mihara et al. (1975) obtained about half as large a Peierls energy for InSb. This reduction of the Peierls relief is the primary reason for dislocations being more mobile in III-V-compounds compared to elemental semiconductors. For screw dislocations in undoped GaAs the abrupt kink model does not fit the measurements of $Q(\tau)$ by Osvenskii et al. (1973a). Rather the data above of 5 MPa can be fitted to the dragging point model (Celli et al. 1963). Here Q should be

(7) $$Q = Q_0 - V\tau$$

We find Q_{0s} = 1.23 eV and V = 12.5 b^3, assuming kink collision Q_0 is the sum of the formation energy of a single kink and half the energy needed to surpass an obstacle by a kink. From V the distance between two obstacles (x^* = 11.5 nm) can be deduced. Because the measurements of $Q(\tau)$ for 60° dislocations and screws have been done with different materials it cannot be decided whether the rate limiting process in fact is different for the two types; the strong increase of Q_s below 5 MPa seems to indicate that the abrupt kink model would apply for screws, too, in pure GaAs.

If this interpretation, which admittedly is based on limited experiments, is correct the formation of kink pairs (KP) of critical length is rate limiting. Then any change of Q (by doping, illumination etc.) affects KP nucleation, too. Maeda (1988) made some suggestions for REDG. The analysis of the Patel effect given by Hirsch (1979) conforms with this need if one takes into account the levels of α and β dislocations in the energy gap (Jones et al. 1981, Maeda 1985).

6. REFERENCES

Alexander H and Haasen P 1968 in *Solid State Phys. ed.* (F Seitz et al.) 22 27
Alexander H 1989 in press: *Rad. Eff. and Cryst. Def.*
Androussi Y, Vanderschaeve G and Lefebvre A 1987 *Inst. Phys. Conf. Ser.* 87 291
Blom G M and Woodall J M 1988 *J. Electr. Mat.* 17 391
Bourret E D , Tabache M G and Elliot A G 1987 *Appl. Phys. Lett.* 50 1373
Boivin P Dr. *Thesis* 1988 Univ. Poitiers
Burle-Durbec N, Pichaud B and Minari F 1987 *Phil. Mag. Lett.* 56 173
Celli V, Kabler M V, Ninomiya T and Thomson R 1963 *Phys. Rev.* 131 58
Chaudhuri J R, Patel J R and Rubin L G 1962 *J. Appl. Phys.* 33 2736
Choi S K, Mihara M and Ninomiya T 1977 *Jap. J. Appl. Phys.* 16 737
De Cooman B C and Carter C B 1987 *Inst. Phys. Conf. Ser.* 87 259
Ehrenreich H and Hirth J P 1985 *Appl. Phys. Lett.* 46 668
Erofeeva S A and Osipyan Yu A 1973 *Sov. Phys. Sol. St.* 15 538
Feuillet G 1982 *Thesis Master of Science*, Univ. Oxford, UK
Fujita S, Maeda K and Hyodo S 1988 *phys. stat. sol.* a 109 383
Gomez A M and Hirsch P B 1978 *Phil. Mag.* 24 1383
Gottschalk H, Patzer G and Alexander H 1978 *phys. stat. sol.* a 45 207
Gottschalk H, Kaufmann K and Alexander H 1987 *Inst. Phys. Conf. Ser.* 87 287
Haasen P 1962 *Z. Phys.* 167 461
Hirsch P B 1979 *J. de Phys.* 44 C6-117
Hirsch P B, Pirouz P, Roberts and Warren P D 1985 *Phil. Mag.* B 52 759

Hirth J P and Lothe J 1968 *Theory of Dislocations* (New York, McGraw-Hill)
Höche H R and Schreiber J 1984 *phys. stat. sol.* a 86 229
Hurle D T J 1979 *J. Phys. Chem. Sol.* 40 613
Jimenez-Melendo M, Djemel A, Riviere J P and Castaing J 1986 *Mat. Sci. Forum* 10 - 12 791
Jimenez-Melendo M, Djemel A, Riviere J P and Castaing J, Thomas C and Duseaux M 1988 *Rev. Phys. Appl.* 23 251
Jones R 1981 *Inst. Phys. Conf. Ser.* 60 45
Jones R, Öberg S and Marklund S 1981 *Phil Mag* 43 839
Küsters K H and Alexander H 1983 *Physica* 116 B & C 594
Küsters K H, de Cooman B C and Carter C B 1986 *Phil. Mag.* A 53 141
Lagowski J, Gatos H C, Ayoma T and Lin D G 1984 *Appl. Phys. Lett.* 45 680
Lefebvre A, Francois P and di Persio J 1985 *J. de Phys.* 46 L 1023
Lefebvre A, Androussi Y and Vanderschaeve G 1987a *Phil. Mag. Lett.* 56 135
Lefebvre A, Androussi Y and Vanderschaeve G 1987b *phys. stat. sol.* a 99 405
Maeda K and Takeuchi S 1981 *Jap. J. Appl. Phys.* 20 L165
Maeda K and Takeuchi S 1985 in *Dislocations in Solids* ed H Suzuki et al. (Tokyo, Tokyo Univ. Press) 433
Maeda K 1985 in *Dislocations in Solids* ed H Suzuki et al. (Tokyo, Tokyo Univ. Press) 425
Maeda K 1988 in *Defects in Crystals* ed E Mizera (Singapore, World Scientific) 163
Matsui M and Yokoyama T 1986 *Inst. Phys. Conf. Ser.* 79 13
Mdivanyan B E and Shiksaidov M Sh 1988 *phys. stat. sol.* a 107 131
Mihara M and Ninomiya T 1975 *phys. stat. sol.* a 32 43
Nakada Y and Imura T 1987a *phys. stat. sol.* a 102 625
Nakada Y and Imura T 1987b *phys. stat. sol.* a 103 85
Ninomiya T, Thomson R and Garcia-Moliner F 1964 *J. Appl. Phys.* 35 3607
Ninomiya T 1979 *J. de Phys.* 40 C6-143
Osvenskii V B and Kholodnyi L P 1973a *Sov. Phys. Sol. St.* 14 2822
Osvenskii V B, Kholodnyi L P and Mil'vidskii M G 1973b *Sov. Phys. Sol. St.* 15, 661
Paufler P 1989 private communication
Petroff P and Hartmann R L 1974 *J. Appl. Phys.* 45 3899
Ponce F A, Anderson G B, Haasen P and Brion H G 1986 *Mat. Sci. Forum* 10 - 12 776
Rabier J, Garem H, Demenet J L and Veyssiere P 1985 *Phil. Mag. Lett.* A 51 L67
Sato M, Takebe M and Sumino K 1985 in *Dislocation in Solids* ed H Suzuki et al (Tokyo, Tokyo Univ. Press) p 429-32
Sato F, Matsui M and Chikawa J 1986 *Inst. Phys. Conf. Ser.* 79 193
Schreiber J and Leipner H S 1988 *Sov. Phys. Quantum Electr.* 15 2304
Snigireva I I, Khodos I I, Shiksaidov M Sh and Ushakova A P 1988 in *Defects in Crystals* ed E Mizera (Singapore, World Scientific) pp 218
Steinhardt H and Haasen P 1978 *phys. stat. sol.* a 49 93
Takebe T, Murai S, Tada K and Akai S 1986 *Inst. Phys. Conf. Ser.* 79 283
Tanaka M and Jouffrey B 1984 *Phil. Mag.* A 50 733
Tohno S, Shinoyama S, Katsui A and Takaoka H 1986 *Appl. Phys. Lett.* 49 1204
Wattenbach M, Krüger J, Kisielowski-Kemmerich C and Alexander H 1989 *Mat. Sci. Forum* 38 - 41 73
Wessel K and Alexander H 1977 *Phil. Mag.* 35 1523
Yonenaga I and Sumino K 1987 *J. Appl. Phys.* 62 1212
Yonenaga I and Sumino K 1989 *J. Appl. Phys.* 65 85

Inst. Phys. Conf. Ser. No 104: Chapter 3
Paper presented at Int. Symp. on Struct. Prop. Disloc. Semicond., Oxford, 1989

293

Dislocations in indented GaAlAs

P Charsley and R Haswell

Department of Physics, University of Surrey, Guildford, Surrey, GU2 5XH, UK

ABSTRACT: Dislocations formed by low load indentation at room temperature have been observed by TEM, for GaAlAs semiconductor alloys. Al concentrations of 10 at % and 30 at % have been studied. The results indicate an increasing tendency for stacking faults to be formed with increasing concentration of Al.

1. INTRODUCTION

Dislocations in III-V semiconductor materials can have important effects on the optoelectronic behaviour. In devices currently being developed III-V semiconductor alloys, such as GaAlAs, GaInAs and GaInAsP, have become increasingly used. The properties of dislocations in these alloys is important for device behaviour as well as having an intrinsic interest. In this paper we present TEM studies of dislocations in $Ga_{1-x}Al_xAs$ with x=0.3 and 0.1 after indentation on {001} faces at room temperature. These results are compared with previous studies by other workers (Hoche and Schreiber (1984) and Lefebvre et al (1987)) on GaAs.

This earlier work on GaAs has shown that when a crystal is indented on the (001) surface the dislocations in the perpendicular rosette arms are very different. This corresponds to the asymmetry of the rosette arms as observed by etch pit techniques (e.g. Warren, Pirouz and Roberts (1984)) and also to the cracks produced, when compared with elemental semiconductors such as Ge. The rosette arms parallel to [110] (in the notation used by Warren et al (1984)), which are defined by the As(g) planes (or B(g) planes as we shall refer to them) are long and contain dislocations with no observable splitting. Along the [$\bar{1}$10] directions, defined by the A(g) planes (i.e. Ga(g) in the notation of previous authors) extensive stacking faults or microtwins are observed. This direction is also the <110> direction which exhibits crack formation.

In a published paper (Haswell and Charsley (1989) we have shown that for GaAlAs with 30 at % of Al the dislocations observed show extensive splitting along both of the rosette arm directions. Nevertheless the extent of the splitting remains asymmetric. In this paper we will consider the effects of changing the Al concentration. It should be pointed out that we have not yet fully identified the A(g) and B(g) planes, but preliminary results using a convergent-beam technique (Liliental-Weber and Parechanian-Allen (1986)) strongly suggest that the identifications made in this paper are correct. These ideas are supported when the results are compared with those for GaAs.

2. EXPERIMENTAL TECHNIQUES

The specimens were in the form of epitaxial layers of (001) orientation between 1 and 2μm thick grown by MOCVD on GaAs substrates. The specimens studied were GaAℓAs with compositions 10% Aℓ, undoped; 30% Aℓ Si doped and 30% Aℓ of P-type doped with Zn. Indentations were at room temperature with a 5g load, using a Vickers pyramid indenter, with the diagonals parallel to the <110> directions. Samples were not annealed after indentation. They were thinned from the substrate side by mechanical polishing and chemical thinning, using a bromine/methanol etch in the ratio 1:9. Specimens were studied using a JEOL 2000 FX microscope operated at 200 kV.

3. EXPERIMENTAL RESULTS

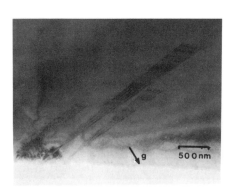

Figure 1.   Bright field g=2$\bar{2}$0. Indented Ga$_{0.7}$Aℓ$_{0.3}$As (n-type) showing stacking faults and dissociated dislocations parallel to [110].

Figure 2.   Indented Ga$_{0.7}$Aℓ$_{0.3}$As (n-type); g, 3g weak beam image, g=220.

Figures 1 and 2 show dislocations and stacking faults in the two rosette directions of an indented GaAℓAs specimen, with 30 at % Aℓ which is Si doped. The most extended stacking faults are assumed to lie along [$\bar{1}$10], however the dislocations along [110] are also widely split, in closely parallel groups, contrary to results on GaAs.

Observations of n-type 30 at % Aℓ (doped with Si) and p-type (doped with Zn) show essentially similar results with no cracks observed in either case. The values for the Vickers Hardness have approximately equal values for both n- and p-type 30% Aℓ. This value is reduced in the 10 at % Aℓ specimens Cracks are not observed in any of the 30% Aℓ specimens and the Vickers hardness values remain unchanged within experimental error at 1,000 ± 20 Kg mm^{-2}.

The undoped 10 at % Aℓ exhibits a reduced hardness of 900 ± 20 kg mm^{-2}; in addition cracks are observed parallel to a single set of {110} planes. These cracks, which again illustrate the 2-fold symmetry, will be assumed to be parallel to the [$\bar{1}$10] directions as reported by Warren et al (1984). The cracks were not observed until the specimens were thinned (from the un-indented side).

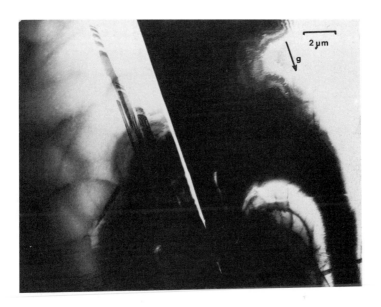

Figure 3. Bright field g=220. Indented $Ga_{0.9}Al_{0.1}As$ showing dissociated dislocations and stacking faults parallel to [1$\bar{1}$0].

Figure 3 shows an area at low magnification containing the two rosette arms. The dislocations in the rosettte arm parallel to the crack are shown in figures 4 and 5, at higher magnifications and different reflections; groups of dislocations which are widely split into partials with overlapping stacking faults can be seen. The perpendicular rosette arms show predominantly perfect dislocations dominated by very long screw dislocations. However the arrow indicates the presence of extended faults in limited regions.

4. CONCLUSIONS

When the extent of the formation of stacking faults is compared for the specimens of GaAlAs with published results for GaAs, indented with similar loads at the same temperature, we conclude that Al enhances their formation. This enhancement increases with the Al content but is not greatly affected by doping. From the results obtained so far it appears that a concentration of 10 at % is needed before any significant effect is observed. We think that the effect can be understood through a reduction in the mobility of the trailing partials with an increasing Al content. Nevertheless at this stage we cannot be certain of the identification of the A(g) and B(g) slip planes without further work. The effect of increasing the Al concentration is also to increase the flow stress as measured by the hardness value but to reduce the tendency to cracking. Whereas the first effect is not unexpected the reduction in the tendency to cracking is surprising and some further work in this direction is in progress.

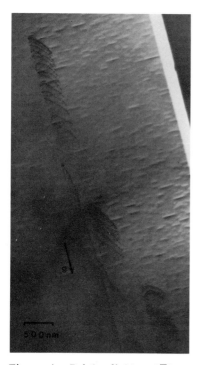

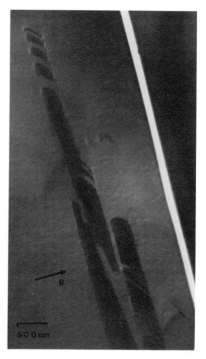

Figure 4. Bright field g=2$\bar{2}$0. Indented Ga$_{0.9}$A$\ell_{0.1}$As showing dissociated dislocations parallel to [1$\bar{1}$0]

Figure 5. Same area as Fig 4 showing stacking faults and dissociated dislocations parallel to to [1$\bar{1}$0], g=220.

ACKNOWLEDGEMENTS

We would like to thank STC Technology (Harlow) for their assistance with the provision of specimen material

5. REFERENCES

Haswell R and Charsley P, 1989, Philos. Mag. Lett. **59**, No.4
Hoche H R and Schreiber, 1984, Phys. Stat. Sol.(a) **86**, p229
Liliental-Weber Z and Parechanian-Allen L, 1986, Appl. Phys. Lett. **49**, pp1190-1192
Lefebvre A, Androussi Y and Vanderschaeve G, 1987, Phys. Stat. Sol.(a) **99**, pp405-412
Warren P D, Pirouz P and Roberts S G, 1984, Philos. Mag. A **50**, ppL23-L28

Inst. Phys. Conf. Ser. No 104: Chapter 3
Paper presented at Int. Symp. on Struct. Prop. Disloc. Semicond., Oxford, 1989

297

On the nature of the asymmetry effect in dislocation mobility in Si single crystals

B Ya Farber and Yu L Iunin

Institute of Solid State Physics Academy of Sciences of the USSR, 142432 Chernogolovka, Moscow district, USSR

ABSTRACT: To reveal a mechanism determining essential differences (to an order of magnitude) in the dislocation velocities of broadening and narrowing half-loops the motion of individual dislocations has been investigated under the action of loading pulse sequence. Dependences of the mean dislocation displacements upon pulse duration, pulse separation and the frequencies of sinusoidal loading are presented. The results obtained are analyzed in terms of the diffusion and drift of the kinks in an inhomogeneous potential relief stipulated by point defects.

1. INTRODUCTION

At present the results of dislocation velocity measurements in semi-conductors are interpreted in the framework of the Peierls mechanism of dislocation glide (Hirth and Lothe 1968). In this case the velocity of a long enough dislocation $[L > L_k = (2v_k/J)^{1/2}]$ is determined as:

$$v_d = a\, c_k\, v_k = a\, (2v_k J)^{1/2} \qquad (1)$$

where a is the lattice parameter, J is the rate of the kink pairs formation, c_k is the kink density, and v_k is the kink drift velocity. It should be noted that the theory did not involve any predictions of the dislocation velocity dependence on the direction of its glide. There are no grounds for such predictions in the structure of elemental semiconductors either. Nevertheless the asymmetry of individual dislocation mobility was observed in Si and Ge (Nikitenko et al, 1981, 1982, 1984) during the reverse motion of the dislocation along one slip plane. This effect shows up in the fact that the velocities of the half-loop narrowing are by 1 or 2 orders of magnitude higher than the velocities of their broadening. As seen from (1) this may be caused by an increase of the kink pair formation rate, or by kink drift velocity or by both factors at once. To distinguish between these possibilities we measured the dislocation mobility in the swept part of the slip plane as well as in the as grown material using the method of periodic pulse loading proposed by Nikitenko et al (1985).

2. EXPERIMENTAL

The investigated samples were rectangular prisms with the edge orientation {110}, {112}, {111} measuring 35x4x1.5 mm³. They were cut from dislocation-free single crystals of n-type Si grown by the float-zone method, doped with

P up to a specific conductivity $\sim$ 150 Ohm cm. The initial procedure of
the dislocation introduction was similar to that described by Nikitenko et
al. (1984). Individual dislocations were forced to move in the direction
of the half-loop broadening ("forward" motion) or narrowing ("backward"
motion) depending on the sign of the load applied.

The samples were deformed by four point bending around the [112] axis.
Three kinds of load were used, viz: static, periodic pulse and alternating
sinusoidal ones. In the second case the load was a sequence of unipolar
pulses with the duration t_i separated by pauses with duration t_p when the
applied stress σ = 0. The active loading time, i.e. the sum of the stress
pulse durations, was equal to the static loading time over which the
dislocations moved a distance l_{st} $\sim$ 20-30 μm. The alternative loading
was performed using a bending jig with eight supports, where the four
outermost ones were fixed and the internal four supports were moved
reversibly producing a sinusoidal load in the frequency range 1-100 Hz.
Magnetic hanging was used to eliminate the action of the weight of the
loading system elements during the pulse separation in both periodic pulse
loading and sinusoidal loading.

3. RESULTS

Study of the asymmetry of the dislocation mobility in Si (Nikitenko et al
1984) during static loading has shown that there is a linear dependence
between average dislocation displacements and loading duration, which means
that dislocation motion in the forward and backward directions can be
characterized by a steady-state velocity.

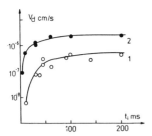

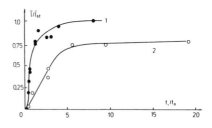

Fig. 1. Average 60° dislocation
velocities vs pulse duration
(t_i = t_p) for the forward (1)
and backward (2) dislocation
motion T = 600°C, σ = 7 MPa.

Fig. 2. Normalized mean dislocation
displacements vs the relative pulse
duration at (t_i = t_p). 1 - forward
motion, 2 - backward motion. T = 600°C,
σ = 7 MPa, Σt_i = t_{st} = 7200 s (1) and
1500 s (2).

Fig. 1 shows the dependence of the 60° dislocation velocities on the pulse
duration for the condition of loading pulse sequence with t_i = t_p, T = 600°C,
σ = 7 MPa. It is seen that dependences on t_i for the forward (curve 1) and
backward (curve 2) dislocation motion are qualitatively similar. Both the
curves have ranges of strong (at short t_i) and weak (at long t_i) dependence
on t_i. It should be noted that the range of the strong dependence on t_i for
the backward motion ceases at a shorter pulse duration than the one corresponding to the forward motion.

More pronounced differences between forward and backward motion, revealed

in the pulse experiments, can be seen in Fig. 2 where the same data are
presented in normalized units. Fig. 2 shows the dependences of the
average 60° dislocation displacements on the relative pulse duration for
forward (curve 1) and backward (curve 2) dislocation motion. Mean
dislocation displacements are normalized to the corresponding average
dislocation displacements measured during static loading. The pulse
durations are presented in the units of the average time (t_a) for dislocation transfer to an adjacent valley of the potential relief: $t_a = a/v_d$.
It is seen that, as in Fig. 1, the range of a weak dependence $l(t_i)$ is
followed by a strong one as t_i decreases for both types of dislocation
motion. There are inflection points on the curves 1, 2, where the
standard deviation for the average dislocation displacements reaches its
maximum value. The inflection point on curve 1 corresponds to $t_i^*/t_a \approx 0.5$
($\sim$ 30 ms) and for curve 2 $t_i^*/t_a \approx 1$ ($\sim$ 10 ms). The l values approach the
static ones (l_{st}) at $t_i/t_a \approx 6$-8 for the forward motion and do not reach l_{st}
up to $t_i/t_a \approx 20$ for the backward motion. A transition from strong to weak
$l(t_i)$ dependence corresponds to $t_i/t_a \approx 1$ for the forward motion and t_i/t_a
≈ 5 for the backward one.

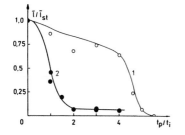

Fig. 3. Normalized mean dislocation
displacements vs pause duration
(unit t_p/t_i)
(1) Forward motion: $t_i/t_a = 1.8$ = const.
(2) Backward motion: $t_i/t_a = 3$ = const.
The other conditions are the same as
in fig. 2.

Fig. 3 shows the dependences of the normalized mean dislocation displacements
on the duration of pauses at $t_i/t_a \approx 1.8$ for the forward motion and t_i/t_a
≈ 3 for the backward one. It is seen that an increase of the pause duration
reduces monotonically the dislocation displacements from l_{st} down to zero for
the forward motion and to $1/l_{st} \approx 0.06$ for the backward one. In spite of the
fact that relative pulse duration for the forward motion is smaller than that
for the backward one the critical pause duration required for an essential
decrease of the dislocation displacements is much greater.

4. DISCUSSION

We are going to analyze the results obtained in the framework of the model
proposed by Farber et al (1986) and Nikitenko et al (1987) for the treatment of the results measured under the action of loading pulse sequences.
It is supposed that dislocation motion proceeds due to the formation and
spreading of stable kink pairs with the kinks moving along the dislocation
line in the regime of diffusion and drift. During the pulse action stable
kink pairs can be formed and spread along the dislocation line. If during
the pulse mutual annihilation of the kinks of opposite sign from adjacent
pairs does not occur then during the pulse separation (when $\sigma = 0$) the
kink pairs formed became unstable and begin to shrink towards the formation
center. If the pauses are long enough the kink pairs will collapse and
the dislocation cannot move irrespective of the number of load cycles.
If the kink annihilation takes place during single pulse action the pauses
will not affect the kink motion and dislocation displacement will

correspond to that for static loading. Due to these assumptions the inflection point on the curves $l(t_i)$ may be attributed to the mean time of the diffusive expansion of the kink pair up to a stable configuration. The kink pair is stable against collapse upon reaching the critical size.

$$l_c = x' + x^*, \text{ where } x^* = (\alpha/\sigma_{ef}ab)^{1/2}, \text{ and } x' = kT/\sigma_{ef}ab \qquad (2)$$

Here α is the constant of the elastic interaction between kinks forming a kink pair, b is the Burgers vector and σ_{ef} is the apparent stress acting on the kinks. Petukhov (1982) has shown that $\sigma_{ef} = \sigma - \sigma_s$, where σ_s is the starting stress for dislocation motion. Using these assumptions we can estimate the kink diffusivity

$$D_k = (x')^2/4t_i^* \qquad (3)$$

Under specific experimental conditions $\sigma_s = 1.5$ MPa, $\sigma = 7$ MPa, $x' = 45b$, $(t_i^*)_f = 30$ ms, $(t_i^*)_b = 10$ ms, one obtains from (3) $(D_k)_f = 1.9 \times 10^{-11}$ cm^2/s and $(D_k)_b = 7.7 \times 10^{-11}$ cm^2/s for the forward and backward motion respectively. So dislocation motion in the backward direction is characterized by a high value of the kink diffusivity. Since D_k determines the kink drift velocity:

$$v_k = (D_k/kT)\sigma_{ef}ab \qquad (4)$$

this should result in an increase of v_k. However simple estimations show that by taking into account only the increase of v_k one fails to explain the results obtained. Indeed, from (1)-(4) it follows that $t_i^*/t_a \sim c_k x'$. This means that at a constant kink density the critical relative pulse duration would be the same for the forward and backward motion which contradicts experimental facts shown in Fig. 2. So the asymmetry of the dislocation mobility is at least partly determined by a change in the kink density on the dislocation line during its glide in the backward direction.

The following experimental data support this assumption. Fig. 2 shows that dislocation displacements during pulse experiments do not reach the static values even at $t_i \gg t_a$ and only 6% of the kink pairs do not collapse with rising pause duration as seen in Fig. 3 where $t_i = 3t_a$. This strongly differs from the results obtained for forward dislocation motion, where $\bar{l}$ approaches the $\bar{l}_{st}$ value at $t_i \approx t_a$ and, when $t_i = 1.8\ t_a$, the pause duration required for a collapse of the kink pairs,is very long. Some process is therefore occurring which prevents the mutual annihilation of the kinks from adjacent kink pairs. Strong obstacles on the dislocation line which impede the kink motion can play such a role. In this case the dislocation moves due to superkink formation and overcomes the obstacles when the height of the superkink (h) reaches its critical value (h_c). When $h < h_c$ the superkink is unstable and during the pulse separation it will collapse, as is indeed observed in Fig. 3. It is reasonable to suppose that the h_c value should be correlated with transition from strong to weak $\bar{l}(t_i)$ dependence in Fig. 2. It takes place at $t_i \approx t_a$ for the forward motion and at $t_i \approx 5t_a$ for the backward one. So $h_c \approx 5a$.

The average separation between strong obstacles may be estimated as the distance covered by a stable kink pair during t_a.

$$L = l_c + 2v_k t_a \qquad (5)$$

Under our specific experimental conditions one obtains from (4), (5)
L $\approx$ 90 b. It is known that when the kink free path is restricted by
strong obstacles the stationary dislocation velocity can be estimated
as: v_d = aJL (Alexander 1986). Using measured values for the $v_d \approx$
3 x 10^{-6} cm/s and $v_d \approx$ 4 x 10^{-7} cm/s for the backward and forward motion
respectively one can estimate the rates of stable kink formation in these
two cases. It yields $J_f \approx$ 2.4 10^6 cm^{-1}s^{-1} and $J_b \approx$ 2.6 10^{7}cm^{-1}s^{-1}. It
is seen that $J_b \gg J_f$.

So one of the main reasons for the increase of the backward dislocation
velocity should be attributed to the increase of the kink pair formation
rate. Although the kink mobility during the process of a stable kink pair
formation is also increased, the apparent kink drift velocity turns out to
be even smaller (compared with the case of the forward dislocation motion).
This is due to the fact that most of the time ($\sim$ 80%) a stable kink pair
has to wait to overcome the obstacles.

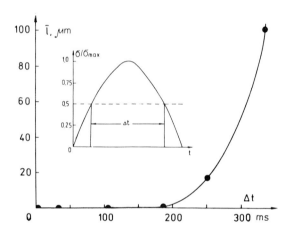

Fig. 4. Dependence of
the mean 60° dislocation
back displacements on the
pulse duration during
sinusoidal loading
T = 650°C, σ_{max} = 10 MPa,
t = 900 s. The inset
shows the scheme of the
pulse duration determin-
ation.

Alternating loading gives rise to some possibilities for the investigation
of kink dynamics. Due to an essential difference between forward and
backward dislocation velocities one may predict macroscopic shrinkage of
the half-loop under sinusoidal loading. However experiments in the
frequency range 1-100 Hz, at T = 600°C and σ_{max} = 10 MPa did not reveal
any dislocation motion. Raising the temperature to 650°C we observed the
shrinkage of the dislocation half-loop at f < 1.3 Hz. Fig. 4 shows the
dependence of the mean back dislocation displacement on the duration of a
single pulse of sinusoidal load. It is seen that the onset of the
dislocation motion takes place at Δt = 250 ms. During this time the
dislocation covers a distance of $v_d\Delta t \approx$ 150Å (where v_d = 2 x 10^{-5} cm/s
is the stationary dislocation velocity measured at t = 650°C and
σ = 7 MPa). The reversible character of the dislocation motion under
the action of sinusoidal loading may be controlled by anchoring points
on the dislocation line. From the equation obtained by Seeger (1981) one
can estimate the mean separation between these pinning points as:

$$L = (2E_o v_d \Delta t/b\sigma)^{1/2} \tag{6}$$

where $E_o \approx G b^2/2$ is the line energy of the dislocation. Under our specific experimental conditions this yields $L \approx 1000$ b.

5. CONCLUSIONS

Detailed analysis of the results concerning the investigation of the asymmetry of the dislocation mobility in Si and Ge during static loading has shown (Nikitenko et al. 1982, 1984) that the most probable reason for this effect should be attributed to a change in the point defect states in the crystal volume adjacent to the swept part of the slip plane.

The data presented here give further information about the state of these defects and their influence on the micro-mechanism of dislocation motion. From the discussion presented it may be concluded that point defects in the swept part of the slip plane are distributed more inhomogeneously than in the as grown material. This results in the increase of the kink pair formation rate as well as in a deviation of the dislocation from a straight line during its motion over the swept part of the slip plane in the backward direction. It should be noted that analogous changes of the dislocation form take place during forward dislocation motion in the as grown crystal but at much lower stress level (Nikitenko and Farber 1989).

6. REFERENCES

Alexander H 1986 Dislocations in Solids ed F R N Nabarro (Amsterdam: North Holland) pp 115-234
Farber B Ya, Iunin Yu L and Nikitenko V I 1986 Phys.Stat.Sol.(a) 97 469
Hirth J and Lothe J 1968 Theory of Dislocations (New York: McGraw Hill)
Nikitenko V I, Farber B Ya and Yakimov E B 1981 JETP Letters 34 233
Nikitenko V I, Farber B Ya and Bondarenko I E 1982 Sov.Phys.JETP 55 891
Nikitenko V I, Farber B Ya and Yakimov E B 1984 Cryst.Res. & Techn. 19 295
Nikitenko V I, Farber B Ya and Iunin Yu L 1985 JETP Letters 41 124
Nikitenko V I, Farber B Ya and Iunin Yu L 1987 Sov.Phys.JETP 66 738
Petukhov B V 1982 Fiz.Tverd.Tela. 24 439
Seeger A 1981 J. Physique 42 C5-201
Nikitenko V I and Farber B Ya 6 Int.Conf. on Structure and Properties of Dislocations in Semiconductors this volume

Inst. Phys. Conf. Ser. No 104: Chapter 3
Paper presented at Int. Symp. on Struct. Prop. Disloc. Semicond., Oxford, 1989

303

Excitation-enhanced dislocation mobility in semiconductors: a microscopic mechanism

N Maeda and S Takeuchi

Institute for Solid State Physics, University of Tokyo,
Roppongi, Minato-ku, Tokyo 106, Japan

ABSTRACT: The observed excitation enhancement of dislocation motion in semiconductors can be interpreted in terms of the reduction in activation energy by non-radiative recombination of injected carriers at the straight-dislocation sites and/or the kink sites. Formulae for the enhanced mobility have been derived based on the abrupt kink model for the possible three cases and have been compared with experimental results.   The observed reduction in activation energy possibly corresponds to the electronic energy level in the bandgap associated with the straight-dislocation sites.

1. INTRODUCTION

It has been revealed in recent years that dislocation motion in covalent semiconductors is remarkably enhanced by irradiation by electron beam or laser light. This irradiation effect was measured quantitatively first in GaAs (Maeda and Takeuchi 1981, Maeda et al 1983), later in InP (Maeda and Takeuchi 1983), Si (Kusters and Alexander 1983, Maeda et al 1987) and GaP (Maeda and Takeuchi, to be published). These experiments showed that the mobility enhancement is observed below a critical temperature obeying another Arrhenius relation with the characteristic parameters: (i) the pre-exponential factor which is smaller than that in darkness by several orders of magnitude is almost proportional to the excitation intensity (Maeda K and Takeuchi S 1983), and (ii) the activation energy which is independent of the excitation intensity is considerably smaller than that in darkness. These features of the enhanced mobility have been interpreted in terms of recombination enhanced defect motion (REDM) (Weeks et al 1975, Sumi 1984); dislocation motion is enhanced with the assistance of the energy released upon non-radiative recombination of electron-hole pairs at the dislocation.

On the other hand, it is well-known that the dislocation mobility in semiconductors is affected by doping of electrically active impurities (for example, Patel and Chaudhuri 1966). An alternative interpretation of the irradiation effect, therefore, is that it is a modification of the doping effect, since crystal excitation changes the effective Fermi level (quasi-Fermi level).   From this point of view, Belyavskii et al. (1985) have proposed a mechanism which accounts for both the irradiation and doping effect. Thus the basic interpretation of the irradiation effect is still controversial. In this paper, however, we formulate the enhanced dislocation velocity from the standpoint of recombination enhancement, and analyze the microscopic mechanism of the enhanced mobility by comparing these formulations with the experimental results.

2. NON-RADIATIVE RECOMBINATION AT DISLOCATION

In this section, we consider what influences non-radiative recombination at dislocations have on the elementary processes of dislocation motion. Successive elementary processes of dislocation motion consist of two parts: the smallest initial unit kink-pair formation process and the kink diffusion process along the dislocation line. Since these two processes are both accompanied by the bond-switching process, a travelling dislocation is expected to experience a rather high potential barrier at every step of the processes.

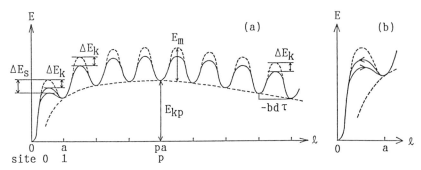

Fig.1   (a) Changes of the energy of a kink-pair under crystal excitation in the processes of initial kink-pair formation and subsequent kink migration (E:energy, ℓ:distance between kinks, a:lattice periodicity).   (b) Potential barriers between site 0 and site 1 (a part of (a)).

A dislocation line consists of two kinds of elementary sites, straight-dislocation sites and kink sites. Since they have entirely different atomic structures, the electronic states of them are generally different from each other. Consequently, non-radiative recombination at the two kinds of sites are generally not equivalent; thus we have to deal with them separately. We assume that both the electronic levels associated with the straight-dislocation site and the kink site exist in the bandgap of the crystal, hence non-radiative recombination of injected carriers occurs at both of these electronic levels. Let ΔE_s and ΔE_k be the energies released upon non-radiative capture of excited carriers at the straight-dislocation site and the kink site, respectively. We assume   that dislocation motion is enhanced with the help of ΔE_s and ΔE_k. Figure.1 shows the effect of ΔE_s and ΔE_k on the elementary processes of dislocation motion. In this figure, potential barriers for dislocation motion under crystal excitation, which are drawn by a solid line, are shown to be reduced by ΔE_s or ΔE_k compared to barriers in darkness which are drawn by a dotted line. Characteristic features of the influence of ΔE_s and ΔE_k are as follows:
(i) ΔE_s assists the initial unit kink-pair formation; namely the potential barrier is effectively reduced by ΔE_s only for the process of site 0 → site 1.
(ii) ΔE_k assists the process of kink migration; namely the potential

barriers are effectively reduced by ΔE_k for the processes of site i $\leftrightarrow$ site(i+1) (i=1,2,$\cdots$) and site 0 $\leftarrow$ site 1.

It is noted that, as shown in Fig.1(b), the potential paths of the process of site 0 $\rightarrow$ site 1 and that of site 0 $\leftarrow$ site 1 are not identical, since ΔE_s and ΔE_k are generally unequal. We shall consider in the following sections what changes are brought about in dislocation motion when potential barriers are subjected to such modification as shown in Fig.1.

3. GENERALIZED FORMULA FOR DISLOCATION MOTION

The dislocation motion has been discussed in detail by Hirth and Lothe (1968) on the basis of the abrupt kink model. The dislocation velocity is generally written in two different forms according to whether the kink-kink collision occurs or not:

$$v = \begin{cases} 2d\sqrt{Jv_k} & (X \ll L) \qquad (1) \\ dJL & (X \gg L) \qquad (2) \end{cases}$$

Here J is the frequency of kink-pair formation per unit length of the dislocation, v_k is the lateral velocity of a kink along the Peierls valley, d is the interval of Peierls valleys, X is the mean free path of migration of a kink, and L is the segment length of a dislocation. The condition X$\ll$L and X$\gg$L, respectively, correspond to the dislocation motion with and without kink-kink collisions.

Since Eqs.(1) and (2) are valid in general cases, our problem is reduced to formulate J and v_k. The present authors have derived J directly from a set of rate equations which represent the equiblium jump frequency of kinks at each kink site (i.e. site 0,1,2,$\cdots$). The result of J is written in the form

$$J = v_s \frac{bd\tau}{kT} \exp[-\frac{1}{kT}\{\sum_{i=0}^{p-1}(\Delta E_{i+}-\Delta E_{(i+1)-})+E_{dif}\}], \qquad (3)$$

where v_s is the trial jump frequency of straight-dislocation sites, b is the Burgers vector, τ is the shear stress component in the direction of b, and E_{dif} is the effective activation-energy barrier to diffusive kink migration. As for the summation term in Eq.(3), p is the site number at which the energy of a kink-pair assumes the maximum value (see Fig.1(a)), ΔE_{i+} and $\Delta E_{(i+1)-}$ denote the potential barriers to the processes of site i $\rightarrow$ site (i+1) and site i $\leftarrow$ site (i+1), respectively. If the potential path is identical everywhere for the forward and backward jumps, which is the case of the dislocation motion in darkness, the summation term in Eq.(3) gives the energy of the kink-pair at the site p which is indicated by E_{kp} in Fig.1(a). The lateral kink velocity v_k is obtained by means of the Einstein relation in diffusive phenomenon:

$$v_k = v_k \frac{bd\tau}{kT}a^2 \exp[-\frac{1}{kT}E_{dif}], \qquad (4)$$

where v_k is the trial jump frequency of kinks.

Substituting Eqs.(3) and (4) into Eqs.(1) and (2), we obtain the generalized formula for dislocation motion:

$$
v = \begin{cases}
\sqrt{\nu_s \nu_k}\dfrac{2\,a\,b\,d^2\,\tau}{kT}\exp\left[-\dfrac{1}{kT}\{\dfrac{1}{2}\sum_{i=0}^{p-1}(\Delta E_{i+}-\Delta E_{(i+1)-})+E_{dif}\}\right] & (X<<L) \quad (5)\\[4ex]
\nu_s\dfrac{b\,d^2\,\tau}{kT}L\exp\left[-\dfrac{1}{kT}\{\sum_{i=0}^{p-1}(\Delta E_{i+}-\Delta E_{(i+1)-})+E_{dif}\}\right] & (X>>L) \quad (6)
\end{cases}
$$

For the dislocation motion in darkness, taking account of the relations $\nu_s \simeq \nu_k \simeq \nu_D$ (Debye frequency), the summation term $= E_{kp}$ and $E_{dif} = E_m$, we obtain the same formula as Hirth and Lothe derived:

$$
v = \begin{cases}
\nu_D\dfrac{2\,a\,b\,d^2\,\tau}{kT}\exp\left[-\dfrac{1}{kT}(\dfrac{1}{2}E_{kp}+E_m)\right] & (X<<L) \quad (7)\\[4ex]
\nu_D\dfrac{b\,d^2\,\tau}{kT}L\exp\left[-\dfrac{1}{kT}(E_{kp}+E_m)\right] & (X>>L) \quad (8)
\end{cases}
$$

4. DISLOCATION VELOCITY UNDER CRYSTAL EXCITATION

As remarked in Section 2, the electronic states of straight-dislocation sites are generally different from those of kink sites, so that the following three cases have to be considered:

(i) Case I: First, we consider the case in which recombination enhancement occurs at both straight-dislocation sites and kink sites. In this case, ν_s and ν_k in Eqs.(5) and (6) are, respectively, given by $\nu_s \simeq \eta_s r_s$ and $\nu_k \simeq \eta_k r_k$; where r_s (r_k) is the non-radiative carrier-capture frequency at straight-dislocation sites (kink sites), and η_s (η_k) is the efficiency factor combining the carrier capture process with the process of initial unit kink-pair formation (kink migration). On the other hand, the summation term and E_{dif} in Eqs.(5) and (6) are, respectively, given by $\Sigma(\Delta E_{i+}-\Delta E_{(i+1)-}) = E_{kp}-\Delta E_s+\Delta E_k$ and $E_{dif} = E_m-\Delta E_k$. Substituting these relations into Eqs.(5) and (6), we obtain the formula for dislocation velocity:

$$
v = \begin{cases}
\sqrt{\eta_s r_s}\sqrt{\eta_k r_k}\dfrac{2\,a\,b\,d^2\,\tau}{kT}\exp\left[-\dfrac{1}{kT}\{\dfrac{1}{2}E_{kp}+E_m-\dfrac{1}{2}(\Delta E_s+\Delta E_k)\}\right] & (X<<L) \quad (9)\\[4ex]
\eta_s r_s\dfrac{b\,d^2\,\tau}{kT}L\exp\left[-\dfrac{1}{kT}(E_{kp}+E_m-\Delta E_s)\right] & (X>>L) \quad (10)
\end{cases}
$$

(ii) Case II: Secondly, we consider the case in which recombination enhancement occurs only at straight-dislocation sites. In this case, the substitution of the relations $\nu_s \simeq \eta_s r_s$, $\nu_k \simeq \nu_D$, $\Sigma(\Delta E_{i+}-\Delta E_{(i+1)-}) = E_{kp}-\Delta E_s$ and $E_{dif} = E_m$ into Eqs.(5) and (6) yields the following formula:

$$
v = \begin{cases}
\sqrt{\eta_s r_s \nu_D}\dfrac{2\,a\,b\,d^2\,\tau}{kT}\exp\left[-\dfrac{1}{kT}(\dfrac{1}{2}E_{kp}+E_m-\dfrac{1}{2}\Delta E_s)\right] & (X<<L) \quad (11)
\end{cases}
$$

$$\left| \eta_s r_s \frac{bd^2\tau}{kT} L \exp\left[-\frac{1}{kT}(E_{kp}+E_m-\Delta E_s)\right] \right. \qquad (X \gg L) \qquad (12)$$

(iii) Case III: The last case is the one in which recombination enhancement occurs only at kink sites. In this case, the substitution of the relations $v_s \simeq v_D$, $v_k = \eta_k r_k$, $\Sigma(\Delta E_{i+} - \Delta E_{(i+1)-}) = E_{kp} + \Delta E_k$ and $E_{dif} = E_m - \Delta E_k$ into Eqs.(5) and (6) yields the following formula:

$$v = \begin{cases} \sqrt{v_D \eta_k r_k} \dfrac{2abd^2\tau}{kT} \exp\left[-\dfrac{1}{kT}(\dfrac{1}{2}E_{kp}+E_m-\dfrac{1}{2}\Delta E_k)\right] & (X \ll L) \qquad (13) \\[3mm] v_D \dfrac{bd^2\tau}{kT} L \exp\left[-\dfrac{1}{kT}(E_{kp}+E_m)\right] & (X \gg L) \qquad (14) \end{cases}$$

5. DISCUSSION

The features of the above results of Eqs.(9)~(14) are summarized in the Table. The result of Eq.(14) may seem inconsistent with our intuition, but it is a natural consequence of irreversible potential paths between site 0 and site 1 as shown in Fig.1(b).

Table   Features of enhanced dislocation mobility. (SD: straight-dislocation site, K: kink site, I: excitation intensity)

segment length	site of recombination	pre-exponential factor	reduction in activation energy
X≪L	SD + K	$\propto I$	$(1/2)(\Delta E_s + \Delta E_k)$
	SD	$\propto \sqrt{I}$	$(1/2)\Delta E_s$
	K	$\propto \sqrt{I}$	$(1/2)\Delta E_k$
X≫L	SD + K	$\propto I$	ΔE_s
	SD	$\propto I$	ΔE_s
	K	no effect	no effect

The most important problem is that to which case among I, II and III the observed velocity enhancement actually corresponds. We shall consider this problem comparing Eqs.(9)~(14) with the following experimental facts: (i) as remarked in Section 1, the pre-exponential factor is proportional to the excitation intensity I, and (ii) some observed reductions in the activation energy ΔE are larger than the half value of the bandgap energy E_g (i.e. $\Delta E > (1/2)E_g$). If the condition is X≪L, the case I is the only possibility from the experimental results of both the pre-exponential factor and the activation energy. This indicates that, if X≪L, non-radiative recombination should have occurred at both straight-dislocation sites and kink sites, and that the observed reduction in the activation energy is interpreted as the average value of the energies released at both kinds of these sites. On the other hand, if the condition is X≫L, both of case I and case II are the possible cases; in either of these cases the observed reduction in the activation energy should correspond to the energy released at the kink sites. Even if the enhancement of kink migration occurs, it does not contribute to the mobility enhancement. Accordingly, it is concluded that, if X>>L, the enhancement of the initial unit kink-pair formation process is ascribed to the enhancement of dislocation mobility.

There still remains a problem of which condition of X<<L and X>>L is actually realized. As for this problem, the authors presume that the condition of X>>L is realized under usual experimental conditions. The reasons are : (i) straight portions of the moving dislocation or the distances between geometrical kinks are expected to be rather short, probably less than 1μm, due to the dislocation curvature which is caused by internal stress in the crystal, and (ii) recent investigations by electron microscopy (Hirsch et al 1981, Louchet 1981, Louchet and Thibault-Desseaux 1987, Louchet et al 1988; 1981a, b, Fnaiech 1987) have shown that dislocation velocity in semiconductors is proportional to the dislocation length up to about 1μm. Thus, the present authors come to the conclusion that the observed reduction in the activation energy corresponds to the electronic energy level associated with the straight-dislocation sites.

ACKNOWLEDGEMENT

The authors would like to thank Prof. K. Maeda, Department of Applied Physics, University of Tokyo, for valuable discussions.

REFERENCES

Belyavskii V I, Darinskii B M and Sviridov V V 1985 Sov. Phys.-Sol. State 27 658
Fnaiech M, Reynaud F, Couret A and Caillard D 1987 Phil. Mag. A55 405
Hirsch P B, Ourmazd A and Pirouz P 1981 Microscopy of Semiconducting Materials 1981, edited by A G Cullis and D C Joy , Inst. Phys. Conf. Ser.No.60 (London, Bristol: The Institute of Physics) p 29
Hirth J P and Lothe J Theory of Dislocations (McGraw Hill, New York, 1968) p 531
Küsters K H and Alexander H 1983 Physica 116B 594
Louchet F 1981a Microscopy of Semiconducting Materials 1981, edited by A G Cullis and D C Joy , Inst. Phys. Conf. Ser. No.60 (London, Bristol: The Institute of Physics) p 35; 1981b Phil. Mag. A43 1289
Louchet F, Cochet-Muchy D, Brechet Y and Pelissier J 1988 Phil. Mag. A57 327
Louchet F and Thibault-Desseaux J 1987 J. Phys. Appl. 22 207
Maeda K, Sato M, Kubo A and Takeuchi S 1983 J. Appl. Phys. 54 161
Maeda K and Takeuchi S 1981 Jpn. J. Appl. Phys. 20 L165; 1983 Appl. Phys. Letts. 42 664; 1983 J. Phys. 44 C4-375
Maeda N, Kimura K and Takeuchi S 1987 Bulletin of the Academy of Sciences of the USSR Physical Series Vol.51 No.4 93
Patel J R and Chaudhuri A R 1966 Phys. Rev. 143 601
Sumi H 1984 Phys. Rev. B29 4616
Weeks J D, Tully J C and Kimerling L C 1975 Phys. Rev. B12 3286

Inst. Phys. Conf. Ser. No 104: Chapter 3
Paper presented at Int. Symp. on Struct. Prop. Disloc. Semicond., Oxford, 1989

309

Mechanical properties and dislocation dynamics of compound and alloy semiconductors

Ichiro YONENAGA and Koji SUMINO

Institute for Materials Research, Tohoku University, Sendai 980, Japan

ABSTRACT : The mechanical properties of GaAs and GaP compound and GaAsP alloy semiconductors are investigated by means of compressive deformation. The dynamic characteristics of dislocations are deduced from the results of strain rate cycling tests. The dynamic behaviour of dislocations in the compound and alloy semiconductors is found to be similar to that in elemental semiconductors such as Si etc. The flow stress of an alloy semiconductor has a component that is temperature-insensitive which is absent in other types of semiconductors.

1. INTRODUCTION

III-V compound semiconductors are attracting keen interest as the materials which can be used for high-speed devices and various optoelectronic devices. Alloy semiconductors are especially useful as the materials for such kinds of devices since their lattice parameter and the wavelength of emitted light can be controlled by the chemical composition of the alloy.

It is known that dislocations cause inhomogeneities in electrical and optical properties of a semiconductor crystal. Nevertheless, far less is known of the dynamic properties of dislocations or even the mechanical properties of these kinds of semiconductors in comparison with elemental semiconductors. It is also interesting from a fundamental viewpoint to clarify the rate-determining processes of deformation and dislocation motion in these crystals.

This paper reports mechanical properties of GaAs and GaP compound and GaAsP alloy semiconductors and deduces the dynamic properties of dislocations in them.

2. EXPERIMENTAL

Specimens were prepared from GaAs, GaP and GaAs$_{0.8}$P$_{0.2}$ crystals grown by the liquid encapsulated Czochralski technique (Hibiya *et al* 1987). The main impurities in the GaAs, GaP and GaAsP crystals were Si at a concentration of 1×10^{16}, S of 3.3×10^{17} and Si of 7×10^{16} cm^{-3}, respectively. The densities of grown-in dislocations in the crystals were in the range of $10^4 - 10^6$ cm^{-2}. They were finished by mechanical lapping followed by chemical polishing into a rectangular parallelepiped shape, approximately $2.7 \times 2.7 \times 10.6$ mm^3 in size. The compressive axis was parallel to the [$1\bar{2}3$] direction and the side surfaces parallel to the (111) and ($\bar{5}41$) planes. Compressive tests were carried out at constant strain rates with the use of an Instron-type machine at various temperatures in a high purity argon gas atmosphere.

The effective stress which is required to make dislocations in motion keep a certain velocity via thermally activated processes was determined by means of strain-rate cycling.

3. RESULTS

Figures 1 (a) to (c) show stress-strain curves of GaAs and GaP compound and GaAsP alloy crystals, respectively, at various tempratures under a shear strain rate of 2×10^{-4} s^{-1}. The changes in the effective stress with strain are also shown in the figures.

The curves of all the crystals are characterized by a remarkable stress drop after yielding followed by a gradual increase in the stress with strain due to work-hardening.   The stress drop becomes small or even absent as the temperature increases.   Such characteristics in the stress-strain curves are common to all kinds of semiconductors so far investigated such as Ge, Si and InSb (Kojima and Sumino 1971, Yonenaga and Sumino 1978, Shimizu and Sumino 1975).

The effective stress in any crystal is constant with respect to the strain in the deformation stage after the lower yield point, which means that the mean velocity and the density of moving dislocations are both constant with strain in this deformation stage.   This dynamic state of moving dislocations is termed *the steady state of deformation* (Sumino 1974, Suezawa *et al* 1979) and has already been found to be realized in Ge and Si (Kojima and Sumino 1971, Sumino *et al* 1974, Yonenaga and Sumino 1978).   Such a steady state is seen here to be realized also in both III-V compound and III-V alloy crystals.

The flow stress and the effective stress both depend sensitively on the deformation conditions in all of GaAs, GaP and GaAsP as in Si and Ge. We thus know that the deformation of both the III-V compound and III-V alloy semiconductors is controlled by the same dislocation processes as those operating in elemental semiconductors.

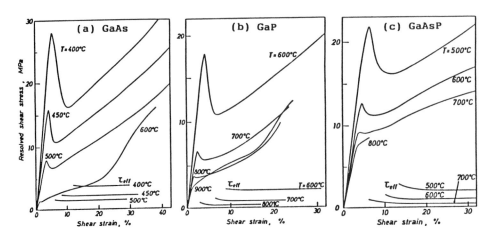

Fig. 1 Stress-strain curves of crystals at various temperatures for deformation at a shear strain rate of 2×10^{-4} s^{-1}. (a) GaAs, (b) GaP, and (c) GaAs$_{0.8}$P$_{0.2}$

The stress-strain behaviour of all of the compound and alloy semiconductor crystals does not depend systematically on the density of grown-in dislocations, contrary to the case of elemental semiconductors. This implies that grown-in dislocations in the former are immobile. It seems that the surfaces of crystals of both III-V compound and alloy semiconductors involve some irregularities that act easily as effective generation centers for dislocations.

A GaP crystal shows the Portevin-LeChatelier phenomenon on the stress-strain curve for deformation at high temperatures and low strain rates, which has been interpreted to be due to the dynamic interaction of dislocations with impurities in the cases of Si, Ge and GaAs highly doped with certain kinds of impurities (Siethoff 1970, Brion *et al* 1971, Yonenaga and Sumino to be published). It must be noted, however, that the concentration of the main impurity (S) in the GaP specimens used is more than two orders of magnitude lower than the concentration of impurities which give rise to the same phenomena in other semiconductors. The origin of such a strong interaction between dislocations and impurities in the GaP crystal is a problem to be clarified in future.

Figure 2 shows the magnitudes of the upper yield stress and the steady state values of the effective stress in GaAs, GaP and GaAsP plotted against reciprocal temperature of deformation at a strain rate of 2×10^{-4} s^{-1} together with those in Si of the same density of dislocations prior to deformation. It is seen that the upper yield stress of GaP in the temperature range $600 - 800$ °C and the upper yield stress of GaAs in the temperature range $450 - 600$°C are approximately equal to that of Si in the temperature range $800 - 1100$°C. It is noticeable that the strength of GaAsP alloy at low temperatures lies between those of GaAs and GaP compounds but that it lies above those of GaAs and GaP at high temperatures. The reason will be discussed in the next section.

The upper and lower yield stresses and the effective stress, denoted here collectively as τ, of any crystal are well expressed as a function of temperature T and strain rate $\dot{\varepsilon}$ by the following empirical equation :

$$\tau = A \, \dot{\varepsilon}^{\,1/n} \exp \left(U / k T \right) \qquad (1)$$

where k is the Boltzmann constant and A is a constant that depends on the

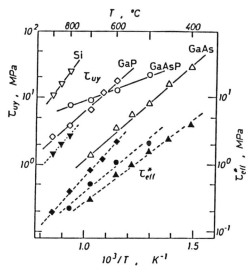

Fig. 2 Magnitudes of the upper yield stress τ_{uy} and the steady state values of the effective stress τ^*_{eff} in GaAs, GaP, GaAsP and Si plotted against the reciprocal temperature $1/T$ for the deformation under a shear strain rate of 2×10^{-4} s^{-1}.

material.   The magnitudes of parameters n and U for various crystals are given
in Table I.

Table I   Magnitudes of n and U for the upper and lower yield stresses, τ_{uy}
and τ_{ly} and the steady state values of the effective stress τ^*_{eff} for various
compound and alloy semiconductors

	τ_{uy}		τ_{ly}		τ^*_{eff}	
	n	U (eV)	n	U (eV)	n	U (eV)
GaAs	2.4	0.55	3.1	0.45	3.7	0.40
GaP	2.8	0.68	2.9	0.49	3.0	0.74
GaAs$_{0.8}$P$_{0.2}$	4.8	0.20	8.2	0.18	3.5	0.48

4. DISCUSSION

In a crystal having the sphalerite structure, motion of α, β and screw dislocations
all contribute to plastic defromation.   It is commonly observed in III-V compound
semiconductors that α dislocations have a mobility higher than those of other
types of dislocations by orders of magnitude. In such cases the deformation of
a crystal is rate-controlled by the motion of screw dislocations, and the strain
rate $\dot{\varepsilon}$ is given by the following equation :

$$\dot{\varepsilon} = 2 N v_s b \tag{2}$$

where N is the total density of moving dislocations, v_s the velocity of screw
dislocations and b the magnitude of the Burgers vector of the dislocations.

It is empirically known that the velocity of a dislocation of any type in a
semiconductor crystal is, in general, well described by the following equation as
a function of the stress τ and temperature T :

$$v = v_0 \, (\tau / \tau_0)^m \, \exp \, (-Q / k T) \tag{3}$$

where $\tau_0 = 1$ MPa, and v_0, m and Q are constants. The magnitudes of v_0,
m and Q for the screw dislocations that are rate-controlling in deformation can
be deduced from the magnitudes of n and U for the effective stress in the steady
state shown in Table I, together with the absolute its magnitude, according to
the theory of Sumino (1974).

The magnitudes of v_0, m and Q determined for various materials are given in
Table II.

Table II   Magnitudes of v_0, m and Q

	v_0 (m/s)	m	Q (eV)
GaAs	4100	1.7	1.5
GaP	130000	1.0	2.2
GaAs$_{0.8}$P$_{0.2}$	39000	1.5	1.7

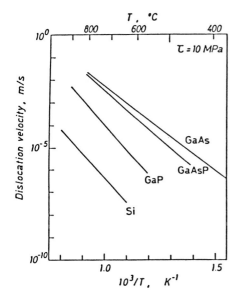

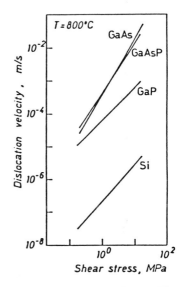

Fig. 3 Dislocation velocities in GaAs, GaP, GaAsP and Si under a shear stress of 10 MPa plotted against the reciprocal temperature.

Fig. 4 Dislocation velocities in GaAs, GaP, GaAsP and Si at 800 ℃ plotted against the shear stress.

Figure 3 shows the velocities of screw dislocations under a shear stress of 10 MPa plotted against the reciprocal temperature as deduced from Eq. (3) and Table II in GaAs, GaP and $GaAs_{0.8}P_{0.2}$. The velocity in GaAs deduced with the above procedure is in good agreement with that determined directly by the etch pit technique (Yonenaga and Sumino 1989). From the results in Table II and Fig. 3, the dynamic characteristics of dislocations in GaAsP alloy are intermediate between those in GaAs and GaP crystals, being closer to that in GaAs rather than that in GaP. This seems to reflect the bonding character of the alloy. Figure 4 shows the velocities of screw dislocations at 800 ℃ plotted against the stress, deduced in the same way.

It is now well accepted that the flow stress of a crystal in any deformation stage is divided into two components of different origins. One is the effective stress, mentioned in the preceding section, which depends on the temperature. The other is the athermal stress which is the component needed to overcome any resistance not thermally surmountable and which depends weakly on the temperature. The variation in the upper yield stress with temperature in the GaAsP alloy seen in Fig. 2 seems to show that the flow stress of the GaAsP alloy has an athermal stress component that is absent in GaAs and GaP compounds.

Several origins for athermal stress are conceivable in an alloy semiconductor. First, short-range order which is observed in most III-V ternary alloy semiconductors is expected to bring about an extra athermal resistance to the motion of dislocations. According to the calculation by Ichimura and Sasaki (1988), the stress increment due to the short-range order is of the order of 2 MPa for the equi-composition of $GaAs_{0.5}P_{0.5}$. The experimental magnitude of the athermal stress of about 7 MPa is far larger than this value. The short-range order is, thus, thought not to be the dominant origin of the concerned athermal stress. Secondly, since the lengths of a Ga-As bond and a Ga-P bond are different by about 4 %, local fluctuations of the alloy composition can induce a fairly large stress field which a dislocation can not surmount with thermally activation. Thirdly, the dynamic development of solute atmospheres around dislocations during deformation may lead to an apparently athermal nature of resistance against the motion of dislocations.   Either or both of the latter two are thought to be the cause of the athermal stress in the GaAsP alloy crystal.

5. CONCLUSION

Mechanical properties and dislocation dynamics have been investigated in GaAs and GaP compounds and $GaAs_{0.8}P_{0.2}$ alloy.   The deformation of the III-V compounds and the alloy is controlled by dislocation processes which are essentially the same as those rate-determining in elemental semiconductors such as Si. The velocities of dislocations rate-determining in the deformation of the semiconductors have been deduced as a function of stress and temperature. The flow stress of an alloy semiconductor seems to have a component that is temperature-insensitive and which is absent in other types of semiconductors.

ACKNOWLEDGEMENT

The authors are grateful to Mr. H. Watanabe and Dr. J. Matsui of NEC Corporation for supply of a GaAsP crystal.

REFERENCES

Brion H G, Haasen and Siethoff H 1971 Acta Metall. **19** 283
Hibiya T, Watanabe H, Ono H, Matsumoto T and Iwata N 1987 J. Electrochem. Soc. **134** 981
Ichimura M and Sasaki A 1988 Jpn. J. Appl. Phys. **27** L176
Kojima K and Sumino K 1971 Crystal Lattice Defects **2** 147
Shimizu H and Sumino K 1975 Phil. Mag. **32** 123
Siethoff H 1970 Phys. Stat. Sol. **40** 153
Suezawa M, Sumino K and Yonenaga I 1979 Phys. Stat. Sol.(a) **51** 217
Sumino K, Kodaka S and Kojima K 1974a Mater. Sci. Eng. **13** 263
Sumino K 1974b Mater. Sci. Eng. **13** 269
Yonenaga I and Sumino K 1978 Phys. Stat. Sol.(a) **50** 685
Yonenaga I and Sumino K 1989 J. Appl. Phys. **65** 85

Effect of surface charge on the dislocation mobility in semiconductors

J.T.Czernuszka

Department of Metallurgy and Science of Materials, University of Oxford,
Parks Road, Oxford OX13PH.

ABSTRACT: Indentations   on {1Ī00} surfaces of zinc oxide produce
arrays of basal dislocations. Careful etching experiments have   shown
that, under ambient conditions, Zn(g) dislocations are more mobile
than O(g) dislocations. By varying the surface charge it was   found
possible to make the O(g) dislocations more mobile than the Zn(g)
dislocations. The depth to which this effect is observed is of the
order of the Debye length. These observations allow some insight into
the well-known "chemomechanical effects".

1.   INTRODUCTION

It has been widely observed and recognised that the   near-surface
plasticity of ionically bonded solids is markedly dependent on the
chemical environment in which they are tested.   These variations,
originally investigated by Rebinder (1928), are termed "chemo-mechanical
effects" and have been reviewed by Westwood et al. (1981). This paper
examines under what conditions the environment may affect the mobility of
dislocations in semiconductors.

Zinc oxide was chosen as a model material for a number of reasons
including: when immersed in an electrolyte it follows simple Schottky
barrier theory with no intrinsic surface states (Dewald 1960), thus, an
external bias will cause near surface bending of the energy levels; it is
a relatively "soft" material so that dislocations travel a distance from
the indentation site sufficient to observe distinct "arrays" by chemical
etching; this system has been previously studied with respect to its
hardness behaviour by Ahearn et al (1980) and so should allow a ready
comparison of the results.

2.   EXPERIMENTAL PROCEDURE

Single crystals of vapour phase grown zinc oxide were obtained as
hexagonal prisms with the long axis parallel with the [0001] direction and
with {1Ī00} side faces; and were approximately 20mm long and 2mm wide.
These crystals were etched in phosphoric acid at 80 C for about 1 minute
to remove any handling damage. Zinc oxide is naturally n-type owing to an
excess of interstitial zinc (Thomas 1957).

To examine the effect of surface charge on near-surface plasticity, the
arrangement shown schematically in Figure 1 was used. A 1M KCl solution,
buffered with borate buffered to the required pH (9 and 9.5), was used as
the electrolyte. The choice of such a concentrated solution ensured that
the solution double layer (the Guoy layer) is eliminated leaving a layer
wholly of the Helmholtz type. In this situation the zeta-potential is
effectively zero. Oxygen was removed by continuously bubbling nitrogen
through the electrolyte for at least 24 hours prior to the start of the
experiment. The ZnO was fixed to a brass plate with indium to form an
ohmic contact. All of the connecting leads and metal surfaces were masked
with insulating lacquer. Indentations were formed by a Vickers indenter at
a load of 5g. The bias voltage was set to the most negative value and
subsequent indentations made at progressively less negative values. Three
indentations were made at each voltage, and the cell allowed to
equilibriate for approximately 30 seconds before the next set of
indentations. The sweep rate was 60 mV/min. This procedure was chosen to
correspond as closely as possible to that of Ahearn et al (1980) since it
has been shown that the direction of voltage sweep and sweep rate could
affect the final results, implying that adsorption kinetics are important.

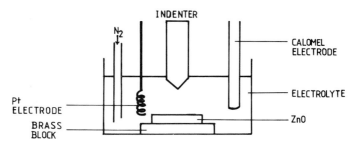

Fig.1 A schematic diagram of the experimental arrangement

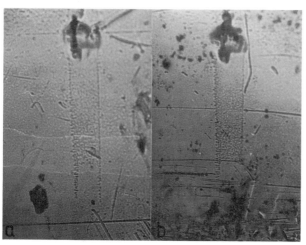

Fig.2 Etched indentations
made under (a) ambient
conditions and (b) at a
bias of -200mV, pH=9.5.

10 μm

3. RESULTS AND DISCUSSION

A reagent to etch the positions of dislocations has been developed consisting of 2 HF, 5 HNO$_3$, 20 H$_2$O, with a few grains of silver nitrate, for approximately 2 minutes. Figure 2a shows an etched indentation. The dislocation rosettes lie on the basal plane. For crystals with the wurtzite structure, basal dislocations possess a polar character. As can be seen from Figure 2a, one side of the rosette is longer than the other. The sample was cleaved along the basal plane, and the two halves etched in 20% HNO$_3$ to determine the polarity (Mariano and Hanneman, 1963). If it is assumed basal dislocations lie with their extra half planes towards each other then the longer arm consists of Zn(g) dislocations and the shorter arm O(g) dislocations. Thus, the behaviourof both types of dislocation with applied voltage could be studied.

There was no consistent variation in length of (each type of) rosette with voltage. But, by measuring the differences in length between the Zn(g) and O(g) dislocations a consistent trend was observed. The results are plotted in Figure 3. At pH=9.5 an anodic bias causes Zn(g) dislocations to move further from the indentation site than the O(g) dislocations. For a slightly cathodic bias, i.e. more negative than about 50mV, the distances travelled by the respective dislcations are reversed. Figure 2b shows an example of an indentation made at a bias of -200mV. At pH=9.0 a parallel behaviour is observed; now the rosette lenghth reversal occurs at about -250mV. Thus, the application of an external bias can alter dislocation mobility, as inferred from the rosette lengths; this has been called an "electromechanical effct". Also, a simialr situation can be achieved by varying the pH of the environment – a "chemomechanical effect". In both cases, the charge transfer across the environment:specimen interface will have induced near surface bending of the energy levels to preserve electrical neutrality, that is the Fermi level remains at the same level in the bulk and at the surface.

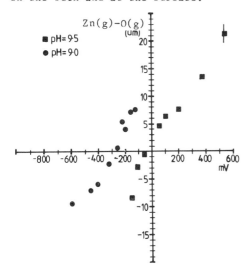

Fig.3 Plot of the difference in rosettelength of Zn(g) and O(g) dislocations against applied voltage at two values of pH.

Hirsch (1980) has outlined a theory which suggests that the dislocation velocity in semiconductors is related to the energy difference between the Fermi level and a kink level associated with a dislocation. A greater energy difference implies a greater concentration of charged kinks, and hence (in certain circumstances) a higher dislocation, velocity. Now, any levels within the band gap will follow the conduction and valence bands. Thus, the energy difference between a kink level, say, and the Fermi level may be altered by an applied voltage, resulting in a change in dislocation velocity. This effect would extend into the crystal only as far as the band bending occured, that is of the order of the Debye length. There is an alternative theory due to Ahearn et al (1980), which suggests that, for ZnO at least, the change in dislocation velocity with applied voltage is due to a change in interaction of a charged dislocation with charged zinc ions. Bending of the energy levels would alter the charge asssociated with the ions. Both theories could account for the surface specificity of the chemomechanical and electromechanical effects.

To verify this surface specificity, a specimen was indented at a bias of -200mV, pH=9.5 (and a load of 20g). Under these conditions the O(g) moved further than the Zn(g) dislocations. This specimen was etched to reveal the positions of the dislcations and their extent measured. Layers, approximately, 2um thick, were removed by chemically polishing in hot orthophosphoric acid and the specimen re-etched. A plot of the variation of rosette length with depth is shown in Figure 4. It can be seen that as the surface is progressively removed the extent of the Zn(g) and O(g) dislocations reverts to the normal case (ie that of Figure 2). The depth to which band bending occurs is given by the Debye length, $L = \sqrt{(K\epsilon kT/ne^2)}$ ϵ is the prermettivity of free space, K that of the solid, e the electronic charge and n the doping level ($= 10^{14} cm^{-3}$ from Hall efect measurements). The calculated value of L is approximately 300nm. This value is consistent with the results of Figure 4, although much smaller than could be removed chemically.

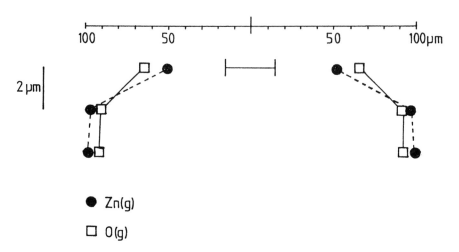

100   50   50   100µm

2 µm

● Zn(g)

□ O(g)

Fig. 4 Plot of rosette length as a function of depth into the bulk.

Another semiconductor was examined to test further the general criteria for observing a "surface" effect. n-type GaAs produced much smaller dislocation arrays around indentations than ZnO; consequently, the hardness itself was measured. There was no variation in hardness, within experimental error, with applied voltage or pH at loads of 5-50g . This behaviour can be explained in terms of near surface band bending. For GaAs it is known that the energy levels are pinned at the surface by surface states (eg Weider 1980), so an applied voltage will produce little or no band bending. This was indeed confirmed by capacitance:voltage measurements.

4. CONCLUSIONS

1) Under ambient conditions $Zn(g)$ are more mobile than $O(g)$ basal plane dislocations in n-type ZnO.
2) In an electrolyte of pH=9.5 an applied voltage of -50mV is sufficient to make the mobility of both types of dislocation the same. A more negative bias made the $O(g)$ dislcoations more mobile.
3) Similar behaviour could be obtained at pH=9.0, but a voltage of about 250mV was necessary to make the mobility of the dislocations equal.
4) The results show that chemomechanical and electromechanical effects are related.
5) The effects occur to a depth of the order of the Debye length.

5. ACKNOWLEDGEMENTS

The Venture Research Unit of BP plc is thanked for the provision of a Research Fellowship. Mr P. Robertson, RSRE, Malvern provided the ZnO specimens. Dr. S. Haywood, Clarendon Laboratory, Oxford kindly performed the Hall effect and Capacitance:Voltage experiments

6. REFERENCES

Ahearn J S, Mills J J Westwood A R C and Kalivoda D A 1980 "Fundamentals of Tribology" ed N P Suh and N Saka (MIT Press) p 295-315
Dewald J F 1960 Bell System Tech. J. 39 615
Hirsch P B 1980 J. Microscopy 118 3
Mariano A N and Hanneman R E 1963 J. Appl. Phys. 34 384
Rebinder P A 1928 Proc. 6th Physics Congr. (Moscow:State Press)
Thomas D G 1957 J. Phys. Chem. Solids 3 229
Weider H H 1980 J. Vac. Sci. Technology 17 1009
Westwood A R C, Ahearn J S and Mills J J 1981 Colloids Surf. 2 1

{001} ⟨110⟩ slip in GaAs

C-D Qin and S.G. Roberts

Department of Metallurgy and Science of Materials,
University of Oxford, Parks Road, Oxford, OX1 3PH, U.K.

ABSTRACT: Dislocations of a new slip system, {001}⟨110⟩, have been
found around indentations on the {001} surface of n-type GaAs. This
mode of slip was studied by polish/etch techniques and by TEM, and was
found to have a polar character, analogous to that of the normal
{111}⟨1$\bar{1}$0⟩ slip.

1. INTRODUCTION

In GaAs (as in most f.c.c. semiconductors), the primary slip system is
{111}⟨1$\bar{1}$0⟩ (Alexander and Haasen 1968). Dislocations of this type have
been intensively studied, and much is known about their structure and
mobility. In particular, As(g) and Ga(g) dislocations have very different
mobilities, which depend strongly on doping (Choi et al 1977, Ninomiya
1979). To date, however, there has been no evidence for the existence of
{001}⟨110⟩ slip in f.c.c. semiconductors.

Indentation on {001} planes of GaAs produces extensive slip patterns along
⟨110⟩ directions, corresponding to the {111}⟨1$\bar{1}$0⟩ primary slip mode (e.g.
Warren et al 1984, Hoche and Schreiber 1984, Lefebvre and Vanderschaeve
1988). However, close examination of both low (20°C) and high temperature
(400°C) indentations revealed slip lines in ⟨100⟩ directions, which would
correspond to slip on {110} or {100} planes. This slip type was examined
in detail for n-type GaAs (10^{18} Te cm^{-3}) by etching and by TEM.

2. ETCHING EXPERIMENTS

Figure 1 shows etch features around an indentation made at 400°C using a
120° diamond cone with a 200g load on a {001} face. Lines along ⟨100⟩ are
visible even on the indented surface at this temperature, but can be more
clearly seen (as here) if a layer of material is first stripped from the
surface. This layer contains rosette and recovery slip (Warren et al 1984,
Roberts et al 1986) which obscures the visibility of the unusual ⟨100⟩
slip traces. The layer was removed by a chemical polish (1HF:1HNO$_3$:2H$_2$O),
which removed ~15um in ~15 seconds and also acted as a dislocation etch.

3. ELECTRON MICROSCOPY

Specimens for electron microscopy were prepared by making an array,
spacing 100um, of 10g conical indentations, and back polishing and
thinning to perforation. Specimens showing the unusual slip mode clearly
could only be easily prepared of room temperature indentations, as

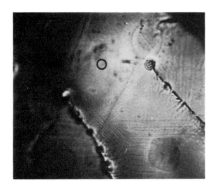 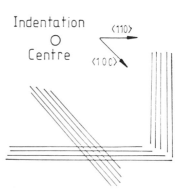

Figure 1 Optical micrograph (Nomarski D.I.C.) of etched features beneath an indentation. Linear etching features can be seen in ⟨110⟩ directions and ⟨100⟩ directions.

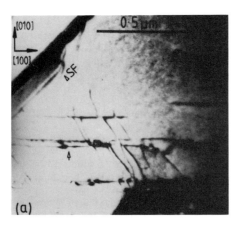 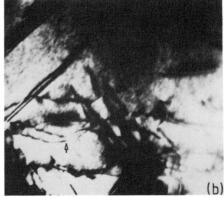

Figure 2 TEM images of dislocations near one corner of an indentation: (a) beam down ⟨001⟩ showing the dislocations along [100] as straight lines; (b) beam down ⟨112⟩ showing the dislocations as loops. The indentation centre is at the bottom left.

extensive "rosette" slip obscured the area of interest at the higher temperatures (see above). Figure 2 shows bright-field images of the slip bands around an indentation. In fig. 2a, the incident beam is normal to the foil (i.e. down [00$\bar{1}$]), and dislocations in the slip band along [100] (arrowed) are seen as straight lines. In fig. 2b, the electron beam is along [1$\bar{1}$2], and the dislocations in this band can be clearly seen as loops. Thus the slip must be in the {010} plane normal to the surface, and not in the inclined {110} planes which could have given the equivalent surface etch features.

Figure 3 shows a Burgers vector analysis of the dislocations in the (010) plane (arrowed in fig.2). The contrast effects are analysed in the Table. The dislocations on the (010) plane are out of contrast only in (d). This slip system is therefore (010)[101]. Generally, in any one of the {010} planes, dislocations are all of one <101> vector of the possible two. Note the extinction of the stacking faults (denoted as 'SF') on convergent slip planes associated with the As(g) dislocations, on the (1Ī1) plane in (b) and (d) (g.R integral). The polarity of these dislocations was determined by a chemical etching method (Warren et al 1984), to be As(g) type.

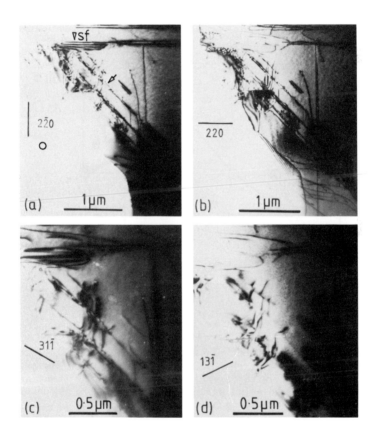

Figure 3 g.b analysis of the new slip mode: (a) g = 2Ī0; (b) g = 220; (c) g = 31Ī; (d) g = 13Ī (all bright field images). Note the dislocations are out of contrast only in (d). "0" shows the position of the indentation centre.

Table: g.b analysis (fig. 3)

b	g	2Ī0 (a)	220 (b)	31Ī (c)	13Ī (d)
1/2[Ī101]		1	1	1	0
1/2[10Ī]		1	1	2	1
(R) 1/3[1Ī1]		4/3	0	1/3	-1

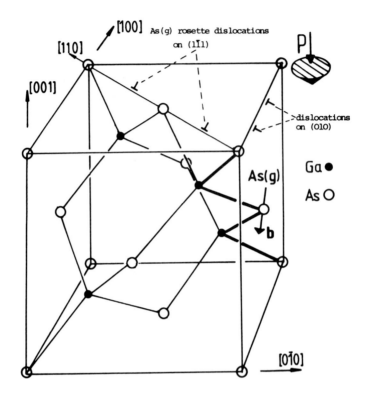

Figure 5 Identification of the type of {001} dislocations. The
geometry corresponds to that in figs. 2 and 3. The dislocations on
the (010) plane, with extra half planes to the right of the slip
plane, correspond to As(g) dislocations if the core structure is that
in fig.4. The indentation position is shown as 'P'.

Following the above assumptions and assuming the observed {001}
dislocations are of the faster As(g) type, fig.5 shows the orientation of
the Burgers vectors of $(0\bar{1}0)[\bar{1}0\bar{1}]$ dislocations, relative to a unit cube,
oriented so that these $(0\bar{1}0)[\bar{1}0\bar{1}]$ As(g) dislocations correspond to those
seen in figs. 2 and 3. The terminating atoms are different for $(0\bar{1}0)[\bar{1}0\bar{1}]$
and $(0\bar{1}0)[\bar{1}01]$ dislocations; those corresponding to the dislocations seen
in figures 2 and 3 are of the As(g) type. Also, dislocations on the $(1\bar{1}1)$
planes with extra-half planes inside the indentation must be of the As(g)
type (this is the faster type of dislocations in n-type GaAs, Warren et al
1984). Moreover, they correspond to the {111} rosette dislocations seen in
figs.2 and 3 (denoted as 'SF'). The geometry coincidence between fig.2,
fig.3 and fig.5 confirms that the {001}<110> dislocations seen in figs.2
and 3 are of the As(g) type given the above assumptions. However, as the
core structure of the {001}<110> dislocations is uncertain, this
identification can only be very tentative.

4. DISCUSSION

Consideration of the geometry of edge-type dislocations on the {010} planes indicates that, like those on the {111} planes, they can have a polar character. However, unlike the {111} dislocations, the character depends not only on the orientation of the extra-half plane of the dislocations, but also on the orientation of the Burgers vector, as these two vectors are not equivalent in a crystal of the $\overline{4}$3m point group, such as GaAs (see below). The two possible <101> vectors of dislocations of the same sign on a particular {010} plane correspond to dislocations of opposite character. We might expect to see dislocations only of the faster type, if (like those on the {111} planes) the As(g) and Ga(g) dislocations have different mobilities (Ninomiya 1979).

It is possible to assign "absolute" As(g) or Ga(g) character to these dislocations, if certain assumptions are made:
1) The extra half-plane is on the side of the slip plane nearer to the indentation centre; this seems reasonable from earlier studies on GaAs and other materials.

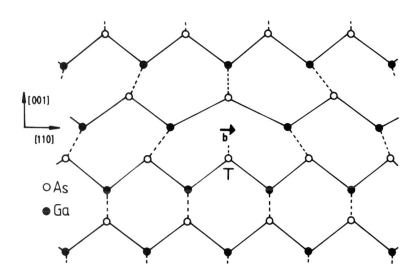

Figure 4 Schematic core structure of an {001}<110> edge dislocation, viewed down the dislocation line. Bonds not in the plane are shown dotted; it is bonds of the dotted type that seem more likely to be broken for dislocations of the indicated Burgers vector; i.e. this dislocation is of the As(g) type.

2) Two types of bonds can be broken to form a dislocation. The broken bonds at the dislocation core are assumed to be those normal to the direction of Burgers vector rather than those at about 35.3^{o} angle; this seems reasonable from 'ball and stick' models (see fig.4), but needs further investigation.

5. CONCLUSIONS

A new slip system, {001}<110>, has been shown to be active beneath microhardness indentations in n-type GaAs. Results from etching (400°C, 200g indentation) and TEM experiments (20°C, 10g indentation) show that this type of slip is activated over a wide range of experimental conditions. Dislocations of this type have a polar nature, which depends on both the sense of the slip and on the direction of the Burgers vector. Tentative models of the core structures identify the faster type as As(g).

REFERENCES

Alexander H and Haasen P 1968 Solid State Physics 22 27
Choi S K Mihara M and Ninomiya T 1979 Jap. J. Appl. Phys. 16 737
Hoche H-R and Schreiber J 1983 Phys. Stat. Sol.(a) 86 229
Lefebvre A and Vanderschaeve G 1988 Phys. Stat. Sol. 107 647
Ninomiya T 1979 J. Phys. (Paris) 40 C6-143
Roberts S G, Warren P D and Hirsch P B 1986 J. Mat. Res. 1 162
Warren P D, Pirouz P and Roberts S G 1984 Phil. Mag. A50 L23

Acknowledgement

Financial support provided by the Chinese Government and the British Council (C-DQ) and the SERC (SGR) is gratefully acknowledged.

Inst. Phys. Conf. Ser. No 104: Chapter 4
Paper presented at Int. Symp. on Struct. Prop. Disloc. Semicond., Oxford, 1989

The plasticity of GaAs at medium and low temperatures

J Rabier

Laboratoire de Métallurgie Physique, URA 131 CNRS
Faculté des Sciences, 86022 Poitiers Cedex, France

ABSTRACT: At medium and low temperature plastic deformation of GaAs is controlled by the Peierls mechanism and the doping dependence of the macroscopic plasticity is clearly observed.In the medium temperature range yield stress measurements as a function of doping as well as TEM observations are in agreement with the individual dislocation mobilities measurements. At lower temperature the effect of electronic doping on the yield stress is different. These results are related to the individual velocity of perfect and partial dislocations. Consequences on dislocation multiplication and strain hardening are discussed as a function of stress and temperature.

1. INTRODUCTION

Extensive studies have been conducted on the mechanical behaviour of elemental semiconductors such as Si and Ge. In comparison only a small number of works have been performed on plasticity of compound semiconductors as GaAs crystals (for recent reviews see Alexander 1986, Rabier and George 1987). In the past few years plasticity of GaAs compounds has been studied in the medium ($T<0.6T_m$) and low temperature ranges: beside usual deformation experiments, efforts have been devoted to extend the range of deformation down to lower temperatures using suitable deformation techniques. As a consequence it has been possible to investigate the high stress regime, which is also of interest to understand the nucleation of dislocations in highly stressed epilayers (De Cooman and Carter 1987). Most of these recent experiments were associated with systematic observations of the deformation substructures at the TEM scale.

Extending the temperature range of plastic deformation down to low temperatures has several consequences: it allows the study of the deformation regime controlled by pure Peierls mechanisms and allows one to get well defined dislocation substructures. The analysis of thermal activation processes of plastic deformation requires also an investigated domain of temperatures as large as possible. Furthermore, at the higher stresses different deformation mechanisms can be involved: extended stacking fault nucleation and twinning. The effects of electronic doping on plastic deformation can be also observed unambiguously since the effects of diffusion which lead to impurity precipitation on dislocations are minimized and the electronic effects due to the Fermi level displacement are expected to be more important at low temperature.

The deformation domain controlled by Peierls forces and the influence of electronic doping on the plastic behaviour of GaAs have been studied using various techniques in the medium temperature range: constant strain rate deformation tests in tension or compression (Sazhin *et al* 1966, Swaminathan and Copley 1975, Laister and Jenkins 1973, Hobgood *et al* 1986, Nakada and Imura 1987, Yonenaga *et al* 1987), creep tests (Steinhardt and Haasen 1978), flexion (Laister and Jenkins 1973). Micro indentation tests were used to study the low temperature range (T<400°C) (Warren *et al* 1984, Hirsch *et al* 1985). However modelling the stress field in an indentation test proved to be very difficult so that the comparison between the data of these two sets of experiments in the whole range of temperature is not reliable.

Two types of deformation technique can be used to obtain a macroscopic plastic deformation at low temperatures: pre-straining in order to get a fixed dislocation density sufficiently high to assure the imposed strain rate at low temperature, and deformation superimposing a hydrostatic pressure to the applied stress to prevent nucleation and propagation of cracks in the usual brittle regime (Veyssière *et al* 1985). These two techniques have been used for silicon (Demenet *et al* 1984, Omri *et al* 1987, Castaing *et al* 1981, Rabier *et al* 1983), and discussed elsewhere (Rabier and George 1987).

The aim of this paper is to discuss and compare the recent results obtained on macroscopic plastic deformation at medium and low temperatures, with reference to:
(i) the influence of electronic doping on the macroscopic plastic behaviour of GaAs (isoelectronic impurities are not taken into account)
(ii) the thermal activation regime of dislocation glide and the relation with dislocation velocities.
(iii) the mechanisms controlling plastic flow at the microscopic level.

2.   MEDIUM TEMPERATURE PLASTICITY

2.1.Plastic deformation results

Most of the published results are relevant to the plasticity of as-grown crystals having a low initial dislocation density. These tests can be performed in a limited range of temperature which depends on electronic doping: usually 350°-650°C. Examples of stress-strain curves are shown in figure 1.

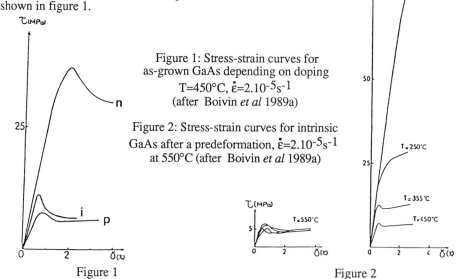

Figure 1: Stress-strain curves for as-grown GaAs depending on doping
T=450°C, $\dot{\varepsilon}$=2.10^{-5}s^{-1}
(after Boivin *et al* 1989a)

Figure 2: Stress-strain curves for intrinsic GaAs after a predeformation, $\dot{\varepsilon}$=2.10^{-5}s^{-1} at 550°C (after Boivin *et al* 1989a)

Figure 1   Figure 2

The elastic limits which were obtained show large scatter from study to study: this comes mainly from different experimental conditions (differences in compression axis, strain rate, concentration of impurities). In the same deformation conditions,the plastic domain of n type samples appear to be very narrow, whereas p type samples have about the same plastic properties as undoped crystals (Nakada and Imura 1987, Boivin *et al* 1989a cf figure1). These tests show that it is impossible to deform as-grown n type single crystals below 450°C whereas in the same conditions undoped and p type have a brittle-ductile transition at 350°C. This precludes any comparison of the different doping influence on plasticity over a large temperature range.The presence of a yield drop is due to the lack of a sufficient dislocation density at the begining of the test. As soon as the applied stress is large enough to increase the dislocation velocity, the rapid multiplication process increases rapidly the dislocation density which yields a diminution of the applied stress down to the lower yield stress. This minimum results from the subsequent hardening due to the excess of dislocations introduced during the rapid multiplication. In these conditions the stress at the lower yield point τ_{lyp} can be written as:

$\tau_{lyp}=\tau_{eff}+\tau(\rho)+\tau_\mu$ (Escaig *et al* 1982),where τ_{eff} is the effective stress acting on dislocation, $\tau(\rho)$ is an internal stress term which results from hardening and depends explicitly on the dislocation density $(\alpha(\rho)^{1/2})$, and τ_μ is an internal stress term corresponding to long range interactions usually taken as the athermal stress.

The extension of the yield drop prevents a test being conducted at low temperature since samples may undergo stresses larger than their fracture limit, in the transient yielding regime.The effect of a pre-deformation is to reduce the yield drop amplitude, and consequently the internal stress term $\tau(\rho)$. Ideal predeformation conditions are those for which $\tau(\rho)=0$. This explains the observed softening, as compared to direct tests. Pre-straining has also the advantage that it induces homogeneous dislocation densities in the different samples. Figure 2 shows examples of stress-strain curves $\tau(\gamma)$ obtained for intrinsic GaAs after a pre-straining in the athermal plateau.

Using pre-straining, undoped crystals can be deformed down to 250°C (Astié *et al* 1986) and 150°C (Boivin *et al* 1989a). By abrading a surface, Karmouda (1984) succeeded in deforming undoped crystals down to 260°C. p type samples can be deformed down to 150°C, and n type samples down to 350°C (Boivin *et al* 1989a). After pre-straining in the athermal plateau, instead of having in the subsequent deformation a yield point amplitude which increases when temperature decreases, this amplitude decreases. Futhermore a parabolic stage is found after the apparent elastic limit (Astié *et al* 1986, Boivin 1988, Boivin *et al* 1989a). The same features were also found for GaSb (Omri 1987). This is in contradistinction with what is found for silicon (Omri *et al* 1987). Indeed thermal activation being less important, dislocation velocities decrease so that the dislocations introduced by pre-deformation are no longer sufficient to accomodate the imposed strain rate. If dislocations multiply rapidly a yield point is expected whose amplitude increases with decreasing temperature. This does not seem to apply to GaAs.

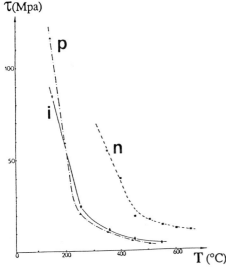

τ(Mpa)

Figure 3: Yield stresses as a function of temperature for intrinsic, p and n type GaAs, n=2.2 10 ^{18} cm^{-3}, i (n=7 10 ^{13} cm^{-3}), p=2 10 ^{18} cm^{-3} .
($\dot{\varepsilon}$=2.10^{-5}s^{-1}, after Boivin *et al* 1989a)

Figure 3 shows the variation of yield stress with temperature for different dopings in the wor
of Boivin *et al* 1989a. n type samples have larger yield stresses than p type and intrinsi
crystals. Moreover these two latter types of samples show quite a similar plastic behaviour,
type samples being more plastic than undoped ones between 200°C and 500°C, whereas th
reverse is observed below 200°C. The curves τ(T) exhibit a rapid increase of the yield stres
below about 250°C for undoped and p type GaAs and 450°C for n type: below thes
temperatures, dislocation glide is strongly thermally activated. This is also observed in weakl
doped n type GaAs by Karmouda (1984) (T<400°C), and in undoped GaAs by Astié *et a*
(1986) (T<350°C). This feature can be compared to the variation of the yield stress of bc
metals at low temperature for T<0.2T_m.

2.2.Thermal activation analysis of plastic deformation

Experimental results shows a domain of strong thermal activation for any type of sample, whe
deformation has been sucessfully conducted at low temperatures. This domain can be analyze
using both the theory of thermally activated processes originally developed by Schoeck (1965
and the formalism of Haasen (Alexander and Haasen 1968) which was introduced to describ
the yield drop of elemental semiconductors.

In the first analysis, starting from the Orowan relation, $\dot{\gamma}=\rho bv$, $\dot{\gamma}$ can be expressed as:

$$\dot{\gamma}=\dot{\gamma}_0 \exp{-\Delta G(\tau,T)/kT} \quad (1)$$

where $\Delta G(\tau,T)$ is the variation of the Gibbs energy which is necessary for the dislocation t
overcome the obstacle. The pre-exponential term is independent of τ. Taking into account th
work done by the effective stress acting on the mobile dislocation segment,

$$\Delta G_0(T)=\Delta G(\tau,T)+\tau_{eff}V \quad (2) \quad \text{where} \quad V=-(d\Delta G(\tau,T)/d\tau)_{T,ss} \quad (3).$$

The quantities ΔG_0 (Peierls barrier at 0K) and V (activation volume) characterize th
mechanism. By integrating the curves V(τ), ΔG(T) and $\Delta G(\tau)$ can be obtained (Omri *et a*
1987). In figure 4 are reported results of Karmouda (1984), Astié *et al* (1986) and Boivin *et a*
(1989a) on intrinsic GaAs or weakly doped crystals. In Karmouda's work a plateau is foun
for T>350°C on the curve ΔG(T), whereas it does not exist in the other results. This deviatio
can be explained mainly by a different estimate of τ near the athermal plateau. V(τ) value
measured by stress relaxation technique are small (of the order of b^3) for the lowes
temperatures. These values are characteristic of a Peierls mechanism. The values of τ and V fo
intrinsic GaAs are very close in the different experiments for T<400°C. This proves that at lo
temperature, the differences in experimental conditions are not critical. Extrapolating $\tau_{eff}=\tau-\tau$
equal to zero leads to values of ΔG_0 (table 1). These curves are very similar for undoped and
type crystals. The values of $\Delta G(\tau-\tau_\mu)$ for n type are always larger, for the same effectiv
stress.

In order to compare macroscopic tests with dislocation velocities usually expressed as:

$$v=B\tau^m \exp{-Q/kT} \quad (4),$$

the phenomenological equation:

$$\dot{\gamma}=A\tau^n \exp{-Q/kT} \quad (5)$$

can be employed. Expression (5) is also obtained from the model of Haasen (Alexander an
Haasen 1968) which describes the beginning of the stress strain curve i.e. the yield drop i

elemental semiconductors. Q is the activation energy of dislocation velocity and n=m+2. Unlike the Schoeck formalism, stress appears only in the pre-exponential term, activation energy Q being independent of τ. The activation volume measured is here an apparent activation volume and is expressed as:

$$V_a = kT \, (d\ln \dot{\gamma}/d\tau)_T = nkT/\tau \quad (6)$$

n can be determined by the slope of $V_a(kT/\tau)$ and for constant strain rate tests Q/n is the slope of $\ln\tau = f(1/T)$.

The stress exponent n is found to be constant only at low temperature (T<250°C (Boivin 1989a), T<450°C (Karmouda 1984, Astié *et al* 1986) for undoped, T<250°C for p type and T<450°C for n type (Boivin 1989a)). Stress exponents deduced at low temperature are high (see table 1). The values obtained for Q can be compared with those measured by Steinhardt and Haasen (1978) in creep tests in the temperature range [300°C, 650°C], at low stress and similar doping conditions. The agreement is only good for intrinsic and n type in the work of Boivin *et al* (1989a). Other values of Q are greater than those found by Steinhardt and Haasen (1978). In the same table are reported the ΔG values obtained for an effective stress τ=10 MPa. All the activation energy values are systematically higher than those reported in individual dislocation velocity measurements (see George and Rabier 1987).

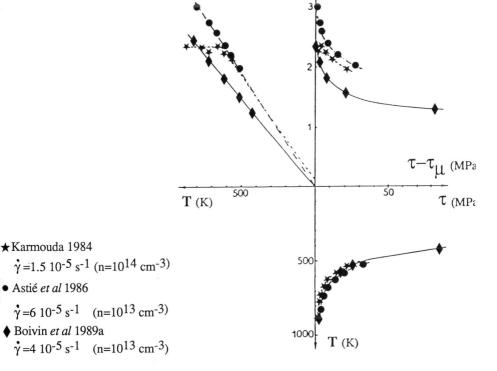

★ Karmouda 1984
   $\dot{\gamma}$ =1.5 10^{-5} s^{-1} (n=10^{14} cm^{-3})

● Astié *et al* 1986
   $\dot{\gamma}$ =6 10^{-5} s^{-1} (n=10^{13} cm^{-3})

◆ Boivin *et al* 1989a
   $\dot{\gamma}$ =4 10^{-5} s^{-1} (n=10^{13} cm^{-3})

Figure 4: Cross-diagram for intrinsic or undoped GaAs showing the variations of ΔG as a function of T and τ_{eff}=τ-τ_μ and τ (T).

Table 1:

type	n	U/n	U eV	ΔG_0 eV	$\Delta G(\tau_1)$ eV	$U_c(\tau_1)$ eV	
undoped	5.5	0.27	1.48	2.5	1.7	1.5	Boivin *et al* 1989a
	8	0.27	2.2	3.	2.2		Astié *et al* 1986
	6.6	0.31	2.04	2.25	2.35		Karmouda 1984
p type	5.5	0.32	1.76	2.5	1.8	1.3	Boivin *et al* 1989a
n type	5.5	0.37	2.03	2.8	2.1	2.2	id

n: stress exponent, U: activation energy deduced from the formalism of Alexander and Haasen

U_c: activation energy for creep reported by Steinhardt and Haasen 1978 (τ_1=10MPa),

ΔG: activation energy determined using the formalism of thermally activated processes for

$\tau_1=\tau-\tau_\mu$=10MPa, ΔG_0: Peierls stress at 0K.

Applying these two analyses to the thermally activated deformation of GaAs suffers several limitations.The Schoeck analysis seems to be difficult to apply to semiconductor materials with a low initial dislocation density since the rapid multiplication of dislocations during the yield drop leads to large internal stresses. Using a pre-deformation allows the plastic domain range to be extended down to lower temperatures and to obtain experimental conditions better suited to determine the activation parameters, since the yield point amplitude can be reduced or suppressed. Nevertheless the range of temperature studied is still narrow compared to those studied for bcc metals: in the case of pure iron (Peyrade 1978) deformation has been studied down to $0.02T_m$ instead of $0.28T_m$ for GaAs. This limited range of temperature can introduce large errors in the ΔG value. On the other hand, the value of the slope of $\Delta G(T)$ ($34<\alpha<36.5$) appears always larger than those reported and accepted for bcc metals ($\alpha \cong 25$). This value can be due to the occurrence of several deformation mechanisms in the investigated domain of temperature. The model of Haasen describes the yield point in elemental semiconductors with low initial dislocation density. This yields a relation between ($\tau,\dot{\gamma},T$) valid only at the lower yield point. Futhermore the value of the stress exponent measured (m=3.5) is larger than that predicted by the model ($1\leq m\leq 2$); in addition this value has been found to depend on temperature in a different manner to the model's hypotheses.

2.3.TEM observations of deformation substructures

Deformation at high temperature up to the lower yield stress induces an isotropic dislocation substructure, i.e. without any preferential orientation, mainly consisting of dipoles and multipoles and with very few screw dislocations. The same features are observed in silicon at the lower yield point after high temperature deformation (Oueldennaoua *et al* 1988). Deformation at low temperature stopped close to the macroscopic yield stress leads to dislocation substructures characterized by a large increase in the screw dislocation density in the primary glide plane compared to the other dislocation segments. The transition between substructure where screw dislocations are nearly absent and one where they are the major part of the substructure (which is correlated to the absence of a yield drop at low temperature), has been observed by Astié *et al* (1986) and Boivin *et al* (1989a) in undoped GaAs and Omri (1987) in GaSb.

Comparing the observations made at 350°C for the different dopings, it can be seen that the microstructure of n type GaAs is already characterized by long screw dislocations, whereas those of intrinsic and p type GaAs have typically a high temperature character. This is consistent with velocity measurements since n doping strongly diminishes the mobilities of the different types of dislocations. Not all the observed dislocations are mobile dislocations which contribute to plastic flow, but numerous dipoles, multipoles and by-products are seen, some of them coming from pre-deformation. Few isolated 60° α or β dislocations are seen in samples deformed at low temperature, so that it is difficult to compare mobilities of the different types of dislocation. In addition, dislocation configurations are often unstable under the action of the electron beam. Consequently, during observation, internal stresses can be relaxed together with out-of-equilibrium dislocation configurations, unless dislocations are stopped by interactions with nearby configurations. The large majority of screw dislocations seen at low temperature indicates that they are slow and that they control deformation in any type of doping. This is also true for p type crystals although small differences in mobilities between α,β and screw dislocations were deduced from direct velocity measurement. Nevertheless, when it is possible to observe the α (or β) segments on a glide loop, the observations are in qualitative agreement with velocity data (Ninomya 1979): longer 60° segments are found in p type crystals, compared to intrinsic ones (Boivin *et al* 1986, Boivin *et al* 1987). This is also reported for experiments in which dislocations generated at higher temperature were then submitted to a high stress at lower temperature (De Cooman 1987, De Cooman and Carter 1987). Dislocations appear generally to be dissociated in the glide plane. The intrinsic stacking fault energy determined from isolated configurations in weak beam conditions yields a stacking fault energy which does not vary significantly with the nature of electronic doping (Boivin 1988). This shows that the effect of doping on the plastic behaviour of GaAs cannot be explained by differences in the core structures of dislocations.

3. LOW TEMPERATURE PLASTICITY

Plastic deformation tests can be performed at low temperatures below the usual brittle-ductile transition using a confining pressure. This technique, initially used for geological applications (Griggs and Kennedy 1956), prevents cracks from nucleating and propagating and allows macroscopic plastic deformation of GaAs down to room temperature (Rabier *et al* 1985, Rabier and Garem 1985, Lefebvre *et al* 1985).

3.1.Deformation tests under confining pressure

Examples of stress-strain curves obtained under confining pressure are shown in figure 5, for intrinsic, p type and n type GaAs.

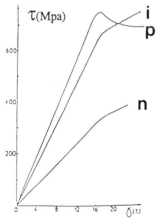

Figure 5 : Stress-strain curves under confining pressure for intrinsic, n type and p typeGaAs ($\dot{\varepsilon} = 2.10^{-5}$ s^{-1}).

At room temperature no yield point appears on the stress-strain curves for intrinsic and n type GaAs, whereas a small one is found for p type.Tests conducted using the same technique show that yield points found at higher temperature disappear when the temperature decreases (Boivin 1988, Boivin *et al* 1989b François *et al* 1988). This confirms what was found in the medium temperature range (§2.1): dislocation multiplication becomes more difficult at low temperature. n doping decreases the yield stress of GaAs whereas p doping causes hardening as compared to the intrinsic crystal (Rabier *et al* 1985). The doping effect on flow stress at room temperature is very important, especially in the case of the softening of n type GaAs. The mechanical behaviour of p type and intrinsic GaAs appears to be rather similar. The electronic doping effect on plastic deformation is thus found to be reversed at room temperature as compared to higher temperatures (Sazhin *et al* 1966, Laister and Jenkins 1973, Swaminathan and Copley 1975, Steinhardt and Haasen 1978, Nakada and Imura 1987, Boivin *et al* 1989a). This confirms the trends found using standard deformation techniques (cf fig 3) when the temperature decreases. Deformation at room temperature appears to be inhomogenous and concentrated in very dense glide bands parallel to the (111) glide plane. The observation of wavy slip lines demonstrate that cross slip is active at room temperature whatever the type of samples.

Deformation tests under atmospheric pressure, after pre-straining at high temperature, and under confining pressure, can be performed in overlapping temperature ranges. Comparison of these two sets of tests is discussed in an other paper (Demenet *et al* 1989).

3.2. TEM observations of deformation substructures

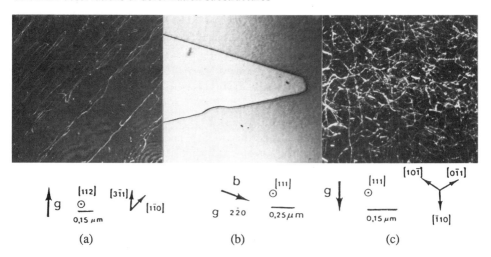

(a)   (b)   (c)

Figure 6 : Dislocation substructures at the macroscopic yield stress after deformation at room temperature (see figure 5)
    (a) intrinsic GaAs: screw dipoles
    (b) n type : asymmetric behaviour of screw segments
    (c) p type: distribution of various dislocation segments

TEM observations show that different deformation mechanisms from those reported at higher temperatures act at room temperature: coexistence of weakly dissociated dislocations and microtwins is observed. In n type and intrinsic GaAs, perfect dislocations consist mainly of screw segments. In p type GaAs, screw and 60° mixed segments are found in equal proportion at the lower yield point : the corresponding mobilities of these segments seem to be quite similar (figure 6). Furthermore for the different dopings, twin nucleation induced by the glide of uncorrelated partial dislocations appears to be a deformation mechanism at room temperature whose importance increases with the plastic strain. Twinning seems to be more important at comparable strains in p-type GaAs than in intrinsic and n-type GaAs. Since perfect dislocations seem to control the beginning of the deformation, twinning will not be discussed here. Microtwinning has been studied in semi insulating crystals by Androussi *et al* (1987).

Cross-slip is observed at room temperature in n type and intrinsic GaAs and to a lesser extent in p type GaAs. Numerous extended 30° segments are observed mainly in n type GaAs, which result from the projection of 60° segments belonging to the cross-slip plane. Some isolated configurations observed in n-type GaAs have been reported to show asymmetric behaviour between opposite screw segments: $(30\beta,30\alpha)$ screw segments exhibit a higher ability to cross-slip whereas $(30\alpha,30\beta)$ opposite screw segments can easily be dissociated in the glide plane. Deformation substructures are also characterized by a very high density of dislocations at the yield stress. A pre-plastic stage can be assumed, similar to b c c metals, which corresponds to the movement of the fastest α (or β) 60° dislocations.

In the high stress regime recovery mechanisms, such as cross slip, are required at low strains due to the very high density of dislocations necessary to induce plastic deformation. In contrast to the medium temperature range, where deformation is controlled by weakly dissociated dislocations, at high stress the mobility asymmetry between the partial dislocations of a segment becomes an important parameter. The ability to nucleate extended stacking faults and to induce cross slip depends on this asymmetry.

4. PLASTIC DEFORMATION AND DISLOCATION MOBILITIES

Because of the existence of three types of dislocations with different mobilities, the plastic strain rate of GaAs can be expressed as:

$$\dot{\varepsilon}=b(N_\alpha v_\alpha+N_\beta v_\beta+N_s v_s) \quad (7)$$

where N and v are the density and velocity of the different dislocations. Steinhardt and Haasen 1978 using a multiplication law $\dot{N} \propto \tau\rho v$ obtained:

$$\dot{\varepsilon}=bN\sqrt{1/2}(v_\alpha+v_\beta)v_s \quad (8)$$

For T<200°C TEM observations show that screw dislocations control plasticity.The strong activation domain compares well to that of b c c metals. This suggests that a common analysis can be relevant (Louchet and Kubin 1979, Yonegana *et al* 1987), which is also confirmed by the absence of yield point at low temperature. This type of analysis with $\dot{N}_s=2(\dot{v}_\alpha+v_\beta)$, $\dot{N}_\alpha=\dot{N}_\beta=2v_s$ and $v_\alpha>>v_\beta,v_s$, $N_s>>N_\alpha,N_\beta$ yields:

$$\dot{\varepsilon}=2bN_s v_s \quad (9)$$

These two analyses give rather good agreement between measured yield stresses and the average dislocation mobilities as a function of electronic doping for temperatures larger than 200°C since $\tau_n>\tau_i>\tau_p$ ($v_n<v_i<v_p$ (Ninomya 1979)). However this effect seems to be different at lower temperatures (T<150°C), and the extrapolation of velocity measurement at low temperature and high stress does not explain this effect.

Since screw dislocations control plastic deformation at low temperature, how their properties are affected by a high stress regime should be investigated. At medium temperatures screw dislocations exhibit a mobility nearly equal to that of the slowest dislocation segments α or β. Indeed their mobility is controlled by the slowest partials, since the perfect screw dislocation mobility μ_s can be written as:

$$1/\mu_s= 1/\mu_{30\alpha}+1/\mu_{30\beta} \quad (10)$$

One can reasonably assume that $\mu_{30\alpha}\cong\mu_\alpha$ and $\mu_{30\beta}\cong\mu_\beta$ (Jones *et al* 1981) so that the mobility ratio between α and β partial dislocations can be assumed to vary in the same way as the mobility ratio of perfect α and β perfect dislocations, as a function of electronic doping. This μ_α/μ_β ratio for perfect dislocations has been found to decrease from n to p type (see for example Ninomya 1979).

In a glide loop the two screw segments ($30\alpha/30\beta$) and ($30\beta/30\alpha$) consist of partial dislocations having asymmetrical mobilities. The dissociation width (d) of each of the screw segments can be calculated using the formalism of Wessel and Alexander (1977), from the ratio $R^*=R_1/R_2$ between the lattice friction forces acting on the leading (R_1) and trailing partials (R_2):

$$d= d_0 / (1+b\tau/2\gamma (f- ((1-R^*)/(1+R^*))) \quad (11)$$

where f=1/6 in the case of a <123> compression axis and d_0 is the equilibrium splitting width at zero stress. In the case of n type GaAs, where the mobility asymmetry between partial dislocations is assumed to be greater, one of the screw segments possesses a fast leading partial of α type, the other one a slow leading partial of β type; R^* is then larger than 1 on the ($30\alpha/30\beta$) segment and lower than 1 (=$1/R^*$) on the ($30\beta/30\alpha$) segment. This yields (for a high applied stress) to a widening of the splitting width on the ($30\alpha/30\beta$) segment and a contraction on the ($30\beta/30\alpha$) segment.

The activation energy for cross slip can then be calculated in the framework of the Escaig model (1968). Figure 7 shows the calculated activation energy as a function of the mobility ratio of partial dislocations (Rabier and Boivin 1989). When the stress increases (τ>100 MPa), the activation energy for cross slip decreases rapidly as a function of R^*. For τ=500 MPa and R^* ranging between 5 and 10, W is of the order of 1.5 eV, a value which is typically of the order of magnitude of that reported for dislocation glide (see George and Rabier 1987). Calculations of the activation volume as a function of stress and R^*, give small values for high stresses of the order of magnitude of those reported for a glide process controlled by a Peierls mechanism. These calculations of activation parameters for cross slip show that pure glide and cross slip can be concurrent mechanisms at low temperature and high stress in GaAs.

Figure 7: Calculated activation energy for cross slip as a function of the mobility ratio of partial dislocations.

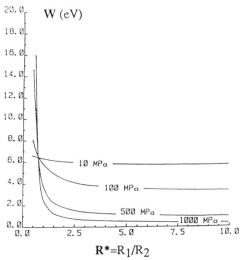

These calculations show that:
-at high temperature, cross slip is thermally activated and can occur as a recovery process when the dislocation density is important.
-at medium temperature and low stress, cross slip is difficult.
-at low temperature cross slip is made easier by the applied stress and the asymmetric mobilities of 30α and 30β partial dislocations.

Dislocation multiplication and strain hardening are directly related to cross slip so that plastic deformation results can be discussed as a function of screw dislocation cross slip characteristics.

When the temperature decreases, dislocation multiplication appears to be more difficult than at high temperature. The yield point which is characteristic of a rapid multiplication is replaced at medium temperature by a parabolic strain hardening stage without any maximum: long screw dislocations are left behind by the glide of faster dislocations α (or β) and lead to the formation of screw pile-ups which prevent dislocation sources from operating normally. There is futhermore a micro- or pre-plastic stage which corresponds to α or β dislocation glide analogous to that observed in bcc metals where edge segments are more mobile than screw ones at low temperatures. Then few α or β mobile dislocations can be observed by TEM compared to the high density of screws reported in samples deformed at medium temperature when the observations are made at the macroscopic yield stress (Astié *et al* 1986, Boivin 1988, Boivin *et al* 1989a).

At low temperature and high stresses, dislocation mobilities are too low to accomodate the deformation so that the dislocation density increases rapidly in the primary plane. In these conditions cross-slip, which is assisted by the applied stress, is a recovery mechanism acting as soon as the macroscopic yield stress is reached. The ease of cross slip depends on the $30\alpha/30\beta$ partial dislocation mobility asymmetry. This asymmety increases from p type to n type: cross slip is more and more easy from p type to n type and this favours the annihilation of screw pile-ups. In this scheme the dislocation sources can be activated at lower stresses in n type crystals. Multiplication of dislocations also depends on the possibility for a dislocation segment to create a source, i.e. for screw segments to develop a super jog, by cross slip, which can act as a pinning point for a pole mechanism. For these two reasons, dislocation multiplication is likely to be easier in n type GaAs than in intrinsic and p types. This is in agreement with the measured values of yield stresses at room temperature, as well as the observations of the cross slip and dissociation asymmetries in n type GaAs (Boivin *et al* 1989b).

5.   CONCLUSION

The effect of electronic doping on GaAs plasticity, using standard deformation techniques, is in qualitative agreement with individual dislocation velocity measurements for temperatures larger than 200°C. However this effect seems to be different at lower temperatures (T<150°C). TEM observations show that plastic deformation at low temperature is controlled by a Peierls mechanism on screw dislocations, the glide loop shape providing evidence of the velocity asymmetry between α,β and screw dislocations as a function of electronic doping. The determination of the fine structure of dislocations for the different dopings confirm that rather than being a core effect, the effect of doping is of an electronic nature. Stress-strain curves provide the evidence that dislocation multiplication appears more difficult when the temperature decreases. This is correlated with an increase in the screw dislocation density in the primary glide plane and in the cross slip plane.

Deformation tests at constant strain rate under a confining pressure can be performed down to room temperature on extrinsic and intrinsic single GaAs. The electronic doping effect on the macroscopic yield stresses is reversed as compared to higher temperatures (T>200°C). In the high stress regime recovery mechanisms, such as cross slip, are required at low strains due to the very high density of dislocations necessary to induce plastic deformation. In contrast to the medium temperature range, where deformation is controlled by weakly dissociated dislocations, at high stress the mobility asymmetry between the partial dislocations of a same segment has important consequences. On this asymmetry depends the cross slip ability of screw dislocation. A cross slip model shows that dislocation multiplication is likely to be easier in n type GaAs than in intrinsic and p type GaAs.

ACKNOWLEDGMENTS

The author thanks his co-workers, Drs P Boivin, J L Demenet and H Garem for their contributions to this work.

REFERENCES

Alexander H and Haasen P 1968 Solid State Phys **22** 26
Alexander H 1986 Dislocations in Solids, Nabarro FRN edt, North Holland **7** 113
Androussi Y, Vanderschaeve G and Lefebvre A 1987 Inst. Phys.Conf. Ser. **8** 291
Astié P, Couderc J J, Chomel P, Quélard D and Duseaux M 1986 Phys. Stat. Sol.(a) **96** 225
Boivin P 1988 Thèse d'Université, Poitiers
Boivin P, Rabier J and Garem H (a) submitted to Phil. Mag.
Boivin P, Rabier J and Garem H (b) submitted to Phil. Mag.
Boivin P, Rabier J, Garem H and Duseaux M 1986, Defects in semiconductors, Von Bardeleben H.J edtr., Mat. Sci. Forum **10-12** 781
Boivin P, Rabier J and Garem H 1987 Inst. Phys. Conf. Ser. **87** 297
Castaing J, Veyssière P, Kubin L P, and Rabier J 1981 Phil. Mag. A **44** 1407
De Cooman B C 1987 Thesis, Cornell
De Cooman B C and Carter C B 1987 Inst. Phys. Conf. Ser. **87** 259
Demenet J L, Desoyer J C, Rabier J and Veyssière P 1984 Scripta Met. **18** 41
Demenet J L, Boivin P and Rabier J 1989 this issue
Escaig B 1968 J. Phys. Paris **29** 225,
Escaig B, Ferré D and Farvacque J L 1982 Phys. Stat. Sol.(a) **71** 329
François P, Lefebvre and Vanderschaeve G 1988 Phys. Stat. Sol.(a) **109** 187
George A and Rabier J 1987 Revue Phys. Appl. **22** 941
Griggs D T and Kennedy G C 1956 Amer. J. Sc. **254** 712
Hirsch P B, Pirouz P, Roberts S G and Warren P D 1985 Phil.Mag B **52** 759
Hobgood H M, McGuigan S, Spitznagel J A and Thomas R N 1986 Appl. Phys. Lett **24**,1654
Jones R Oberg S and Marklund S 1981 Phil. Mag. B **43** 839
Karmouda M 1984 Thèse de troisième cycle, Lille
Laister D and Jenkins G M 1973 J. of Mat. Sci **8** 1218
Lefebvre A, François P, and Di Persio J 1985 J. Phys. Lett. Paris **46** 1023
Louchet F and Kubin L P 1979, Phys Stat Sol a **56** 169
Nakada Y and Imura T 1987 Phys. Stat. Sol. (a) **103** 435.
Ninomya T 1979 J. Phys. Paris **40** C6-143
Omri M, Tête C, Michel J P and George A 1987 Phil. Mag A **55** 601
Omri M. 1987 Thèse d'état Nancy
Oueldennaoua A, Allem R, George A and Michel J P 1988 Phil. Mag A **57** 51
Peyrade J P 1978 Thèse d'état Toulouse
Rabier J, Veyssière P and Demenet J L 1983 J. Phys. Paris **44** C4-283
Rabier J, Garem H, Demenet J L and Veyssière P 1985 Phil. Mag. A **51** L-67
Rabier J and Garem H 1985, Materials Under Extreme Conditions, Ahlborn H et al edts, MRS-Europe, 177
Rabier J and George A 1987 Revue Phys. Appl.**22** 1327
Rabier J and Boivin P 1989 submitted to Phil. Mag.
Sazhin N P, Mil'vidskii M G, Osvenskii V B and Stolyarov O G 1966 Sov. Phys. Solid State **8** 1223
Schoeck G 1965 Phys. Stat. Sol. **8** 499
Steinhardt H and Haasen P 1978 Phys. Stat. Sol.(a) **49** 93
Swaminathan V and Copley J M 1975 J. Amer. Ceram. Soc. **58** 482
Veyssière P, Rabier J, Jaulin M, Demenet J L and Castaing J 1985 Rev. Phys. Appl. **20** 805
Warren P D, Pirouz P and Roberts S G 1984 Phil. Mag. A **50** L-23
Wessel K and Alexander H 1977 Phil. Mag. 35 1523
Yonenaga I, Onose U and Sumino K 1987 J. Mat. Res. **2** 252

Inst. Phys. Conf. Ser. No 104: Chapter 4
Paper presented at Int. Symp. on Struct. Prop. Disloc. Semicond., Oxford, 1989

339

High temperature plasticity of III–V semiconductors

A. ZOZIME and J. CASTAING

Laboratoire de Physique des Matériaux
C.N.R.S. Bellevue, 92195 Meudon cedex, France

ABSTRACT : High temperature plasticity of GaAs, InSb and InP has been investigated. It shows many similarities with that of fcc metals. Impurity alloying largely increases the flow-stress of semiconductors, even at very high temperature.

1. INTRODUCTION

The interest in high temperature mechanical bahaviour of semiconductors is twofold (i) device processing implies that materials undergo high temperatures under stresses (crystal growth, diffusion anneals ...) (ii) electronic properties influence the dislocation behaviours e.g., the plasticity is affected by doping up to some temperature (for reviews, see Rabier and George 1987, George and Rabier 1987). Silicon properties at low temperatures are well documented for doping levels up to 10^{18}-10^{19} cm^{-3}, the temperature to have 10^{18}cm^{-3} intrinsic carriers being 730 K (0.43 T_M ; T_M = melting temperature). The situation is opposite for III-V's which have larger gaps than Si, with temperatures for 10^{18}cm^{-3} carriers of 1400 K (0.93 T_M) and 1335 K = T_M for GaAs and InP respectively. On can therefore expect that dopant concentrations in the 10^{18}cm^{-3} range can affect properties of III-V semiconductors, even in the vicinity of T_M .

In this paper, we discuss recent results on the plasticity of III-V for T > 0.5 - 0.6 T_M , which has been studied in relation with dislocation multiplication during crystal growth (Djemel et al. 1989). The density of as-grown defects is inversely correlated to the high temperature strength which displays a substantial spread for the various semiconductors (see Table 1). Mechanical data are particularly abundant for GaAs, InSb and InP that are described and discussed in the following.

TABLE 1 - Melting temperatures and CRSS at 0.65 T_M for III-V semi-conductors after Gottschalk et al. (1978)

	GaAs	Gap	GaSb	InAs	InP	InSb	Si
melting temperature T_M (K)	1510	1740	980	1215	1335	805	1685
τ_{LY} (MPa) at 0.65 T_M and $\dot{\epsilon} \sim 4 \times 10^{-5} s^{-1}$	1.9	4.0	15.8	0.7	1.8	5.0	4.5

2. PLASTIC DEFORMATION
2.1. Mechanical tests

Plastic deformation experiments were generally carried out on single crystals grown by the liquid encapsulated Czochralski technique, using liquid B_2O_3. The initial dislocation density obtained by etch-pit counting, was typically between $10^3 cm^{-2}$ and $10^5 cm^{-2}$ for undoped materials, having carrier concentrations up to n $\sim 10^{15}$-$10^{16} cm^{-3}$. Most of the papers reported on plastic deformation carried out in compression at constant strain rates with or without a protective atmosphere. Both orientations, <123> and <100> of the long axis of the crystal specimens were used with strain rates between 10^{-5} to $10^{-2} s^{-1}$. Much less work has dealt with high temperature creep (Völkl, Müller and Blum, 1987).

The materials were doped with different impurities, mainly In in GaAs (isoelectronic), S and Zn in InP (respectively n and p doping), with atomic concentrations ranging between 10^{17} to 10^{20} cm^{-3}.
Most of the authors give the resolved shear stress versus shear strain $\tau(\gamma)$ curves and not the engineering stress-strain $\sigma(\epsilon)$ curves. Experiments carried out on undoped InSb between about 0.65 and 0.8 T_M (Siethoff and Schröter, 1984) and on undoped InP at 0.75 T_M (Siethoff et al., 1988) showed that the curves are characterized by five stages, which are easily distinguishable on the curves giving the work hardening coefficient $d\tau/d\gamma$ as a function of stress (Fig. 1). These stages are : (I) easy glide, (II) strong work hardening, (III) first dynamical recovery, (IV) work hardening, (V) second dynamical recovery.

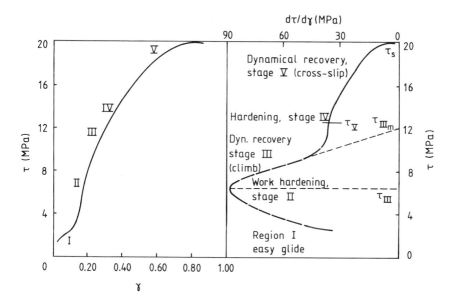

Fig 1. Resolved shear stress τ versus shear strain γ curve and the corresponding work hardening coefficient $d\tau/d\gamma$ as a function of the shear stress γ for undoped <123> InP ; T = 1003K (0.75T_M) ; strain rate = 2.1 x 10^{-4}s^{-1}. The yield point at the beginning of deformation is omitted for clarity. (Siethoff and Schröter 1984).

Load drops are observed (Fig. 2) for undoped InP deformed at 0.67 and 0.71 T_M (Brown, Cockayne and MacEwan, 1980). This figure shows that the upper yied point and the load drop depend on the strain rate. The load drop increases in magnitude for increased strain rate and can disappear for low strain rate (10^{-5}s^{-1}).

Fig. 2 shows also that less pronounced load drops are observed for the multiple slip orientation <100> which also enhanced greatly the work hardening rate. On the other hand, the magnitude of the upper yield stress is much greater for the single slip orientation <123>. For this orientation discontinuities are also present in the curve at a strain rate of 10^{-3}s^{-1}.

GaAs single crystals have been deformed at temperatures as high as 0.9 T_M by Djemel et al. (1986), Tabache et al. (1986), Guruswamy et al. (1987) (for a review see Djemel et al., 1989) Some results are displayed in Fig. 3 where the σ (ϵ) curves for σ // <123> show yield drops and the curves for σ // <001> are parabolic, with large flow stresses. This behaviour has been observed by various authors for GaAs, InP, as well as for other crystals. High temperature deformations have to be performed in a medium which prevents the decomposition of the semiconductor compounds. Liquid B_2O_3 is of widespread use ; Djemel et al. (1986) found that it softens undoped GaAs with respect to tests performed in in air and argon. An analogous observation was made by Imhoff et al. (1985) for CdTe.

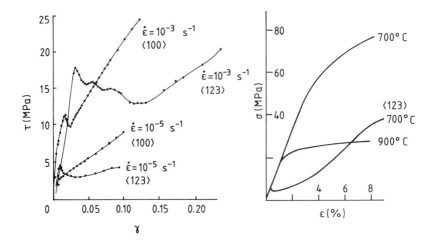

Fig 2. Resolved shear stress as a function of the shear strain for undoped InP, for two values of the strain rate ($10^{-3} s^{-1}$ and $10^{-5} s^{-1}$) and two orientations :
    <123> deformed at 0.67 T_M = 874 K
    <100> deformed at 0.71 T_M = 948 K
(Brown, Cockayne and MacEwan, 1980).

Fig 3. Engineering stress-strain $\sigma(\epsilon)$ curves for GaAs at various temperatures. σ // <123> is indicated close to the curve. The two other curves correspond to σ // <001>. After Guruswamy et al. (1987).

2-2 Influence of doping

Most of the studies on the influence of impurities have been perfomed at low temperature to relate the plastic deformation to electron or hole concentrations.

The influence of doping on the mechanical properties of InP was studied by many authors (Brasen and Bonner, 1983), (Brown, Cockayne and MacEwan, 1983), (Müller, Rupp, Völkl, Wolf and Blum, 1985), (Völkl et al., 1987), (Gall and Peyrade, 1987). The experimental results from Müller et al. (1985) with Zn doping are given in Fig 4. These curves show that increasing dopant concentration increases the levels of the upper and the lower yield points.

There are a great number of studies on the influence of isoelectronic doping in Ga As, such as In doped Ga As, which were induced by its use to grow dislocation-free single crystals (Djemel at al 1989). Indium has been shown to harden single crystals, even at temperatures as high as 1100°C (0.91 Tm) ; the yield stresses are increased by indium, but the flow stresses at large strains are largely increased (Fig 5).

Indium diffusion to dislocations has been revealed by static and dynamic strain ageing. Djemel et al (1988 a) reported serrated flow in the 800-1000°C range and yield drops when straining is resumed after an interruption of the deformation tests below 800°C. The latter phenomena was found consistent with elastic impurity-dislocation interactions (Djemel et al., 1988a).

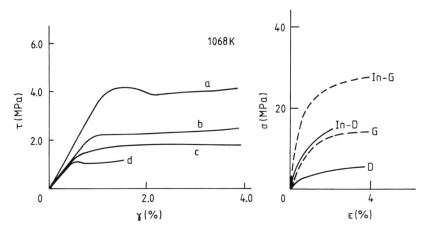

Fig 4. Resolved shear stress τ as a function of the shear strain γ for InP at a constant strain rate ($\dot{\epsilon}$ = $10^{-4}s^{-1}$) for T = 0.8 T_M, and different S doping concentrations :
a) 1.0 x $10^{19}cm^{-3}$
b) 5.0 x $10^{18}cm^{-3}$
c) 1.3 x $10^{18}cm^{-3}$
d) undoped
(Müller et al., 1985).

Fig 5. Engineering stress-strain σ (ϵ) curves for undoped and In-doped GaAs (curves labeled In). $\sigma//<001>$; T = 1100°C. Curves D and G, respectively from Djemel et al. (1986) and Guruswamy et al. (1987) who used strain rates of 2 x $10^{-5}s^{-1}$ and $10^{-4}s^{-1}$ respectively.

2-3 Critical resolved shear stress

The value of the lower yield point τ_{LY} is often used to define the critical resolved shear stress (CRSS). The curves τ_{LY} (T) for InP, published by Müller et al. (1985) and Gall et al. (1987) are given in Fig 6 with the Si curve for comparison (Omri et al., 1987).

These curves show that the CRSS is thermally activated in undoped InP and in S or Zn doped InP. The athermal regime does not appear clearly on the curves. However, the change around 700 K = 0.56 T_M of the slope of the a' curve (undoped InP) is regarded as a transition to a low temperature zone where the stress varies rapidly with the temperature, the athermal regime being reached for T > 1023 K = 0.76 T_M. The transition zone is wider in S doped InP than in undoped InP. On the other hand, the "low temperature zone" of Zn doped InP is translated to higher temperatures, a least until 1068 K = 0.80 T_M.

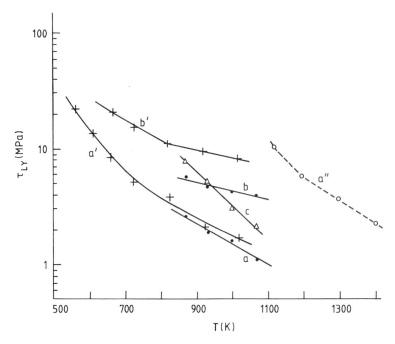

Fig 6. Magnitude of the lower yield point (CRSS) of InP, as a function
of the temperature, at a constant strain rate $\dot{\epsilon} = 10^{-4} \mathrm{s}^{-1}$

a) undoped InP   ⎫   (Müller, Rupp, Völkl,
b) S doped InP $(10^{19} \mathrm{cm}^{-3})$ ⎬   Wolf and Blum 1985)
c) Zn doped InP $(2.2 \times 10^{18} \mathrm{cm}^{-3})$ ⎭

a') undoped InP   ⎫   (Gall and Peyrade, 1987)
b') S doped InP $(\sim 10^{19} \mathrm{cm}^{-3})$ ⎬
                                               ⎭

a") Si $(n \sim 10^{15} \mathrm{cm}^{-3})$ (Omri et al, 1987)

Fig.6 shows clearly that Zn or S doping hardens InP. The effect of
lattice hardening by Zn is strongly temperature dependent. Moreover,
this dependence is increased by Zn concentration (Müller et al.,
1985). The temperature dependence of the CRSS is much weaker for S
doping and does not markedly change with S concentration, but the
results are rather scattered (Fig.6). It can be seen that the lattice
hardening effect of Zn dopant becomes negligible for high temperatures
(near T_M) whereas the S dopant has less influence than Zn at lower
temperatures (870K = 0.65 T_m) and more at T ~ T_M.

The CRSS for Ga As are summarized in Fig.7, where we have plotted
average data values for various stress orientations and for strain
rates in the same range. Values for <123> compression are lower than
for <100>. Indium doping is found to increase the CRSS by factors up
to five, except at T ≤ 600 K where it was found to soften GaAs
(Fig.7). The addition of indium raises the CRSS of GaAs above the ones
reported for other III-V semiconductors (Fig.7 and Table 1).

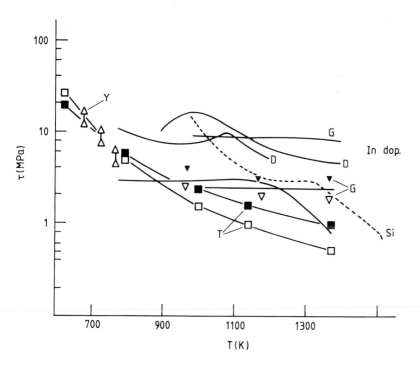

Fig 7. Resolved flow stress τ for GaAs at high temperatures. The lines correspond to tests with σ //<100>. The points are deduced from test with σ //<123>; open symbols for undoped crystals, full symbols for In doped crystals. G = Guruswamy et al. (1987) ; D = Djemel et. al. (1986, 1988a, 1989) ; T = Tabache et al. (1986) ; Y = Yonenaga et al. (1987). The dotted line is for silicon (Omri et al. (1987)).

2-4 Microstructural observations

In order to determine the mechanisms involved in the deformation of materials, it is necessary to perform microstructural observations. For GaAs, cathodoluminescence in a scanning electron microscope permitted to reveal the formation of subgrains of sizes in the 10 μm range (Djemel et al., 1988 b). Transmission electron microscopy has confirmed the existence of the deformation induced polygonization for undoped and In doped Ga As ; its formation seems more difficult for In doped Ga As than for undoped one (Jimenez-Meléndo et al. (1988)). Weak beam dark field was used to examine the dissociation of dislocations ; stacking fault energies x were deduced for In doped Ga As (x = 27 mJ/m^2) which were smaller than for undoped Ga As (x = 45 mJ/m^2) (Jimenez, Melendo et al (1988). This decrease in x was observed for large enough concentrations of indium, viz. 4 x 10^{22} cm^{-3}.

3 . DISCUSSION
3.1. Strain mechanism

Compression tests have been the most commonly used alhough it suffers from many limitations described, for instance, by Bretheau et al. (1979). This may explain part of the scatter of the mechanical data. The use of <123> or <100> compression axis should not influence the CRSS, which corresponds to the initial slip ; however, the difficulty to detect the first event of deformation leads to use the lower yield stress when there are drops in the σ (ϵ) curves and a larger proof stress $\sigma_{0.2}$ for ϵ = 0.2 %, when the σ (ϵ) curves are parabolic (Fig. 2 and 3). The first situation corresponds to single <110> {111} slip and the second one to multiple slip, which gives rise to many dislocation intersections, then to work-hardening contributing to $\sigma_{0.2}$.

Besides the motion of dislocations, deformation can be achieved by twinning. Twins have often been observed in as-grown compound semiconductors and after low temperature deformation under hydrostatic pressure (Rabier, this conference). Brown et al. (1980) reported deformation of InP by twinning in the temperature range 0.56 T_M < T < 0.75 T_M .

3.2 Rate controlling mechanisms

At low temperatures, the obstacles to dislocation glide are the Peierls potential, i.e. the strength of the chemical bond, which is very sensitive to the electronic characteristics (Fermi Level). Plastic properties are then directly related to the electronic doping (see Rabier, this conference)

For high temperatures, such a mechanism is not any more involved, although electronic doping effects are still possible in III-V semiconductors. In most cases, impurities are believed to be associated with point defects, forming complexes able to have strong elastic interactions with dislocations. This, together with internal stresses is believed to be the kind of mechanisms controlling the deformation at small strains, i.e. explaining the CRSS values.

At large strains dislocation multiplication and annihilation play a dominant role. Similar to silicon and germanium, InSb and InP show in their response to dynamic compression two recovery stages, characterized by the parameters τ_{III} (beginning of stage III), τ_{III_m} (steady state condition for stage III), τ_V (beginning of stage V), τ_s (saturation stress), (Fig. 1). Stage V appears at high temperatures (> 0.75-0.8 T_M) so that, as in Fig. 1, it is sometimes difficult to characterize.

Dynamical recovery was interpreted using the theoretical models of steady-state creep (Siethoff, 1983a, 1983b, Siethoff and Schröter, 1984). It was shown that the power law creep model can be fitted to the experimental data of τ_{III} (γ_{III}, T) or τ_{III_m} (γ_{III_m},T). This power law creep model relates the steady-state creep of a material to the diffusion controlled climb of edge dislocations. The cell structure created during the work hardening stage II, disappears partially in stage IV, leaving mainly screw dislocations arranged in tangles (Brion and Haasen, 1985). Let us note here that the power law creep models are purely empirical in their origin and strictly spoken, not related to the climb (or any other) rate-controlling mechanism.

The dynamical recovery in stage V has to be interpreted with a different steady state creep model. Assuming the rate to be controlled by the cross slip of screw dislocations, the fit of Escaig's model to the experimental data τ_V ($\dot{\gamma}_V$,T) is quite good.

It is worthwhile to mention that with decreasing temperature and increasing strain rate, the stress level τ_V approaches τ_{III}. In this case the power law characteristic of diffusion controlled recovery breaks down and a cross slip mechanism can account for the experimental data.

These descriptions of deformation at large strains agree with the general ideas on plasticity, however it would be worthwhile to check that no plastic instability, appearing at $\gamma \sim 0.5$-0.8, affects the strength of the specimens.

3.3. Influence of doping

Isovalent as well as heterovalent dopants have been shown to dramatically decrease the dislocation densities in as-grown crystals. This has been ascribed to hardening effects of impurities which have complex interactions with dislocations (Djemel et al., 1989).

Basically, elastic interactions seem to dominate the influence of both heterovalent and isovalent impurities on high temperature deformations, giving rise to strain ageing in the case of In in GaAs. Impurity-point defect complexes can be centers of strong distortions which interact with dislocations. This explains an increase in the flow stress from the beginning of the deformation. In the case of In doped GaAs, the work-hardening rate is larger than that of undoped crystals, giving larger steady state flow stresses (Fig. 5). This suggest a more difficult recovery process in doped GaAs than in undoped ones. With the classical assumption that cross-slip is the rate controlling mechanism, this has been ascribed to the observed increases in dissociation width of dislocations. Such a process delays the appearance of polygonization during deformation (Jimenez et al., 1988), as well as during crystal growth (Djemel et al., 1989).

4. CONCLUSIONS

Electronic doping can influence III-V semiconductor properties even close to the melting point. However, the high temperature deformation process is dominated by mechanisms based on elastic interactions already described for other materials, such as metals and compounds. The details of the influence of impurities on dislocation mobilities seem to be useful in order to understand the phenomena involved in the multiplication of defects during the growth of single crystals, a problem of obvious technological interest.

REFERENCES

Bretheau T, Castaing J, Rabier J and Veyssière P 1979 Adv. Phys. **28** 829
Brasen D and Bonner W A 1983 Mat. Sc.Eng. **61** 167
Brion H G, Haasen P 1985 Phil. Mag. A **51** 879
Brown G T, Cockayne B and MacEwan W R 1980 J. Mat. Science **15** 1469
Brown G T, Cockayne B, MacEwan W R and Ashen D J 1983 J. Mat. Sc. Letters **2** 667
Djemel A and Castaing J 1986 Europhys. Lett. **2** 611
Djemel A, Castaing J and Duseaux M 1988a Phil Mag. A **57** 671
Djemel A, Castaing J and Chevallier J 1988b Revue Phys. Appl. **23** 1337
Djemel A, Castaing J, Burle-Durbec N and Pichaud B 1989 Revue Phys. Appl., to be published
Gall P and Peyrade J P 1987 J. Cryst. Growth **82** 689
George A and Rabier J 1987 Revue Phys. Appl. **22** 941
Gottschalk H, Patzer G and Alexander H 1978 Phys. Stat. Sol a **45** 207
Guruswamy S, Rai R S, Faber K T and Hirth J P 1987 J. Appl. Phys. **62** 4130
Imhoff D, Gelsdorf F, Pellissier B and Castaing J 1985 Phys. Stat. Sol. a **90** 537
Jimenez-Melendo M, Djemel A, Rivière J P, Castaing J, Thomas C and Duseaux M 1988 Revue Phys. Appl. **23** 251
Müller G, Rupp R, Völkl J, Wolf H and Blum W 1985 J. Cryst.Growth **71** 771
Omri M, Tete C, Michel J P and George A 1987 Phil. Mag. A **55** 601
Rabier J and George A 1987 Revue Phys. Appl. **22** 1327
Siethoff H 1983a J. Physique C4 supplément n°9 44 217
Siethoff H 1983b Phil. Mag. A **47** 657
Siethoff H and Schröter W 1984 Z. Metallk. **75** 475
Siethoff H, Ahlborn K, Brion H G and Völkl J 1988 Phil. Mag. A **57** 235
Tabache M G, Bourret E D and Elliot A G 1986 Appl. Phys. Lett **49** 289
Völkl J, Müller G and Blum W 1987 J. Cryst. Growth **83** 383
Yonenaga I, Onose V and Sumino K 1987 J. Mater. Res **2** 252

Inst. Phys. Conf. Ser. No 104: Chapter 4
Paper presented at Int. Symp. on Struct. Prop. Disloc. Semicond., Oxford, 1989

349

Plastic deformation of Si and Ge bicrystals

A George (*) A Jacques (*), X Baillin (x), J Thibault-Desseaux (+) and
JL Putaux (+)

(*) Laboratoire de Physique du Solide, CNRS URA n°155, Ecole des Mines de
Nancy, Parc de Saurupt, 54042 Nancy, France
(x) CEA, IRDI, DMECN, Département de Métallurgie, Centre d'Etudes
Nucléaires, 85 X, 38041 Grenoble, France
(+) CEA, Département de Recherche Fondamentale, Service de Physique,
Centre d'Etudes Nucléaires, 85 X, 38041 Grenoble, France

ABSTRACT : Dislocation-grain boundary (GB) interactions have been
studied in bicrystals with symmetric tilt coincidence boundaries. After
deformation at moderate temperatures, the regions close to the GB are
seen to be in an advanced state of deformation characterized, soon
after the yield point, by multiple slip, Lomer-Cottrell barriers and
cross-slip. Direct GB crossing by dislocations was observed in situ
during straining experiments in the HVEM. The process required large
stresses. Shockley partials were transmitted first. Successive
transmission of several dislocations from one pile-up tip was
conditioned by the elimination of residual GB dislocations. HREM
provided a straightforward determination of the Burgers vector and
core structure (including associated steps) of the deformation induced
extrinsic GB dislocations.

1. INTRODUCTION

The interest of the materials scientist in studying grain boundary (GB) -
dislocation interactions in semiconductors is manyfold.

(i) GB-dislocation interactions are probably one of the keys to solve the
long-standing problem of predicting the mechanical strength of polycrystals from the known behaviour of the single crystalline material. As
for single crystal work, elemental semiconductors can be viewed as model
substances since they can be obtained in the form of large, highly
perfect, bicrystals with well-defined GBs.

(ii) From a crystallographer's point of view, it is now recognized that
GBs have ordered structures and that their reactions with lattice
dislocations must obey precise topological laws, the first being that the
total Burgers vector must be kept constant and a second, that grain
boundary dislocations (GBDs) usually have associated steps in order to
preserve the GB structure on either side of them. Experimental
confirmations are, however, difficult to obtain and here again Si and Ge
are good candidates for HREM observations of selectively introduced
defects.

(iii) if the electrical properties of dislocations are considered, the physicist can study two-dimensional objects with the possibility to change their defect content (dangling bond density) by plastic deformation.

Present authors'concern was restricted to mechanical and crystallographic aspects and the present paper reports on observations that have been made in deformed Si and Ge bicrystals with the three following questions in mind.

(i) How are dislocation arrangements modified near GBs, compared with the grain interior ?
(ii) Can slip be directly transferred from one crystal to the other, i.e. how and in which conditions are dislocations able to cross a given GB ?
(iii) How is the structure of a GB altered by plastic deformation ?
The emphasis is put on the second question. The reader is referred to J. Thibault (this conference) for a more thorough treatment of the third question.

2. BICRYSTALS

The $(1\bar{2}2)$, $\Sigma = 9$ grain boundary was chosen from currently available bicrystals. This symmetrical tilt boundary corresponds to a misorientation of 38,94° around the [011] axis. The boundary plane is indexed $(1\bar{2}2)_I$ in lattice I and $(1\bar{2}2)_{II}$ in lattice II (superscripts are used to designate the crystal to which a vector or plane belongs, whilst subscripts designate the cubic basis in which vector components are expressed). Bicrystals were pulled from seeds cut from natural 2nd order twins for both Ge (50 Ω cm) and Si (1 Ω cm). The deviation from exact coincidence orientation is less than 5×10^{-3} degrees. Ge bicrystals contain typically 10^3 cm cm^{-3} dislocations and Si bicrystals are dislocation free.

The rotation matrix R, the coincidence lattice (CL) and the DSC lattice are useful to deal with dislocation-GB reactions. Their basis vectors, expressed in the lattice I cubic basis are respectively

$$
\begin{array}{ccc}
\text{R} & \text{CL} & \text{DSCL} \\[4pt]
\dfrac{1}{9}\begin{bmatrix} 7 & -4 & 4 \\ 4 & 8 & 1 \\ -4 & 1 & 8 \end{bmatrix} &
\dfrac{1}{2}\begin{bmatrix} 4 & 2 & 3 \\ 1 & -4 & -1 \\ -1 & 4 & 2 \end{bmatrix} &
\dfrac{1}{18}\begin{bmatrix} 4 & 2 & 3 \\ 1 & -4 & 3 \\ -1 & 4 & 6 \end{bmatrix}
\end{array}
$$

$$\uparrow \quad \uparrow \quad \uparrow$$
$$dsc_1 \quad dsc_2 \quad dsc_3$$

The atomic structure of this boundary in elemental semiconductors can be described as a periodic arrangement of L and L' structural units made up of five-and seven- membered atom rings. L and L' units correspond to each other in a glide $(1/4[41\bar{1}]_I)$ mirror $((1\bar{2}2)_I)$ symmetry. The structure (Krivanek et al. 1977, d'Anterroches and Bourret 1984) has no dangling bond and each L or L' unit can be considered as the core of a perfect Lomer dislocation.

The special features of [011] tilt bicrystals should be emphasized as far as plastic deformation is concerned :
(a) One Burgers vector, 1/2[011], is common to each crystal.

(b) Four (111) slip planes (two in each grain) meet along the [011] direction in the boundary plane.

A few experiments were done with Σ = 25 Si bicrystals. This boundary corresponds to a misorientation of 16.26° around the [001] axis. The boundary plane is indexed $(7\bar{1}0)_I/(710)_{II}$. Since only dislocation arrangements near the GB have been studied in these bicrystals, the rotation matrix, Cl and DSCL need not be detailed here.

3. DEFORMATION EXPERIMENTS AND IMAGING CONDITIONS

Two kinds of experiments were performed : (a) compression tests and subsequent observations of dislocations at the GB by TEM and HREM in Si samples. (b) in situ straining experiments in the HVEM with Si and Ge bicrystals.

Compression tests at constant strain rate ($\dot{\varepsilon}$ = 6 x $10^{-6}s^{-1}$) in the temperature range 993 K-1123 K were conducted under reducing atmosphere. Tests were stopped in the yield region and samples cooled down under load in order to freeze the dislocations arrangements for TEM observations. The bicrystals were stressed with stress axes lying in the boundaries, so that symmetric compatible loading was assured, and corresponding to single slip in each grain (Martinez-Hernandez et al. 1987). For HREM observations, a few samples were deformed at higher temperatures ($\approx$ 1500 K) or up to larger strains.

Conventional TEM work was done with Philips EM300 and JEOL 200 CX microscopes, HREM work with a JEOL 200 CX microscope.

In situ straining was done in the 1 MeV electron microscope of the CENG, operated at 400 kV and using the high temperature deformation stage of Pelissier et al. (1980). Thanks to the two tilt axes available, convenient diffracting conditions could be obtained for each crystal simultaneously. Different stress axes were chosen depending on the transmission reaction which was looked for in any case. Pre-straining was critical. The amount of pre-strain was fixed to introduce few dislocations at the GB plane, so that in a given area, they were usually confined in one and the same grain. Samples for in-situ straining were cut from pre-strained ones, with the same stress axis. Thinning of the area close to the GB was done by ion milling. For more details see Baillin et al. (1987, 1989). In situ experiments were carried out at rather low temperature (0.5-0.6 T_m) in order to be able to measure the velocity of dislocations in real time, which provided a direct estimate of the actual shear stress, using the data of Schaumburg (1972) for Ge and George and Champier (1979) for Si. Stresses of 50-100 MPa are typical, not including pile up stresses. The applied stress in HVEM experiments was about one order of magnitude larger than in compression tests.

In situ experiments were interrupted prior to failure and samples, cooled down under load, could be later observed by conventional TEM in order to determine the Burgers vectors and details of the slip geometry.

4. ENHANCED WORK HARDENING NEAR GRAIN BOUNDARIES

A GB is a two-dimensional obstacle for most dislocations. The blocking of slip by the boundary creates in bicrystals a pile up situation that could not form in single crystals and dislocation rearrangements in order to minimize the total stress lead to typical configurations which were never observed in single crystals deformed under similar conditions (Martinez-Hernandez et al.1987, George 1988). Only two examples are given here, to demonstrate that even for a symmetrical bicrystal subjected to symmetrical loading, i.e. "plastically compatible", the inhomogeneity of plastic deformation at the microscopic scale creates quite involved dislocation arrangements near the GB. Figure 1 shows a dislocation network in a $\Sigma = 25$ Si bicrystal deformed between the upper and lower yield points. Secondary dislocations were blocked at the GB before being intersected by primary dislocations. The resulting attractive reaction forms Lomer-Cottrell locks (Martinez-Hernandez et al. 1987, Korner et al. 1987). In such a network, primary dislocations are stretched out between the LC segments, the secondaries, which acted as obstacles, appearing as steps of the ladder structure.

Figure 2 shows another typical configuration : prismatic dislocations formed by repeated cross-slip. Cross-slip was very frequently observed during in situ straining experiments as well as after compression tests. It provided an efficient way of homogeneizing the distribution of crystal dislocations trapped in the GB (Baillin et al. 1987).

Lomer-Cottrell barrier formation and profuse cross-slip appear in single crystals at stages II and III, respectively, of the stress-strain curve. They are observed to occur near GBs at the yielding of bicrystals. It can be said that the neighbourhood of the GB is in an advanced state of deformation. Very similar configurations were observed in $\Sigma = 9$ and $\Sigma = 25$ bicrystals, which suggests that they do not depend much on the type of the GB.

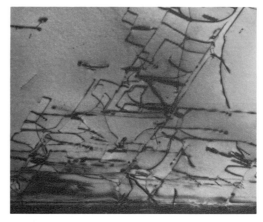

Fig.1. Dislocation network in a $\Sigma = 25$ Si bicrystal deformed at 923 K, $\tau=46$ MPa, nominal strain 6.3×10^{-3} .See text for details

Fig.2. Prismatic dislocations created by repeated cross-slip. Primary and cross-slip planes viewed nearly end-on. $\Sigma = 25$ Si bicrystal deformed at 1023K, $\tau = 11$ MPa, nominal strain 7×10^{-3}

5. DISLOCATION TRANSMISSION BY THE $\Sigma = 9$ GB

Let us consider a b^I Burgers vector dislocation, gliding in a crystal I P^I plane. If no image force (which could stem from non symmetrical GBs in anisotropic crystals) prevents it entering the GB, it can decompose into interfacial dislocations (GBDs), with $b^I = \Sigma\, b_i^{(gb)}$ (1)

$b_i^{(gb)}$ must be vectors of the DSC lattice to leave the GB structure unchanged on each side of them.

The dislocation may also be transmitted into a crystal II P^{II} plane, with a b^{II} Burgers vector. Such a process is favoured in symmetrical bicrystals since there is always a P^{II} plane which has the same intersection as P^I with the GB plane. In most cases, b^I is not a crystal-II lattice vector, $b^{II} \neq b^I$, and the Burgers vector preservation can only be achieved by creating a GBD.

$$b^I = b^{II} + b^{gb} \qquad (2)$$

Since the energy per unit length of a dislocation is proportional to $|b|^2$, and $|b^I|^2 = |b^{II}|^2$, reaction 2 is not energetically favourable and must be helped by the applied stress. Earlier in situ experiments using synchrotron X-ray topography and HVEM (Baillin et al. 1987, Jacques et al. 1987) have confirmed that only the common Burgers vector 1/2[011], parallel to the tilt axis, can be easily transmitted by $\Sigma = 9$ GBs. The two experiments reported below succeeded to induce direct transmission of dislocations with non common Burgers vector because they had been designed to combine several favourable parameters : (i) matching of P^I and P^{II} planes at the GB, (ii) small Burgers vector of the residual GBD, (iii) maximum efficiency of the pile-up stress to help the reaction. When dealing with real situations, the above scheme must be complicated to account for the dissociation of glide dislocations into Shockley partials and the possibility that residual GBDs dissociate into unit DSC GBDs according to eqn.(1). As a matter of fact, El Kajbaji and Thibault-Desseaux (1988) have shown that dislocations are still dissociated at a pile-up tip against the GB plane and in situ straining experiments prove that dislocations are transmitted in the form of single Shockley partials before complete 1/2<110> dislocations can eventually be reformed in P^{II}.

Figure 3 is made of frames selected from the video tape during straining of a Ge bicrystal at ≈ 690 K, $\tau \approx 150$ MPa. The pile up of interest, with primary dislocations $[101]_{II} (\bar{1}11)_{II}$ has formed against the line labelled A in Figure 3a, coming from a source located out of the camera field of view. The arrival of the 5th and 6th dislocations induced a progressive alteration of the contrast ahead of the pile up. A dark line appeared, which slowly expanded in the $(\bar{1}11)_I$ plane, symmetric to the pile up plane with respect to the GB (Figure 3b). The contrast behind this bulging line suggests it is a partial dislocation trailing a stacking fault, an assumption supported by the second configuration to be described below. Immediately after a 7th dislocation had joined the pile up, a second partial developed in grain I and repelled the first one (Figure 3c). The sequence continued with new perfect dislocations added to the pile up and new partials transmitted in grain I. The chronology of the sequence is sketched in Figure 3d. It may be noticed that in most cases the time interval between the arrival of a new dislocation and the next transmission was just the time needed for all pile up dislocations to take their new equilibrium positions, as if any transmission was triggered by the stress level reached at the pile up tip.

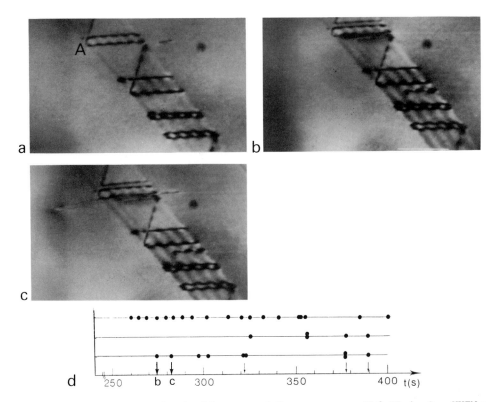

Fig.3. Transmission of Shockley partials across a Σ=9 GB in Ge. HVEM experiment at 690K. Marker 1 μm. Times in seconds from the beginning of the recording. (a) to (c) selected frames from the video tape (d) Chronology of significant events. Upper line, arrivals of dislocations in the pile up. Middle line, cross slip from the pile up plane. Lower line, emission of partials in the opposite grain

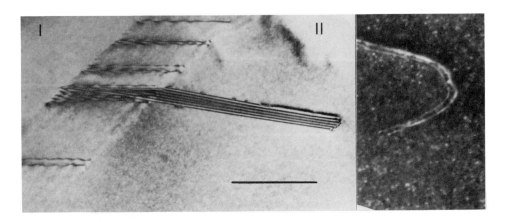

Fig.4. Transmission of a pair of Shockley partials across a Σ=9 GB in Ge. Marker 1 μm

Figure 4 shows the simplest configuration observed after in situ straining at $\approx$ 803 K, $\tau \approx$ 85 MPa had been interrupted immediately after one crossing event had been observed. A pile up of 4 dislocations with b = 1/2[110]$_I$ gliding in $(\bar{1}11)_I$ has formed against the GB. Transmitted dislocations developed in the symmetrical $(\bar{1}11)_I$ slip plane and are still linked to the GB by an extended stacking fault. Weak beam micrographs prove that there are two Shockley partials at the front of the fault, both with the same Burgers vectors 1/6[211]$_{II}$. The fringe contrast of the large fault is that of an extrinsic stacking fault as expected if each partial has trailed an intrinsic fault.

The two geometries of Figures 3 and 4 are equivalent. Let us describe the process in the sense crystal I $\rightarrow$ crystal II. Assuming that only dislocation glide is possible in the investigated range of temperature, the transmission could proceed as follows. The leading partial of the first dislocation b^{lI} of the pile up enters the GB. It can either decompose into a transmitted Shockley partial b^{lII} in grain II, leaving a b^r residue according to eqn(2), or dissociate in the GB, giving one DSC dislocation with b^g = dsc$_1$ = 1/18[41$\bar{1}$]$_I$, glissile in $(1\bar{2}2)_I$, and a residual b'^r. Next the trailing partial of the first dislocation of the pile up, b^{tI}, enters the GB, reacts with the residue and decomposes... and so on. Since the p^{lI} slip plane is known, there are only three possibilities for b^{lII}. Figure 5 shows the forces which are exerted on the Shockley partials and b^g dislocations, respectively. Equal shear-stress curves (τ = 5σ) in slip systems of crystal II were calculated for an arbitrary value of the applied stress σ (40 MPa) and a pile up of 20 dislocations. This allows consideration of only the sign of b for which the dislocation is repelled from the GB. As a first approximation, the energy barrier for the transmission or dissociation was estimated by the Σb^2 criterion in a two-dimensional model, neglecting also the energy of GB steps associated with b^g or b^r and the stacking fault energy.

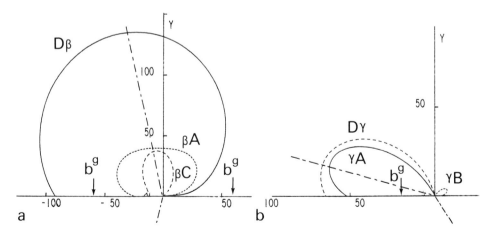

Fig.5. Equal shear stress curves on partial dislocations in crystal II at the tip of a pile up of 20 dislocations in crystal I (σ = 40 MPa). Distances are expressed in units of the lattice parameter. The GB is horizontal. The origin is at the meeting point of the GB and pile-up plane, which are viewed end-on, as the slip plane considered in crystal II (a) Ge samples, (b) Si samples

Results are presented in Table 1 where only the most probable reaction is presented. Two $1/2[101]_I$ dislocations could be converted into one b^g dislocation and three identical Shockley partials with $\bar{b} = 1/6[211]_{II}$, leaving the GB free of any residual GBD at the pile up tip, since the $\Sigma = 9$ coincidence relationship exactly gives :

$$2 \ DC^I = b^g + 3 \ D\beta^{II}$$

in Thompson's notation. Figure 5 confirms that the $D\beta^{II}$ partial is the most stressed at the pile up tip. It must be noticed that this scheme does not lead to the transmission of perfect dislocations, unless some rearrangement occurs later in the triplets of $D\beta^{II}$ dislocations. Baillin et al. (1989) remarked that such a triplet is formally equivalent to two DA + DC dislocations superimposed in neighbouring planes, if the third $D\beta$ partial closes the two-layer (extrinsic) fault created by the emission of the first two partials trailing each a one-layer (intrinsic) fault. The core structure of the third $D\beta$ can be viewed as the superposition of βA and βC cores on two adjacent planes (Amelinckx 1979). It is not yet clear whether the arrangement of slip planes is the right one, and whether the separation of the two potential dislocations is possible.

step n°	next dislocation of the pile-up	dislocation + at the GB plane	new dislocation ⇒ at the GB plane	emitted + dislocation	Energy balance (xb_p^2)
1		$1/6[211]_I$	⇒ $r_1 - 1/9[112]_I$	$+ \ b_g - 1/18[41\bar{1}]_I$	- 0.22
2	$1/6[12\bar{1}]_I$	$+ \ r_1$	⇒ $r_2 - 1/18[581]_I$		0.21
3		r_2	⇒ $r_3 - 1/54[172]_I$	$+ \ 1/6[211]_{II}$	- 0.54
4	$1/6[211]_I$	$+ \ r_3$	⇒ $r_4 - 1/54[19,16,11]_I$		0.41
5		r_4	⇒ $r_5 - 1/54[5,\bar{1},10]$	$+ \ 1/6[211]_{II}$	- 0.26
6	$1/6[12\bar{1}]_I$	$+ \ r_5$	⇒	$1/6[211]_{II}$	- 0.26

Table 1. Successive dislocation reactions at the GB plane in Ge bicrystals

Figure 6 presents another example of dislocation transmission observed in Si at 840 K, $\tau \approx$ 100 MPa. A full description of this rather intricate configuration is given elsewhere by Baillin et al. (1989). The salient features are as follows. Pile ups of $1/2[101]_I$ dislocations gliding in $(11\bar{1})_I$ b planes have formed at the GB. Dislocations were emitted in $(11\bar{1})_{II}$ planes of grain II. Burgers vector analysis by conventional TEM after interruption of in situ straining prove that they were perfect dislocations with $\bar{b} = 1/2[101]_{II}$. However, near the pile up tip, extended stacking faults trailed by Shockley partials could also be identified. An additional interesting feature was the observation of dislocations with $\bar{b} = 1/2[0\bar{1}1]_I$ gliding in $(111)_I$ planes in crystal I. These loops expanded from the pile up tip and immediately followed dislocation transmission events. Figure 5 shows the respective magnitudes of the forces on the possible Shockley partials.

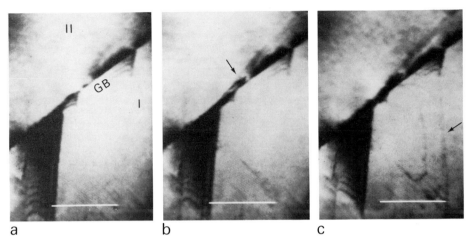

a   b   c

Fig.6. Transmission of dislocations across a $\Sigma = 9$ GB in Si. HVEM experiment at 840 K. Selected frames from the video tape. Marker 1 μm (a) initial state (b, c) after several transmission events

Table 2 gives the simplest explanation, which however is not unique. (A b^9 dislocation could have been emitted by the first leading partial that has entered the GB and the emission in grain II of the r^3 residue, which is a Shockley partial, was disregarded since it would have trailed an extrinsic stacking fault). Furthermore the emission of $CA^I = 1/2[0\bar{1}1]_I$ dislocations implies that they glide first in the $(100)_I$ cube plane before being transferred on to (111), planes, where they were observed. At the end of step n°5, one perfect dislocation has been transmitted with $\bar{b} = D\bar{A}^{II} = \gamma A^{II} + D\gamma^{II}$, as observed, a b^9 GBD has been created and a perfect CA^I dislocation has been re-emitted in a (111) plane of crystal I. This has removed one perfect dislocation plus the leading partial of the second one from the pile up, which means that a cycle is not achieved. Next possible steps and alternatives were examined by Baillin et al. (1989) with the result that, for this geometry, all the transmission processes appear to be difficult since some reactions present high energy barriers. This might explain why in these experiments, the transmission always stopped rapidly, after a few dislocations in the best cases.

step n°	next dislocation of the pile-up	+ dislocation at the GB plane	⇒ new dislocation at the GB plane	+ emitted dislocation	Energy balance (xb_p^2)
1		$1/6[2\bar{1}1]_I$	⇒ $r_1 - 1/27[\bar{2}55]_I$	+ $1/6[2\bar{1}1]_{II}$	0.44
2	$1/6[112]_I$	+ r_1	⇒ $r_2 - 1/54[5,1,\bar{2}8]_I$		0.22
3		r_2	⇒ $r_3 - 1/18[\bar{2}55]_I$	+ $1/6[112]_{II}$	0.33
4	$1/6[2\bar{1}1]_I$	+ r_3	⇒ $r_4 - 1/9[2\bar{4}4]_I$		0.66
5		r_4	⇒ $1/18[41\bar{1}]_I$	+ $1/2[0\bar{1}1]_I$	0.66
6		+ $1/6[112]_I$	⇒		

Table 2. Successive dislocation reactions at the GB plane in Si bicrystals

6. HREM OBSERVATION OF DISLOCATIONS AT GRAIN BOUNDARIES

HVEM and conventional TEM observations provide a precise determination of incoming and transmitted dislocations. Samples could also be designed in order to make b^9 dislocations visible during in situ straining. Anyway those techniques do not allow one to visualize residual GBDs, the situation at a pile-up tip being so intricate that it cannot be resolved even by weak-beam TEM. HREM is probably the only method to directly determine the residues, in spite of obvious limitations : the orientation of dislocations at the GB plane must be parallel to an electron beam axis allowing atomic columns to be resolved. Referring to the above observations, HREM allows study of the Si case, but not the Ge case. To be observed, configurations must be stable enough to have a reasonable probability of being frozen-in. Only high-energy-barrier steps, requiring thermal activation to be overcome and thus long waiting times, fulfil this requirement. Any intermediate configuration that would decompose spontaneously cannot be detected. To be resolved, the configurations must be 2-dimensional over the foil thickness ($\approx$ 10 nm), whereas it is highly probable that saddle-point configurations are 3-dimensional in essence. Last, it is practically impossible to perform HREM observations on the foils used for HVEM experiments.

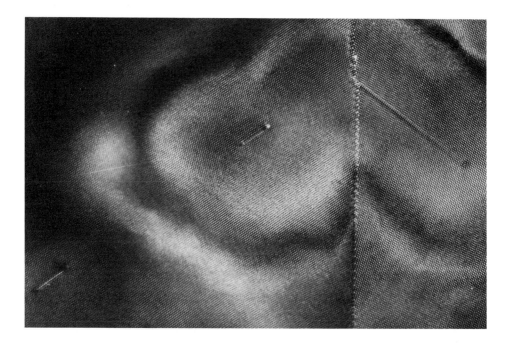

Fig.7. Emission of a 30° partial at the tip of a pile-up. The leading pile-up dislocation has totally entered the GB, the next dislocation has not yet reached it. The transmission in the slip plane symmetric to the pile-up plane is not likely to develop further since it is against the applied stress and is only locally favoured by the pile-up stress. (Micrograph El Kajbaji)

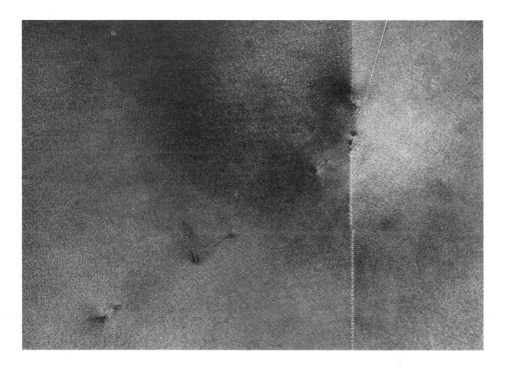

Fig.8. Partial dislocation transmission in $(11\bar{1})_{II}$ ahead of a pile up in $(11\bar{1})_{I}$. The transmitted partial has gone away; only the stacking fault is seen in grain II. Note the local alterations of the GB structure and the barrier formed in the pile up by reaction with a secondary dislocation.

In spite of these difficulties, HREM has supplied a number of interesting results (El Kajbaji et al. 1987, El Kajbaji and Thibault-Desseaux 1988, Thibault-Desseaux et al. 1989)

(i) The three kinds of GBDs with unit DSC Burgers vectors were observed and their core structure as well as associated atomic steps in the GB plane could be determined. They include : $\pm$ b^9 dislocations, but also $\pm$ b^c = $\pm$ dsc$_2$ = $\pm$ 1/9[1$\bar{2}$2]$_I$ and $\pm$ Dγ^I (or Bγ^I), $\pm$ Bα^{II} (or Dα^{II}), which are 30° partials of the $(11\bar{1})_I$ and $(1\bar{1}1)_{II}$ planes and whose screw component cannot be distinguished by HREM.

(ii) Two other types of GBDs were termed "imperfect" or "partial". The first one is a dislocation located in the GB but still attached to a well defined stacking fault extended in one grain. Its core structure and also the associated step are different from those of an isolated GBD with the same Burgers vector. The second type is a GBD with a broad core, which can be viewed as two non DSC dislocations, with a "GB stacking fault" in between. Such a configuration always remains very compact.

(iii) Several other non-unit DSC GBDs were observed. Their associated step is often a signature, which allows determination of the reaction between lattice and GB dislocations which produced them. The emission of a b^9 dislocation from a trapped γA^I Shockley partial was directly observed.

(iv) Non-unit GBDs can decompose not only by emitting glissile b^g dislocations but also by climb. Such climb decomposition was observed to take place at fairly low temperature ($\approx$ 1000 K in Si).

(v) The dislocations in a pile-up are distributed over parallel slip planes at a few interplanar spacings. This means that successive dislocations do not reach the GB at the same point, which can reduce the repulsion between residues and the next dislocation.

(vi) Although transmission events were very rare in samples deformed at low stresses, lattice dislocations were observed in a compact form in the GB after having just emitted one b^g dislocation. This means that the trailing partial can recombine with the residue left after the leading partial has emitted a b^g dislocation. According to the Σb^2 criterion this recombination is nearly as difficult as most of the postulated reactions listed in Tables 1 and 2.

Exceptionally, transmission events could be identified. Two examples are given, which are still too complex to be detailed here. Figure 7 shows the head of a pile up in $(11\bar{1})_I$ and a 30° partial, possibly αB^{II}, gliding on $(1\bar{1}1)_{II}$ and still connected with the GB by an intrinsic fault. This reaction is favourable according to the Σb^2 criterion but cannot lead to effective dislocation transmission since αB^{II} is driven towards the GB by the applied stress. Yet, stress calculations show that this partial is repelled from a pile-up tip. Figure 8 looks very similar to configurations which are expected in the case of transmission from $(11\bar{1})_I$ to $(11\bar{1})_{II}$ observed by HVEM. Unfortunately the transmitted partial in grain II in front of the stacking fault has gone out of the field of view and the present foil was too thick to obtain truly good images.

7. REFERENCES

Amelinckx S 1979 *Dislocations in Solids* ed by F.R.N. Nabarro (Amsterdam : North-Holland) vol 2 p 67-460
Baillin X, Pelissier J, Bacmann J J, Jacques A and George A 1987 *Phil. Mag. A* **55** 143
Baillin X, Pelissier J, Bacmann J J, Jacques A and George A 1989 *Phil. Mag.* to be published.
D'Anterroches C and Bourret A 1984 *Phil. Mag. A* **49** 783
El Kajbaji M, Thibault-Desseaux J, Martinez-Hernandez M, Jacques A and George A 1987 *Rev. Phys. Appl.* 22 569
El Kajbaji M, Thibault-Desseaux J 1988 *Phil. Mag. A* **58** 325
George A and Champier G 1979 *Phys. Stat. Sol.(a)* 53 529
Jacques A, George A, Baillin X and Bacmann J J 1987 *Phil. Mag. A* **55** 165
Korner A, Martinez-Hernandez M, George A and Kirchner H O K 1987 *Phil. Mag. Letters* **55** 105
Krivanek O L, Isoda S and Kobayashi K 1977 *Phil. Mag. A* **36** 931
Martinez-Hernandez M, Korner A, Kirchner H O K, Michel J P and George A 1987 *Phil. Mag. A* **56** 641
Pelissier J, Lopez J J and Debrenne P 1980 *Electron. Microsc.* 4 30
Schaumburg H 1972 *Phil. Mag.* 25 1429
Thibault-Desseaux J, Putaux J L, Bourret A and Kirchner H O K 1989 *J. de Phys.* to be published

Inst. Phys. Conf. Ser. No 104: Chapter 4
Paper presented at Int. Symp. on Struct. Prop. Disloc. Semicond., Oxford, 1989

361

The effect of oxygen and hydrogen on the brittle–ductile transition of silicon

P. Haasen, M. Brede and T. Zhang
Institut für Metallphysik and SFB 126, Universität Göttingen, FRG

ABSTRACT

The BDT of silicon as a model substance has been investigated by several groups in recent years since the first and singular publication of St. John in 1975. The results do not always agree; neither do recent interpretations. We have used a mode I fracture mechanics experiment while the Oxford group uses bending tests. They obtain a much lower BD temperature T_c but also a much smaller K_{IC} at the BDT. It is argued that a plateau in $K_{IC}(T)$ observed in Göttingen below T_c leads to ductility in the Oxford specimens while the Göttingen specimens still fracture - although now with a plastic zone at the crack tip. To pin down dislocation mobility rather than dislocation nucleation as a mechanism for the BDT n-doped silicon is used. There is also a very interesting correlation between increased dislocation mobility in a hydrogen plasma and decreased T_c. Oxygen i.e. CZ vs FZ material also makes a strong effect on the BDT. This depends on the state of oxide precipitation as large particles evidently act as dislocation sources. The symposium being one on dislocations and not on fracture the paper will center on chemical aspects of dislocation motion which tries to screen the crack tip stresses.

1. INTRODUCTION

Silicon has a pronounced transition (BDT) from brittle to ductile behaviour at about $0.6\ T_m$ (melting temperature). This was established clearly in a classic, single-shot paper by St. John (1975) using fracture mechanics specimens and x-ray topography in the plastic range. His conclusion that dislocation motion to blunt the crack, not the nucleation of these dislocations at the crack tip determined the BDT largely contradicted the calculations by Rice and Thomson (1974). Haasen (1983) further worked out the ideas of St. John and proposed to test these by comparing the BDT of intrinsic silicon with that of strongly n-doped material. It is well known from dislocation velocity measurements (Patel and Chaudhuri 1966) that its activation energy decreases from 2.1eV to 1.7e V with extrinsic n-doping, depending somewhat on the stress range. This should clearly show up in a 100K decrease of the BDT and was implemented in the subsequent experimental study of Brede and Haasen (1988). All of these experiments were done on initially dislocation-free silicon to ensure dislocation nucleation at the crack tip itself. Indeed Michot and George (1982, 1986) have shown in static loading experiments that dislocations are generated at the crack tip near 1000K at applied loads much lower than those which initiated cleavage at room temperature. The generation of dislocations at a crack tip in a thin foil was observed directly by Chiao and Clarke (1989).

These authors (1985) also compared CZ-grown silicon with 3.5 x 10^{17} oxygen/cm^3 with FZ silion (<10^{15} oxygen/cm^3) and found a 0.4ev higher activation energy for the BDT of the former. This is in qualitative agreement with dislocation velocity measurements (Imai and Sumino 1983) as there is a finite starting stress of τ_c=5...15MPa in CZ-silicon below which dislocations do not move at all. Brede and Haasen (1988) found that the activation energy of the BDT in highly n-doped material does not always follow the high stress behaviour of the dislocation velocity but can be explained by the data at small stresses. This is in disagreement with recent computer simulations of pre-indented fracture experiments in bending by the Oxford group (Hirsch *et al.* 1989) which leads to extremely high stresses in the plastic zone at the crack tip near the indentation. Unfortunately the experimental situation widely differs in the mode I fracture mechanics specimens and the pre-indented bent bars as do the temperatures at the BDT. The discussion of these differences is still going on and will not be the subject of this symposium on dislocations in semiconductors. We rather want to elaborate on recent BDT experiments on Cz-silicon in which the oxygen was precipitated into oxide particles of various sizes (Behrensmeier *et al.* 1987) and on silicon which was hydrogenated during fracture testing in a plasma (Zhang and Haasen 1989).

2. SUMMARY OF RESULTS OF MODE I FRACTURE TESTS ON N-DOPED SPECIMENS

From {112} slices of Si crystals tapered double cantilever beam specimens were cut as described by St. John (1975) and Brede and Haasen (1988). They were sliced, precracked at room temperature and the crack kept open during heating by a small load. The stress intensity K_I produced by the tensile load (P) was rather independent of the crack length (a) due to the fracture mechanics guided geometry. The crack front was parallel to <112> in a {111} plane

The materials investigated are listed in table 1

Table 1. The materials: (FZ:float zone; CZ: Czochralski)

Material	Doping [at/cm^3]	Growth	Oxygen [at/cm^3]	Reference
A	$1.6 \cdot 10^{18}$P	FZ	$<10^{15}$	
B	$6.5 \cdot 10^{18}$P	FZ	$<10^{15}$	Brede and
C	$2.9 \cdot 10^{18}$As	CZ	$6.7\ 10^{17}$	Haasen (1988)
D	$1.2 \cdot 10^{18}$P	CZ	$\cong 10^{17}$	
E	$\approx 10^{16}$P	FZ	$<10^{15}$	Michot and
F	$\approx 10^{16}$P	CZ	$3.5\ 10^{17}$	George (1985)
G	$3.0 \cdot 10^{14}$B	?	?	St. John (1975)
I		CZ	$\approx 10^{17}$ 750°C 48h	Behrensmeier
II	10Ωcm p		950°C 48h	et al. (1987)
III			750°C 48h + 950°C 48h	

The specimens were tensile loaded in mode I at a constant rate $\dot{\delta}$ at temperatures between 670 and 1200 K in vacuum. Two types of load-crack opening curves were observed. On the very low temperature side of the BDT, temperature T_c, the samples failed by brittle fracture (Figure 1, curve 1) at an applied stress intensity K_{co}=0.9MPa$\sqrt{m}$. In a small T range around T_c the toughness increased sharply by a factor 5-8 corresponding to the upper yield stress τ_{uy} (Figure 1, curve 2). On the high T side the maximum applied load decreased with increasing T as usual for τ_{uy}. In the P(δ) curves rising to the upper yield a dent was often observed at K_{co} corresponding to a finite crack jump by Δa, as seen on the etched crack surface (see Brede and Haasen 1988). In a 80°C range below T_c the toughness K_{IC} was found to be higher than K_{co}, typically by a factor 1.7 (but up to 2.5). In this range dislocations generated at the crack front were observed - whether there was a crack jump at K_{co} or not.

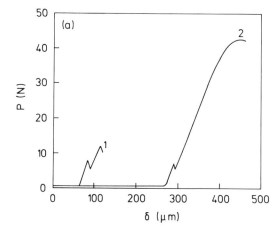

Fig. 1 Different types of load-crack opening curves (1 below T_c, 2 above). First dent corresponds to K_{co}, second to fracture.

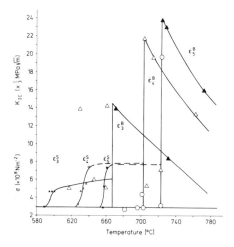

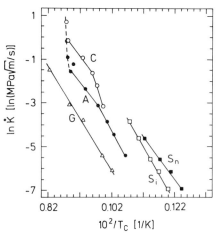

Fig. 2. Measured bdt's by Brede ($\dot{\varepsilon}^B$) and Samuels ($\dot{\varepsilon}^S$) at three opening rates for intrinsic Si and mat. A.

Fig. 3. Arrhenius plots for different materials of Brede, (A,C), St. John (G) and Samuels (S_i, S_n).

Figure 2 shows $K_{IC}(T)$ curves measured by Brede for three different $\dot{\delta}$ in mode I tension and points of Samuels measured on similar material at comparable loading rates. While the uppermost experimental points at $K_{IC} \approx 2K_{co}$ of Samuels correspond to plastic behaviour of the bent specimens, the points of Brede in mode I tension (large triangles) at similar K_{IC} indicate still brittle behaviour, though with some dislocation activity at the crack tip. The line connects both data sets to a plateau at $(2-3)K_{co}$, but is below the BDT of the Brede specimen which leads to much higher K_{IC} at T_c. This difference in BDT could well be due to the different states of stress, specimen geometry and pretreatment etc. Figure 3 shows Arrhenius plots of both data sets (Brede A, Samuels S_i) and data from St. John (perhaps low p-doped FZ material G). The temperature shift is evident as is the curvature in the Brede data on both ends which we will turn to below. In the middle temperature range of Brede, as well as for the Samuels data the slope corresponds to activation energies of 1.6-1.7eV (intrinsic) and 2.1-2.3eV (n-doped material) in good agreement with those measured for dislocation velocities.

At low stresses and low temperatures the representation of dislocation velocity measurements by the eq (see Alexander 1986)

$$v = \left(\frac{\tau}{\tau_0}\right)^m B_0 \exp\left(-\frac{U}{kT}\right) \tag{1}$$

with $m \approx 1$ (τ_0=1MPa) is no longer valid. Here higher values of m, B and U are often observed, dislocations move more slowly than eq (1) predicts down to immobility at the starting stress τ_c which is high particularly in the presence of oxygen impurity. We believe that such a low stress effect is visible in the Arrhenius plot of Michot and George's (1985) BDT data on materials E(FZ) vs. F (Cz), the latter having the higher starting stress, therefore a higher T_c, Figure 4. The same applies to materials E vs G. as phosphorus doping results in a higher τ_c than boron doping. At the low temperature end of the BDT data on materials A and C, figure 3, the dislocation velocity decreases faster, so the BDT moves to higher temperature. The higher activation energy there (curve C in Figure 4) is in good agreement with that of the dislocation velocity at small stresses. This we use in modelling a near equilibrium plastic zone, an almost fully shielded crack to explain the BDT in our mode I tensile specimens.

3. THE INFLUENCE OF PRECIPITATED OXYGEN ON THE BDT

The comparison of BDTs in materials E (FZ) and F (Cz) by Michot and George (1985) (Fig. 4) shows already a shift of T_c to higher temperatures due to dissolved oxygen as well as an increased activation energy. This is also found with dislocation velocities - but only at low stress (Imai and Sumino 1983). The heat treatments of a slightly p-doped Cz material of similar oxygen content as F produced the microstructures of specimens I and of II/III (table 1, Behrensmeier *et al.* 1987). In the former amorphous SiO_2 platelets on {100} planes of Si are expected, in III (and also II) these platelets are associated with extrinsic stacking faults on {111} planes due to precipitated silicon. They show up by etch pits and act as dislocation sources (Figure 5)

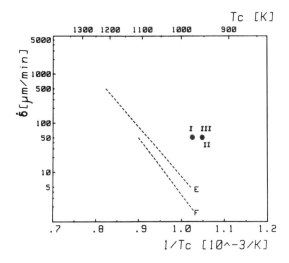

Fig. 4. Arrhenius plot
for silicon of different
oxygen content and heat
treatments of table 1

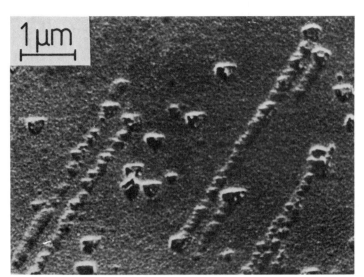

Fig. 5. SEM of mat III, stressed to 1.4 MPam$^{1/2}$
above T$_c$, cooled and fractured; crack advanced from
l to r

Figure 6 shows the stress intensity at fracture (open symbols)
or yield (full symbols) for materials I and III. The first
crack advance is marked by crosses. The loading rate was held
constant at $\dot\delta=50\mu m/min$; otherwise the results are similar to
the ones by Brede, discussed above. Figure 4 allows the
comparison of the effect of oxygen precipitation on BDT at
this $\dot\delta$: It is a pronounced shift of T$_c$ to lower temperatures,

$\Delta T_c(F{\Rightarrow}I)=-140°C$, which is in our opinion an effect of precipitation hardening, leading to a higher τ_c, higher dislocation velocities. Further aging in specimens II/III activates dislocation sources, from the 3.4 x 10^9 oxide platelets/cm^3 in a 20-50 μm wide zone near the crack front as evidenced by etch pitting; this further lowers T_c by only ΔT_c=-20°C. The platelet diameter of about 1μm leads to Frank-Read stresses of 19 MPa which is also the mean stress calculated in the BCS model for the observed mean spacing within the dislocation trains emanating from crack tips stressed to 1MPa$\sqrt{m}$ and cooled under load (Behrensmeier *et al.* 1987). Of course the etch pits do not catch the unrelaxed dislocation configuration around a dynamically loaded crack. The noted stress values and oxygen effects in our opinion support however the assumption of a low stress situation in the plastic zone near a dynamically loaded mode I crack.

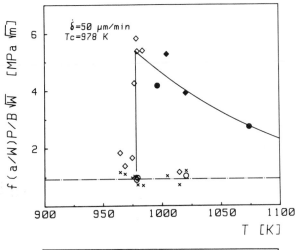

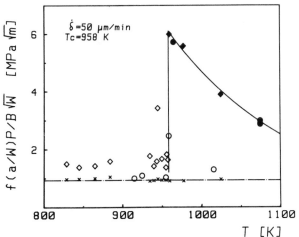

Figs. 6. Stress intensity at fracture (open symbols) or upper yield (closed) vs temperature. Crosses mark crack advance
top: mat I
bottom: mat III

4.   THE EFFECTS OF A HYDROGEN PLASMA ON THE BDT

Hydrogen is known to passivate dangling bonds in semiconductors which are often associated - if only in a transitory state - with dislocations. The passivation has been observed by electrical measurements (Pohoryles 1981). In metals hydrogen embrittlement is often observed and related to two possible causes: A decrease of the surface energy (Zhang 1985) and/or a change in plastic deformation associated with fracture (Tabata and Birnbaum 1984). In silicon these effects on the BDT can be separated and were therefore studied by Zhang and Haasen (1989). For the first time an influence of ionized hydrogen on dislocation mobility in a semiconductor could be clearly identified.

The material used was that called "D" in table 1, i.e. Cz with 10^{18} phosphorus/cm^3. The plasma was produced between a Pt coil and the specimen at 1kV potential in several mbars hydrogen gas pressure at two power levels (3 and 40 W). Since the plasma is also a heating source the temperature of the specimen was accurately measured and controlled to ±1K.

Three aspects of the results of mode I tensile testing under H^+ will only be summarized here (Zhang and Haasen 1989)

a) the influence of hydrogen on the threshold stress intensity (K_E) for dislocation emission from the crack tip.

The specimen was loaded at 1123 K to a stress intensity K, held therefor 30 min, then cooled to room temperature under load and fractured at room temperature. In this experiment K_e is determined as the minimum K applied at high T to increase the toughness at 300K beyond K_{co}. The results are shown in Figure 7 with and without hydrogen charging. First both sorts of specimens have the same K_{co} at small preloadings K. This will be elaborated upon below. Second K_e for charged and uncharged specimens clearly differ, i.e. 0.5 and 0.7 MPa√m, respectively. Therefore hydrogen eases dislocation emission by $\Delta K_e =0.2$MPa√m.

b) The effect of hydrogen on the surface energy of silicon

According to Griffith the low temperature fracture toughness of silicon is related to the surface energy. The effect of H^+ was tested in two ways: (i) specimens were charged at 1173K for 2 hrs without load, rapidly cooled and fractured at 300K. (ii) Specimens were charged at 573K for 6 hrs under a load of 0.8 MPa√m or for 30 hrs under 0.7 MPa√m, cooled and fractured. No change in K_{co} was ever found neither any crack advance observed by optical microscopy. Thus we conclude that the surface energy of silicon is not affected by hydrogen, surprising as this may seem. The hydrogen content of our specimens at 1000K should be at least $6 \cdot 10^{17} cm^{-3}$ (Zhang and Haasen 1989).

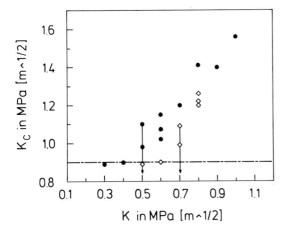

Fig. 7. Fracture toughness K_c at 300K after preloading to K at high T for charged (full symbols) and uncharged silicon (open). (Similar results on FZ Si were obtained by Michot *et al.* (1982))

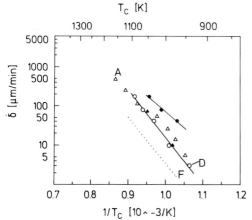

Fig. 8. Arrhenius plots for charged (full symbols) and uncharged Si (open) together with those of mat. A, D, F (table 1)

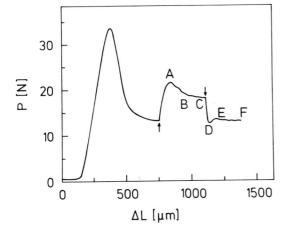

Fig. 9. The influence of hydrogen on the flow stress. Arrows show switching off (upward) and on (downward) a 40W plasma at 994K

c) The effect of hydrogen on the BDT

We do not discuss the features of $P(\delta)$ and K_{IC} here in detail as they are similar to results reported above. We just show an Arrhenius plot for the BDT of charged and uncharged specimens in Figure 8. There is a considerable ductilization effect of H^+ in Si, corresponding to a maximum $\Delta T_c = -62K$ measured for $\delta = 41\mu m/min$. Also the activation energy changes from $2.4\pm0.6eV$ for the uncharged to $1.6\pm0.4eV$ for the charged material. (Also shown in Figure 8 are the Brede results for similarly P-doped specimens A of FZ Si and those of Michot and George (1986) for similarly oxygen containing material F without P-doping). As the surface energy of Si does not change with hydrogen, the ductilization must be related to the dislocation mobility. This was investigated next.

d) Hydrogen effect on the lower yield stress of silicon

Precracked specimens were mode I tensile deformed in the plastic range, for $T > T_c$, beyond τ_{uy} under an 40W plasma. At the first arrow upwards in Figure 9, in the lower yield (τ_{ly}) range, the plasma was turned off and the stress increased to A, then settled to a new plateau BC; At C the plasma was turned on again with the stress dropping to D and settling at EF. The change in flow stress by the plasma was quite reversible. Since the plasma is also a fast heating source the temperature of the specimen decreased by shutting off the plasma and vs.vC before the heating control could correct this (in the plateaus!). The change in flow stress by the measured drop in temperature of 11K between A and B is used to estimate the activation energy of the plastic flow. The load change DE is clearly smaller than AB - due to the presence of hydrogen - although the absolute temperature change is the same. The delayed approach AB to steady state is not due to delayed disappearance of hydrogen as the overshooting at point D shows.

For estimating the activation energy U of the dislocation velocity from $\Delta\tau_{ly}$ above we use Haasen's (1962) theory.

According to

$$\tau_{ly} = C_m \dot{\varepsilon}^{\frac{1}{m+2}} \exp\left(\frac{U}{(m+2)kT}\right) B_0^{-\frac{1}{m+2}} \tag{2}$$

one can calculate U at given strain rate and known m=1.7 for silicon from $\Delta\tau_{ly}(\Delta T)$, the changes AB and DE respectively. The result is U=2.5 eV without hydrogen, 1.7 eV with hydrogen. These are in reasonable agreement with the Arrhenius plot (Fig. 8) of the BDT. Thus the change in ductility with hydrogen is just due to faster dislocations.

The change from the plateau EF to BC at known ΔU also gives the change of the preexponential B_0 of $v(\tau,T)$, eq (1) to $B_0/B_0(H)=3600$.

5. IMPLICATIONS FOR A THEORY OF THE BDT IN DISLOCATION-FREE SILICON UNDER TENSION

The experimental observations described above on the effects of various dopants and impurities on the BDT of mode I tensile specimens point towards a low stress situation for the plastic zone which develops ahead of the crack tip and screens as well as blunts it. This is in contradiction to a model presented by Hirsch *et al.* (1989) for the indented and bent fracture specimens where the authors calculate extremely high stresses (of the order 100 MPa, i.e. large compared to τ_c) for the dislocations which move from the indentation's plastic zone to "sensitive parts" of the crack tip. If the effective stress there never reaches K_{CO} then general plasticity is found when the applied $K \approx 2K_{CO}$. This is however only just the plateau in the K_{IC} vs T curves for the mode I tensile tests (Fig. 2) where the specimens still fracture brittly if though with some visible dislocation activity. It takes still 80K higher temperatures, faster dislocations, to overcome brittleness in our specimens, and then $K_{IC} \approx$ (5-8)K_{CO}, not just 2 K_{CO}! There is no computer simulation for this later situation yet, where certainly no dislocations are available before the crack emits them (whether this occurs at $K_E=0.95K_{CO}$ as assumed by Hirsch *et al.* (1989) to explain a sharp BDT in their bending experiments or at $K_E \approx 0.5 - 0.7K_{CO}$ as shown in Figure 7 is of no importance at this point).

A quantitative theory for the BDT in mode I tensile testing of Si has been formulated by Brede and Haasen (1988) on the basis of the following three equations

(I) If the normal stress σ_y at the crack tip reaches E/10 then fracture occurs. σ_y is related to the local stress intensity k_{tip} by the local tip radius ρ

$$\sigma_y = \frac{2k_{tip}}{\sqrt{\pi\rho}} \tag{3}$$

(II) The tip radius increases by dislocation emission with velocities $v(\tau_c,T_c)$ at spacings Δ at effective stresses of the order τ_c according to

$$\dot{\rho} = \dot{n}b_\perp = \frac{v(\tau_c,T_c)}{\Delta}b_\perp \tag{4}$$

$b_\perp$ is the normal component of the Burgers vector on the crack plane (which is blunting)

(III) According to the quasi-static theory (Thomson 1983) of the plastic zone of length d which is screening a mode III crack k_{tip} is related to the applied stress intensity K_a by

$$k_{tip} = K_a \frac{3}{2\pi} \sqrt{\frac{\Delta}{d}} \left(\ln \frac{4d}{\Delta} + \frac{4}{3} \right) \qquad (5a)$$

$$\text{where } d = \frac{K_a^2}{\tau_c^2} \frac{1}{18\pi} \qquad (5b)$$

It is shown in Brede's and Zhang's papers that the resulting T_c (δ) relation fits their experiments for the experimental stiffness of the specimen, $\tau_c \approx$ 3-6MPa, K_c=1.7 K_{c0} (the "plateau") and $\Delta \approx$ 1μm (which is larger than actually observed).

REFERENCES

Alexander H 1986 *Dislocations in Solids* vol. 7 ed F R N Nabarro (Amsterdam: North Holland) p 113
Behrensmeier R, Brede M and Haasen P 1987 *Scripta Met.* 21 1581
Brede M and Haasen P 1988 *Acta Met.* 36 2003
Chiao Y H and Clarke D R 1989 *Acta Met.* 37 203
Haasen P 1962 *Z. Physik* 167 461
Haasen P 1983 *Atomistics of Fracture* NATO Conf. Series VI (New York: Plenum Press) p 707
Hirsch P B, Roberts S G and Samuels J 1989 *Proc. Roy. Soc.* A421 1, 25
Imai M and Sumino K 1983 *Phil. Mag.* A49 599
Michot G and George A 1982 *Scripta Met.* 16 519
Michot G and George A 1985 *Proc. ICSMA VII* ed H S McQueen *et al.* (Oxford: Pergamon) p 1187
Michot G and George A 1986 *Scripta Met.* 20 1495
Patel J R and Chaudhuri A R 1966 *Phys. Rev.* 143 601
Pohoryles B 1981 *Phys.Stat. Sol. (a)* 67 K75
Rice J and Thomson R 1974 *Phil. Mag.* 29 73
St. John C 1975 *Phil. Mag.* 32 1193
Tabata T and Birnbaum H K 1984 *Scripta Met* 18 231
Thomson R 1983 *Physical Metallurgy* vol. 2 eds R W Cahn and P Haasen (Amsterdam: North Holland) p 1487
Zhang T 1985 Thesis Beijing Univ. Iron Steel Techn.
Zhang T and Haasen P 1989 *Phil. Mag.* in press

Inst. Phys. Conf. Ser. No 104: Chapter 4
Paper presented at Int. Symp. on Struct. Prop. Disloc. Semicond., Oxford, 1989

373

Dislocation dynamics and the brittle–ductile transition in precracked silicon

P.B. Hirsch, S.G. Roberts, J. Samuels and P.D. Warren
Department of Metallurgy and Science of Materials
University of Oxford, Parks Road, Oxford OX1 3PH.

ABSTRACT: A model which simulates the **dynamic** behaviour of shielding dislocations near a crack tip under an increasing stress has been developed. The model reproduces well all the features of the BDT in silicon, provided that dislocation sources are activated only very close to K_{Ic}, and the dislocations travel a distance related to the crack size before the crack can be fully shielded. T_c can then be predicted, assuming that the dislocation sources are formed after the motion of dislocations from near the specimen's surface to the crack tip. Dislocations generated near T_c are highly stressed, and are in a continuously developing "inverse pile-up". The experimental results are described, and the characteristics and predictions of the model discussed.

1. INTRODUCTION

The Brittle-Ductile transition (BDT) in precracked Si tested at constant strain rate is characterised by the facts that (1) it is sharp, (2) at the BDT cleavage occurs after plastic deformation, at a stress higher than that for cleavage at low temperatures, (3) the strain-rate dependence of the transition temperature (T_c) is controlled by the activation energy for dislocation velocity (St.John 1975; Brede and Haasen 1988; Hirsch, Roberts and Samuels 1987, 1988; Samuels and Roberts 1989). The increase in fracture stress at the BDT is associated with plastic relaxation processes induced by the crack tip stress field. The conditions for unstable quasi-brittle fracture can be written:

$$f_c = 2 (\gamma + \gamma') \tag{1}$$

and

$$\frac{df_c}{da} > 2 \frac{d (\gamma + \gamma')}{da} \tag{2}$$

where f_c is the local force on the crack determined by the derivative of the total energy of the system, γ is the surface energy and γ' an effective surface energy term representing other forms of energy dissipation during cleavage, mainly through plastic work. The fracture stress can increase either because of plastic flow occurring during the propagation of the cleavage crack (*e.g.* Burns and Webb 1970), (*i.e.* γ' increases), or because of the reduction in the driving force f_c due to plastic relaxation around the crack during loading (*e.g.* Liu and Bilello 1977), either up to crack initiation, or following some stable crack growth. In silicon, there is no evidence for significant dislocation generation during cleavage at the BDT, and no stable crack growth has been observed (Samuels and Roberts 1989).

The BDT is therefore associated with a reduction in f_c due to plastic flow during loading, before crack initiation. This leads directly to a strain-rate dependence of T_c, resulting from the strain rate dependence of plastic flow, through the dislocation velocity, stress, temperature relationship. Two theories have been proposed to explain the BDT in Si, by Brede and Haasen (1988) and Hirsch, Samuels and Roberts (1987, 1988, 1989) respectively. Very broadly, in both treatments the BDT occurs when the dislocations emitted from the crack during loading shield and blunt the crack so that the local stress or stress intensity factor is not sufficient to cause cleavage. However, in the

Brede and Haasen theory, it is assumed that the plastic zone is saturated at all times, *i.e.* it approximates to the static equilibrium solution of the dislocation distribution for a given value of the applied stress-intensity factor K, while in the Hirsch *et al* theory the BDT is controlled by the dynamic evolution of the dislocation array, far from the equilibrium situation.

In the present paper the existing mode III dislocation dynamics model will be reviewed briefly and extended to a mode I model, experiments designed to test the predictions of the model concerning the effects of dislocation density and prestressing treatments described, and the results of the dislocation dynamics model compared and contrasted with the Brede and Haasen treatment.

2.THE DISLOCATION DYNAMICS MODEL

2.1 Background - Experimental Results

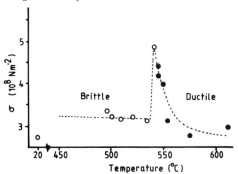

Fig. 1 Typical "BDT curve". Stress to either fracture or bending is plotted as a function of temperature. Note the abrupt transition between fracture at a constant stress in the low-temperature régime to fracture at an elevated stress, and then bending, as temperature is increased.

The model has been designed to simulate the experiments carried out by Samuels and Roberts (1989), on specimens of Si, with semicircular surface cracks introduced by room temperature indentation, and tested in 4-point bending at different strain rates and temperatures. Fig. 1 shows a typical set of results of fracture stress as a function of temperature at a given strain-rate. The transition is very sharp; the width is less than 10°C. At the transition the specimens still fail by cleavage, but at stresses much greater than the low temperature fracture stress. Above T_c the specimens fail by plastic bending.

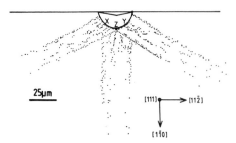

Fig. 2 Dislocation array around the precursor flaw on the fracture face of a "transition" specimen. Dislocations have originated at points X and Y. and move out in well-defined arrays, shielding the deepest point on the crack profile, Z.

At T_c the dislocation arrangement (revealed by etching) takes on a characteristic form shown in fig. 2, for a specimen which has failed at $K = 1.4K_{Ic}$. The pronounced trains of dislocations emanating from the points X,Y where inclined {111} glide planes are tangent to the (111) cleavage plane should be noted in particular. Analysis of the resolved shear stresses on the available slip systems around the crack indicate that stresses are highest at these sites (and on the slip systems with the same Burgers vector, though different slip planes) (Hirsch, Roberts and Samuels 1989).   Further etch pitting experiments show that at T_c dislocation activity at the crack tip does not begin till the applied stress intensity factor $K = 0.9K_{Ic}$, where K_{Ic} is the low temperature critical stress intensity factor. No evidence of dislocation activity is observed at temperatures less than a few degrees below T_c.

Experiments in which the strain rate is varied allow an activation energy for the process controlling the BDT (U_F) to be determined, assuming that

$$\dot{K} = B \exp(-U_F/kT_c) \tag{3}$$

where $\dot{K}$ is the rate of increase of the stress intensity factor and B is a constant. Activation energies thus determined, for silicon of various dopings, by St. John, Samuels and Roberts, and Brede and Haasen were found to be very similar to those for dislocation glide; *i.e.*, writing the dislocation velocity v in the form:

$$v = A\,\tau^m \exp(-U/kT) = \tau^m v_0 \tag{4}$$

where τ is the resolved shear stress on the dislocation, U is the activation energy for dislocation velocity, m is a parameter that varies slowly in the temperature regime considered (George and Champier 1979), A is a constant and v_0 is the temperature-dependent part of the velocity expression, the experiments on the strain-rate dependence of T_c show that

$$\dot{K}/v_0(T_c) = \text{constant} \tag{5}$$

2.2 Dislocation shielding and the conditions for fracture

Following the treatment given by Thomson (1986) the total local force on the crack is given by

$$f_c = \frac{K_{eIII}^2}{2\mu} + \frac{(1-\nu)}{2\mu}(K_{eI}^2 + K_{eII}^2) \tag{6}$$

where K_{eI}, K_{eII}, K_{eIII} are the local (effective) stress intensity factors in modes I, II, III respectively, μ is the shear modulus, and ν = Poisson's ratio. In the experiments described in section 2.1, $K_{eI} = K_I - K_{DI}$, $K_{eII} = -K_{DII}$, $K_{eIII} = -K_{DIII}$, where K_I is the applied mode I stress intensity factor and K_{DI}, K_{DII}, K_{DIII} are the modes I, II, III dislocation shielding terms. Thus

$$f_c = \frac{K_{DIII}^2}{2\mu} + \frac{(1-\nu)}{2\mu}\left((K_I - K_{DI})^2 + K_{DII}^2\right) \tag{7}$$

Lin and Thomson (1986) give the following expressions for K_{DI}, K_{DII}, K_{DIII}, where θ is the angle between the slip and crack planes, and b_e and b_s are the Burgers vector components along the edge and screw directions in the slip plane:

$$K_{DI} = \frac{\mu\, b_e}{2\sqrt{(2\pi r)}\,(1-\nu)}\left(3\sin\theta\,\cos\frac{\theta}{2}\right) \tag{8a}$$

$$K_{DII} = \frac{\mu\, b_e}{2\sqrt{(2\pi r)}\,(1-\nu)}\left(2\cos\theta\,\cos\frac{\theta}{2} - \sin\theta\,\sin\frac{\theta}{2}\right) \tag{8b}$$

$$K_{DIII} = \frac{\mu\, b_s}{\sqrt{(2\pi r)}}\cos\frac{\theta}{2} \tag{8c}$$

For the particular geometry in the Si experiment, $\theta = 70°$ and $K_{DII} = 0$. Also, comparing relevant terms in (7), $(1-\nu)K_{DI}^2/K_{DIII}^2 \approx 8.5$, so that for shielding calculations at the BDT when K_I may be much greater than K_{DI}, say $K_I \gtrsim 2K_{DI}$, the K_{DIII} term may be neglected. Thus (7) can be rewritten:

$$f_c\frac{2\mu}{(1-\nu)} = (K_I - K_{DI})^2 \tag{9}$$

The experiments (fig. 2) suggest that dislocations are generated at discrete points on the crack (*e.g.* X and Y), and that shielding at the point Z, midway between the sources, will only become important when the dislocations from X,Y have moved past Z. In the model (fig. 3) the shielding at

Z (which is much less than at X and Y) is considered critical. The curved crack is replaced by a straight crack; the velocity of the dislocations is assumed to be the same along and normal to the crack. Fig. 4a shows the projections normal to the straight line crack; for a dislocation parallel to the crack edge the edge and screw components are: $b_e = b\sqrt{3}/2$, $b_s = b/2$, where b is the length of the perfect Burgers vector.

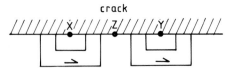

Fig. 3 Simplified model of crack front and dislocation loops, which expand from X & Y and eventually shield Z.

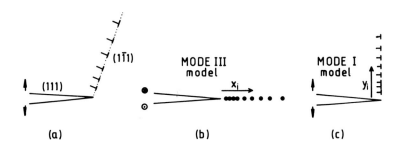

Fig. 4 Approximations to crack - dislocation array arrangements: (a) edge dislocations emitted at 70° to the crack plane; (b) the "Mode III" model; (c) the "Mode I" model.

The necessary condition for fracture is then taken as

$$K_{ez} \equiv (K_I - K_{DI})_z = K_{Ic} \qquad (10)$$

where K_{ez} is the local stress intensity factor at Z and K_{DI} is evaluated at Z. The crack (length a) will be unstable, without any stable crack growth, if:

$$\frac{df_c}{da} > 0 \qquad \text{or, using (9) \& (10),} \qquad \frac{dK_{ez}}{da} > 0 \qquad (11)$$

Now, $$\frac{dK_{ez}}{da} = \left(\frac{\partial K_{ez}}{\partial \sigma}\right)_a \frac{d\sigma}{da} + \left(\frac{\partial K_{ez}}{\partial a}\right)_\sigma = \left(\frac{\partial K_{ez}}{\partial \sigma}\right)_a \frac{d\sigma}{da} + \frac{K_I}{2a} - \left(\frac{\partial K_{DI}}{\partial a}\right)_\sigma \qquad (12)$$

where σ is the stress. Assuming that the shielding is entirely due to edge dislocations on the plane at 70° to the crack plane, and replacing the array by a superdislocation with Burgers vector $n\,b_e$ at R, it can be shown that $(\partial K_{DI} / \partial a)_\sigma = 1/6 \; K_{DI} / R$, and with (9):

$$\frac{dK_{ez}}{da} = \left(\frac{\partial K_{ez}}{\partial \sigma}\right)_a \frac{d\sigma}{da} + \frac{K_I}{2}\left(\frac{1}{a} - \frac{1}{3R} \right) + \frac{K_{Ic}}{6R} \qquad (13)$$

Now, $a = 13\mu m$; fig. 3 shows that at Z there will be a dislocation-free zone $\approx XZ$; therefore $R \geq XZ \approx 7.5\mu m$. With these values for a and R, since $(\partial K_{ez} / \partial \sigma)_a$ is either positive or zero, the instability condition is satisfied. It should be noted, however, that in principle for other dislocation arrangements, stable crack growth is possible if $(\partial K_{ez}/\partial a)_\sigma < 0$. The cleavage crack then advances according to:

$$\frac{da}{d\sigma} = \frac{u}{\dot{\sigma}} = -\left(\frac{\partial K_{ez}}{\partial \sigma}\right)_a \bigg/ \left(\frac{\partial K_{ez}}{\partial a}\right)_\sigma \tag{14}$$

where u = crack velocity, and K_I increases in stable crack growth, as expected from a "resistance curve". When $(\partial K_{ez}/\partial a)_\sigma$ becomes positive after some stable crack growth, cleavage could ensue. Such an effect occurs for example in the experiments of Hull, Beardmore and Valintine (1965) in precracked single crystals of tungsten at sufficiently high temperatures. The amount of stable crack growth increases with increasing temperature.

2.3 Mode III model

In order to calculate the shielding due to an array of dislocations generated at and moving away from the crack tip, the dislocation interactions must be taken into account. For a slip plane inclined to the crack plane these interactions for edge dislocations are very complex (Thomson 1986, eqn. 16.16). To simplify the model a mode III model was initially chosen, in which the loops have Burgers vectors parallel to the crack front, and are elongated in the Burgers vector direction. Under these conditions the dominant dislocation-dislocation interaction is assumed to be that between straight parallel screw dislocations, and the curvature at the ends of the loops is taken into account by the line tension term. The model then becomes one-dimensional (see fig. 4b), and the shear stress at any dislocation x_i is given by:

$$\tau_{x_i} = \frac{K_{III}}{\sqrt{(2\pi x_i)}} - \frac{\alpha\mu b}{x_i} + \frac{\mu b}{2\pi} \sum_{j\neq i} \left(\frac{x_j}{x_i}\right)^{\frac{1}{2}} \frac{1}{(x_i - x_j)} \tag{15}$$

where x_j is the position of the jth dislocation, b the Burgers vector, μ the shear modulus, ν Poisson's ratio, and α the line tension/image stress parameter which depends on the shape of the loop (taken as 1/4 in most of the calculations). The first term in eqn. (1) is the crack tip field stress, the second a combined configurational/image stress term, the third the dislocation-dislocation interaction term.

Substituting from (4), for a constant strain rate test (*i.e.* $\dot{K}_{III}$ constant), (15) becomes:

$$\left(\frac{\dot{K}_{III}}{v_0}\right)^{\frac{1}{m}} \left(\frac{dx_i}{dK_{III}}\right)^{\frac{1}{m}} = \frac{K_{III}}{\sqrt{(2\pi x_i)}} - \frac{\alpha\mu b}{x_i} + \frac{\mu b}{2\pi} \sum_{j\neq i} \left(\frac{x_j}{x_i}\right)^{\frac{1}{2}} \frac{1}{(x_i - x_j)} \tag{16}$$

It is then assumed that dislocation loops are nucleated at the tip when the net stress at a critical distance $x=x_c$ is sufficient to expand the loop, *i.e.* the right-hand side of eqn. (16) is equal to zero at $x_1=x_c$. The first dislocation is emitted at a critical value of $K_{III} \equiv K_N$, where

$$K_N = \alpha\mu b \sqrt{(2\pi/x_c)} \tag{17}$$

For a given $\dot{K}_{III}$ the evolution of the dislocation array can then be calculated from eqn. (4), as a function of K_{III}. Since the dislocation positions are known, the dislocation shielding term K_{DIII} can be calculated. In the experiments considered, dislocation loops are emitted only from particular points on the crack front. The local stress intensity K_{ez} at a point Z remote from the source is assumed to be:

$$K_{ez} = K - K_{DIII} = K - \sum_{j>j_0} \frac{\mu b}{\sqrt{(2\pi x_j)}} \tag{18}$$

summing only over all dislocation loops which have moved past this point Z. The value of $K_{III} \equiv K_F$ at which brittle fracture occurs is given by eqn. (10), modified for mode III. The model predicts that abrupt transitions of the type observed in Si occur only if crack tip sources are nucleated at an applied stress intensity factor $K_{III} \equiv K_0$ just below K_{Ic}, and if these sources then continue to operate at $K_{III} \equiv K_N \ll K_0$. Fig. 5 shows K_F and d_F (the extent of the dislocation array) as a function of temperature, for various values of K_0, for $K_N = 0.21\ K_{Ic}$, demonstrating that K_0 must be $= K_{Ic}$. Other computations show that, for the transition to be sharp, $K_N \ll K_0$ (Hirsch *et al* 1989).

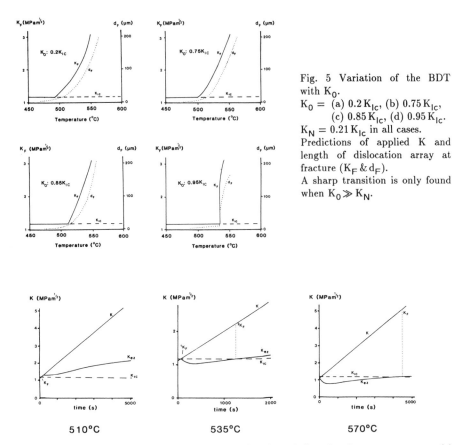

Fig. 5 Variation of the BDT with K_0.
$K_0 =$ (a) $0.2\,K_{Ic}$, (b) $0.75\,K_{Ic}$,
     (c) $0.85\,K_{Ic}$, (d) $0.95\,K_{Ic}$.
$K_N = 0.21\,K_{Ic}$ in all cases.
Predictions of applied K and length of dislocation array at fracture (K_F & d_F).
A sharp transition is only found when $K_0 \gg K_N$.

Fig. 6. Characteristics of a sharp BDT. K_{ez} as a function of time for three temperatures: (a) 510°C - brittle - K reaches K_{Ic} before dislocations cross d_{crit}; (b) 535°C - transition - K_{ez} diverges from K exactly at K_{Ic}; (c) 570°C - ductile - K_{ez} reaches K_{Ic} at at high value of applied K. (intrinsic: $K_0=0.95K_{Ic}$, $K_N=0.21K_{Ic}$)

The basic reason for the step in K_F at the BDT is illustrated in fig. 6 which shows K_{ez} as a function of time at three temperatures around T_c. At 510°C K_{ez} reaches K_{Ic} before dislocations pass Z and the behaviour is brittle. At $T_c = 535$°C, K_{ez} drops initially below K_{III} where $K_{III} = K_{Ic}$, and later rises to reach K_{Ic} when $K_{III} \leq 2K_F$. At 570°C K_F is very high and the associated stress level is above that for general yielding. The key result of these computations is that in order to explain the sharp transition crack tip sources have to be nucleated at K_{III} just below K_{Ic}. This is consistent with the etch pit observations which show that significant dislocation activity begins at $K_I \approx 0.9\,K_{Ic}$ at T_c. The model accounts satisfactorily for the observed transition temperatures.

The observed strain-rate dependence (3) follows directly from (16), since all dislocation positions and therefore the shielding are the same for $(\dot{K}_{III}/v_0) =$ constant. A further important result of the calculations is that the stress in the dislocation array is very high, ≈ 50-100MPa. The array at the BDT is therefore very far from equilibrium.

2.4 Mode I model

A mode I model for edge dislocations gliding on a plane passing through the crack front at 90° to the crack plane (fig. 4c) has been developed. In this case the dislocations and their images are coplanar, and the problem becomes again one-dimensional. Using the expression for the crack tip stress field, the forces between edge dislocations and the image forces given by Lin and Thomson (1986) and in a different form by Lakshmanan and Li (1989), we find that the shear stress on an edge dislocation at y_i in the slip plane (at right angles to the crack plane) in the Burgers vector (b) direction is given by:

$$\tau_{y_i} = \frac{K_I}{4\sqrt{(\pi y_i)}} - \frac{\alpha\mu b}{4\pi(1-\nu)y_i}$$

$$+ \frac{\mu b}{4\pi(1-\nu)} \sum_{j\neq i} \left\{ \left(\frac{y_j}{y_i}\right)^{\frac{1}{2}} \frac{1}{(y_i - y_j)} + \frac{8 y_i y_j^2}{(y_i+y_j)^3 (y_i-y_j)} \right\} \qquad (19)$$

where y_j is the position of the jth dislocation.

Using eqn. (4), as before, τ_{y_i} is replaced by $(\dot{K}_I/v_0)^{1/m} (dy_i/dK_I)^{1/m}$. Comparison of eqn. (19) with eqn. (16) shows that apart from numerical factors, which tend to make the dislocation interactions stronger relative to the crack tip stress field, there is also one additional interaction term. The value of K_I at which the first dislocation is emitted becomes:

$$K_N = \alpha\mu b \left\{ (1-\nu) \sqrt{(\pi y_c)} \right\}^{-1} \qquad (20)$$

and the local mode I stress intensity factor (from eqn. 38.1 in Thomson (1986)) becomes:

$$K_{ez} = K_I - K_{DI} = K_I - \sum_{j>j_0} \frac{3\mu b}{4(1-\nu)\sqrt{(\pi y_i)}} \qquad (21)$$

Compared with eqn. (18), for the same dislocation position the shielding (for mode I) from edge dislocations in the orthogonal plane is about 40% more efficient than that from screws (for mode III) in the crack plane.

Figs. 7a,b show K_F, d_F, N_F (the number of dislocations in the array) for $K_0 = 0.95 K_{Ic}$, $K_N = 0.2 K_{Ic}$ for modes III and I respectively. For mode I the velocity data for 60° dislocations were used; the transition temperature T_c is similar to that found in the mode III calculation, but the height of the step in K_F at T_c is much smaller, and fewer dislocations are emitted. The height of the step in K_F increases as K_0 approaches K_{Ic}; fig. 8 shows K_F, d_F, N_F as a function of temperature for $K_0 = 0.98 K_{Ic}$, and $K_N = 0.2 K_{Ic}$.

The number of dislocations expected in the array when the specimen fractures at the transition at $\approx 570°C$ and $K_F \approx 1.5 MPa\sqrt{m}$ is about 150, and d_F is $\approx 40 \mu m$. This compares with the experimentally observed values of ≈ 100 and $\approx 100 \mu m$ respectively at $T_c = 540°C$ and $K_F = 1.4 K_{Ic}$ (see fig. 2) (Samuels and Roberts 1989). Although for the mode I calculation the predicted values of N_F, d_F are closer to the experimental values, the predicted increase in cleavage stress at T_c is less than that observed.

The next step in the modelling is to take the screw component of the dislocations into account; this introduces another term in (19), and is expected to give results lying between those for the mode III and I calculations. These computations remain to be done. However, it is clear also from the mode I calculations, as it is from the mode III calculations, that sharp transitions are only predicted if $K_0 \approx K_{Ic}$. Furthermore, in both modes, the stresses in the dislocation arrays are in the range 50-100MPa.

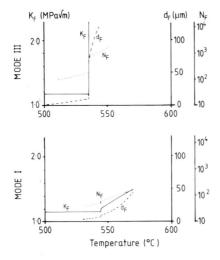

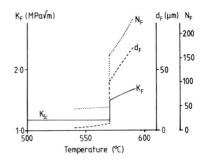

Fig. 7 Results of Mode III and Mode I simulations for $K_0 = 0.95K_{Ic}$, $K_N = 0.2K_{Ic}$. The variation with temperature of applied K (K_F), length of the dislocation array (d_F) and the number of dislocations emitted (N_F), at fracture.

Fig. 8 Mode I simulation with increased K_0. The sizes of the steps in K_F, d_F and N_F increase. ($K_0 = 0.98K_{Ic}$, $K_N = 0.2K_{Ic}$)

3. EFFECT OF EXISTING DISLOCATIONS

In the Samuels and Roberts (1989) experiments, there is a plastic zone under the indentation, but separated from the crack. The value of K_0 is identified with that at which dislocations from this plastic zone have moved to the crack tip and transformed into sources by cross-slip. It can then be shown that:

$$K_0^{m+1} - K_d^{m+1} = \frac{2\,(m+1)\,8\pi^{m/2}r_0\,(1 + m/2)}{(m+2)\,f^m} \left(\frac{\dot{K}_I}{v_0}\right) \tag{22}$$

where K_d is the value of K_I at which the stress at a dislocation source in the plastic zone under the indentation at distance r_0 from the crack is sufficient for it to emit a dislocation, and f is an orientation factor. In terms of loop length l,

$$K_d \simeq \frac{(8\pi r_0)^{1/2}}{f} \frac{\mu b}{l} \tag{23}$$

Independent estimates of K_d, from electron microscope observations, and static loading experiments, suggest values of K_d lying between $\simeq 0.3$ and 0.46 MPa√m. Equation (23) can be used to predict $(\dot{K}/v_0)$ at T_c, in good agreement with experiment (Hirsch *et al* 1989). The model predicts that as r_0 increases, v_0 should increase for the same $\dot{K}_I$, *i.e.* T_c should increase. This effect has been observed in experiments using a larger crack size arising from a larger indentation.

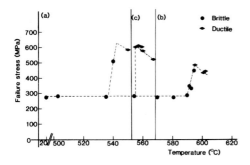

Fig. 9 Effects of surface deformation on the BDT.
(a) "Control" specimen: $T_c = 540°C$;
(b) Specimen with dislocated zone (top $4\mu m$) removed by chemo-mechanical polishing: $T_c = 595°C$;
(c) As (b), then polished with $6\mu m$ diamond paste:
$T_c = 550°C$.

The importance of existing dislocations in controlling T_c has been tested by chemo-mechanically polishing off the top $4\mu m$ of specimens containing cracks with radius $13\mu m$ (Warren 1989). This treatment should remove most of the plastic zone under the indenter. Fig. 9b shows that T_c has increased from 540°C for the control specimen, to 595°C for the polished specimen, in qualitative agreement with expectations. Similar specimens were then abraded with $6\mu m$ diamond paste after removal of the top $4\mu m$. Fig. 9c shows that T_c has shifted to the lower value of 550°C, again in qualitative agreement with expectations since abrasion should introduce surface sources (Warren 1989).

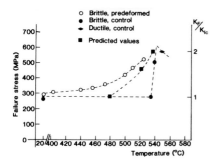

Fig. 10
Effect of pre-deformation on the BDT.
BDT curves (failure stress : temperature) for:
"control" specimens;
specimens predeformed to generate dislocation sources;
predicted curve for $K_0 = K_N = 0.21 K_{Ic}$.

The theory also predicts that if nucleation of crack tip sources does not have to take place, *i.e.* if crack tip sources already exist, the transition should become gradual with crack tip induced plasticity starting at lower temperatures. Fig. 10 shows the results of experiments in which specimens were prestressed at 2°C above T_c to a stress corresponding to $K = 1.04\,MPa\sqrt{m}$, just below the room temperature K_{Ic}, followed by rapid unloading and cooling to a lower temperature at which reloading occurs. The model predicts that such treatment should either nucleate crack tip sources, or have placed dislocations so close to the crack tip that sources would be formed almost immediately on reloading. The experiments confirm the predictions of a smooth transition (Warren 1989). Fig. 10 also shows the predicted curve for the mode III model, and for the control specimen.These experiments confirm that existing dislocations control the BDT in the experiments of Samuels and Roberts (1989). However, if the value of r_0 and l in eqn. (23) are such that $K_d > K_{Ic}$, existing dislocations cannot play any role. This is likely to be the case in the experiments of Chiao and Clarke (1989), in which dislocations have been observed to be nucleated from crack tips in thin foils of Si, in the absence of any existing dislocations.

4. COMPARISON WITH THE BREDE-HAASEN RESULTS AND THEORY

4.1Background

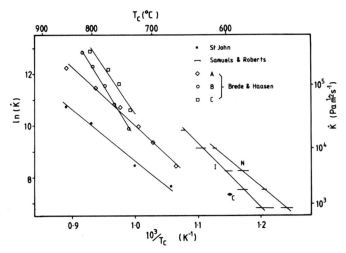

Fig. 11 Comparison of data from the experiments of Samuels and Roberts (1989), Brede and Haasen (1988) and St. John (1975). (for doping levels see refs.)

Brede and Haasen (1988) used the tapered double cantilever geometry originally used by St.John (1975). In this geometry the value of K is approximately constant in the middle of the specimen. The sharp cracks were produced by cutting a narrow slit, and driving in a wedge into the open end of the slit. The crack profile was approximately parallel to <112>. Fig. 11 compares the Oxford results of $\log_e \dot{K}$ against $1/T_c$ for intrinsic and n-type material with those from Brede and Haasen for similar n-type material (line A) and with those from St.John for intrinsic material. It should be noted that above ˜700°C, silicon doped to the levels used in the Brede-Haasen experiments has carrier concentrations dominated by intrinsic thermal activation, rather than by the extrinsic doping.

The values of T_c found by Brede and Haasen and St.John are considerably higher than in the Oxford experiments, and the strain-rates are also generally much greater. For comparable strain rates ($\dot{K} = 10^{-2}$ MPa√ms^{-1}) the Brede-Haasen values of T_c are about 70°C greater than in the Oxford experiments. Many of the Brede-Haasen specimens exhibited a crack jump, below T_c, and there is evidence that dislocations are nucleated at the crack tip when the crack arrests. Such specimens subsequently fail at temperatures below T_c at stresses considerably greater than the room temperature fracture stress.

4.2) Application of the dynamic model to the Brede-Haasen experiments

The sharpness of the transitions found in the Brede-Haasen and St.John experiments, suggests that in these cases also crack tip sources are nucleated at K just below K_{Ic}. The edge pit arrays published by Brede and Haasen and by Behrensmeier, Brede and Haasen (1987) show rows of pits along slip lines, which indicate the operation of discrete crack tip sources or emission from special sites.

We have therefore attempted to apply our dynamical model to the Brede-Haasen experiments. The transition temperature predicted by the model is extremely sensitive to the precise value of K_0 very close to K_{Ic}. However, mode I and mode III modelling of their results (for type "A" material) give reasonable values for T_c at $K_0 = 0.95 - 0.99K_{Ic}$ only with d_{crit} values of 20-30μm - much larger than the source spacing ($<˜1\mu$m) apparent from the micrographs in Brede and Haasen (1988). A set of d_{crit}, K_0 and K_N fitting one $T_c/\dot{\delta}$ combination then is found to fit all such combinations for a given material.

In the dynamic model, we distinguish between shielding at the sources (X & Y) and at the midpoint (Z); the dislocation-free zone (DFZ) length at X and Y is much less than the distance between X and Y (*e.g.* Fig 8 of Hirsch *et al* 1989). The local stress intensity factor at the source, K_{es}, is determined by the interactions between the dislocations emitted by that source; *i.e.* the interactions with the dislocations from the neighbouring source are neglected. In the Brede-Haasen experiments the source spacing appears to be $<\tilde{}1\mu m$, of the same order as the length of the DFZ; the interactions between dislocations emitted from neighbouring sources must be taken into account. The result will be an increased backstress at the sources for for a given K, causing a reduction in the dislocation emission rate. On the other hand, at any point Z, any dislocations which have passed Z, not only those from the nearest source, will cause shielding. The net result may not be very different from that predicted on a model which assumes fewer sources and in which such interactions are neglected (although a more detailed model remains to be developed), explaining the successful simulation of the Brede-Haasen results described above.

Both the mode III and the mode I simulations indicate that as for the Samuels-Roberts experiments under dynamic conditions, dislocation sources become activated only at an applied K value very close to K_{Ic}, though the sources thus formed then operate at $K_N \ll K_{Ic}$. We must now seek a mechanism to explain K_0. An obvious possibility is that, as in the Oxford experiments, existing dislocations in the crystal move to the crack and form sources by cross-slip. Brede and Haasen (1988) rule this out, because the density of potential source sites is considered to be too low. However, this possibility cannot be entirely eliminated, and would be consistent with the development of plastic zones under static loading conditions at $K \approx 0.3\,K_{Ic}$ (Michot and George 1986).

Under dynamic conditions, K_0 has to be very close to K_{Ic}, which suggests a mechanism in which the crack advances slightly and arrests, dislocation emission occurring during the arrest process. Brede and Haasen (1988) have clear evidence for dislocation activity at the arrested crack, even below T_c. Dislocation nucleation by the Rice-Thomson mechanism (1974) might then occur preferentially at kinks parallel to traces of inclined slip planes in the arrested crack front, where the transient stresses and/or the orientation may be particularly favourable for nucleation. (The Oxford experiments (Samuels and Roberts 1989) show that emission occurs preferentially where the glide plane is parallel to the crack front, *i.e.* along [110]). Emission at such preferred sites might continue if $K_N < K_{Ic}$; alternatively once a loop has been nucleated the screw components can form single ended sources by cross-slip. The loop evolution is likely to be similar to that described by Hirsch *et al* (1989, fig. 5). K_N for the nucleated source will depend simply on the length of the screw dislocation. The density of special sites and resulting sources will control T_c which is the temperature at which the dislocation velocity is high enough, for a given source density, for K_{ez} to drop below K_{Ic}, as in fig. 16. Since K_0 is effectively constant, the strain-rate dependence of T_c will still be determined by the activation energy controlling dislocation velocity, provided the source density is independent of temperature and strain-rate.

Finally, it should be noted that Brede and Haasen assume a mode III model in which: (a) the dislocation array is saturated at all times, *i.e.* the dislocations are in static equilibrium, with the net stress balanced by a small friction stress ($\approx$3MPa); (b) the local stress intensity at the source, K_{es}, controls the BDT; (c) $K_N \approx 0.3\,K_{Ic}$. The dynamic simulations for this model, however, show that at the BDT the dislocation array is far from equilibrium, the stresses being 50-100MPa. Also, using the criterion for the BDT that $K_{es} = K_{Ic}$, the predicted values of T_c are 200-300°C below those observed. Thus models based on dislocation statics do not adequately describe the actual dynamic situation at the BDT.

5. CONCLUSIONS

(a) A dynamic model has been developed in which dislocations are emitted from a discrete number of crack tip sources. At the BDT the dislocations are emitted and move sufficiently rapidly to shield the points between the sources, such that the local stress intensity factor remains below K_{Ic} for values of K above K_{Ic}.

(b) The abrupt nature of the transition is due to the fact that dislocation sources are nucleated at $K_0 \approx K_{Ic}$, and that these sources then operate at $K_N << K_{Ic}$.

(c) In the Samuels-Roberts experiments in which dislocations from existing sources can be activated at $K_d < K_{Ic}$, K_0 is controlled by the time taken for a dislocation to travel to the crack tip and to form a crack tip source.

(d) In the Brede-Haasen experiments in which not sufficient suitable external dislocations are available, it is suggested tentatively that crack tip sources may be formed following crack advance at $K = K_{Ic}$; loops may be nucleated by the Rice-Thomson (1974) mechanism when the crack arrests, and sources can then be nucleated by cross-slip of screw segments.

(e) The stresses acting on the dislocations at T_c are very high (50-100 MPa), and the arrays are very far from equilibrium.

(f) The strain-rate dependence of T_c is controlled by the activation energy for dislocation velocity; this is predicted by the dislocation dynamics model.

(g) The model predicts a dependence at T_c on dislocation density and loop length, and of the shape of the transition curve on prestress. Both these effects have been found experimentally.

REFERENCES

Behrensmeier R, Brede M and Haasen P 1987 *Scripta Metall.* 21 1581
Brede M and Haasen P 1988 *Acta Metall.* 36 2003
Burns S J and Webb W W 1970 *J. Appl. Phys.* 41 2078
Chiao Y H and Clarke D R 1989 *Acta Metall.* 37 203
George A and Champier G 1979 *Phys. Stat. Sol.* 53a 529
Hirsch P B, Roberts S G and Samuels J 1987 *Scripta Metall.* 21 1523
Hirsch P B, Roberts S G and Samuels J 1988 *Revue Phys. Appl.* 23 409
Hirsch P B, Roberts S G and Samuels J 1989 *Proc. R. Soc. Lond.* A 421 25
Hull D, Beardmore P and Valintine A P 1965 *Phil. Mag.* 12 1021
Lakshmanan V and Li J C M 1989 *Acta Metall.* in press
Lin I-H and Thomson R 1986 *Acta Metall.* 34 187
Liu J and Bilello J C 1977 *Phil. Mag.* 35 1453
Michot G and George A 1986 *Scripta Metall.* 20 1495
Samuels J and Roberts S G 1989 *Proc. R. Soc. Lond.* A 421 1
St.John C 1975 *Phil. Mag.* 31 1193
Thomson R 1986 *Solid State Physics* eds H Ehrenreich and D Turnbull (New York: Academic Press) 39 1
Warren P D 1989 *Scripta Metall.* in press.

Inst. Phys. Conf. Ser. No 104: Chapter 4
Paper presented at Int. Symp. on Struct. Prop. Disloc. Semicond., Oxford, 1989

385

Fracture and crack tip plasticity in silicon and gallium arsenide

G Michot and A George

Laboratoire de Physique du Solide, CNRS URA N°155, Ecole des Mines de Nancy, INPL, Parc de Saurupt, 54042 Nancy, France
LURE, CNRS, Université de Paris-Sud, 91405 Orsay, France

ABSTRACT : K_{Ic} measurements have been done at room temperature in Si for $\{111\}$ and $\{110\}$ cleavage planes and in GaAs. Crack tip plasticity and the brittle-ductile transition (BDT) were investigated in situ by synchrotron X-ray topography. The kinetics of growth of the plastic zone (PZ) allows a determination of the stress in the PZ. The very sharp BDT can be related to dislocation velocities and explained by a shielding effect of dislocations, which decreases the effective K preventing it from reaching K_{Ic} in the ductile range.

1. INTRODUCTION

In brittle materials which contain cracks, stress concentrations arise at crack tips. With the crack tip at the origin, the stress field at a point of coordinates r, φ can be expressed :

$$\sigma_{ij} \, (r, \varphi) = (K/\sqrt{2\pi r}) \; f(\varphi) \tag{1}$$

K is the stress-intensity factor, which can be calculated from the applied load P and specimen geometry. In the absence of plastic deformation , if the load is increased, brittle fracture will happen at $K = K_c$, the critical stress intensity factor, which is a material property for a given mode of loading and, in single crystals, a given type of cleavage planes. According to Griffith's classical analysis, the cleavage strength is related to the surface energy, γ and Irwin and Wells (1965) and Eshelby (1968) have shown that :

$$2\gamma = K_{Ic}^2 \, (1 - \nu^2) \, / \, E \tag{2}$$

(plane strain conditions are assumed, ν : Poisson's ratio, E : Young's modulus, the subscript I is for mode I loading or opening mode).

When temperature increases, dislocations may gain a sufficient mobility to move to -or to be generated at- the crack tip. The "plastic zone" (PZ) thus formed relieves the stresses and increases the critical load for crack propagation. Dislocations may cause crack-tip blunting and/or shielding. The shielding effect can be rationalized, attributing to PZ dislocations a K_D value such that the effective stress intensity factor is now (Majumdar and Burns 1981) :

$$K_{eff} = K - K_D \tag{3}$$

where K is the value that would correspond to the externally applied load
in the absence of dislocations. It is possible that generalized plastic
deformation takes place before K_{eff} reaches K_{IC}. The material has become
ductile.

Obviously, semiconductors must present such a brittle-ductile transition
(BDT) because of the large activation energy which characterizes
dislocation glide. Semiconducting materials and especially silicon can
even be taken as model substances since they offer the unique possibility
of a quantitative analysis in terms of dislocation dynamics.

The paper reports on investigations that have been performed both in the
brittle range and about the BDT. Brittle fracture was studied in Si and
GaAs. In Si, K_{IC} measurements are given not only for $\{111\}$ but also for
$\{110\}$ cleavage. In GaAs, the influence of the In content was investigated
and an environment sensitive subcritical crack growth was observed. The
BDT was studied in Si, mainly using in situ observations of the formation
of the PZ in dynamical loading, by X-ray topography with the synchrotron
radiation of LURE-DCI.

2. EXPERIMENTAL DETAILS

Tapered double-cantilever-beam (DCB) samples (Figure 1) were used. Crack
initiation and arrest at room temperature was obtained following the
technique of St John (1975). Pre-cracked samples were loaded in mode I by
means of the two pin-holes. K_{IC} measurements and some BDT experiments were
performed in an Instron machine at a constant deflection rate, which
allowed several crack starts and arrests to be measured with the same
sample. In situ observations in Si were done with the high temperature
deformation stage described by George and Michot (1982) but with an
improved loading system which can be driven so that the applied stress
intensity factor increases at constant rate $\dot{K}$. This stage was mounted on
the two-axis spetrometer of the topography beam port of LURE-DCI. The beam
was first monochromatized ($\lambda \approx 0.08$ nm) by a 220 reflection on a Ge
crystal. Permanent control of the Bragg angle was used to keep the
interesting area in selective reflection when plastic deformation took
place. X-ray topographs were recorded at given time intervals on Ilford L4
nuclear plates (exposure ≈ 20 s) or directly on a video tape when
dislocation velocities were too high.

High temperature experiments were conducted under reducing atmosphere
(10 % H_2, 90 % N_2). GaAs crystals were tested in various atmospheres,
oxidizing (air), neutral (argon) or reducing.

Samples which did not break during in situ loading were cooled down with
the load applied and subsequently loaded up to fracture in the Instron
machine in order to measure the influence of a "static" PZ on the fracture
load in the brittle range (warm prestressing effect).

3. BRITTLE FRACTURE IN Si AND GaAs

In Si, cleavage usually occurs between $\{111\}$ planes, which correspond to
the smallest ratio E/d of Young's modulus over plane spacing. K_{IC}^{111} was
measured in samples cut from $\{112\}$ wafers with the notch parallel to a
<110> direction. We obtained :

$$K_{IC}^{111} = 0.93 \pm 0.05 \text{ MPa}\sqrt{m}$$

in close agreement with the value of 0.96 MPa√m̄ produced by St John (1975) and with the results of Brede and Haasen (1988) (0.9 MPa√m̄) and of Samuels and Roberts (1989), Hirsch et al. (1987, 1989), who found a slightly higher $K_{Ic} \simeq 1.17$ MPa√m̄ with a different specimen geometry and testing technique.

{110} cleavage planes could also be obtained in samples cut from {001} wafers, again with a notch parallel to <110>, with a critical stress intensity factor :

$$K_{Ic}{}^{110} = 0.89 \pm 0.04 \text{ MPa√m̄}$$

very close to the value observed for {111} cleavage (Michot 1987).

Figure 2 shows typical crack surfaces for these two types of cleavage, as observed by X-ray topography. Interestingly, {110} cleavage looks much more perfect than {111}. Only Moiré fringes due to a slight closure mismatch are observed for the former, {110} surfaces being very smooth,

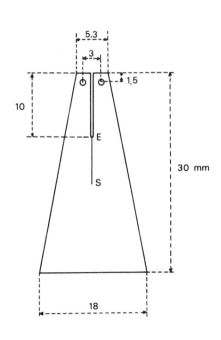

Fig.1. DCB samples
(typical thickness : 0.7 mm)

Fig.2. X-ray topographs of {111} (a) and {110} cleavage planes (b)

while {111} cleavage surfaces exhibit many ledges. This observation could explain the discrepancy between experiments and theory which arises when K_{Ic} values are converted into surface energies using eqn.2.

$$\gamma^{111} = 2.70 \text{ J.m}^{-2}$$
$$\gamma^{110} = 2.25 \text{ J.m}^{-2}$$

with a $\gamma^{111}/\gamma^{110}$ ratio larger than unity though theoretical estimates (e.g. Gilman 1959) give ≈ 0.7. It is well recognized that cleavage techniques give values in excess of the true surface energy, due to their large sensitivity to crack surface defects and parasitical effects such as energy dissipation by sound waves. It is believed that, because of the higher perfection of {110} cleavages, γ^{110} is closer the true value than is γ^{111}. Michot (1989) has proposed the following explanation for the poorer quality of {111} cleaved surfaces. Cracks are spread on several parallel planes from the beginning. They tend to propagate in <110> orientations. Such a direction of propagation is unique in {110}, while the competition between the three available directions in {111} leads to an exaggeration of initial defects.

Fewer measurements were performed in GaAs (Michot et al. 1988). Due to the polarity of {111} surfaces, easy cleavage planes are {110}. K_{IC} was observed not to depend on the gaseous environment and the indium content up to 1 at. per cent.

$$K_{IC} = 0.44 \pm 0.04 \text{ MPa}\sqrt{m}$$

corresponding to a surface energy,

$$\gamma = 0.76 \text{ J.m}^{-2}$$

Contrary to Si, a regime of slow crack growth was observed near K_{IC}. The crack velocity was found to strongly depend on the atmosphere, being much larger in hydrogen than in air and in air than in argon or helium.

4. THE BRITTLE-DUCTILE TRANSITION - EXPERIMENTS

At a given loading rate $\dot{K}$, samples failed by brittle fracture or were plastically deformed, depending whether the temperature was lower or higher than T_c, defined as the BDT temperature (Figure 3). This transition is very sharp and T_c is defined within ≈ 5 K, the experimental uncertainty on T in the present in-situ loading experiments. Fracture was always observed at $K \approx K_{IC}$ contrary to Brede and Haasen (1988) who reported brittle fracture at $\dot{K}$ significantly larger than K_{IC}. The Arrhénius plot of Figure 3 suggests a correlation between the transition temperature and the loading rate through an exponential relation :

$$\dot{K} \approx \exp(-E/kT_c) \tag{4}$$

with a measured activation energy of 2.4 ± 0.2 eV, close to -although somewhat higher than- the activation energy for dislocation glide determined by many authors to be 2.2 ± 0.1 eV. (For convenience, the loading rate $\dot{K}$ was scaled to $\dot{K}_0 = 310 \text{ Pa}\sqrt{m} \text{ s}^{-1}$ in the rest of the paper).

At temperatures far below T_c no dislocations appeared prior to fracture. Closer to T_c, in several cases, fracture occurred at $K_I \approx K_{IC}$ although dislocations had clearly started to develop from the crack tip. Figure 4 shows an example of such a beginning PZ which has not prevented brittle fracture. Of course the dislocation density remained small, when compared with that observed in the fully developed PZ of samples loaded at $T > T_c$, as can be seen on the selected "snapshots" of Figure 5.

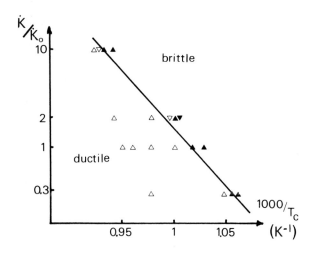

Fig.3. Arrhénius plot of the loading rate versus the inverse of critical temperature T_c ($\dot{K}_o$ = 310 Pa√m s^{-1}, full symbols correspond to fractured samples)

As reported in earlier work (Michot et al. 1980), dislocation loops formed at or near the crack tip -this cannot be decided from X-ray topographs- and glide in those {111} planes which intersect the crack front. In (1̄1̄2) samples with a (111) crack, the PZ consists of five distinct areas or lobes : one -called E- in the (11̄1) plane parallel to the crack propagation direction, the other four in the oblique planes (1̄11) and (11̄1), two in each, with the one ahead and the other behind the crack tip. These lobes are called A, B, C and D as shown in Figure 5.

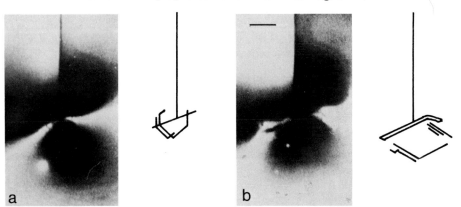

Fig.4. Plastic zone developed at $\dot{K}$ = 0.3 $\dot{K}_o$ prior to brittle fracture at $T \simeq T_c$ = 946 K. Synchrotron topographs. (a, b) K/K_{IC} = 0.6 and 0.79 respectively. Marker : 200 μm

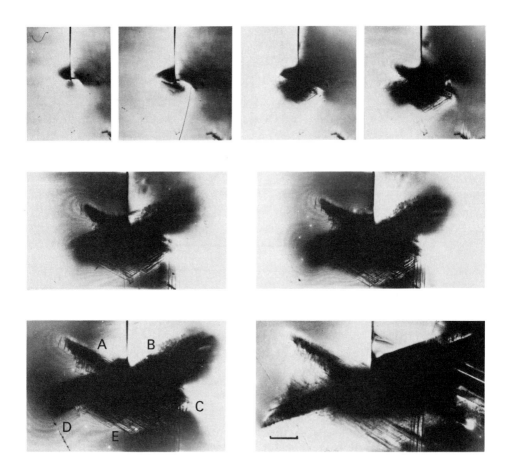

Fig.5. Plastic zone developed at T = 1053 K (T_c=988 K), $\dot{K}=\dot{K}_o$=310 Pa√m s^{-1}. Synchrotron topographs. K/K_{IC} = 0.36, 0.49, 0.63, 0.77, 1.03, 1.18, 1.32, 1.68. Marker : 1 mm

In the ductile range, the growth kinetics of each lobe can be appreciated from Figure 6 where the lobe size d is plotted as a function of K/K_{IC} = ($\dot{K}$t + K_r)/K_{IC} (K_r is a residual stress intensity factor applied permanently in order to avoid crack healing) : the sample was loaded under $\dot{K} = \dot{K}_o$ = 310 Pa√m s^{-1} at T = 998 K ($\approx$ 10 K above T_c). It was difficult to measure accurately the starting of the first dislocation loops, since they were in most cases masked by the strong black-white contrasts caused by elastic strains around the crack tip. While the E lobe appears first and then grows at the smallest rate, the other four lobes do not always start growing simultaneously but grow at the same rate, which is observed to be constant up to large PZ sizes in a given experiment. This means that, for a fixed temperature, outer dislocations of the PZ glide at constant velocity -i.e. under a constant effective stress- when K increases at a constant rate. (In the following, the PZ size d averaged over lobes A to D is used).

Extrapolating the linear PZ growth rate to d = 0 gives an estimate of the value K_{em} at which the first dislocation could have been emitted. K_{em} ranges from 0.2 to 0.6 K_{IC}. The minimum value, 0.2 K_{IC}, is fairly close to the $K_{min} \approx 0.3$ MPa√m, determined by Michot and George (1986) as the threshold value for dislocation generation at crack tips under constant load (creep conditions). When the testing temperature approaches T_c at constant $\dot{K}$, the general trend is a decrease of the difference $K_{IC}-K_{em}$. Yet, K_{em} is still far from K_{IC} at a temperature as close to T_c as experimentally possible.

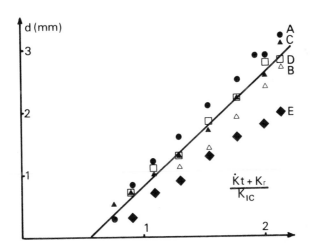

Fig.6. Plastic zone size versus K/K_{IC} ($\dot{K} = \dot{K}_o = 310$ Pa√m s^{-1}, T = 998 K, each symbol for one lobe)

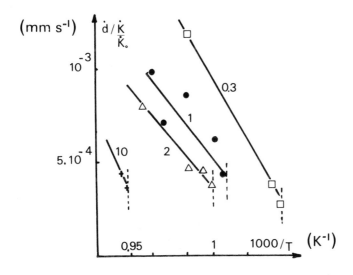

Fig.7. Plastic zone growth rate divided by $\dot{K}/\dot{K}_o$ as a function of 1/T. ($\dot{K}/\dot{K}_o$ = 0.3, 1, 2 and 10, each symbol for one loading rate)

Figure 7 is a plot of the PZ growth rate $\dot{d}$, divided by $\dot{K}$, as a function of the reciprocal temperature. At constant $\dot{K}$, the decreasing values of $\dot{d}/\dot{K}$ can be extrapolated to $T = T_c$. It can be checked that to a good approximation $\dot{d}/\dot{K}$ is constant at $T = T_c$ over a 150 K interval. It can be said that, at the BDT, the PZ grows at a rate which is proportionnal to the loading rate.

The velocity of the outer dislocations of the PZ has been measured and the shear stress τ acting on them was calculated, using the phenomenological description of the data of George and Champier (1979) :

$$v(mm/s) = 2,9 \times 10^6 \ (\tau/1 \ MPa)^{1.2} \ \exp \ (-2.2 \ eV/kT) \tag{5}$$

Shear stresses ranging from 5 to 16 MPa are found (table 1). At a given $\dot{K}$, τ increases when T decreases towards T_c. For a given $T-T_c \approx 5$ K, the stress varies by 60 % only for $\dot{K}$ varying by a factor of 30. It may be noticed that such estimates are subject to some scatter because of the experimental uncertainty on T.

$\dot{K}/\dot{K}_o$	0.3		1					2				10	
T(K)	1023	955	1053	1043	1023	998	992	1063	1023	1010	1003	1083	1078
$\dot{d} \times 10^4$ (mm/s)	3.5	1.1	9.9	7.0	8.5	6.2	4.2	16	9.2	9	7.5	44	37
τ (MPa)	5.6	9.5	7.4	6.6	11.6	15	12.5	9	12.5	16	15.8	14.6	13.6
T_c (K)	946		988					1000				1075	
τ_c (MPa)	10		14.6					16.3				(13.1)	

Table 1. Shear stresses exerted on the leading PZ dislocations

Table 2 gives the results of fracture tests at room temperature on samples previously loaded above T_c. High temperature loading was stopped at $K = K_{HT}$, comprised between 1.5 and 2.6 K_{IC}. The fracture load at room temperature was converted into K_c^*. Those results are analyzed as follows. Room temperature fracture occurs when the effective K equals K_{IC}. Thus, according to eqn.(3) :

$$K_c^* = K_{IC} + K_D \tag{6}$$

where K_D represents the shielding effect of dislocations. (Following Michot and George (1985), we assumed that blunting is negligible for our specimen geometry). Since dislocations are immobile at 20°C, if they are frozen by cooling the sample under load, K_D is assumed to be the same that prevailed when the high temperature test was interrupted. Eqns(3) and (6) give a straightforward determination of the effective K at this point :

$$K_{eff} = K_{HT} - K_D = K_{HT} + K_{IC} - K_c^* \tag{7}$$

K_{eff} values vary from $\approx 0.2 \ K_{IC}$ to $0.5 \ K_{IC}$. Surprisingly to us, no correlation could be observed between K_{eff} and the proximity from T_c.

$\dot{K}/\dot{K}_0$	0.3		1					2				10	
$T-T_c$ (K)	77	9	65	55	35	10	4	63	23	10	3	8	3
K_{HT}/K_{IC}	1.51	1.24	1.76	1.39	1.76	2.14	1.64	1.94	2.16	2.07	2.32	1.96	2.59
K_c^*/K_{IC}	2	1.86	/	2.06	/	1.69	1.22	/	2.77	/	3.11	2.55	/
K_{eff}/K_{IC}	0.51	0.38	/	0.33	/	0.55	0.58	/	0.39	/	0.22	0.41	/

Table 2. K_{eff} derived from the warm prestressing effect. The values of K_c^* are not given for samples which failed by cleavage on planes distinct from the pre-cracked plane

5. THE BRITTLE-DUCTILE TRANSITION - DISCUSSION

A simplified view of the BDT can be given as follows, starting from the situation at some temperature $T_E \gg T_c$ where perfect shielding can be realized. Then, a quasi-equilibrium is established at the crack tip with $K_{eff} = K_{min}$, which means that $\dot{K}_{eff} = \dot{K} + \dot{K}_d = 0$ (Michot and George 1985). Let τ_y be the stress in the PZ, the PZ growth rate depends on τ_y and T_E through eqn.(5). The mean dislocation nucleation rate $\dot{N}_E$ is constant and proportionnal to $K_{min}\dot{K}$ (Michot 1989)(*). Such a situation is very similar to the static loading conditions for which the equilibrium distribution of PZ dislocations, forming an inverted pile-up ahead of the crack tip, can be determined (see for example Michot 1986) provided that the PZ is sustained by a static friction stress opposing dislocation movement.

If, at the same K, the temperature is decreased, the loss of dislocation mobility can be recovered only by an increase of the stress level. It appears reasonable to assume that a higher stress in the PZ stems from a higher K_{eff}. (It is implicitly assumed that the stress is constant in the PZ or, at least, that there is a monotonic relationship between the stress level close to the crack tip and at the PZ border). Such an increase of the PZ stress has indeed been observed (Table 1), although it is not sufficient to keep the PZ growth rate constant.

At $T > T_c$, K_{eff} must saturate before reaching K_{IC}. Possibly, a steady state regime of the PZ growth can be realized at some K_{eff} and τ values, so that the increasing applied K is again exactly counterbalanced by shielding. At $T = T_c$, the maximum value of K_{eff} is K_{IC}, causing brittle failure. The sketch of Figure 8 summarizes this intuitive analysis. It must be pointed out that only the most vulnerable points of the crack tip are concerned, on the dislocation source points K_{eff} cannot exceed K_{min}.

Warm prestressing experiments should, in principle, allow to measure K_{eff} as explained above. A continuous variation from K_{min} to K_{IC} is expected when the temperature decreases from T_E to T_c. This is in clear contradiction with the results of Table 2. We feel that the reason for this discrepancy is that samples had been loaded too far beyond the BDT

(*) As pointed out by Hirsch et al (1988), dislocation nucleation is a discrete process and K_{eff} should have a zig-zag variation. Perfect shielding assumption means that K_{eff} never exceeds K_{min}

and that K_{eff} is likely to go through a maximum before decreasing when generalized yielding starts, rather than keeping a steady-state value (Figure 9). Similar experiments, with the high temperature loading interrupted earlier at $K_I \approx K_{Ic}$, are planned to check this important point. (It has to be recognized, however, that K_{eff} should not decrease while the PZ growth rate remains constant if the above assumption of a monotonic relationship between these two parameters is valid).

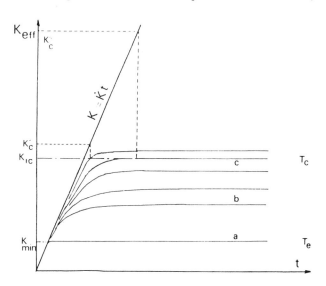

Fig.8. Schematic evolution of K_{eff} with time at different temperatures for a given $\dot{K}$ value, assuming that some steady-state regime sets in (a : $T = T_E$, b : $T_c < T < T_E$, c : $T = T_c$)

It is clear from these results but also from the pioneering work of St John (1975) and more recent studies by Michot and George (1985), Brede and Haasen (1988), Hirsch et al. (1988, 1989) Samuels and Roberts (1989) that the BDT is related to the dislocation velocity. At a given $\dot{K}$, the minimum velocity required to prevent fracture is reached at $T \approx T_c$. To a first approximation, if $\dot{K}$ is doubled, the same unstable situation is obtained with a twice-as-fast velocity :

$$\dot{K} \alpha \ v(T_c) = \tau_c^m \ \exp(-Q/kT_c) \qquad (8)$$

This is experimentally well verified : a linear dependence is observed between $\dot{K}$ and the extrapolated values $v(T_c)$ derived from Figure 7. The corresponding τ_c are derived through eqn.(5). The apparent activation energy derived from eqn.(4) includes the stress contribution explicitly given by eqn.(8), which explains that it amounts to 2.4 eV instead of 2.2 eV, the right value for dislocation glide.

Allowing a continuous recording of dislocations, in situ straining experiments should, in principle, answer the still unsolved questions about the PZ formation and growth in relation to the BDT. Although our experiments have been designed for such purpose, present results are still far from conclusive on many points. A main difficulty arises becauses of

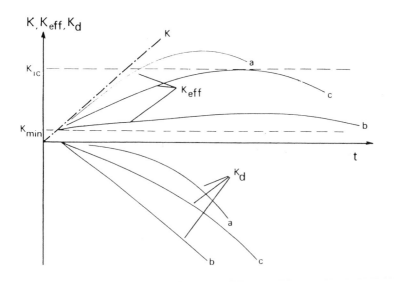

Fig.9. Possible evolution of K_{eff} and K_d with time at different temperatures : $T < T_c$ curve a (in fact interrupted for $K_{eff} = K_{IC}$), $T > T_c$ curve b, $T = T_c$ curve c. The dashed line corresponds to the applied K variation.

the poor spatial resolution of the imaging technique. X-ray topography cannot resolve dislocation loops with a diameter smaller than ca.20 μm. In the highly strained region around a loaded crack tip, the situation is much worse. In addition, with the presently available synchrotron radiation facilities, times of exposure are still of several seconds for high-resolution plates and the resolution much poorer in case of real-time TV recording. A consequence is that few can be said about the starting of the PZ growth, which is probably critical.

Present results confirm that no dislocation activity can be detected before brittle failure up to ≈ 5 K below T_c. Hirsch et al. (1989) have recently produced simulations which show that, if dislocations were emitted at $K_{min} \approx 0.3\ K_{IC}$, the BDT should be more progressive, with extended PZ at $T \ll T_c$. In their model, an abrupt transition was obtained assuming that the nucleation of crack tip dislocations starts at $K \approx K_{IC}$ only, and then can continue at $K \approx K_{min}$, producing a dislocation avalanche. No evidence for such an avalanche could be obtained during our observations, although the lack of data for small PZ sizes does not allow us to conclude that it does not exist. However, dislocations were clearly seen to start at $K \approx 0.6\ K_{IC}$ at $T \approx T_c$, in a sample which failed soon after at K_{IC}. This last observation, on the other hand, is in agreement with another statement of Hirsch that dislocations are not necessarily efficient for shielding the crack tip as soon as they have been emitted : they have also to move to the "most vulnerable" points of the tip.

This brings back to the question of the first dislocation sources. Hirsch has proposed that crack tip sources are provided by existing dislocations that have first to move to the tip. This is likely when cracks are introduced by room temperature micro-indentations, which are known to create also dislocations, but cannot work in present samples since the wedge technique for pre-cracking does not create any dislocation near the crack tip. The earlier proposal by us that swirl defects could act as bulk sources has been rejected by Brede and Haasen (1988) who have estimated a much too low swirl density. The possibility remains that dislocations are created directly at the crack tip following the mechanism proposed by Rice and Thomson (1974) revised by Thomson (1986). It would be valuable to know whether such nucleation occurs at some preferred points along the crack tip, e.g. at the surfaces. Here again X-ray topography is probably not the convenient technique. It is hoped that further information on that point will be obtained by etch pit observations on broken samples. New experiments are also planned with samples of different orientations to see to what extent the BDT is sensitive to the type of cleavage plane and the orientation of available slip planes with respect to the crack tip.

6. REFERENCES

Brede M and Haasen P 1988 *Acta Metall.* **36** 2003
Eshelby J D 1968 *ISI Publications* **121** 13
George A and Champier G 1979 *Phys. Stat. Sol.(a)* **53** 529
George A and Michot G 1982 *J. Appl. Cryst.* **15** 412
Gilman J J 1959 *Fracture* ed B L Averbach, D K Felbeck, G T Hahn and D A Thomas (Cambridge : MIT Press) pp 193-224
Hirsch P B, Roberts S G and Samuels J 1988 *Revue Phys. Appl.* **23** 409 ; 1989 *Proc. R. Soc. Lond.* **A421** 25
Irwin G R and Wells A A 1965 *Metall. Rev.* **10** 223
Majumdar B S and Burns S J 1981 *Acta Metall.* **29** 579
Michot G 1986 *Fracture Control of Engineering Structures (Proc. ECF6)* ed. H C Van Elst and A Bakker EMAS pp 101-110 ; 1987 *Surf. Sci.* **186** L-561 ; 1989 *Crystal Properties and Preparation - Trans. Tech. Publ.* **Vol.17-18** pp.55-98
Michot G, Badawi K, Abd el Halim A R and George A 1980 *Phil. Mag.* **42** 195
Michot G and George A, *1982 Scripta Metall.* **16** 519 ; 1985 *Strength of Metals and Alloys (Proc. ICSMA 7)* ed H J Mc Queen, J P Bailon, J I Dickson, J J Jonas and M G Akben (Oxford:Pergamon Press) pp 1187-1192 ; 1986 *Scripta Metall.* **20** 1495
Michot G, George A, Chabli-Brenac A and Molva E 1988 *Scripta Metall.* **22** 1043
Michot G, George A and Champier G 1982 *Fracture and the Role of the Microstructure (Proc. ECF4)* ed K L Maurer and F E Matzer EMAS pp 30-35
Rice J and Thomson R 1974 *Phil. Mag.* **29** 73
St John C F 1975 *Phil. Mag.* **32** 1193
Samuels J and Roberts S G 1989 *Proc. R. Soc. Lond.* **A421** 1
Thomson R 1986 *Solid State Physics* (ed H Ehrenreich and D Turnbull) (New-York : Academic Press vol 39 pp 1-129)

Inst. Phys. Conf. Ser. No 104: Chapter 4
Paper presented at Int. Symp. on Struct. Prop. Disloc. Semicond., Oxford, 1989

397

Dislocation emission from crack tips and the macroscopic brittle-to-ductile transition

D. R. Clarke and R. F. Cook

IBM Research Division, Thomas J. Watson Research Center, Yorktown Heights, NY 10598

Following a review of the principal results of an in-situ transmission electron microscopy study of dislocation emission from crack-tips in silicon under combined applied stress and temperature, we introduce the fracture mechanics framework necessary to formulate a dislocation model for the brittle-to-ductile transition. In many respects the brittle-to-ductile transition is a model system for developing the full crack resistance curve (including transient effects) for toughening by a zone shielding mechanism.

1. INTRODUCTION

Recent studies by Brede and Haasen (1988) at Gottingen, Hirsch, Roberts and co-workers (1989) at Oxford, Michot and George (1986) at Nancy, and ourselves (Chiao and Clarke, 1989) at IBM, have all sought to extend St John's (1975) pioneering studies of the brittle-to-ductile transition in silicon. The Oxford and Gottingen groups have relied on etching studies after performing fracture tests in order to reveal the presence and possible role of any dislocations, whereas the Nancy group (using X-ray topography) and ourselves (using transmission electron microscopy) have made direct, in-situ observations of dislocation emission from crack tips. From the work reported to date there is a consensus that a) the transition is mediated by dislocations emitted from the tip of a brittle crack and b) the similarity between the apparent activation energies for the strain rate dependence of the transition and those for dislocation glide imply that the transition is controlled directly or indirectly by dislocation mobility. Furthermore, there is also agreement that the crack-tip stress intensity factors can be calculated in terms of standard dislocation models. There remains disagreement about the detailed mechanism of dislocation nucleation, whether dislocations are nucleated at the crack tip, or indeed whether there is even a nucleation barrier to dislocation emission at the transition temperature. Further experimentation will undoubtedly clarify these, and related, issues.

However, the central issue remains how to relate the observed dislocation emission process to the macroscopic brittle-to-ductile transition. This is of importance not only to an understanding of the brittle-to-ductile transition in silicon but also as a test of our general understanding of toughening due to crack-tip process zones. Toughening by the emission of dislocations represents possibly the simplest example of a toughening phenomenon that is amenable to a complete analysis such that an entire crack resistance curve (R-curve) can be computed. This is in contrast to other forms of process zone toughening, such as microcrack toughening or (zirconia) transformation toughening, where the exact form of the constitutive relations on which the R-curve calculations are based are not, as yet, known.

After a brief summary of the results of a recent in-situ TEM study of dislocation emission from crack tips in silicon, we distinguish between the necessary and sufficient conditions for crack extension, and then discuss the fracture and dislocation mechanics associated with dislocation emission. An outline of a model consistently incorporating the dislocation interactions with the mechanics of fracture is then introduced and illustrated with a couple of examples of how stress intensity factors change as the first dislocations are emitted and the crack advances. In the conference presentation we will also show a video-tape recording of dislocation emission from crack tips in silicon.

2. IN-SITU TEM OBSERVATIONS

In this section we present some of the principal findings of the recently published (Chiao and Clarke, 1989) in-situ transmission electron microscopy studies of dislocation emission from crack tips in silicon. The essential features of the experiment were the application of a stress to short cleavage cracks (introduced by indentation) in thin foils of silicon at elevated temperature and real-time observation of crack growth and any dislocation activity. Simultaneously recorded video-tape provided a permanent record of the experiment, enabling subsequent analysis of the dislocation configurations and velocities. The primary motivation for the work was to seek direct experimental evidence for the processes associated with the brittle-to-ductile transition and resolve some of the issues involved.

1. At low temperatures, abrupt extension of the crack would occur without any evidence of dislocation emission.

2. At temperatures associated with the brittle-to-ductile transition perfect dislocation loops having $\bar{b} = 1/2 <110>$ were seen to be emitted from the crack tip on the close-packed (111) planes having the highest Schmid factor. Only one dislocation source appeared to operate at any crack front and in no instance were dislocations seen to migrate to the tip from sources behind the crack-tip.

3. Dislocations loops emitted from (111) cracks under near mode-I loading nucleated and rapidly grew to a distance from the crack and then abruptly stopped (figure 1).

4. Offsets in the Moire fringe pattern formed from the overlapping crack faces revealed the presence of 1/3[111] ledges, presumably left as blunting ledges after emission of those dislocation loops (figure 2).

5. No dislocations were seen to be nucleated away from the crack tip or from regions in the immediate vicinity (~10 nm) of the tip (such as by the activation of some pre-existing source in the crack-tip stress field).

6. In each experiment in which a dislocation was emitted a distinct dislocation free zone was observed. Such a zone is in agreement with the proposals of Ohr (1985) and contrary to the dislocation model of Bilby-Cottrell-Swinden (1963).

7. The coplanar dislocation arrangement observed ahead of a mode III crack adopted an inverse pile-up spacing in agreement with standard linear elastic models for dislocation interaction with a crack-tip in mode-III (figure 3).

8. From the observed dislocation velocities (measured from the video-tape) and the resultant distribution of dislocations, an estimate for the emission condition under mode-I and mode-III loading was made of $K_I = 0.94 \pm 0.06$ MPa$m^{1/2}$ and $K_{III} = 0.17 \pm 0.03$ MPa$m^{1/2}$. The former value is the same as that reported by a number of workers for the room temperature fracture resistance of silicon.

9. Substituting the estimated stress intensity factors (appropriately resolved on to the dislocation glide plane) into the Rice-Thomson (1974) energy analysis leads to the conclusion that spontaneous dislocation emission occurs without a nucleation barrier ($U<kT$).

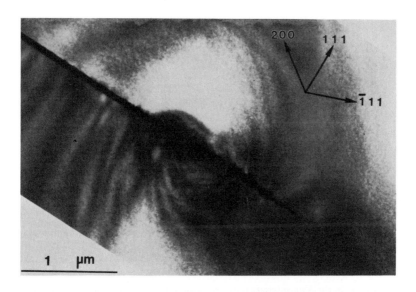

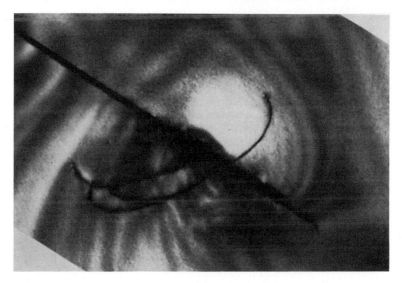

Figure 1. A (111) cleavage crack before (top) and after (bottom) the emission of four perfect dislocation loops. The crack was under approximate mode-I loading and the dislocations, once formed grew and then abruptly arrested at the positions shown.

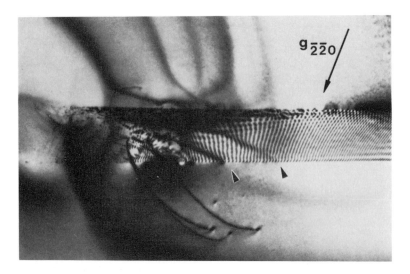

Figure 2. Offsets in the Moire fringe pattern formed from the overlapping crack faces reveal the presence of 1/3 [111] ledges formed after the emission of dislocations.

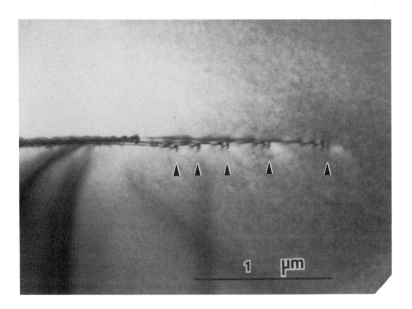

Figure 3. The dislocation arrangement ahead of a mode-III crack. The dislocation positions are in an approximate inverse pile-up arrangement as predicted by linear elastic models for the interaction of dislocations and a crack-tip in mode-III.

3. CONDITIONS FOR CRACK EXTENSION

The dislocation analyses of the brittle-to-ductile transition proposed by Brede and Haasen, Hirsch, Roberts and Samuels, Michot and George, and recently Maeda and Fujita (1989) have all been formulated so as to calculate a stress intensity for the crack, and then relate that to a critical stress intensity factor. The implicit assumption of the works being that when the critical stress intensity factor is attained fracture occurs. However, exceeding the critical stress intensity factor represents only a <u>necessary</u> non-equilibrium condition for fracture. Thermodynamically, a crack is in equilibrium when the strain energy release rate, G, defined as:

$$G = -dU_M/dc$$

and the crack resistance force, R, to fracture

$$R = dU_s/dc$$

are equal. In terms of the (thermodynamic) forces acting on the crack the necessary condition for crack extension is:

$$G > R \tag{1}$$

As in any other thermodynamic system, instability occurs when the second derivative of the total energy is less than zero. This represents a <u>sufficient</u> condition for continued crack extension, namely when the following inequality is satisfied:

$$dG/dc > dR/dc \tag{2}$$

For elastic materials under plane stress conditions, the strain energy release rate and the resistance force may be expressed in terms of stress intensities, K, and material toughnesses, T:

$$K = \sqrt{G E} \quad \text{and} \quad T = \sqrt{R E} \tag{3}$$

Thus, the necessary and sufficient conditions for unstable crack extension are:

$$K_{tip} > T_o \quad \text{and} \quad \frac{dK_{tip}}{dc} > \frac{dT_0}{dc} \tag{4}$$

in the crack-tip frame of reference, where T_0 is the intrinsic fracture resistance of the material, or equivalently in the applied loading system frame of reference, by

$$K_{app} > T \quad \text{and} \quad \frac{dK_{app}}{dc} > \frac{dT}{dc} \tag{5}$$

T in this equation is the (measurable) fracture resistance at crack length c.

For the case of the brittle-to-ductile transition, which the experimental studies clearly indicate is mediated by dislocations, the problem reduces to developing a consistent dislocation model that gives the fracture resistance as a function of crack extension. The latter would be the so-called R-curve for toughening due to dislocation emission. As recognized since the work of St John the problem is compounded by the fracture resistance being strain rate dependent.

4. DISLOCATION AND FRACTURE MECHANICS

It is assumed, and supported by both X-ray topographic studies and our own TEM studies, that silicon is intrinsically brittle at low temperatures and no dislocations are formed during fracture.

In the absence of any dislocations the fracture resistance of the material is denoted by T_0 . Any increase in fracture resistance above this intrinsic toughness, whether in the macroscopically brittle or ductile fracture regime, is given in dislocation models by the shielding (and/or anti-shielding) afforded by the distribution of dislocations ahead and around the sharp crack tip. As the crack is loaded by an externally applied load the applied stress intensity, K_{app} , rises until at some value the crack tip stress intensity equals that at which a dislocation is emitted from the crack tip, K_{el} . In the absence of any other dislocations in the system affecting the stresses at the crack tip, this occurs when:

$$K_{app} = K_{tip} = K_{el} \qquad (6)$$

The Kelly, Tyson and Cottrell (1967) criterion may be generalized in such stress intensity terms to express whether the bonds at the crack tip fail by cleavage or slip. If, as the applied loading on the crack is increased, the stress intensity, $K_{app} = K_{tip}$ attains the intrinsic fracture resistance of the material T_0 first, cleavage fracture will occur. If however, the condition for the emission of a dislocation loop, namely that the energy barrier is less than kT and given by K_{el} , is reached first the bonds will break by shear.

An emitted dislocation contributes to the measured fracture toughness by exerting a back stress intensity onto the crack tip, K_{shd} , thereby shielding the crack tip from the applied loading. The second possible effect of dislocation emission is to alter the shape of the crack tip ("blunting") by the shear associated with the Burgers vector component perpendicular to the crack plane. This contribution is represented by the stress intensity term K_{blt} . The effect of these two terms is to alter the effective crack driving force and hence the net stress intensity acting in the crack-tip frame, K_{tip} , from that applied by an external loading system, K_{app} . Both the shielding and blunting terms tend to reduce the value of K_{tip} , so the crack tip stress intensity may be expressed in equilibrium as:

$$K_{tip} = K_{app} + K_{shd} + K_{blt} = K_{app} + K_\mu = T_o \qquad (7)$$

The term K_μ is introduced for consistency, expressing the "microstructural" contribution (in this case due to dislocations) to the fracture resistance. In mode-II and III where K_{blt} is zero, complete shielding occurs when $K_{shd} = -K_{app}$ leading to a value measured in the crack tip frame stress intensity of $K_{tip} = T_o$, a result derivable from the Bilby-Cottrell-Swinden theory.

Since both the crack shielding and blunting contributions act to lower the crack-tip stress intensity they have the effect of increasing the resistance to crack growth. Thus, alternatively, their contribution may be expressed in terms of the net crack resistance, T, as follows:

$$T = T_0 - K_{shd} - K_{blt} = T_o + T_\mu \qquad (8)$$

Thus, the emission of dislocations from the crack tip increases the applied loading required to propagate the crack and hence has the effect of increasing the measured fracture resistance of the material. However, it should be reiterated that unstable crack propagation will always occur when the inequalities of equation 4 (and equivalently 5) are satisfied, irrespective of the details of the actual processes occurring ahead of the crack tip.

The equations and relationships presented up until this point are completely general and apply equally to whether the crack and dislocations are moving relative to one another or not, or indeed whether the dislocations are in their equilibrium (zero net force) positions. Thus, the equations can be used in developing models that include not only the crack velocity but also the dynamics of the dislocations as they interact with each other and with the moving crack. We have endeavored to develop a fully consistent fracture mechanics model incorporating dislocation interactions with a crack-tip for the brittle-to-ductile transition. A complete exposition of the analysis is beyond the space of this proceedings contribution so only the principal features will be presented here with the fuller analysis submitted for publication elsewhere (Cook and Clarke, 1989).

In previously proposed dislocation models, the crack is assumed to be stationary, dislocations are emitted depending on the applied stress intensity factor and when some critical value is reached the crack is assumed to suddenly extend by cleavage. Here, we assume that both the crack and the dislocations move with velocities dependent on their net driving forces. Thus, for a crack-tip driving force, $G - 2\gamma_o$, (where γ_o is the surface energy per unit area of fracture surface) the crack velocity is

$$v = \omega_o a \exp[-2\Gamma/\eta] \sinh\{\frac{(G - 2\gamma_o)}{\eta}\}$$

where 2Γ is an energy barrier for incremental crack extension a, ω_o is an attempt frequency, and $\eta = k_B T/a^2$. By analogy, the dislocation velocity is given in terms of the acting local force, τb, namely,

$$v = v_o \sinh[\tau b/\eta']$$

To proceed we make two assumptions: 1) the normalized applied loading rate is significantly smaller than the crack extension rate:

$$\dot{\sigma}_a/\mu\sigma_a < < \dot{c}/ac$$

and 2) that the dislocation velocities are greater than the crack velocity at any point in time.

In our model it is recognized that several, discrete regimes of behavior exist associated with the successive emission of dislocations from the crack-tip. Each regime depends on the relative magnitudes of the stress intensities for the emission of each dislocation and the magnitude of the instantaneous microstructural stress intensity (all referred to either the applied loading or the crack-tip loading frame). For instance, the first few regimes may be defined as:

$$K_{e1} > > T_o$$

$$K_{e1} > T_o, K_{e2} > > K_{e1}, K_{\mu1} + K_{e1} > T_o$$

$$K_{e1} > T_o, K_{e2} > > K_{e1}, K_{\mu1} + K_{e1} < T_o$$

$$K_{e2} > K_{e1} > T_o, K_{e3} > > K_{e2}, K_{\mu2} + K_{e2} > T_o$$

(The numerical subscripts here refer to the order of the dislocations emitted).

An indication of the changes in stress intensities and the the crack length that occur in these regimes is presented in the following paragraphs and the accompanying figures 4 and 5. We start by considering a dislocation free solid containing a pre-existing sharp crack of length c_o (as for instance in the experiments of St John and Brede and Haasen) and start to apply a load at a stress rate of $\dot{\sigma}_a$. This corresponds to regime 1 above. The crack is metastably "trapped" until the stress intensity factor at the crack tip reaches a value of T_o at which point the crack becomes unstable to extension. Depending on whether the crack propagation is dominated by the applied loading rate or the crack propagation rate the stress intensity factor in the crack-tip frame of reference will vary with crack length as shown in figure 4. Typically, the compliances are such that some path on the figure is taken for which both the applied stress and crack length are simultaneously varying. Brittle extension of the crack occurs and no dislocations are emitted.

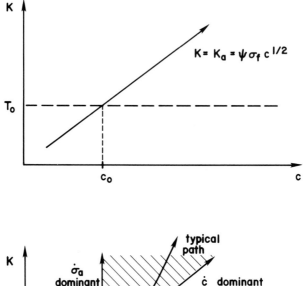

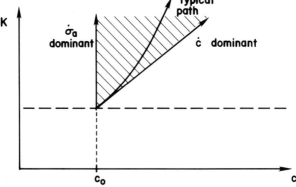

Figure 4. Schematic logarithmic diagram indicating the stress intensity factor for a crack of initial length c_o and how it changes once the intrinsic fracture resistance T_o is exceeded. The diagrams are in the applied loading frame (top) and in the crack-tip frame (bottom). The actual stress intensity factor at the crack-tip depends on the relative magnitudes of the crack growth velocity and the applied loading rate.

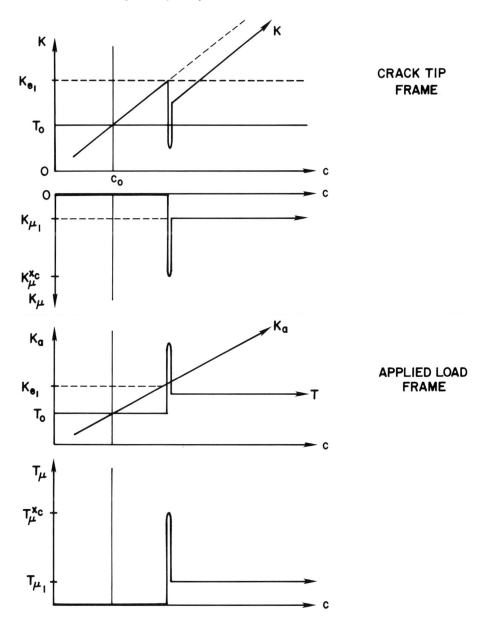

Figure 5. Under the assumption that the crack can grow once it is subject to an excess driving force, the stress intensity factors described in the text vary as shown for the case that $K_{el} > T_o$. The crack system response is shown in both the crack-tip and applied loading frames for clarity.

The next case to consider is when $K_{e1} > T_o$ and $K_{\mu 1} + K_{e1} > T_o$. Under such conditions, the crack is once again metastably "trapped" until T_o is reached at which point it accelerates off as in figure 4. As K_{app} is increased the stress intensity at the crack tip rises until $K = K_{app} = K_{e1}$ at which point the first dislocation is emitted. As it moves away from the crack-tip, the values of K_μ and T_μ simultaneously drop from the maximum value until the dislocation reaches its equilibrium position and the crack is shielded to the extent of $K_{\mu 1}$. As the dislocation velocity is greater than the crack velocity, the dislocation maintains its distance (the dislocation free zone) and hence the degree of shielding remains constant as the crack accelerates off again.

Proceeding with the analysis specific predictions of the crack velocity and dislocation positions can be computed for different parameters in the crack velocity and dislocation velocity equations. Interestingly, even though a constant loading rate may be applied the crack velocity can vary enormously; as the crack extends conditions exist for it to accelerate, to decelerate, to balk or to propagate at a constant velocity. Thus, crack velocity per-se may be a rather unreliable indicator of the overall mode of fracture.

5. DISCUSSION

Perhaps the most important point that this contribution seeks to make is that unstable crack propagation (brittle fracture) can occur even if dislocation emission occurs during crack propagation provided that the overall energy can be reduced. This is in accord with the observation of dislocation etch pits on the fracture surfaces immediately before brittle fracture by Brede and Haasen and the etch pit observations by Samuels and Roberts (1989) of a "transition" specimen (their figure 16). It is also consistent with our in-situ TEM observations. Hirsch et al. have recently come to the same conclusion but based on necessary equilibrium conditions (our equation 1 here) rather on the more stringent sufficient condition. Also, they assume that the dislocations emitted from the operating source have to move a certain distance before they can shield the "most vulnerable" part of the crack.

The loading configuration used by different authors can be expected to influence the possibility of observing dislocations emitted prior to unstable crack growth. For instance, it is notable that Brede and Haasen use a double cantilever beam sample so that whilst controlling the loading rate the crack extends under a constant applied stress intensity. They report that dislocations are emitted at temperatures significantly below the brittle-to-ductile transition. Chiao and Clarke observed dislocation emission in samples that were loaded under essentially displacement control. In contrast, the samples of Samuels and Roberts were loaded at constant strain rates but in a non-stabilizing stress field.

In closing it is worth noting that other than the global energy criterion there is as yet no clear cut criterion for the transition from the macroscopic brittle to ductile fracture behavior in terms of dislocation models. Brede and Haasen assume that the transition occurs when the velocity of the leading dislocation of the plastic zone moves faster than the crack-tip. Hirsch et al. suggest that the transition temperature is primarily controlled by the time it takes for an existing dislocation to reach the crack tip to form dislocation sources. We concur with the conclusion first reached by St John and subsequently reached on the basis of doping effects on the strain rate sensitivity of the transition temperature (by Brede and Haasen and Samuels and Roberts) that the transition is determined by the dislocation velocity but cannot at this time establish a simple dislocation criterion other than that embodied in the total energy equations.

ACKNOWLEDGEMENT

The work was supported in part by the U. S. Office of Naval Research under contract Number N00014-88-C-0176.

6. REFERENCES

Bilby B A, Cottrell A H and Swinden K H 1963 *Proc. R. Soc.* A**272** 304.
Brede M and Haasen P, 1988 *Acta Met.* **36** 2003
Chiao Y-H and Clarke D R, 1989 *Acta Met.* **37** 203
Cook R F and Clarke D R 1989 *Acta Met.* in preparation
Hirsch P B, Roberts S G and Samuels J, 1987 *Proc. R. Soc.* A**421** 25.
Kelly A, Tyson W R and Cottrell A H 1967 *Phil. Mag.* **15** 567.
Maeda K and Fujita S, 1989 *Scripta Met.* **23** 383
Michot G and George A 1986 *Scripta. Met.* **20** 1495.
Ohr S M 1985 *Mater. Sci. Eng.* **72** 1.
Rice J R and Thomson R 1974 *Phil. Mag.* **29** 73.
Samuels J and Roberts S G 1989 *Proc. R. Soc.* A**421** 1.
St. John C 1975 *Phil. Mag.* **32** 1193.

Inst. Phys. Conf. Ser. No 104: Chapter 4
Paper presented at Int. Symp. on Struct. Prop. Disloc. Semicond., Oxford, 1989

TEM investigation of dislocations around indentations on {0001} SiC

C-D Qin and S.G. Roberts

Department of Metallurgy and Science of Materials,
University of Oxford, Parks Road, Oxford, OX1 3PH, U.K.

ABSTRACT: TEM investigation of indentations on $(000\bar{1})$ (carbon) and
(0001) (silicon) faces of 6H SiC single crystals shows that
dislocation slip is by basal partial dislocations ($b = a/3\langle 1\bar{1}00\rangle$). At
500°C, these move as isolated partials. At 1000°C, they move as
paired partials, forming a total b of $a/3\langle 11\bar{2}0\rangle$. The full dislocation
distribution was observed around conical indentations at both 500°C
and 1000°C.

1. INTRODUCTION

Evidence has shown that dislocations can be introduced into 6H SiC
crystals under various deformation conditions (e.g. Adewoye 1981, Fujita
et al 1987, Maeda et al 1987). The stacking fault energy is very low
(Maeda et al 1987), therefore dissociated dislocations are normally
generated during deformation. However, there is still a lack of
agreement among the investigators as to the deformation behaviour of 6H
SiC. In studies of plastic deformation, Amelinckx et al (1960), using
optical and X-ray microscopy, reported high densities of dislocations
having $\langle 11\bar{2}0\rangle$ vectors slipping primarily on the {0001} basal slip plane,
and also on the $\{4\bar{4}01\}$ and $\{1\bar{1}01\}$ planes. Stevens (1972), however,
reported a scarcity of dislocation slip traces but a large number of
stacking faults bounded by partial dislocations in TEM studies.

More recently, hardness anisotropy studies have been used to determine
active slip systems in single crystal 6H SiC (Sawyer et al 1980, Naylor
1981 and Fujita et al 1986). Sawyer et al (1980) and Naylor (1981)
proposed the operation of $\{1\bar{1}00\}\langle 11\bar{2}0\rangle$ and $\{0001\}\langle 11\bar{2}0\rangle$ slip systems
under the hardness indenter so as to fit their hardness anisotropy
measurements, made at temperatures between 20°C and 800°C. They analysed
their results using the modified hardness anisotropy model of Brookes et
al (1971). Fujita et al (1986) calculated the hardness anisotropy using
the model of Brookes et al (1971) with $\{0001\}\langle 1\bar{1}00\rangle$ slip systems
(corresponding to partial dislocations), based on results of Maeda et al
(1987) that dislocations in this material were found to be widely
dissociated; their results fit the actual hardness anisotropy measured
on basal planes. Hardness anisotropy was observed to depend on testing
temperatures, so, to fit the model, the degree of basal dislocation
reconstruction and cross-slip was assumed to be temperature dependent,
and the high mobility of prismatic edge dislocations at low temperatures
was taken into account (Maeda et al 1987). These assumptions, however,
lacked experimental confirmation.

Thus further investigation of dislocation behaviour in SiC is necessary
to understand the plastic flow mechanisms in the temperature range 20°C –
1000°C. The hardness anisotropy models can also be better understood by
taking 6H SiC as a model "ultra-hard" material.

2. EXPERIMENTAL

Two types of 6H SiC single crystals (doping levels unknown, one type
blue-black and the other yellow in colour) with {0001} faces were
mechanically polished, with the final polish being 1 micron diamond
paste. Etching in Na_2O_2 at 450°C for 8 minutes was performed to
distinguish the polarity of "Si" and "C" {0001} faces (Brack 1965).

Conical indentations were made on both (0001) "Si" and (000$\bar{1}$) "C" faces
at 500°C with 50g loading, dwell time of 15 seconds using a Matsuzawa
MHT1 hardness tester equipped with a hot-stage. Vickers indentations made
at 1000°C were also performed using a special high temperature hardness
tester operated in vacuum (manufactured by Stanton Redcroft Ltd).
Thereafter, specimens were back-polished by mechanical polishing and ion
beam milling to perforation. A Philips CM12 transmission electron
microscope was used and operated at 120kV.

3. RESULTS

Dislocation patterns were found to be the same for the two types of
materials (black and yellow), and also the same for indentations on the
two opposite faces (silicon and carbon faces). Typical results are
reported below.

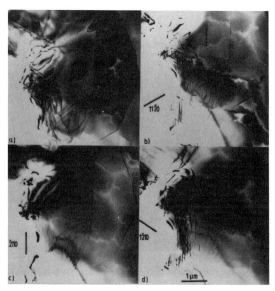

Fig.1 Bright field (BF) TEM of an indentation made at 500°C on the
silicon face of the yellow material. The indentation centre is at
the lower left hand corner of each micrograph. (a) multibeam image
showing the unpaired partial dislocation loops bounding the
stacking faults. (b)-(d) g.b analysis (g vectors are marked).
Burgers vectors are shown for some of dislocations in (b).

3.1 Low temperature (500°C) indentations

Figs. 1 and 2 show that all extended dislocations lie in the basal planes, and are partials with Burgers vectors a/3<1100>, always of the same type in a particular region. They tend to align towards the <1120> and <1100> line orientations, though <1120> directions are the more favoured. No "trailing" partials are observed even close to the indentation centre; these complementary partials may not have been produced. Median/radial cracks are observed lying approximately along the <1100> orientations; the extended partial dislocations are generally connected to median/radial cracks (see fig.2).

The dislocation and median/radial crack distribution had 6-fold symmetry. Fig.2 shows a part of the six-fold dislocation distribution.

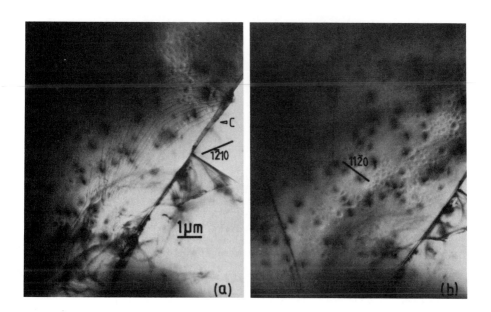

Fig.2 BF TEM of an indentation made at 500°C on the carbon face of the black material. The indentation centre is at the lower left hand corner of each micrograph. (a) g=1210: unpaired partial dislocations are in contrast and are associated with the [1100] crack (Marked "C"). (b) g=1120: these are out of contrast.

3.2 Elevated temperature (1000°C) indentations

Fig.3 shows partial dislocations around indentations made at this high temperature. Partials were found in pairs in this case, though the width of stacking faults is not consistent (one very narrowly paired dislocation marked as "P"), and cannot be used to measure the stacking fault energy accurately. Dislocation Burgers vectors (determined by g.b analysis) are denoted by small bars.

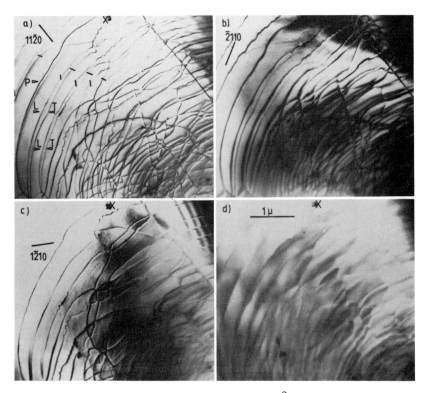

Fig.3 BF TEM of an indentation made at 1000°C on the {0001} face of
the yellow material. The indentation centre is at the lower right
hand corner of each micrograph. 'X' denotes the same point in each
micrograph. (a) g=11$\bar{2}$0: all partials are in contrast. (b) g=$\bar{2}$110:
some of the partial dislocations are out of contrast. (c) g=1$\bar{2}$10:
the other set of partial dislocations are out of contrast. (d)
multibeam image shows the stacking fault contrast. Some paired
partial dislocations are marked as 'L' and 'T'; L and T denote
leading and trailing partial dislocations, respectively.

4. DISCUSSION

4.1 Dislocation slip system

For indentations made at 500°C and at 1000°C, all dislocations seen were
basal partial dislocations. This clarifies previous suggestions about
other possible slip systems for indentation on basal planes (e.g.,
Sawyer et al 1980). From the present work, it is not certain why there
are prismatic dislocations around indentations on {1$\bar{1}$00} planes (Maeda
et al 1987).

The transition from unpaired to paired partial dislocations on
increasing the indentation temperature shows that slip of the trailing
partial dislocations is activated only at elevated temperatures. Maeda
et al (1987) explained this transition as being due to the different
mobilities of the two types of dislocations (Si(g) or C(g)), i.e., they
assumed the bounding partial dislocations to be of different types

(though in their figures the boundary partials shown were of the same type). This will not generally be true because these bounding partials will be the same type of dislocations if they have dissociated from a 60° dislocation rather than from a screw dislocation (assuming that dislocation lines align along ⟨11$\bar{2}$0⟩ directions and bounding partial dislocation lines are parallel to parent dislocation lines).

4.2 Partial dislocations

Compared with other materials (e.g. Ge, Si and GaAs) which show paired dissociated dislocation slip, 6H SiC is different mainly in that it has a high Young's modulus and a very low stacking fault energy (Maeda et al 1987). γ/b is related to the nucleation force for partial dislocation nucleation; Gb^2/γ varies as the ratio of the nucleation energy between the perfect and partial dislocations. Table.1 shows Gb^2/γ for some different materials; Gb^2/γ of SiC is about 100 times greater than that for the others. Thus unpaired partial dislocations can be easily generated.

The ratio for SiC implies direct nucleation of the partial dislocations may be favoured. Thus this means that for determination of hardness anisotropy for 6H SiC using, e.g., the model presented by Brookes et al (1971) the operative slip systems {0001}⟨1$\bar{1}$00⟩ should be used.

Table 1*

	$G(10^4MN/m^2)$	γ (mJ/m^2)	b(A)	$\gamma/b(MN/m^2)$	Gb^2/γ
SiC	18.6	2.5	1.73	144.5	23130
Si	6.37	69	2.22	310.8	200
Ge	5.2	60	2.31	260	200

*Note, G represents the shear modulus, γ the stacking fault energy, b the length of the partial dislocation Burgers vector. The shear modulus of SiC is from Shaffer and Jun (1972), stacking fault energy from Madea et al (1987). Shear moduli for Si and Ge are from Frost and Ashby (1982), and the stacking fault energies are from Hirth and Lothe (1982).

5. CONCLUSIONS

In conclusion, dislocations around indentations on basal planes at 500°C and 1000°C are unpaired and paired basal partial dislocations, respectively. The wide dislocation dissociation results from the very low stacking fault energy in this material. Thus {0001}⟨1$\bar{1}$00⟩ slip systems, instead of {0001}⟨11$\bar{2}$0⟩ and {1$\bar{1}$00}⟨11$\bar{2}$0⟩ slip systems, should be used if the Brookes et al (1971) model is applied to basal plane hardness anisotropy analysis at low temperature. At high temperature, the pairs add up to {0001}⟨11$\bar{2}$0⟩ and these would be valid in the context of a hardness anisotropy model.

These TEM studies did not detect any doping effects or any possible different plastic zone configurations for indentations on the two opposite basal faces.

References

Amelinckx S, Strummane G and Webb W W 1960 J. Appl. Phys. 31 1359
Adewoye O O 1981 Crystal lattice Defects 9 pp 107–11
Brack K 1965 J. App. Phys. 36 3560
Brookes C A, O'Neill J B and Redfern B A W 1971 Proc. Roy. Soc. Lond. 322 73
Frost H J and Ashby M F 1982 Deformation-Mechanism Maps (Pergamon Press)
Fujita S, Maeda K and Hyodo S 1986 J. Mat. Sci. 5 L450
Fujita S, Maeda K and Hyodo S 1987 Phil. Mag. A55 203
Hirth J P and Lothe J 1982 Theory of Dislocations (New York: Mcgraw-Hill), 375
Maeda K, Suzuki K, Fujita S, Ichihara M and Hyodo S 1987 Phil. Mag. 57 573
Naylor M G S 1981 Ph.D. Thesis University of Cambridge
Sawyer G R, Sargent P M and Page T F 1980 J. Mat. Sci. 15 1001
Shaffer P T B and Jun C K 1972 Mat. Res. Bull., 7 63
Stevens R 1972 J. Mat. Sci. 7 517

Acknowledgments

Financial support by the British Council and the Chinese Government (C-DQ) the SERC (SGR) is gratefully acknowledged.

Inst. Phys. Conf. Ser. No 104: Chapter 4
Paper presented at Int. Symp. on Struct. Prop. Disloc. Semicond., Oxford, 1989

Plasticity and dislocation mobilities at low temperature in silicon and gallium arsenide

J L Demenet, P Boivin and J Rabier

Laboratoire de Métallurgie Physique, URA 131 CNRS
Faculté des Sciences, 86022 Poitiers Cedex, France

ABSTRACT : Macroscopic and microscopic features of silicon and gallium arsenide deformed at low temperature and high stress are reported. Similarities and discrepancies are pointed out and analyzed in term of dislocation mobilities. The effect of pressure on dislocation mobility is also discussed.

1. INTRODUCTION

Mechanical properties of semiconductors strongly depend on temperature. The temperature range of interest can be roughly divised into three domains characterized by the experimental procedure to conduct compression tests at a fixed strain rate. In the so-called high temperature range, initially dislocation-free crystals can be deformed in the usual range of strain rates ($\dot{\gamma} \approx 10^{-4}$ s^{-1}) without risk of brittle failure. At medium temperature a plastic deformation can be achieved in a standard Instron-type machine only at low strain rates and after a suitable pre-strain at higher temperature has created a high starting dislocation density. At low temperature, brittle fracture must be prevented by a confining hydrostatic pressure superimposed to the uniaxial applied stress.

These three ranges of temperature and stress correspond roughly to three types of analysis of the mechanical properties. At high temperature, semiconductors exhibit the general features of the high temperature plasticity of fcc metals : stress-strain curves, hardening stages...(Siethoff and Schröter 1984, Schröter and Siethoff 1984). Indeed the covalent character of the bonds is not determining in the dislocation configurations when the temperature is large enough. In this range of temperature dislocation fine structure has the same consequences on semiconductors plasticity as it has on fcc metals plasticity. At medium temperature, the yield behaviour of semiconductors can be explained by their covalent character yielding to low velocities of dislocations compared to those of fcc metals. Yonenaga *et al* (1987) state that stress-strain characteristics of a GaAs crystal in the temperature range 400°-500°C are very similar to those of a Si crystal in the temperature range 800°-900°C, reflecting a comparable average dislocation mobility of both materials in the associated temperature ranges. At low temperature the plastic properties of Si and GaAs are very different : stress-strain curves, effect of doping,.... This is thought to reflect the different behaviour of dislocation fine structure (mobility of the different segments, mobility of the partial dislocations). These points show that when temperature is lowered, macroscopic plastic properties reflect more precisely the intimate structure of dislocations.

In this context, this paper reports the main macroscopic and microscopic features of deformation at low temperature and high stress, under confining pressure, of silicon and gallium arsenide. Most of the experimental results presented in this paper are detailed elsewhere : Rabier *et al* (1983), Demenet (1987) for Si, Boivin (1988), Boivin *et al* (1989) for GaAs. Similarities and discrepancies are pointed out and analyzed in term of dislocation mobilities. In addition, the effect of pressure on dislocation mobility is discussed.

2. EXPERIMENTAL RESULTS

2.1. Plastic deformation

The apparatus for studying plastic deformation at low temperature under confining pressure has been described elsewhere (Veyssière *et al* 1985). Temperature ranges explored for Si and GaAs are 300°-600°C (0.34-0.51 T_m) and RT-200°C (0.2-0.31 T_m) respectively, at strain-rates of 2 x 10^{-5} s^{-1} and 2 x 10^{-6} s^{-1}.

In silicon, deformation curves exhibit the well-known yield drop, in the whole investigated temperature range (figure 1). Variations of the lower-yield point of intrinsic silicon deformed along a ⟨213⟩ axis have been analyzed with Haasen's model. Determination of the n and Q parameters of the phenomenological law :
$$\dot{\epsilon} = A.\sigma^n \exp(-Q/kT)$$
leads to n ≈ 5-5.5 and to Q varying between 1.7 eV and 2.2 eV (fig.2). This result is consistent with exponent and energy values deduced from individual perfect dislocation velocity measurements in the same temperature and stress ranges (Kisielowski-Kemmerich 1982, Küsters and Alexander 1983). The influence of the compression axis has been studied in intrinsic silicon. Rabier *et al* (1983) have shown that compression along ⟨213⟩, ⟨110⟩ and ⟨100⟩ led to nearly the same values of the lower yield stress, while in the high temperature range, the effect of orientation is pronounced (Omri 1981). Doping by phosphorous atoms (n-type conductivity) softens the material. The lower yield stress value is, as in the high temperature range, lowered by comparison to intrinsic silicon (fig.1).

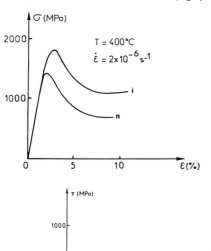

Fig.1. Stress-strain curves of intrinsic and n-type silicon deformed under confining pressure.

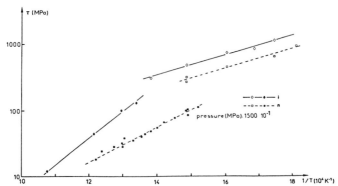

Fig.2. Silicon. Ln τ versus 1/T for tests under confining pressure (1500 MPa) and at atmospheric pressure (0.1 MPa). $\dot{\epsilon} = 2$ x $10^{-6}s^{-1}$.

Stress-strain curves of intrinsic and n-type GaAs single crystals deformed under confining pressure in the same strain-rate conditions as silicon, do not exhibit the yield-drop phenomenon at room temperature (figure 3). This is a typical feature of the deformation at low temperature of these crystals. Examination of lateral faces of samples by optical microscopy reveals   cross-slip activity. On another hand, stress-strain curves of p-type GaAs exhibit a small yield-point. At room temperature, the flow stress of n-type crystal is less than the flow-stress (or the lower yield stress) of intrinsic (or p-type) crystal, in contrast to higher temperature measurements (T > 200°C) (fig.4) (Boivin *et al* 1989). The transition temperature where the effect of doping is reversed is close to 150°C and corresponds to a resolved stress of about 150 MPa. At last, deformation along a ⟨100⟩ axis yields to a significant lowering of the yield stresses (between 130 and 200 MPa at RT depending on the doping) compared to deformation along a ⟨213⟩ axis.

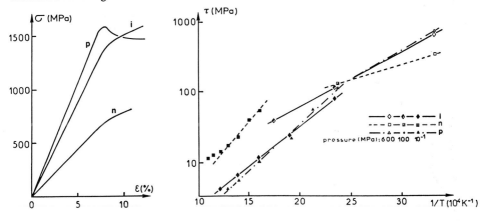

Fig.3. Stress-strain curves of intrinsic, p-type and n-type GaAs deformed under confining pressure (room temperature, $\dot{\epsilon} = 2 \times 10^{-5}$ s^{-1}).

Fig.4. GaAs. Ln τ versus 1/T for tests under confining pressure (600 MPa and 100 MPa) and at atmospheric pressure (0.1 MPa). $\dot{\epsilon} = 2 \times 10^{-5}$ s^{-1}.

2.2. TEM Observations

TEM observations on silicon deformed under confining pressure have been partly described in a recent paper (Demenet *et al* 1987). By compression along ⟨213⟩, deformation occurs by perfect dislocations. At the lower yield point, deformation is not perfectly homogeneous and hexagonal loops are seen with a somewhat marked predominance of screw segments at the edge of slip bands. With increasing deformation rate, dislocation distribution becomes homogeneous and the screw and 60° segments are found to be nearly in the same quantity. Microtwins, twins and cross-slip are also occasionally observed. In the case of silicon deformed along a ⟨110⟩ axis, dislocations are also weakly dissociated and homogeneously distributed. However, after a compression along ⟨100⟩, perfect dislocations are never observed. Microstructure is built with twins and partial dislocations.

TEM observations on GaAs have been mainly performed on crystals deformed at room temperature. In the three kinds of crystals (intrinsic, n-type and p-type), weakly dissociated dislocations and twins or microtwins are simultaneously observed after a compression along ⟨213⟩. In intrinsic and n-type material, perfect dislocations have a very pronounced screw orientation and glide loops are very elongated (Boivin *et al* 1989). In addition, cross-slip, evidenced at a macroscopic level, is confirmed by the observation of numerous segments at 30° from the screw direction. These segments are lying in the cross-slip plane and are seen in projection on the primary glide plane. In p-type deformed crystals, screw and 60° segments are rather homogeneously distributed and cross-slip evidences are much less numerous. Additional experiments have been performed in order to get a more precise insight into the mobility of dislocations. n-type ⟨213⟩ GaAs crystal has been deformed at high temperature (550°C then 250°C) at atmospheric pressure, then submitted to a deviatoric stress of 200 MPa for one hour at room temperature in the confining pressure apparatus. Triangular faults have been evidenced (Boivin 1988, Boivin *et al* 1989). The analysis shows they come from dissociation of 60°α segments and this evidences the higher mobility of 90°α partials compared to 30°α partials, whatever their leading or trailing characters. In addition, the behaviour of screw segments depends on the nature of the leading partial. Screw segments 30°ß/30°α, i.e. with the leading 30°ß partial, show high ability for cross-slip, while 30°α/30°ß segments, i.e. with the leading 30°α partial, dissociate in the glide plane.

3. DISCUSSION

In the temperature range where Si and GaAs are ductile, stress-strain characteristics of both materials are very similar. The corresponding temperature ranges are 800°-900°C for Si and 400°-500°C for GaAs (Yonenaga *et al* 1987). In addition, configurations of dislocations developed by deformation are analogous. When decreasing temperature, the same shift on ductility is observed. With the help of a confining pressure, silicon is deformed down to 300°C and gallium arsenide down to RT. In the whole temperature range, intrinsic, p- and n-type silicon exhibit a yield drop at the beginning of deformation. In addition, the influence of doping is not modified when temperature decreases (fig.2). In contrast, the plastic behaviour of GaAs changes from high to low temperature. The yield drop disappears at low temperature on intrinsic and n-type material (fig.3) and the influence of doping is reversed (fig.4). Are those phenomenons due to a drastic modification of doping effect on the velocity of glide dislocations or to the existence of other(s) deformation mechanism(s) ? TEM observations are required to answer to this question.

In silicon, the distribution of dislocations is rather homogeneous although the screw orientation is somewhat prevalent at the beginning of the deformation. The mobility of segments is nearly the same. This feature is associated with the occurrence of a yield drop. The observed microstructure in samples deformed by compression along ⟨213⟩, ⟨110⟩ and ⟨100⟩ is well explained by the glide force systems acting on partials. By compression along ⟨213⟩ (or ⟨110⟩), the glide force on the trailing partials is 7/5 (or 2/1) of the glide force on the leading partials, and weakly dissociated dislocations are observed. The minimum resolved shear-stress to separate partials can be deduced from the ratio of friction forces R* acting on leading (l) and trailing (t) partials (R* = R_l/R_t) (Alexander *et al* 1980, Demenet 1987). For a resolved shear stress of 1000 MPa, which corresponds to the lower yield stress at ≈ 350°C and $\dot{\epsilon}$ = 2 x 10^{-6} s^{-1} in intrinsic silicon, dissociation of partials could be observed in ⟨213⟩ deformed material if R* < 0.35 and in ⟨110⟩ deformed material if R* <0.25. By compression along ⟨100⟩, the glide force on the leading partials is twice the glide force on the trailing partials. Partials can dissociate for a resolved shear stress varying between 230 and 910 MPa when R* varies between 0 (no friction force on the leading or the trailing is pinned) and 1 (same friction force on the two partials). At low temperature, the stress level in the deformation tests lies in this range and so this explains the extended stacking faults observed. In conclusion it can be claimed that in silicon, no preferential segment is responsible of the deformation at low temperature and high stress and that differences in partial mobilities are not important.

In GaAs, the same force analysis can explain the low dissociation of dislocations produced by compression along ⟨213⟩ and the extended stacking faults created by compression along ⟨100⟩. The lower stacking fault energy leads to lower minimum stress for dissociation. However, another explanation is needed to explain the other features observed in the ⟨213⟩ deformed samples : screw-type dislocations are prevalent at low temperature for intrinsic and n-type GaAs, cross-slip and twinning are observed whatever the doping, screw segments of opposite sign have asymmetric behaviour under high stress in n-type GaAs. A model has been developed to explain these observations (see Rabier, this issue). In contrast with Si, the difference in velocity between α, ß and screw dislocations reported in individual dislocation velocity measurements shows that the density of screw dislocations will be more important than the faster ones. This yields to the formation of screw pile-ups which have to be eliminated. So, multiplication of dislocation can become difficult, a problem which is not encountered in Si. Cross-slip mechanisms are then of importance in understanding the low temperature deformation of GaAs.

Although applying a confining pressure allows to reach low temperature and high stress regimes for Si and GaAs, the comparison between tests conducted using different deformation techniques at the same temperature shows that yield stresses obtained under pressure are always larger (figures 2 and 4). It is of interest to check if those particular deformation conditions can play a role on the dislocation mobility. This point is discussed in what follows. When an external stress σ is applied, dislocations are subjected to the well-known Peach-Koehler force expressed by $\mathbf{F}$ = $(\sigma.\mathbf{b})_x\mathbf{u}$, where

σ is the stress tensor, **b** is the Burgers vector and **u** is the unit vector of the dislocation line (Hirth and Lothe 1982). If dislocations are dissociated into two Shockley partials, as usually in semiconducting materials, each partial (30° or 90° partial, belonging to a screw or a 60° segment) is subjected to a Peach-Koehler force depending on the deformation axis orientation. The Peach-Koehler force on a partial can be divided into two components : one component in the glide plane, the so-called glide force, and one component in the plane normal to the glide plane, the so-called climb force. The climb force can be directed out of the extra-half plane or into the extra-half plane associated to the partial. In the former case, it is a positive climb corresponding to a vacancy emission mechanism, and in the later case, it is a negative climb corresponding to a vacancy absorption mechanism. Forces on partials (30° or 90°) of a pre-existing dislocation loop, according to the compression axis orientation have been calculated and their effect studied by Demenet *et al* (1989) for Si. If a sample is confined in a pressure field defined by a tensor **P**, an additional force is exerted on dislocations. In the case of a hydrostatic pressure, tensor **P** is reduced to its diagonal elements, P. A simple calculation shows that the additional force is always directed into the extra-half plane without any component in the glide plane (Demenet 1987). So, the confining pressure modifies the climb force on partials. Depending on the sense of the climb force due to the uniaxial stress, the resulting climb force is amplified, lowered or reversed with respect to a standard uniaxial stress. An example is given in figure 5 for each segment (30°/30°, 30°/90° and 90°/30°) of a loop with dislocations lying along the ⟨110⟩ directions. The thin arrows represent the Peach-Koehler force on each partial when a σ = 500 MPa stress is applied by compression along a ⟨213⟩ axis. The thick arrows represent the resulting Peach-Koehler force on each partial when a hydrostatic confining pressure P = 1500 MPa is added to the uniaxial stress. The resulting climb force on the two partials of a 60° segment is always negative (negative climb). However, in the case of the screw segment, built with two 30° partials, the resulting climb force on the 30° trailing partial is amplified and it is reversed on the 30° leading partial. A 120 MPa confining pressure is high enough to erase the climb force on the 30° leading partial.

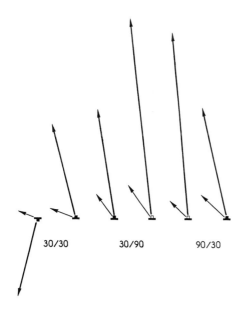
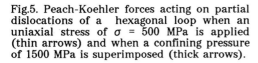

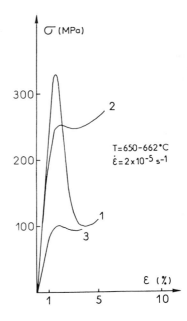

Fig.5. Peach-Koehler forces acting on partial dislocations of a hexagonal loop when an uniaxial stress of σ = 500 MPa is applied (thin arrows) and when a confining pressure of 1500 MPa is superimposed (thick arrows).

Fig.6. Silicon. Stress-strain curves at atmospheric pressure (1), under confining pressure (2), at atmospheric pressure after pressurization (3).

Alexander *et al* (1983) and Grosbras *et al* (1984) have demonstrated, studying dislocation dissociation widths in silicon under well-defined orientations, that the climb force acts on the dislocation mobilities. Particularly a negative climb force amplifies the friction forces acting on 30° partials (Demenet 1987, Demenet *et al* 1989).

Experimental tests have been performed on silicon in the temperature range within which both deformation under confining pressure and at room pressure are possible. Results are reported in fig.6. The yield drop is nearly suppressed and the lower yield stress is much higher for sample deformed under pressure (700 MPa). In addition, the deformation curve of silicon deformed at room pressure after pressurization (0.5 h at T = 450°C and P = 150 MPa) is shown. It can be seen that pressurization introduces dislocations which contribute to reduce the yield drop. It is obvious that the rough hydrostaticity and the unavoidable frictions in the confining solid medium cannot be excluded to explain partly the high value of the lower yield stress, particularly at this "high" temperature (0.54 T_m). In order to check this point, some comparative tests have been performed in a gas machine (Paterson, 1970) with a confining pressure of 300 MPa. The lower yield stress was 10 % less than obtained with a solid confining pressure of the same intensity. So, it seems reasonable to think that confining pressure modifies friction forces on partials. One way could be a change of dislocation core of partials which could lead to modify the mobility of partials, as proposed by Duesbery (1983). Although no effort has been devoted to check if climb forces act on the mobility of dislocations in GaAs, the similarity between results obtained for Si and GaAs with and without pressure can be explained by the existence of such an effect.

ACKNOWLEDGEMENTS

The authors thank Dr P Chopra (Research School of Earth Sciences, Canberra) for performing the experiments on silicon in the gas machine.

REFERENCES

Alexander H, Kisielowski-Kemmerich C and Weber E R 1983 *Physica* 116B 583
Boivin P 1988 Thèse, Poitiers
Boivin P, Rabier J and Garem H 1989 submitted to *Phil. Mag.*
Demenet J L 1987 Thèse d'Etat, Poitiers
Demenet J L, Rabier J and Garem H 1987 *Inst. Phys. Conf. Ser.* 87 355
Demenet J L, Grosbras P, Garem H and Desoyer J C 1989 *Phil. Mag. A* 59 501
Duesbery M S 1983 *Acta Metall.* 31 429
Grosbras P, Demenet J L, Garem H and Desoyer J C 1984 *Phys. Stat. Sol.(a)* 84 481
Hirth J P and Lothe J 1982 *Theory of Dislocations* 2nd edition (New York : Wiley and Sons) pp 288
Kisielowski-Kemmerich C 1982 Thesis, Köhln
Küsters K H and Alexander H, 1983 *Physica* 116B 594
Omri M 1981 Thèse, Nancy
Paterson M S 1970 *Int. J. Rock Mech. Min. Sci.* 7 517
Rabier J 1989, this proceedings
Rabier J, Veyssière P and Demenet J L 1983 *J. de Physique* 44 C4-243
Schröter W and Siethoff H 1984 *Z. Metallkde.* 75 482
Siethoff H and Schröter W 1984 *Z. Metallkde.* 75 475
Veyssière P, Rabier J, Jaulin M, Demenet J L and Castaing J 1985 *Rev. Phys. Appl.* 20 805
Yonenaga I, Onose U and Sumino K 1987 *J. Mater. Res.* 2 252

Inst. Phys. Conf. Ser. No 104: Chapter 5
Paper presented at Int. Symp. on Struct. Prop. Disloc. Semicond., Oxford, 1989

421

Effects of extended lattice defects on silicon semiconductor devices

B.O. Kolbesen, W. Bergholz, H. Cerva, F. Gelsdorf,
H. Wendt and G. Zoth
Siemens AG, Otto-Hahn-Ring 6, 8000 Muenchen 83,
Fed. Rep. Germany

ABSTRACT: Extended lattice defects within the electrically active regions of devices are generally harmful to the device performance and hence, may cause yield problems. Here, we discuss a few typical defects that may occur in micron and submicron silicon technology if processes are not optimized: Dislocations generated in 4M DRAMs with trench memory cell, defects created at implantation mask edges, stacking faults formed in Pt doped power MOS devices and gate oxide defects induced by metal precipitates at the silicon/silicon dioxide interface.

1. INTRODUCTION

The driving force of the progress in silicon semiconductor technology is the continuous reduction of the average costs: in the case of memories each year about 35 % per bit. This cost reduction is based primarily on a continuous decrease of the minimum feature size, increase of the packing density by introducing novel space saving device concepts and the use of silicon wafers with larger diameters (125 and 150 mm are the most frequently used diameters at present). In particular, the decreasing lateral minimum feature size requires also a reduction of the vertical feature sizes (thickness of layers, depth of junctions). The consequence of all this is a drastic increase of process complexity: The fabrication of the 4M DRAM (<u>m</u>egabit <u>d</u>ynamic <u>r</u>andom <u>a</u>ccess <u>m</u>emory) discussed in section 2 requires more than 400 process steps. This in turn calls for a high level of process maturity and stability. To achieve production yields at economical levels defect density levels of the order of 0.1 cm^{-2} per critical mask level for chip sizes of 1 cm^2 are required. Among the variety of random defects, such as particles, pin holes etc. which contribute to yield losses, extended lattice defects in the silicon substrate material and defects in the gate oxide can play a prominent role (Kolbesen et al. 1981, 1985, 1988). In the following we discuss a few typical examples of extended defects that can occur in current silicon technology, if processes are not optimized.

2. "TRENCH-INDUCED DISLOCATIONS" IN 4M DRAMS WITH TRENCH MEMORY CELLS

At present 4M DRAMs are at the leading edge of VLSI (<u>v</u>ery <u>l</u>arge <u>s</u>cale <u>i</u>ntegration) MOS technology and are close to mass production (Beinvogl et al. 1988, Sunami 1985, Küsters et al. 1987). Here, for the first time the threshold to submicron structures has been crossed. Fig. 1 shows a SEM micrograph of a cross section of a 4M DRAM with a trench memory cell. Each cell consists of a storage capacitor and a transfer gate (MOS transistor). The three-dimensional memory cell concept constitutes one of the most important innovations in the 4M DRAM generation. It results in a reduction of the cell size of 50 % compared to the planar 1M DRAM cell. In this cell concept the storage capacitor is folded into a 5 μm deep trench of 1 μm width etched into the silicon substrate. The fabrication of those trenches can be regarded as the masterpiece of today's dry etching technology.

The implementation of the storage capacitor into a trench includes a bundle of process steps such as: doping of the substrate by diffusion from an oxide source, growth of the thermal gate oxide (capacitor dielectric), deposition and doping of a poly silicon layer (capacitor plate), insulation of this poly silicon by a CVD (<u>c</u>hemical <u>v</u>apour <u>d</u>eposition) oxide, refilling of the trench by undoped poly silicon, back etching of the refill poly silicon to form again a nearly planar surface topology, growth of a thermal oxide to insulate the capacitor complex from the transistor complex which is fabricated subsequently.

From the device design perspective the trench memory cell concept is an elegant solution to save area. However, obvious drawbacks are not only a considerable increase in process complexity but also a significantly raised susceptibility to the formation of crystallographic defects, compared to planar cell concepts. It has indeed turned out that trench structures are potential sites of dislocation generation ("trench-induced dislocations" (TIDs)). Fig. 2 shows examples of TIDs delineated by short preferential etching (30 sec Secco etch (Secco 1972)) and imaged in a scanning electron microscope. (Before preferential etching all the layers on top of the silicon substrate are removed by a so-called lift off etch technique in hydrofluoric acid). Such dislocations usually extend only a few microns into the silicon bulk. This is illustrated by Fig. 3 which shows a cross section of trenches after preferential etching. A dislocation intersects the sidewall of a trench. Since these defects are confined to a shallow surface layer, they do not show up in conventional transmission (Lang) x-ray topographs. In extreme cases dislocations were found at about 50 % of all trenches on a 4M chip.

The TIDs degrade the electrical performance of the affected memory cell in the following manner: In DRAMs, the information stored in the capacitor of the memory cell can only be maintained for a limited time. The capacitors lose their charge due to various "natural" leakage current sources (Chatterjee et al. 1979). Therefore, the information has to be refreshed periodically in cycles of 16 ms in the case of the 4M DRAM. (For previous generations of DRAMs shorter cycles were used.) By a special test procedure consisting of alternate write and read cycles with increasing period weak memory cells can be detected and localized. The number of failing memory cells (refresh failures) can be plotted as a function of the refresh delay time and can also be imaged in a so-called fail bit map. It is well known from previous generations of DRAMs that extended lattice defects, such as oxidation induced stacking faults and dislocations, cause predominantly refresh failures if they are located in storage capacitors (Kolbesen and Strunk 1981, 1985, Lemme et al. 1988). The same relationship has been verified for TIDs and refresh failures for the 4M DRAM. Fig. 4 shows the relationship between fail bits (refresh failures) and the refresh delay. The good cells are represented by the steep (intrinsic) branch of the curve corresponding to refresh delay times of > 200 ms at 85°C (extrapolated curve dashed). Cells with lower refresh times fail due to leakage currents by generation centres. The latter was verified by measuring the temperature dependence yielding an activation energy of 0.4 - 0.6 eV for the current generation. A good correlation was found between the number of fail bits and the number of etch pits (corresponding to trench-induced dislocations) counted in the characterized memory cell arrays. Moreover, a good correlation was also obtained between single bit "1" failures localized on a fail bit map and TID etch pits found in the failed cells.

The mechanism of formation for dislocations at the trench edges is not fully understood at present. It is obvious that several sources of considerable mechanical stresses are present during device processing, such as film-edge-induced stresses from patterned oxide, nitride and poly silicon films and temperature gradient induced stresses during the heating and cooling cycles of furnace processes. There is also a variety of potential dislocation sources available in the complex process sequence of the 4M DRAM, such as residual damage from ion implantations and dry etching processes, microdefects originating from a practically unavoidable slight metal contamination etc. It is very likely that the sharp edges and corners of the trench geometry play a prominent role in the dislocation generation process, since in planar capacitor structures that type of defects does not occur. It cannot be decided whether such geometric effects inherent in the 3-D structure or the increase in process complexity are mainly responsible for the dislocation generation. In any case, from a pragmatic device manufacturing point of view the prevention of those defects is an absolute necessity and has been achieved by optimization of the complete process sequence.

3. "SPACER-EDGE-INDUCED DEFECTS"

The scaling down of VLSI devices requires the definition of small doped areas in the silicon substrate. Those are usually fabricated by selfalignment techniques when applying ion implantation of the dopant species. In the case of MOS transistors the poly silicon gate acts as implantation mask for the implantation of the source/drain regions. In order to overcome hot electron problems of micron and submicron size MOS transistors the socalled LDD (lightly doped drain) concept has been conceived (Fig.

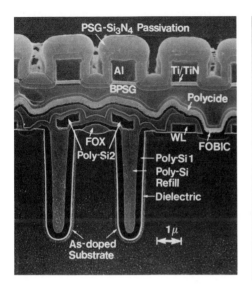

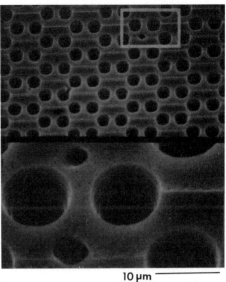

10 µm

Fig. 1: SEM micrograph of a cross section through a 4M DRAM with trench storage capacitors. Also visible: Various deposited dielectric and conductive layers which give rise to thin film edge stresses.

Fig. 2: Plan view SEM micrograph of trench-induced dislocations in a 4M DRAM: delineated as small etch pits between the large holes (= empty trenches). Preparation: Removal of all technology layers by an HF lift-off technique and Secco defect etching of the exposed Si substrate.

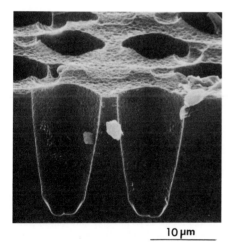

10 µm

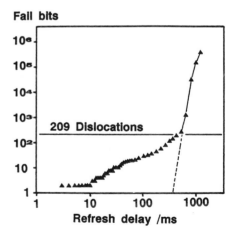

Fig. 3: SEM micrograph of a cross section through the trench capacitors of a 4M DRAM after preparation as in Fig. 2. From the two etch pits it is obvious that the trench-induced dislocation runs from the substrate surface to the wall of the adjacent trench capacitor about 2 µm below the substrate surface.

Fig. 4: Cumulative number of bits in a 512k block of a 4M DRAM that fail a refresh test as a function of the refresh delay time, i.e. the time between successive refresh cycles. Also indicated: the number of trench-induced dislocations detected by defect etching.

5a). In this concept two implantations are necessary: A first phosphorus implantation masked by the poly silicon gate provides the lightly doped drain region, a second arsenic implantation masked by a socalled (oxide) spacer forms the highly doped source/drain contact regions. The implantations are followed by one or several high temperature annealing steps to electrically activate the dopant atoms and to remove the implantation damage. It is known that high dose As implantations produce a buried amorphous layer in the silicon (amorphisation dose $\geq$ 5 x 10 exp 13 atoms/cm^2). During annealing this amorphous layer vanishes by an epitaxial regrowth process restoring the original lattice more or less perfectly. Since the oxide spacers have steep side walls the high dose As implanted regions terminate abruptly at the edge of the spacers.

Cerva and Küsters (1989) have observed that in the case of non-optimal processing extended lattice defects are created in the silicon beneath the spacer edges (Fig. 5a). In an extensive investigation by transmission electron microscopy (TEM) they characterized the subsequent stages of defect formation during the recrystallization process (Fig. 5b,c,d). In the as-implanted state a 70 nm deep amorphous zone is visible that is sharply curved beneath the spacer edge (Fig. 5b). After annealing at 500°C (30 min) half of the amorphous layer has recrystallized, and a notch appears in the amorphous/crystalline interface (Fig. 5c). Annealing at 900°C (60 min) has removed the amorphous layer completely. However, a planar defect has formed under the spacer edge. The defect originated at the position of the notch in the amorphous crystalline interface (Fig. 5d). Such defects always lie on that set of {111} lattice planes which points toward the mask window. The majority of the defects consists of vacancy-type dislocation half loops. Some more complicated defect configurations have been identified as microtwins. Cerva and Küsters (1989) explain the formation of those defects by a model that relies on the different epitaxial regrowth rates for the different substrate lattice plane orientations (Csepregi et al. 1978).

Such spacer-edge-induced defects give rise to enhanced pn-junction leakage currents if they extend into the space charge region of pn-junctions and constitute also a reliability problem for the MOS transistors. By optimizing the implantation and annealing conditions the formation of those defects has been prevented or reduced to a large extent.

4. GATE OXIDE DEFECTS

In an advanced CMOS device a significant fraction of the chip area consists of thin gate dielectrics. In 1M DRAMs the gate oxide amounts to almost 50% of the total chip area of about 45 mm^2. Therefore, with increasing device complexity and a higher level of integration there is a continuous pressure to reduce gate oxide defect densities. During recent years, strong evidence has been compiled that the most important contribution to gate oxide failure by dielectric breakdown stems from crystal defects, and that metal contamination assumes a central role.

In the first instance, this can be demonstrated for gate oxides manufactured under state-of-the-art processing conditions: The density of shallow etch pits (S-pits) which reflects the degree of metal contamination, quite clearly has an impact on the yield of gate oxide test structures (Fig. 6a). S-pit densities in excess of some 10^3cm^{-2} make serious inroads into the yield which is almost 100% for S-pit densities lower than 10^3cm^{-2}. Yamabe et al. (1984) have demonstrated that P-diffusion gettering on the wafer back-surface drastically improves the gate oxide quality of planar 8 mm^2 test capacitors manufactured on the front side of the wafer, i.e. the removal of metal impurities has eliminated the main cause for gate oxide defects.

In order to elucidate the mechanism of gate oxide degradation by metals, intentional contamination experiments have been carried out with Cu, Ni and Pd. The impurities were diffused into wafers with either just 20 nm gate oxide or gate oxide covered by a poly silicon layer from the wafer backside by rapid thermal annealing. Electrical testing was either performed in an electrolytic pin-hole detector (Eisenberg and Brion 1969) or in an automatic tester by a voltage ramp test. It demonstrated a clear spatial correlation between reduced breakdown fields and contaminated areas.

Microscopic examination of such areas led to similar findings for all three impurities studied: For contamination of (100) wafers either before or after poly Si deposition silicide precipitates at the Si/SiO$_2$ interface are observed. For Pd, a 2 nm thinning of the gate oxide by the Pd-Si particle at the Si/SiO$_2$ interface is observed (Fig. 7a,b) (this penetration can be as large as 8 nm in a (111) wafer without poly Si).

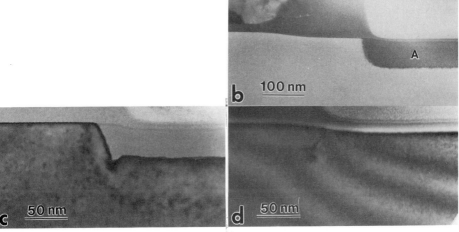

Fig. 5: Schematic cross section through a lightly doped drain (LDD) transistor. Arsenic (1) and phosphorous (2) implanted source/drain regions are indicated as dashed lines. TEM images after successive process steps: b) amorphous Si layer formed by As implantation (50keV, $5 \times 10^{15} cm^{-2}$) (A); c) partial regrowth (500°C/30min); d) complete regrowth after 900°C/ 60min annealing. A defect originates at the position of the notch in the growth front (see c). Imaging conditions: b) brightfield $g = \pm 400$, c), d) <110> aligned brightfield images.

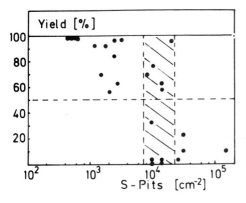

Fig. 6a: Yield of a gate oxide test structure (ramped voltage test, electric breakdown field >8MVcm⁻¹) as a function of the S-pit density (Secco defect etching).

Fig. 6c: SEM image of Cu-Si precipitates under a 20 nm gate oxide. Precipitate B has caused a severe bulge of the oxide whereas precipitate A has actually broken through the oxide. (Anneal: 1200°C, 30s)

Fig. 6b: Cross-sectional TEM micrograph of a Cu-Si precipitate at the SiO_2/(100)Si substrate interface reducing the oxide thickness. (Anneal: 900°C, 60s)

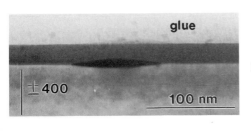

For the case of Ni a gate oxide thinning of only 1 nm is found and the interface appears corrugated (Fig. 8a,b). The situation is similar for Cu contamination after poly Si deposition, whereas the gate oxide can be completely destroyed if no poly Si is present: The Cu precipitates cause the oxide to either bulge out quite severely or ruptures the oxide like a mushroom breaking through the top layer of soil (Fig. 6b,c). In line with this, Cu contamination of the oxide without poly Si can cause gate oxide breakdown at fields as low as $E < 1$ MV, whereas in those cases where the 20 nm oxide is thinned by a few nm (as in the case of Cu with poly Si or Ni and Pd), the breakdown field is reduced by a few MV cm^{-1} (Wada et al. (1987)). A more detailed study of the various precipitation behaviours of different impurities is reported in Cerva et al. (1988, 1989) and Wendt et al. (1989).

For the case of Fe, Honda et al. (1987) have shown that FeSi$_2$ precipitates cause a gate oxide thinning. Furthermore, a quantitative correlation between the Fe concentration and the density of gate oxide defects in large area test capacitors has been established (Fig. 9): From a threshold value of about 10^{12}cm^{-2} - 10^{13}cm^{-2} for the Fe-surface contamination an increase in gate oxide defect density (derived from voltage ramp tests) is found. An important conclusion can be drawn from this result, namely that obviously the effect of lowering the Fe concentration is the growth of <u>fewer</u> precipitates which reach a size sufficient to appreciably thin the gate oxide, and that intentional metal contamination experiments with higher impurity concentrations result in a much higher density of large precipitates but should essentially be a true representation of what happens for lower contamination levels in a device processing line.

To sum up, the following model for gate oxide failure by metal contamination is put forward. Metal contamination results in the formation of metal silicide precipitates near the Si/SiO$_2$ interface, but only a small fraction of these will reach a size sufficient to lower the gate oxide breakdown by an appreciable amount. It follows immediately that the critical impurity concentration depends on the process and the stage in the device process at which the contamination is introduced, since the size distribution of precipitates depends on the thermal history. Furthermore, to attain the lower gate oxide defect densities necessary for larger chip sizes it is obviously necessary to further reduce the metal contamination levels.

For completeness sake we mention that another mechanism of gate oxide impairment by stacking faults has recently been identified (Liehr et al. 1988): It has been found that for annealing in vacuum sites of such defects lead to gate oxide pin-holes by a local enhancement of SiO formation (volatile). However, this mechanism only applies to those process steps with uncovered gate oxides. Furthermore, Lin et al. (1983) have correlated gate oxide breakdown sites with stacking faults decorated by metal impurities, e.g. Cu.

5. METAL CONTAMINATION AND STACKING FAULTS

In the examples mentioned so far the driving force for the formation of extended defects was either mechanical stress or a supersaturation of metal impurities. It is the purpose of this section to demonstrate that the formation of extended defects can be enhanced by metal contamination and that the electrical activity of these defects is mainly due to decoration by metal impurities. In device processing, the most frequent example for such a state of affairs is the formation of oxidation-induced stacking faults (OSFs). It has been realized as early as 15-20 years ago that the occurence of stacking faults is correlated with metal contamination and that these defects can degrade the device performance, even if those devices were a long way from submicrometer dimensions and therefore much more robust.

The following evidence that metal contamination promotes the nucleation of stacking faults has been accumulated over the years:

i)   No OSFs are generated during oxidation provided the process is clean and the Si-surface is damage-free.

ii)   For a "dirty" process the density of formation of stacking faults can be suppressed by gettering, i.e. the removal of metal impurities from the polished wafer surface (Rozgonyi et al. 1975).

iii)   It has been observed that the S-pit density, i.e. the severeness of metal contamination and the density of OSFs are correlated (Murarka et al. (1980)).

iv)   TEM investigations have demonstrated directly that oxidation-induced stacking faults are nucleated on metal precipitates (Shimura et al. (1980)) which otherwise would not be

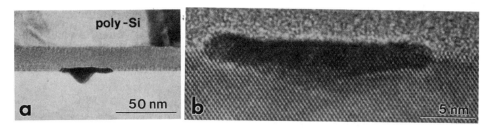

Fig. 7: TEM cross-sections of Pd-Si precipitates at the SiO_2/(100) Si substrate interface.
   a) Brightfield image g = 400, b) <110> aligned lattice image. (Annealing $1200^\circ C$, 30s)

Fig. 8: TEM cross-sections of $NiSi_2$ precipitates at the SiO_2/(100)Si substrate interface.
   a) $NiSi_2$ platelet parallel to a [111] silicon lattice plane extending from the oxide into
   the Si bulk. <110> aligned brightfield image. b) $NiSi_2$ precipitate protruding from the
   Si substrate. <110> multibeam lattice image. (Annealing $1200^\circ C$, 30s)

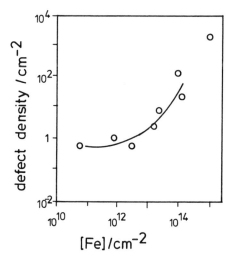

Fig. 9: Gate oxide defect density in a thermally grown 10 nm gate oxide (as determined by a ramped voltage test) as a function of Fe contamination of the wafer surface before the growth of the gate oxide (Takizawa et al. 1987).

generated in spite of a silicon selfinterstitial supersaturation caused by the oxidation process.

In the following example we investigate the effect of stacking faults on MOS power devices (i. e. array of several thousand MOS transistors in parallel): In line with i) - iv) we observed a good correlation between S-pit densities and OSF densities.

The impact of OSFs on the electrical device properties can be quite dramatic for MOS power transistors: In Fig. 10 the spatial distribution of the leakage current per device (area $\sim$ 0.36 cm^2) measured at 400 V reverse bias and the stacking fault density (determined by defect etching and optical microscopy) are compared for one wafer: It is quite obvious that there is an almost 1:1 correspondence between the two quantities, i.e. the stacking faults are the cause of the high leakage currents.

In fact, the leakage current is proportional to the OSF density, as is demonstrated in a double log plot of the leakage current vs. the stacking fault densities (Fig. 11, same data as in Fig. 10). The scatter of about a factor of 5-8 is interpreted to be due to a different degree of metal decoration of the stacking faults. The value of about 1 μA leakage current per OSF is a rather high value: Ogden and Wilkinson (1977) have estimated the leakage current per stacking fault for a number of investigations and found values ranging from picoamps to 500 μA per OSF. Moreover, they found that the leakage current per stacking fault correlated remarkably well with the degree of decoration. Similar findings, combined with EBIC work, have been published by Varker et al. (1974).

Considering the high leakage current per stacking fault and an S-pit density of the order of 10^{5}cm^{-2} in the devices investigated the stacking faults must be regarded as heavily decorated although this point was not investigated by TEM.

It is beyond the scope of the present paper to go into too much detail on the mechanism of recombination/generation at stacking faults and we refer the reader to recent work by Peaker et al. (1988) and Lahiji et al. (1988). The authors showed that the deep levels associated with stacking faults change dramatically if initially clean stacking faults are decorated by e.g. gold or platinum. In fact, mid-gap levels which are necessary for a high generation activity, are observed only after decoration.

The two-fold detrimental action of metal contamination in connexion with stacking faults, namely the lowering of the nucleation barrier for the formation of these defects and the electrical "activation" by decoration can be generalized to other extended defects, e.g. dislocations, oxygen precipitates etc. and emphasizes the necessity for a strict contamination control during device processing.

Acknowledgement
The authors are indebted to Dr. H. Benzinger, Dr. M. Gentsch, G. Götz and S. Sauter for provision of experimental results. This report is based on a project which has been supported by the Minister of Research and Technology of the Federal Republic of Germany under the support no. NT2696. For the contents the authors alone are responsible.

References
Beinvogl W and Hopf E 1988 Advances in Sol. State Physics 28 in print
Cerva H and Wendt H 1988 Mat. Res. Soc. Symp. Proc. 138 in print
Cerva H and Küsters K H 1989 submitted to J. Appl. Phys.
Cerva H and Wendt H 1989 6th Conf. Microsc. Semic. Mats., Oxford April 1989, to be published
Csepregi L , Kennedy E F , Mayer J W and Sigmon T W 1978 J. Appl. Phys. 49 3906
Chatterjee P K , Taylor G W , Tasch A F and Fu H S 1979 IEEE Trans. Electron Devices ED-26 564
Eisenberg P and Brion K 1969 Electronics 42 pp 45-6
Honda K , Nakashini T , Ohsawa A and Toyokura N 1987 J. Appl. Phys. 62 1960
Kolbesen B O and Strunk H 1981 Inst. Phys. Conf. Ser. 57 21
Kolbesen B O and Strunk H 1985 VLSI Electronics: Microstructure Science ed N.G. Einspruch, Vol. 12 "Silicon Materials" ed. H R Huff (Academic Press, New York) p 143
Kolbesen B O , Bergholz W and Wendt H 1989 Proc. Int. Conf. Defects in Semiconductors - 15, Budapest, August 1988, in print

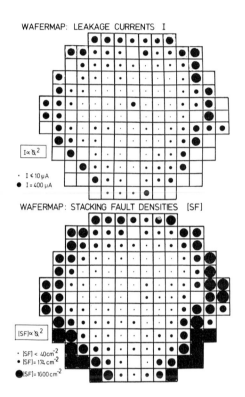

WAFERMAP: LEAKAGE CURRENTS I

$I \propto \varnothing^2$

- $I \leqslant 10\,\mu A$
- $I = 400\,\mu A$

WAFERMAP: STACKING FAULT DENSITIES [SF]

$[SF] \propto \varnothing^2$

- $[SF] < 40\,cm^{-2}$
- $[SF] = 174\,cm^{-2}$
- $[SF] = 1600\,cm^{-2}$

Fig. 10: Wafermap of the distribution of leakage
currents measured at 400 V reverse bias for
MOS power devices and of the density of
stacking faults. The areas of the dots are proportional to the leakage current and the
stacking fault density, respectively. Each
square represents a device of 0.35 cm^2 area.

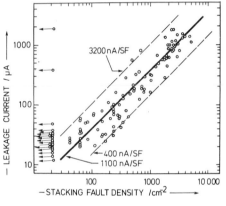

Fig. 11: Double Log plot of the leakage current at
400 V reverse bias vs. the stacking fault
density for MOS-power devices (same data
as in Fig. 10). The arrows at some of the
data points indicate that the stacking fault
densities are below the detection limit
(which is taken as one stacking fault for
the area investigated). The solid line has
the slope 1, i.e. the leakage current is
proportional to the stacking fault density.
The leakage current per stacking fault is
indicated.

Küsters K H , Enders G , Meyberg W , Benzinger H , Hasler B , Higelin G , Röhl S , Mühlhoff H M and Müller W 1987 Symp. VLSI Technology, Digest of Techn. Papers, 93

Lahiji G R , Hamilton B and Peaker T 1988 Electronics Letters 24 134 1340

Lemme R , Gentsch M and Kutzner R 1988 in: "Int. Symp. for Testing and Failure Analysis, Los Angeles 31. Oct. - 4. Nov. 1988, ASM International, Metals Park, Ohio pp 31-9

Liehr M , Bronner G B and Lewis J E 1988 J. Appl. Phys. 52 1982

Lin P S D , Marcus B and Sheng T T 1983 J. Electrochem. Soc. 130 1878

Murarka S P , Seidel T E , Dalton J V , Dishman J M and Read M H 1980 J. Electrochem. Soc. 127 716

Ogden R and Wilkinson J M 1977 J. Appl. Phys. 48 412

Peaker T 1988 private communication

Rozgonyi G A , Petroff P M and Read M H 1975 J. Electrochem. Soc. 122 1725

Secco d'Aragona F 1972 J. Electrochem. Soc. 119 948

Shimura F , Tsuya H and Kawamura T 1980, J. Appl. Phys. 51 269

Sunami H 1985 IEDM Technical Digest 694

Takizawa R , Nakashini T and Ohsawa A 1987 J. Appl. Phys. 62 4933

Varker C J and Ravi K V 1974 J. Appl. Phys. 45 272

Wada K , Ito T and Toyokura N 1987, in "ULSI Science and Technology/1985", S Broydo and C M Osburn eds, Electrochemical Soc. Proc. Vol 87-11, Pennington, N.J. p 119

Wendt H, Cerva H, Lehmann V and Pamler W 1989 J. Appl. Phys. 65 2402

Yamabe K , Taniguchi K , Matsushita Y 1984, in "Defects in Silicon", W M Bullis and L C Kimerling eds, Electrochemical Soc., Pennington, N.J. p 629

Inst. Phys. Conf. Ser. No 104: Chapter 5
Paper presented at Int. Symp. on Struct. Prop. Disloc. Semicond., Oxford, 1989

Observations of dislocation distributions and associated point defects in bulk-grown GaAs

D.J. Stirland

Plessey Research Caswell Limited, Allen Clark Research Centre, Caswell, Towcester, Northants. NN12 8EQ, England.

ABSTRACT: Techniques for the direct observation of macroscopic and microscopic arrangements adopted by dislocations in semi-insulating GaAs are described. The importance of a variety of defects which occur in association with dislocations is discussed in relation to GaAs integrated circuit technology requirements. Methods of determining how this association can be relaxed by thermal treatments of grown ingots are considered.

INTRODUCTION

The crucial requirement for high yield and performance of GaAs integrated circuits fabricated by direct ion implantation is uniformity of the threshold voltage (V_{th}) of metal-semiconductor field-effect transistors (MESFETs) across the semi-insulating GaAs device substrate. The liquid encapsulated Czochralski (LEC) crystal growth technique is now almost universally employed to produce undoped, semi-insulating ingots, but the nature of the process results in comparatively large (10^4-10^5 cm^{-2}) dislocation densities. Fluctuations in V_{th} have been linked to the presence of dislocations and other defects. These factors represent the chief driving forces behind the intensive examinations of the material properties of bulk-grown GaAs in recent years. This contribution is concerned primarily with direct observation techniques for revealing dislocation distributions and their associated defects, and the significance of the results in relation to device parameters.

1. HOW ARE DISLOCATIONS DISTRIBUTED IN UNDOPED GaAs INGOTS?

A quasi-steady state heat transfer/thermal stress model by Jordan et al (1980,1986) assumed that dislocation generation occurred when thermal stresses exceeded the critical resolved shear stresses for the slip systems of the crystal. Fig. 1 shows a transmission X-ray topograph indicating dislocation arrangements in an (001) wafer from the seed end region of a 2" diameter undoped, <001> axis, LEC GaAs ingot. Dislocations are most dense at the periphery, less dense in the central core, and least dense in an annular region between edge and centre (the well known "W-shaped" distribution across a <110> diameter) as shown in Fig. 2. The dislocations are not distributed uniformly in these regions, however. Three types of arrangement are observed. The first is a cellular network (Clark and Stirland 1981) in which dislocations cluster into cell walls encompassing practically dislocation-free cell interiors. Such cells, ~300-500μm in size, are visible in the centre region of Fig. 1. Chen and Holmes (1983) noted the occurrence of a second type of dislocation arrangement in which networks elongated into lineage structures consisting of dislocation lines (up to 1 cm long) oriented approximately along the <110>lying in (001). Examples can be seen in Fig. 1.

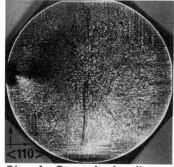

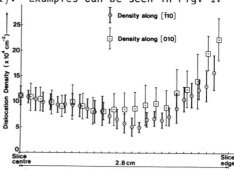

Fig. 1: Transmission X-ray topograph of (001) wafer

Fig. 2. Dislocation density variations across [100] and [110] radii of A/B etched (001) wafer

Linear structures were demonstrated (Stirland et al 1983, 1984) to lie approximately on {110} sheets normal to (001), and often extended several mms down the <001> ingot axis, although Noad (1988) concluded that some of the lineage structures meandered through ingots in roughly the growth direction but followed no specific crystal orientation. Brown et al (1984) established by double crystal X-ray topography that lattice tilts ~0.01° occurred across the crystal regions separated by lineage. The third dislocation arrangement consisted of linear arrays along exact orthogonal <110> directions, forming "tartan" patterns at the <100> edge regions (see Fig. 1). Kitano et al (1986) have suggested that these linear arrays result from dislocations belonging to eight of the twelve <110>{111} slip systems which have yield stress minima values, generated under tangential tensile stress as the ingot periphery cools faster than the interior. The formation mechanisms of the cellular and lineage structures are less certain. Ponce et al (1984) suggested that the cellular structure is due to constitutional supercooling enhanced by deviations from melt stoichiometry, and there is evidence that stoichiometry changes can profoundly alter dislocation densities in low diameter (~1.5 cm) horizontal Bridgman grown <111> axis crystals (Parsey et al 1982, Lagowski et al 1984). Jacob et al (1983) also concluded that not all dislocations could be due to thermal stresses. Transmission electron microscopy (TEM) examinations of cellular and lineage structures should, in principle, enable deductions regarding their formation to be made, but the complex tangles in undoped LEC GaAs make analysis difficult. Two methods of simplifying the situation have been employed. Ono and Matsui (1987) used an alloyed GaAs + 10% atomic phosphorus crystal with a reduced dislocation density ($\sim 10^4$ cm^{-2}), and Ono (1988) an indium doped LEC GaAs crystal essentially dislocation-free, but plastically deformed to generate dislocations. Detailed analyses of dislocation interactions for both specimens showed that sessile Lomer-Cottrell dislocations were formed by reaction between glissile dislocations on e.g. (111) and (1̄11̄) glide planes:

$$\tfrac{1}{2}a\,[10\bar{1}] + \tfrac{1}{2}a\,[011] \rightarrow \tfrac{1}{2}a\,[110] \tag{1}$$

These perhaps indicate initial stages in the formation of lineage and cellular structures.

Fig. 3(a): $\overline{1}1\overline{5}$ reflection X-ray topograph of back-side of wafer

Fig. 3(b): $1\overline{1}\overline{5}$ reflection topograph of device side of same wafer

Fig. 3(c): Right hand lineage superimposed on Fig. 3(b) - traced from Fig. 3(a)

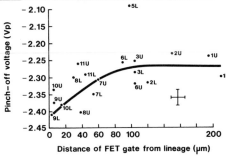

Fig. 4: Variation of FET V_{th} (pinch-off voltage V_p) values with distance from lineage feature on Fig. 3(c)

2. IS THERE AN ASSOCIATION BETWEEN DISLOCATIONS AND MESFET PROPERTIES?

From 1983 onwards evidence gathered to suggest that substrate dislocations affected FET threshold of voltages (V_{th}) (Miyazawa et al 1983, Miyazawa and Hyuga 1986) although some doubts regarding the association were also expressed (Winston et al 1984). These studies were made by preparing arrays of MESFETs as planar devices without gate recesses, measuring V_{th} and other parameters, exposing the device surface to molten KOH and measuring distances of etch pits to FET gates. This destructive technique prevented re-measurement of devices, since the etch removed the metal contacts. X-ray topography offered the possibility of probing individual FETs after the dislocations had been revealed (Stirland 1987). Both reflection and transmission methods were used. Figs. 3(a) and 3(b) show equivalent reflection topographs ($g = \overline{1}1\overline{5}$ and $1\overline{1}\overline{5}$) from the backside and device side respectively of part of a wafer containing two pronounced lineage features, one of which lay close to a line of double FET structures. To assist identification, Fig. 3(c) shows one of these lineage features traced on to the array. Fig. 4 is a graph of V_{th} values for the upper (U) and lower (L) FETs versus distance from the lineage feature. This curve indicates that the lineage, or associated defects, has an influence on V_{th} for distances up to ~40-50µm. It is advantageous to use specially designed MESFET arrays to overcome the experimental difficulties associated with the proximity of metal contact areas and interconnections. Stirland (1987) used the 7 x 8 array of 1µm gate length FETs shown in Fig. 5. A total of 10^5 individual FETs within 2 x 10^3

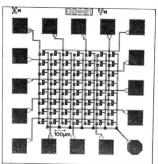

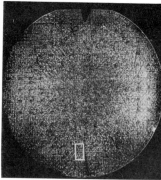

Fig. 6: Transmission X-ray topograph of 2" diameter wafer with 2 x 10³ FET arrays. Every third square is a process control monitor (test chip)

Fig. 5: 7x8 array of 1x10μm FETs multiplexed to 16 pads

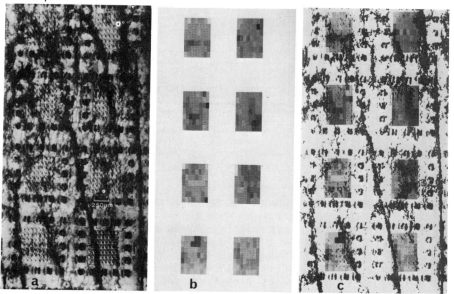

Fig. 7(a)   Part of X-ray topograph (g = 2$\bar{2}$0) showing eight 7x8 arrays
    (b)   gray scale V_p values (100 mV steps) for identical eight arrays
    (c)   overlay of (a) and (b) combined

arrays were available across a 2" wafer. Wafers were processed using a 60 keV Si^{29} implant dose of 8 x 10^{12} cm^{-2} through a 500Å thick silicon nitride encapsulating layer, annealed at 850°C for 15 min. Fig. 6 is a transmission X-ray topograph showing the MESFET arrays on a 2" wafer, and Fig. 7(a) shows the area outlined in white at higher magnification (g = 2$\bar{2}$0). The schematic of Fig. 5 has been superimposed on the lower right hand array. Fig. 7(b) is a grey scale display of V_p (in steps of 100 mV) for the eight arrays and Fig. 7(c) is an overlay to assist correlation. There is close agreement between increased V_p values (darker tones) and proximity to the lineage structures, although the limited resolution of the technique does not permit conclusions regarding the effects of proximity to individual dislocations.

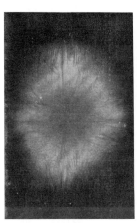

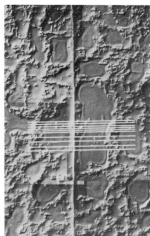

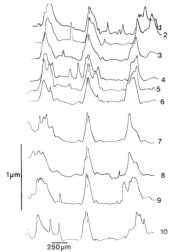

Fig. 8: Transmission infra-red micrograph of (001) wafer 2" diameter

Fig. 9: A/B etched (001) wafer showing lineage and cellular structure

Fig. 10: Surface profiles from scans at positions of lines on Fig. 9.

3. IS THERE AN ASSOCIATION BETWEEN DISLOCATIONS AND EL2?

The main chemical impurity in undoped, semi-insulating, LEC GaAs is carbon, which electrically is a shallow acceptor with concentration usually exceeding the total chemical donor impurities. Concentrations of carbon, [C], measured by radiographic methods (resolution ~15μm Bourret 1987a) are found to be distributed uniformly. The compensating species which produces semi-insulating, high resistivity ($>10^7$ Ω-cm) behaviour is a native deep donor defect known as EL2. It was shown by Holmes et al (1983) that [EL2] increased from 5 x 10^{15} cm^{-3} to 1.7 x 10^{16} cm^{-3} as the As atomic fraction in the melt increased from about .48 to .51, thus indicating that semi-insulating material could only be grown above a critical As concentration for typical [C] ~2.1 - 13 x 10^{15} cm^{-3} obtaining at that time. The identity of EL2 has been the subject of a vast number of investigations (see Bourgoin et al 1988 for a recent survey). It is practically certain that the antisite defect As_{Ga} is involved, possibly complexed with one or more arsenic interstitials. However there is over-whelming evidence that, unlike [C], [EL2] is distributed non-uniformly, in as-grown, semi-insulating GaAs. Martin (1981) showed that EL2 centres exhibited a characteristic absorption band in the near infra-red (~1μm) and calibration of the absorption coefficient α enabled quantitative [EL2] to be measured directly. Martin et al (1981) demonstrated that macroscopic variations in [EL2] closely followed etch pit density varia-tions across <110> diameters of (001) wafers, showing characteristic "W" and "U" distributions. Brozel et al (1983) showed by a high spatial resolution infra-red absorption technique that fine structure variations (~50-100μm) were superimposed on the "W" and "U" variations. The fine structure correlated with regions of high dislocation density in {110} and {001} specimens, corresponding to cellular walls and lineage. Direct transmission images of variations in infra-red absorption across whole wafers were obtained by using an infra-red sensitive silicon vidicon television camera. A typical image is shown in Fig. 8; the similarity to Fig. 1 is apparent.

Examination of A/B etched surfaces provides another method of detecting changes in [EL2] distributions on a microscopic scale. Fig. 9 is a region of an etched wafer similar to that of Fig. 1 showing a <110> lineage sheet and dislocation cells. Nomarski interference contrast indicates that the tightly packed dislocations comprising lineage and cell walls have etched as grooves lying on top of broad mounds. Isolated dislocations inside cells etch as ridges. Brown and Warwick (1986) have established that the A/B etch rate is sensitive to the local Fermi level position: it is lowest for semi-insulating GaAs (Fermi level near mid-gap) and higher for either increased or decreased Fermi levels. Thus the mounds represent regions of enhanced [EL2] and the variation in disloca-tion etching behaviour results from the local environment in which dislo-cations occur. This has been used to quantify the microscopic uniformity of wafers (Stirland et al 1988) by determining the topological variations of surfaces after standard etching procedures. Fig. 10 shows a set of surface profiles at the white lines on Fig. 9 indicating typical magni-tudes of these variations. It has also been possible to demonstrate (Stirland et al 1985a,b) that enhanced [EL2] occurs at single dislocat-ions, by making use of near dislocation-free GaAs (Jacob et al 1983) obtained by doping the melt with $\sim 10^{20}$ cm^{-3} indium. Fig. 11 shows a montage of A/B etched <001> dislocations at the centre of an (001) In:GaAs wafer and Fig. 12 presents four of the etched dislocations at higher magnification. Each dislocation is surrounded by a mound of diameter 50-70µm. Fig. 13 shows transmission infra-red micrographs of the centre of a similar (001) In:GaAs ingot section 3 mm thick. These change with time as the sample was held at 77K under white light illumi-nation. The absorbing spots, which have been directly correlated with dislocation positions (Stirland et al 1985b) show a bleaching of the absorption (Martin 1981) due to transfer of EL2 centres to a metastable state. This photoquenching behaviour is considered to be the most reli-able test for the identification of (neutral) EL2 (Skolnick 1985). It is concluded that individual dislocations are surrounded by cylinders of diameter ~50-70µm consisting of enhanced [EL2] regions. However, it is not possible also to conclude that enhancement of [EL2] per se is there-fore responsible for threshold voltage changes in the vicinity of dislocations.

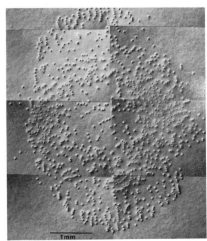

Fig. 11: A/B etched <001> dislocations at the centre of an (001) In:GaAs wafer

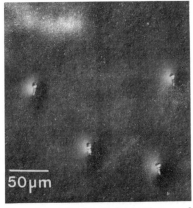

Fig. 12: Four dislocations of Fig.11 at higher magnification

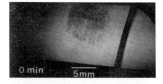
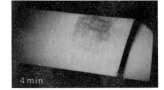

Fig. 14: Lineage and cell wall region of (001) wafer after A/B etching

Fig. 13: Changes in transmission infa-red images of dislocations (EL2 absorption) at the centre of an (001) wafer with time, at 77K under white light

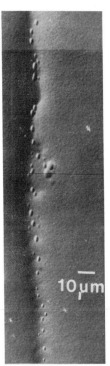

Fig. 15: Single dislocation decorated with precipitates which become etch pits after A/B etching

Increased (negative) V_{th} values result from an increased carrier concentration in the FET channel region, which for a silicon implant is determined by $[Si_{Ga}]-[Si_{As}]$. Miyazawa et al (1988) indicate that this increase could be determined by the $[V_{Ga}]/[V_{As}]$ vacancy ratio in the vicinity of dislocations. Their model requires an increase of arsenic interstitials (As_i) close to dislocations to alter this ratio. It is therefore necessary to look for evidence of other defects in these regions.

4. ARE THERE DISLOCATION-RELATED DEFECTS IN ADDITION TO EL2 CENTRES?

The existence of etch pits on A/B etched GaAs dislocations which could be correlated (by TEM) with etched-out particles was first reported for chromium doped semi-insulating GaAs (Stirland 1977). Identification of one particle as an hexagonal arsenic precipitate was achieved by transmission electron diffraction (Cullis et al 1980) of chromium-doped GaAs, and subsequently a particle in undoped semi-insulating GaAs was also identified as an arsenic precipitate (Stirland et al 1984). Fig. 14 shows a typical lineage and cell wall region of an A/B etched (001) wafer. Almost all the dislocation lines (grooves and ridges) are covered with discrete etch pits. Fig. 15 is a higher magnification area of a single decorated dislocation.

During the past five years a number of publications have confirmed the
original identifications.   The most extensive recent studies of the
particles by high resolution TEM have been made by the Lawrence Berkeley
Laboratory group (Lee et al 1986, 1988, 1989; Bourret et al 1987b).
Arsenic particles were identified at dislocations in peripheral regions
of semi-insulating In:GaAs as well as in horizontal Bridgman grown,
arsenic rich, n-type $(1 \times 10^{17} \text{ cm}^{-3})$ GaAs.   In the latter material the
particles were often associated with dislocation loops lying on $\{111\}$.
For both materials they were tetrahedra in a simple orientation relation-
ship with the GaAs matrix: $(0001)_{As} // (111)_{GaAs}$ with $(11\overline{2}0)_{As} // (\overline{2}02)_{GaAs}$.
Although Cullis et al (1980) did not find a simple relationship between
the arsenic particle and the matrix, or observe a simple shaped particle,
re-examinations of earlier undoped material have found that well-oriented
particles do occur.   Fig. 16 is a stereo-pair of micrographs ($\underline{g}$ = 220)
with foil normal close to $[112]$.   It is conjectured that all the tetrahedral particles were probably associated with two dislocations, although
one of these lay outside the foil surface.   The stereo-image shows that
most of the visible dislocation line lies above the precipitates at the
upper right side, and these lie on a plane almost normal to the foil
surface.   Such observations may indicate that some precipitates are
debris deposited by climbing dislocations, rather than pinning centres on
gliding dislocations.   Either explanation could account for the configuration shown in Fig. 15.

The use of infra-red laser scanning microscopy (Kidd et al 1987a,b) and
scattering tomography techniques (Moriya and Ogawa 1983, Suchet et al
1987) provide essentially non-destructive methods for determining the
distributions of particles such as arsenic precipitates within bulk
material.   Suchet et al (1987) successfully attempted an experimentally
difficult comparison between a 90° tomograph and an A/B etched micrograph
of the same area, and concluded that ~99% of precipitates were pinned at
or very close to dislocations, although ~1% appeared to be isolated in
the matrix.   Typical precipitate densities were ~10^8-10^9 cm^{-3}.   Recently,
Suchet and Duseaux (1987) and Suchet et al (1988) have proposed that two
different types of precipitates can be detected: 'pure' arsenic particles
and contaminated arsenic particles which have gettered impurities.   They
claim that both types are seen by scattering tomography but only the
impure precipitates are detected by the A/B etchant.

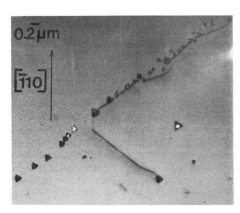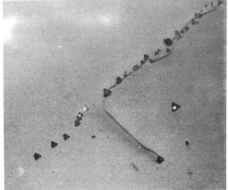

Fig. 16: Stereo-pair of tetrahedral (arsenic) precipitates.   Foil normal
close to $[112]$. (Augustus P D and Stirland D J, 1985 unpublished)

Fig. 17: Dobrilla (1988) structure at cell centre

Fig. 18: Dobrilla (1988) structure associated with precipitates and dislocations?

Fig. 19a: As-grown ingot wafer after A/B etch

Fig. 19b: Post-growth ingot anneal wafer after A/B etch

A new structure appearing in cell centres after A/B etching was reported independently by Dobrilla (1988) and Miyazawa et al (1988). Fig. 17 shows an example of this structure. Miyazawa et al (1988) claim that the roughened structure appeared after annealing (800°C for 20 hrs) but Dobrilla (1988) found the structure in material 'not heat treated after growth'. A characteristic feature noted by Dobrilla, but not by Miyazawa, is the denuded region separating the roughened cell interior from the cell walls (see Fig. 17). Miyazawa suggested that the structure is the indication of "defect agglomeration caused by annealing in essentially dislocation-free regions". It is clear that defects within the cells, which could include arsenic, or arsenic-rich, clusters may show homogeneous nucleation in regions remote from the dislocation sinks in the cell walls. A similar structure also has been seen in close proximity to dislocations (Fig. 18) and is perhaps associated with climb or glide as suggested for Fig. 16. (see also Gall et al 1988 for micro-tomography results on these structures).

5. IS IT POSSIBLE TO DISASSOCIATE DEFECTS AND DISLOCATIONS?

Although it is probable that manufacturers first employed annealing procedures to decrease the number of ingots which cracked or even shattered during wafer cutting, it was soon noted that there were additional advantages. Rumsby et al (1984) showed that "M"-shaped resistivity variations and "W"-shaped mobility variations across $<110>$ wafer diameters were smoothed by whole ingot annealing at 950°C for five hours. In addition, the usual "W"-shaped [EL2] distribution became more uniform, with average [EL2] slightly increased. In particular, annealing has been shown to reduce V_{th} standard deviation values (Packeiser et al 1986, Deconinck et al 1988). However, each manufacturer's annealing "recipe" is commercially confidential and not all procedures are equally effective. For example, the correlation between V_{th} and lineage proximity discussed in section 2 (Figs. 6 and 7) was achieved using wafers from a post-growth annealed ingot supplied by one manufacturer. No correlation was detected when wafers from another manufacturer were used, presumably because in this case the heat-treatments had been optimised for uniformity.

What happens during heat treatments of as-grown ingots? X-ray topography examinations of ingot-annealed wafers show dislocation densities and distributions similar to those of Fig. 1 for annealing temperatures up to ~1000°C (Stirland & Hart 1988). However, although dislocation distributions are essentially unaltered, the etching shows that the environment of the dislocations has changed considerably. Figs. 19(a) and (b) are two A/B etched wafers supplied by the same manufacturer. Fig. 19(a) was from an as-grown ingot, Fig. 19(b) was from an ingot subjected to a post-growth anneal of 920°C for five hours. Both micrographs are displayed with identical Nomarski settings. Surface profiles obtained from the unannealed wafer were similar to those of Fig. 10 whereas profiles from the annealed wafer (not shown) were much smoother (Stirland et al 1988), indicating that the microscopic distribution of [EL2] is more uniform, even though the cellular distribution of dislocations is still evident. The Nomarski contrast reveals that mounds on the unannealed material have become shallow depressions on the annealed material. Close examination at higher magnification shows that the Dobrilla (1988) fine etched structure is now clustered in the cell walls for the annealed sample, in contrast with its location in cell centres for unannealed material as shown in Fig. 19(a). It is apparent that annealing of bulk ingots can significantly modify the association between defects and dislocations in the as-grown materials. Spectacular changes can be achieved by more extreme thermal treatments. Lagowski et al (1986) quenched GaAs specimens (into water) after high temperature anneals at 1100°C-1200°C and found reductions in [EL2] from 2-3×10^{16}cm^{-3} to 1-2×10^{15}cm^{-3}. More significantly the [EL2] could be restored to ~1.5×10^{16}cm^{-3} by subsequent 'conventional' (slow cool) anneals between 750-900°C. Clark et al (1988) have examined quenching effects over the temperature range 700°-1200°C. Their results indicate that, firstly, the uniformity of the EL2 distribution increases throughout with increasing quench temperature. Secondly, [EL2] increases for quenches up to 1000°C and then drops sharply for the higher temperature quenches. Dislocation densities remain unaltered up to 1000°C, but increase above this. Examinations of the dislocation arrangements, after A/B etching, showed that the usual ridge and groove features containing etch pits were replaced by ridge structures alone, with no detectable etch pits present, for specimens quenched from ≥1100°C.

The mechanisms by which the uniformities of [EL2] and therefore electrical behaviour are improved by quenching are almost certainly involved with the dissolution into the GaAs lattice of point defect clusters at elevated temperatures, including arsenic precipitates and other dislocation associated defects such as EL2 centres. Quenching prevents the re-formation of Cottrell atmospheres around dislocations and ensures that the homogeneity of defect concentrations is retained. Quenching followed by carefully controlled conventional annealing to restore [EL2] levels appropriate for particular device requirements is therefore possible.

6. CONCLUDING COMMENTS AND ACKNOWLEDGEMENTS

Simple and direct techniques have been used to describe some of the charateristics of dislocations in undoped and indium doped semi-insulating GaAs. Although enhanced [EL2] and arsenic precipitates are linked with dislocations in as-grown material, the association can be weakened by annealing procedures, thereby offering the possibility of achieving greater substrate unformity for GaAs integrated circuits.

This contribution is by no means a complete review. Inevitably, many important papers have not been cited, for space reasons, and my apologies are offered to their authors. I am grateful to many colleagues with whom discussions have generated ideas and comments, in particular R T Blunt, G R Booker, M R Brozel, S Clark, I Grant, D T J Hurle, P Kidd and R C Newman. D G Hart is thanked for skilled X-ray topography. H Badawi and S Kitching were involved with much of the work described in Section 2 and it is a pleasure to acknowledge their considerable contributions. Figs. 3,4,7,9 and 10 are reprinted by permission of Elsevier Science Publishers and Fig. 2 is reprinted with permission from the IOP Conference Series No. 60, p.339. This work has been supported by the Procurement Executive, Ministry of Defence, (Royal Signals and Radar Establishment).

7. REFERENCES

Bourgoin J C, Von Bardeleben H J and Stiévenard D, 1988, J. Appl. Phys. 64, R65.

Bourret E D, Guitron J B and Haller EE, 1987a, J Crystal Growth, 85 290.

Bourret E D, Elliot A G, Lee B T and Jaklevic J M, 1987b, in Defect Recognition and Image Processing, Ed. E R Weber (Publ. Elsevier) p.95.

Brown G T, Skolnick M S, Jones G R, Tanner B K and Barnett S J, 1984, in Semi-Insulating III-Materials, Eds. D C Look and J S Blakemore (Shiva Publ.), p.76.

Brown G T and Warwick C A, 1986, J. Electrochem. Soc., 133, 2576.

Brozel M R, Grant I, Ware R M and Stirland D J, 1983, Appl. Phys. Lett., 42, 610.

Chen R T and Holmes D E, 1983, J. Crystal Growth, 61, 111.

Clark S and Stirland D J, 1981, Inst. Phys. Conf. Ser. No. 60, 339.

Clark S, Stirland D J, Brozel M R, Smith M and Warwick C A, 1988 in Semi-Insulating III-V Materials, Eds. G Grossmann & L Ledebo (Publ. Adam Hilger) p.31.

Cullis A G, Augustus P D and Stirland D J, 1980, J. Appl. Phys. 51 2556.

Deconinck P, Farges J P, Martin G M, Nagel G, and Löhnert H, 1988, in Semi-Insulating III-V Materials, Eds. G Grossmann and L Ledebo (Publ. Adam Hilger, pg.505.

Dobrilla P, 1988, in Defect Recognition and Image Processing II, Ed. E R Weber (Publ. Elsevier), p.305.

Gall P, Fillard J P, Castagne M, Weyher J L and Bonnafé J, 1988, J. Appl. Phys. 64, 5161.

Holmes D E, Chen R T, Elliott K R, and Kirkpatrick C G, 1983, Appl. Phys. Lett., 43, 305.

Jacob G, Duseaux M, Farges J P, van den Boom M M B and Roksnoer P J, 1983, J. Crystal Growth, 61, 417.

Jordan A S, Caruso R and von Neida A R, 1980, Bell Sys. Tech. J., 59 593.

Jordan A S, von Neida A R, and Caruso R, 1986, J. Crystal Growth, 76 243.

Kidd P, Booker G R and Stirland D J, 1987a, Inst. Phys. Conf. Ser. No. 87, p.275.

Kidd P, Booker G R and Stirland D J, 1987b, Appl. Phys. Lett., 51, 1331.

Kitano T, Ishikawa T, Ono H and Matsui J, 1986, in Semi-Insulating III-V Materials, Eds. H Kukimoto and S Miyazawa (Publ. Ohmsha- North Holland) p.91.

Lagowski J, Gatos H C, Aoyama T and Lin D G, 1984, Semi-Insulating III-V Materials, Eds. D C Look and J S Blakemore (Publ. Shiva) p.60.

Lagowski J, Gatos H C, Kang C H, Skowronski M, Ko K Y and Lin D G, 1986, Appl. Phys. Lett., 49, 892.

Lee B T, Sands T, Gronsky R and Bourret E D, 1986, Inst. Phys. Conf. Ser. No. 83, p.51.

Lee B-T, Gronsky R & Bourret E D, 1988, J. Appl. Phys. 64, 114.
Lee B-T, Bourret E D, Gronsky R and Park I, 1989, J. Appl. Phys. 65 1030.
Martin G M, 1981, Appl. Phys. Lett., 39, 747.
Martin G M, Jacob G, Poiblaud G, Goltzene A and Schwab C, 1981, Inst. Phys. Conf. Ser. No. 59, 281.
Matsuura H, Nakamura H, Sano Y, Egawa T, Ishida T and Kaminishi K, 1985, GaAs IC Symposium (Monterey), IEEE Tech. Digest, p.67.
Miyazawa S, Ishii Y, Ishida S and Nanishi Y, 1983, Appl. Phys. Lett.43 853.
Miyazawa S, Honda T, Ishii Y and Ishida S, 1984, Appl. Phys. Lett. 44 410.
Miyazawa S, 1986 in Semi-Insulating III-V Materials, Eds. H Kukimoto and S Miyazawa, (Ohmsha-North Holland) pg. 3.
Miyazawa S and Hyuga F, 1986, IEEE Trans. Elect. Devices, ED-33, 227.
Miyazawa S, Watanabe K, Osaka J and Ikuta K, 1988, Revue Phys. Appl. 23 727.
Moriya K and Ogawa T, 1983, Jpn. J. Appl. Phys., 22, L207.
Noad J P, 1988, J. Appl. Phys., 64, 1599.
Ono H and Matsui J, 1987, Appl. Phys. Lett., 51, 801.
Ono H, 1988, J. Crystal Growth, 89, 209.
Packeiser G, Schink H and Kniepkamp H, 1986, Semi-Insulating III-V Materials, Eds. H Kukimoto and S Miyazawa (Publ. Ohmsha-North Holland), 561.
Parsey J M, Nanishi Y, Lagowski J and Gatos H C, 1982, J. Electrochem. Soc., 129, 388.
Ponce F A, Wang F-C and Hiskes R, 1984, Semi-Insulating III-V Materials, Eds. D C Look and J S Blakemore (Publ. Shiva) p.68.
Rumsby D, Grant I, Brozel M R, Foulkes E J and Ware R M, 1984, Semi-Insulating III-V Materials, Eds. D C Look and Blakemore J S (Publ. Shiva) p.165.
Skolnick M S, 1985 in Defect Recognition and Image Processing I, Ed. J P Fillard (Publ. Elsevier) p.165.
Stirland D J, 1977, Inst. Phys. Conf. Ser. No. 33a, p.150.
Stirland D J, Grant I, Brozel M R and Ware R M, 1983, Inst. Phys. Conf. Ser. No. 67, 285.
Stirland D J, Augustus P D, Brozel M R and Foulkes E J, 1984, in Semi-Insulating III-V Materials, Eds. Look D C and Blakemore J S (Shiva Publ.) p.91.
Stirland D J, Brozel M R and Grant I, 1985a, Appl. Phys. Lett. 46, 1066.
Stirland D J, Brozel M R and Grant I, 1985b, Inst. Phys. Conf. Ser. No. 76, p.211.
Stirland D J, 1986, in Semi-Insulating III-V Materials, Eds. H Kukimoto and S Miyazawa (Publ. Ohmsha-North Holland), p.81.
Stirland D J, 1987, in Defect Recognition and Image Processing II, Ed. E R Weber (Publ. Elsevier) p.73.
Stirland D J and Hart D G, 1988, unpublished.
Stirland D J, Warwick C A and Brown G T, 1988, in Semi-Insulating III-V Materials, Eds. G Grossmann and L Ledebo (Publ. Adam Hilger) p.93.
Suchet P, Duseaux M, Gillardin G, Le Bris J and Martin G M, 1987, J. Appl. Phys. 62, 3700.
Suchet P, Duseaux M, 1987, Inst. Phys. Conf. Ser. No. 91, p.375.
Suchet P, Duseaux M, Schiller C and Martin G M, 1988, in Semi-Insulating III-V Materials, Eds. G Grossmann and L Ledebo (Publ. Adam Hilger) p483.
Winston H V, Hunter A T, Kimura H, Olsen H M, Bryan R P, Lee R E and Marsh O J, 1984, Inst. Phys. Conf. Ser. No. 74, 497.

Gettering in silicon

Robert Falster

Monsanto Electronic Materials Company, Milton Keynes MK12 5TB UK

ABSTRACT: The gettering effectiveness for transition metals and hybrid solutes such as Au and Pt in a variety of defect structures in silicon has been studied using haze tests, SIMS, RBS and TEM. A saturation is observed regardless of cooling conditions for transition metals in all cases involving surface or "extrinsic" type defect gettering. This is not the case for distributed oxygen related defects where complete gettering of the amounts of transition metals soluble at high temperatures is observed after only the slightest precipitate growth.

1. INTRODUCTION

There are many fast diffusing metallic contaminants common in the environment in which silicon is processed into integrated circuits. Recent advances in surface analytical techniques such as Vapor Phase Decomposition Atomic Absorption or Total Reflection X-ray Flourescence Analysis (Eichinger et al. 1988) and Haze Testing (Graff, 1983 and Bergholz et al., 1987) have revealed a hitherto unknown omnipresence of surface metal contamination. Cleaning technologies are improving but given the process performance demands of current and developing IC technologies, a better understanding of the techniques which can help to minimize the harmful effects of metallic contaminants is required.

Gettering has been proven a powerful tool for coping with contamination. It is achieved in practice by a variety of means which either enhance the solubility or induce the precipitation of the unwanted contaminant at a location in the wafer removed active circuit areas. Several techniques have become popular: phosphorus diffusion, polycrystalline silicon layers, various types of wafer backsurface damage and "intrinsic gettering" which employs the defects which accompany the natural precipitation of supersaturated oxygen in Czochralski (CZ) grown silicon. Given the large variation in nature and distribution of these types of gettering sites: precipitates, stacking faults, grain boundaries, dislocations - as well as in the physical properties of the contaminates, such as the solubility, precipitation behavior, diffusion mechanism and coefficient, it is unlikely that any one model will suffice to explain the whole of gettering phenomenology.

This paper presents results of investigations of the gettering of several transition metals such as Ni, Cu, Fe, Co and Pd as well as such hybrid solutes as Au and Pt.   The cases of polycrystalline silicon films, intentionally induced back surface stacking faults (nucleated by wet sandblasting, called "soft backside damage" or SBD), phosphorus diffusion and oxygen related defects are considered.

2.   "DEFECTLESS" METAL PRECIPITATION

The high diffusivity, low activation energy of diffusion, steep temperature dependence of solubility and ease of silicide formation of the late transition metals usually results in significant precipitation upon cooling from high temperatures even in the absence of obvious crystal defects.   At room temperature after normal cooling, these metals can exist in a variety of forms: precipitated at the sample surface or homogeneously (?) in the bulk; in solution or precipitated at defect sites.   When these defects are intentional it is called gettering, when not, it is a process flaw and potential device failure.   In   the   case   of   the   fastest   diffusers,   Cu   and   Ni, precipitation is always essentially complete. For Cu, Ni and Co, free surface precipitation is usually dominant of the two "defectless" precipitation mechanisms.   This is basis of the "haze test".   Pd usually   results   in   large   numbers   of   both   surface   and   bulk precipitates.

Experiments have shown that free surface precipitation tends to occur at temperatures less than about 600°C (Hill 1981).   Just how strongly surface   metal   precipitation   is   affected   by   the   details   of   low temperature cooling can be seen by the haze patterns in Figure 1.   Two wafers intentionally contaminated by scratching metals on the back surface were given 30 s 1000°C heat treatments in a rapid thermal annealer.   The eyes and eyebrows are stainless steel (the s-pits are due to Ni) and the nose and mouth copper.   The wafer in Figure 1a was quenched to 320°C and held there for 5 minutes prior to removal. It exhibits weak Ni haze and strong Cu haze.   Just the opposite is seen in Figure 1b in which the wafer was quenched to 320°C and removed immediately to room temperature.

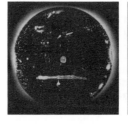

Fig. 1.   Stainless steel and
          copper haze formed
          during (a) quench to
          320°C and (b) to room
          temperature.

Careful dielectric breakdown measurements which can identify metal precipitate related breakdown events show that cooling in the 600°C range plays an important rôle in samples with realistically low levels of contamination as well (Falster 1989).

Usually for a specific metal and cooling treatment, surface precipitates and their etch pits are relatively uniform in size.   This would indicate that the precipitation occured at roughly the same temperature.   This is not always the case however.   The optical micrographs of Secco etch features such as those in Figures 2 and 3 indicate that free surface precipitation can occur over a range of temperatures even for the same metal and cooling treatment. Figure 2 shows a typical large Cu colony surrounded by a dense array of smaller Cu-related etch pits. The large differences in size as well as the gettered ring around the colony indicates clearly that the colony formed first. Ni can also form very large range of free surface precipitate sizes and densities strongly dependent on concentration as is shown in Figure 3.

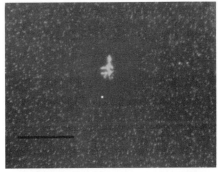

Fig. 2. Cu surface etch pits showing large colony and gettered ring. Bar is 50 µm.

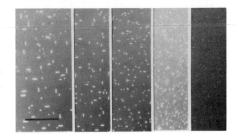

Fig. 3. Distribution of Ni etch features along haze band from center to 1 cm away. Bar is 100 µm.

3.   EXTRINSIC GETTERING OF HAZE FORMING METALS

Extrinsic gettering refers to a broad class of techniques which rely on the application of a more or less uniform layer of crystal disorder to the surface opposite the processed devices (the backside).   If sufficiently dense, nearly any kind of disorder succeeds well in keeping low levels of haze forming metals from precipitating at the undamaged wafer surface after a single heat treatment.   There is a limit to how much precipitation can occur in a region of surface disorder (how much it can getter).   Both haze tests and direct measurements of gettered metal confirm this.   Figure 4 shows the

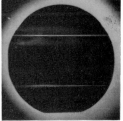

Fig. 4. Ni (top) and Cu haze pattern with (b) and without (a) extrinsic gettering

results of 950°C furnace haze tests on wafers with and without "soft backside damage" in the as-damaged state (prior to stacking fault growth) which have been heavily intentionally contaminated with scratches of Ni and Cu on the wafer frontside. Figure 4a shows the Secco-etched haze patterns typical under these conditions. The effect of the extrinsic gettering is seen in Figure 4b. Significant amounts, but not all, of the metal hazes have been eliminated. This is in sharp contrast to gettering by distributed oxygen related defects as described below. Microscopic examination of the Ni etch pits reveal the small round s-pits to have been eliminated, leaving behind only a distribution of linear features similar to, but smaller than, those of Figure 3. There is essentially no difference in the etch pit patterns between samples contaminated on the damage side. Likewise, variation of the length of time spent at 950°C or the cooling rates in the 700-950°C do not seem to have much influence. Similar results are obtained after the SBD has been grown into stacking faults of a variety of lengths as well as with poly-Si films.

SIMS data of gettering by surface disorder reveal a similar picture. Samples with a variety of surface treatments were sputtered on the opposite side with a small but uniform amount of Cu or Ni. The amount sputtered (ca. 2×10^{15} cm^{-2}, as determined by Rutherford Backscattering Spectrometry), was chosen such that it was slightly below the total amount soluble in silicon at a diffusion temperature of 950°C.   The surface treatments normally reserved for the unpolished backsurface were performed on the polished surface for ease of analysis. Figure 5 shows data typical of Cu comparing the pile up of its precipitates on an undamaged surface (haze) with that of an SBD layer prior to the growth of the stacking faults. The gettering of

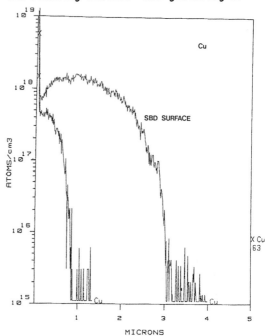

Fig. 5. SIMS profiles of Cu
        gettering by SBD

approximately 50 times more Cu than the free surface is clearly seen. Still, only about a tenth of the total Cu in the system is to be found in the gettered layer. Growth of the damage into stacking faults prior to metal diffusion, longer diffusions, slower cooling do not result in significantly different SIMS profiles. Similar results are obtained with Ni, although significantly less metal can be detected.

Results obtained by the analysis of metal gettering in poly-Si layers are revealing. After deposition, a thin, continuous, oxide layer exists between the single crystal substrate and the CVD poly-Si layer. This layer has been shown to break up or "ball up" under treatments greater than about 1000°C (Jorgenson et al. 1985). Metal diffusions from sputtered sources similar to those described above were performed in poly-Si coated samples (polished side) with either intact or previously (by a 30 min 1100°C heat treatment) "balled-up" interfacial oxide layers. The poly-Si films were 1.4 μm thick and original interfacial oxide layer was 1.7 nm - as determined SIMS measurements. SIMS and RBS analysis of the metal gettered in the layers after various indiffusion treatments at 900°C show the presence of the oxide layer to have a strong effect on the resultant gettered profile of all metals studied.

Figure 6 is a SIMS profile of Ni concentration in a poly-Si layer with an intact oxide after a 15 min. 900°C furnace diffusion from a limited source ($2\text{X}10^{15}$ cm^{-2}) on the opposite side of the 575 μm thick sample. A large peak in concentration is found at the interface, in addition to the just under $1\text{X}10^{18}$ cm^{-3} in the poly-Si layer itself. This interfacial peak is not present in samples whose oxides

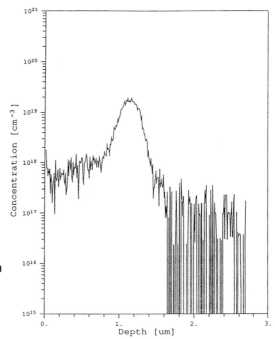

Fig. 6. SIMS profile of Ni gettering by a 1.4μm thick poly-Si layer with an interfacial oxide layer.

have been balled-up by the short high temperature pre-treatment prior to metal diffusion. Similar, if less pronounced, results are observed

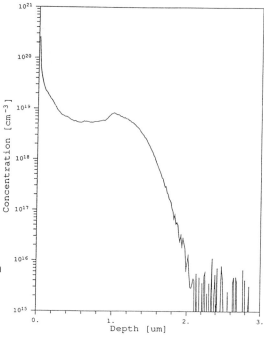

Fig 7.   SIMS profile of Cu
gettering in a 1.4μm
thick poly-Si with an
interfacial oxide.

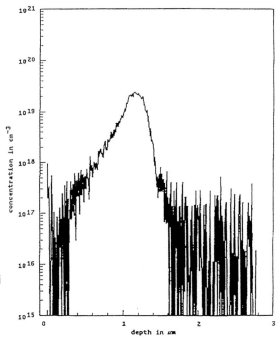

Fig. 8. SIMS profile of Ni
gettering in poly-Si
film after 43 hours
at 900°C.

with Cu (Figure 7). Again, only a fraction of the total metal in the system is gettered.

One important difference is observed in the gettering behaviour of Cu and Ni. Performing the metal diffusions for an extended period of time changes the resultant gettering profiles. Data was taken on samples diffused with Cu or Ni for 43 hours at 900°C. Figure 8 shows the Ni profile following this diffusion. An identical interfacial peak to that of the short diffusion in Figure 6 is seen. Much less Ni is found in the film near the surface. In the case of Cu, no gettering is observed in the film or at the interface after this extended treatment. It is assumed that the gettering of Cu takes place exclusively upon cooling. In the case of the extended treatment it is more effectively gettered by the oxygen related defects which developed in the mean time. For the short treatment, no such defects would be present. The same would seem to be true for the Ni - except for the fraction gettered at the interfacial oxide layer. It was apparently gettered immediately - and remained in a form which was stable throughout the long heat treatment.

The maximum amount of haze forming metal gettered in a disordered surface layer is on the order $5X10^{14}$ cm^{-2}. With Pd, some additional indirect gettering can in some cases be observed however. Figure 9 shows a etched cross section of a sample diffused for 7 min at 950°C with Pd. The micrograph is taken near a surface containing about $1X10^7$ cm^{-2} 0.5 micron long stacking faults resulting from SBD and a 45 min dry oxidation at 1000°C. During the Secco etch, 8 microns of material were removed. A greatly enhanced bulk Pd precipitation is observed on the stacking fault side to a depth of about 50 μm. No such enhancement is seen near the surfaces as-damaged or undamaged samples. The local absorption by the stacking faults of the excess Si self-interstitials produced the Pd precipitation is presumed to be be the cause of this effect.

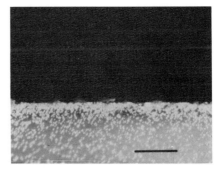

Fig. 9.   Cross section of bulk Pd
          s-pits in the vicinity of
          surface stacking faults.
          Marker is 100 μm.

4.   EXTRINSIC GETTERING OF HYBRID SOLUTES

Hybrid solutes in silicon such as Au or Pt can exist on either fast diffusing, low solubility interstitial sites (M_i) or slower diffusing, higher solubility substitutional sites (M_s). Their diffusion properties are best modelled by the "kick-out" mechanism (Gösele et al. 1980 and Stolwijk et al. 1984).

$$M_i \rightleftharpoons M_s + I$$

The build up of Au or Pt to substitutional equilibrium values in the silicon bulk is controlled by the diffusion and annihilation of silicon self-interstials (I) rather than by the intrinsic properties of substitutional solutes themselves.

Gettering techniques which result in the massive injection of silicon self-interstials such as phosphorus diffusion have a very strong enhancing effect on the gettering of these metals by effectively suppressing the solubility of the substitutional component (Falster 1985, Falster et al. 1985 and Bronner et al. 1987).

Techniques which do not result in the injection of silicon self interstitials, such as SBD or poly-Si layers, do getter hybrid solutes, but without this enhancement.   This point is well illustrated by the case of poly-Si gettering.   Figure 10 shows a SIMS profile of Au gettered into a 1.5 μm thick layer at 1150°C in 1 hour.   A very uniform

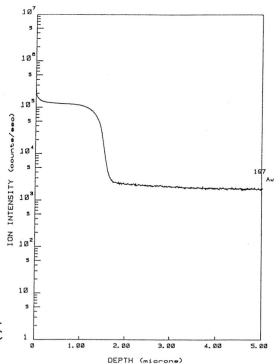

Fig. 10.   SIMS profile of gettered into a 1.5 μm thick poly-Si layer at 1100°C

concentration of Au is found in the film.   RBS calibration places the Au concentration at about 4×10^{18} cm^{-3} in the film.   This is well above the solubility in single crystal silicon.   Cross-sectional TEM analysis revealed no precipitates in the film.

At temperatures below 1100°C, the gettering of significant amounts of Au by "passive" techniques such as poly-Si can take quite a long time. Figure 11 shows the build up of Au concentration in a 1.5 μm layer over time from a limited sputtered source of about 8×10^{14} cm^{-2} located at the other side of the 375 μm thick specimens. Gettering took place until approximately all the Au was absorbed by the film. Measureable gettering at 900°C is not observed until after approximately 24 hours of diffusion.   This should be compared with the gettering of 8×10^{13} cm^{-2} Pt co-diffused with phosphorus after only 2 hours at 860°C (Falster 1985) or 6×10^{14} cm^{-2} Au after a 15 min corona discharge treatment (Falster et al. 1985).

The rate at which the Au is absorbed into the film is approximately equal to the rate at which it can be absorbed into the bulk of the silicon. In all these experiments, sum of the amounts of Au measured on the front and back surfaces remained approximately constant. To illustrate this point further, also plotted on the gettering rate curves of Figure 11 are calculations following Stolwijk et al. (1984) of the calculated square root of time build up of bulk substitutional Au concentration - the total converted to a resultant concentration if entirely contained in a 1.5 μm film.

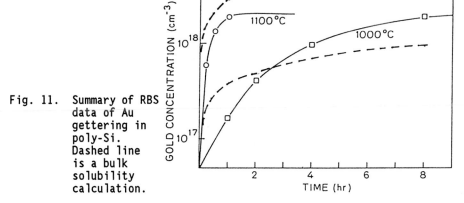

Fig. 11. Summary of RBS data of Au gettering in poly-Si. Dashed line is a bulk solubility calculation.

The presence of an interfacial oxide film enhances the gettering of Au just as it does the transition metals. Figure 12 shows the Au profile

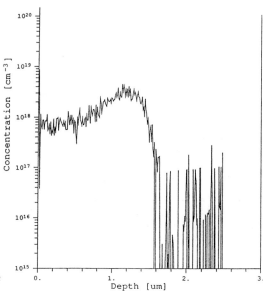

Fig. 12. Au gettering in poly-Si film with an interfacial oxide layer. SIMS data.

obtained in a poly-Si film with no balling-up pretreatment after 40
hours of diffusion at 900°C. The balling up of the interfacial oxide
prior to the Au diffusion eliminates the interfacial peak.

5.   GETTERING BY OXYGEN RELATED DEFECTS

The gettering of transition metals by oxygen related defects has
recently been studied by Falster and Bergholz (1988). Rapid thermal
haze gettering tests of Cu, Ni, Fe, Co and Pd were performed on silicon
slices of a wide variety of oxygen contents both FZ and CZ with initial
interstitial oxygen contents, O_i, ranging from 5.6×10^{17} to
9.0×10^{17} cm^{-3}. Cu and Ni were found to be the easiest to getter
with Co and Pd progressively more difficult. Co was found to interact
with the other metals and affect their ability to be gettered. A
gettering "threshold" for the amount of precipitated oxygen was
observed which varies from metal to metal but is in all cases very low
and very steep. Complete gettering of all metals tested was observed
in wafers with initial O_i of 6.45×10^{17} cm^{-3} after only a portion
of the CMOS simulation was complete. Only 0.25×10^{17} cm^{-3} had
precipitated during this treatment. This is well below the level at
which oxygen precipitates are visible by etching techniques or x-ray
topography. Above this level of precipitation, complete gettering is
always observed. (In fact, after only the slightest precipitate
growth, no amount of metal could be applied in scratch tests which
could not be gettered.) Wafers with marginally less initial oxygen
showed gettering of Ni and Cu only after a complete CMOS cycle. As
grown CZ wafers do not exhibit gettering behavior. FZ wafers showed
no gettering under any circumstances or cycle of heat treatments. No
indication was found that bulk stacking faults are necessary for
effective gettering.

Grown-in oxygen related defects (precipitate precursors) do not then
result in the gettering of transistion metals. It seems however that
only the slightest growth (the rate of which is concentration
dependent) of these precipitates are required to achieve extremely
effective gettering. The case of Fe is however still under
investigation by other techniques as it does not lend itself well to
the haze test. Newly developed scanning infrared microscopy
techniques (Laczik et al. (1989) for the analysis of bulk
precipitation in silicon will be of great use toward the better
understanding of gettering by distributed oxygen related defects.

Of all the gettering systems studied, only distributed oxygen related
defects can result in the complete gettering of the amounts of
transistion metals soluble at high temperatures. Experiments varying
the cooling rates during extrinsic gettering indicate that diffusion
limitation is probably not the reason for this. Furthermore, the total
number of "intrinsic" defect sites is roughly comparable to that of
extrinsic techniques. The stacking fault density of the SBD studied
was about 2×10^7 cm^{-2}. The oxygen related defect density can be
assumed to be roughly 1×10^9 cm^{-3}, corresponding to 5×10^7 cm^{-2}
in a 500 μm thick wafer. In spite of the superiority of distributed
oxygen defects to surface disorder techniques, time is required to
develop intrinsic sites (although perhaps not as much as generally
believed). Extrinsic gettering sites, very effective in gettering low
levels of contaminates, are thus required in addition to the use of
oxygen.

ACKNOWLEDGEMENTS

The author gratefully acknowledges the collaboration of Werner Bergholz and his group at Siemens AG, Munich, on the work regarding gettering by oxygen related defects. Chris Hill of Plessey, Caswell performed the quenched Rapid Thermal haze test heat treatments of Figure 1. Many thanks also to Peter Eichinger of GeMeTec, Munich and Sukanta Biswas of Cascade Scientific, Uxbridge.

REFERENCES

Bergholz W, Zoth G, Wendt H, Sauter S and Asam G 1987 Siemens Forsch. u. Entwickl. Ber. 16 241
Bronner G B and Plummer J D 1987 J. Appl. Phys. 615286
Eichinger P, Rath H J and Schwenke H 1988 Semiconductor Fabrication: Technology and Metrology, ASTM STP 990 (Amer. Soc for Testing and Materials)
Falster R 1985 Appl. Phys. Lett. 46 737
Falster R J, Modlin D N, Tiller W A and Gibbons J F 1985 J. Appl. Phys. 57 554
Falster R 1989 to be published
Falster R and Bergholz W 1989 to be published
Gösele U, Frank W and Seeger A 1980 Appl. Phys. 23 361
Graff K 1983 Aggregation Phenomena of Point Defects in Silicon ed Sirtl E, Goorissen J and Wagner P (Pennington N J: The Electrochemical Society) pp 121-133
Hill D E 1981 Semiconductor Silicon 1981 ed Huff H R and Kriegler (Pennington N J: The Electrochemical Softbound Proceedings Series) pp 354-360
Jorgensen N, Barry J C, Booker G R, Ashburn P, Wolstenholme G R, Wilson M C and Hunt P C 1985 Inst. Phys. Conf. Ser. No. 76 471
Laczik Z, Booker GR, Falster R and Astles M, this conference
Stolwijk N A, Schuster B and Hölzl J 1984 Appl. Phys. A 33 133

Inst. Phys. Conf. Ser. No 104: Chapter 5
Paper presented at Int. Symp. on Struct. Prop. Disloc. Semicond., Oxford, 1989

455

The effect of geometrical and material parameters on the stress relief of mismatched heteroepitaxial systems

R. Beanland & R.C. Pond,
Department of Materials Science and Engineering, The University of Liverpool, P.O. Box 147, Liverpool L69 3BX, UK.

ABSTRACT

It is apparent from experimental evidence that mismatched epilayers grown on vicinal surfaces may have significantly different stress relief mechanisms compared to those grown on surfaces with exact crystallographic indices. One of the effects of growth on a vicinal surface is a misorientation of the epilayer with respect to the substrate. We briefly review the models of misorientation effects, and, using the Hornstra and Bartels model of coherently strained mismatched epilayers, attempt to clarify the differences between (100) and (011) crystal interfaces.

INTRODUCTION

The growth of mismatched semiconductor epilayers is a topical subject, as the potential devices which may become available through epitaxy have novel and technologically important properties. Traditionally, heteroepitaxial growth has been performed upon exact (100) and (111) substrates, and it is only recently that the possibilities of deliberately offcut (vicinal) substrates have been explored. It has often been found that an improvement of crystalline quality is obtained when an offcut is used, although the reasons for this remain unclear. Systematic studies of mismatched epilayers on vicinal substrates have indicated a dependence of the state of strain of the epilayer on the misfit and vicinal angle, which, although well described for coherent epilayers, is poorly understood for semicoherent epilayers. The effect is often described in terms of a 'misorientation', as double crystal X-ray measurements indicate that the low index planes closest to the interface normal in epilayer and substrate are not coplanar. In this paper, we attempt to clarify the misorientation mechanisms in semicoherent epilayers by considering the nucleation of dislocations in a coherent epilayer.

THEORY

A simple and elegant model of coherently strained cubic epilayers was developed by Hornstra and Bartels (1978) with the purpose of measurement of coherently strained epilayer lattice constants. The model, based on anisotropic elasticity theory, allowed prediction of the change in interplanar spacing and orientation of any plane in an epilayer grown on any cubic substrate surface. The angle between equivalent planes in the epilayer and substrate, $\Delta\theta$, is given by

$$\sin\Delta\theta = \underline{n}.\underline{a} \sin\theta_m \tag{1}$$

where $\underline{n}$ is the unit normal to the interface, θ is the angle between the plane in question and $\underline{n}$, and $\underline{a}$ is a 'strain vector' such that the strain tensor of the epilayer is given by

$$A_{ij} = a_i\, n_j + \epsilon_\parallel \delta_{ij} \tag{2}$$

where $\epsilon_\parallel$ is the "misfit", i.e. $(a_e - a_s)/a_e$, and δ_{ij} is the Kroenecker Delta.

Applying this to vicinal epilayers, it is easily seen that the misorientation due to coherency strains, $\theta^c{}_m$ is given by

$$Sin(\theta^c{}_m) = \underline{n.a}\; Sin(\theta_v) \tag{3}$$

where θ_v is the vicinal angle. This model is now in standard use in measurements of coherent epilayers. However, measurements of dislocated epilayers frequently show deviations from equation (3), e.g. Igarishi (1976), Olsen and Smith (1975), and Nagai (1974).

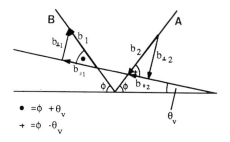

Figure 1.Burgers vector components parallel and perpendicular to the interface for dislocations on inclined glide planes for vicinal epilayers.

A deviation from equation (3) is not surprising when the geometry of stress relief by 'misfit' dislocations is considered. In the diamond and sphalerite structures, glide of $1/2<110>$ type dislocations takes place on $\{111\}$ planes, imposing restrictions on the strain fields which can be totally relieved. It is well known that most 'misfit' dislocations in mismatched semiconductor epilayers glide on inclined $\{111\}$ planes to lie in the interface. These dislocations have a Burgers vector component normal to the interface, and consequently a misorientation of the epilayer will occur unless the population of dislocation types is such that this normal component sums to zero. For exact $\{100\}$ or $\{110\}$ interfaces, this will only occur if the populations of dislocations with opposite tilt-producing components are equal. For vicinal surfaces, equal populations of such dislocations will in general produce a tilt, as the normal components do not cancel. This is shown schematically in Figure (1) for a vicinal angle of θ_v.

The components of the dislocation types parallel and perpendicular to the interface are:

$$b_{\parallel 1} = b\, Cos\,(\varnothing + \theta_v), \quad and \quad b_{\parallel 2} = b\, Cos\,(\varnothing - \theta_v) \tag{4a}$$

and

$$b_{\perp 1} = b\, Sin\,(\varnothing + \theta_v), \quad and \quad b_{\perp 2} = b\, Sin\,(\varnothing - \theta_v) \tag{4b}$$

where $\varnothing$ is the angle between the glide plane normal $\underline{m}$ and the on-axis (i.e. not offcut) direction. If the dislocation populations are equal, each dislocation of one type may be paired with another of the opposite type. For complete relief of misfit, the spacing between each pair of dislocations is $(b_{\parallel 1} + b_{\parallel 2})/\epsilon_\parallel$, and misorientation of the epilayer, $\theta^\perp{}_m$, is then given approximately by

$$Tan\,\theta^\perp{}_m \approx (b_{\perp 1} - b_{\perp 2})/d = (b_{\perp 1} - b_{\perp 2})\epsilon_\parallel/(b_{\parallel 1} + b_{\parallel 2}) = Tan\,\theta_v.\epsilon_\parallel \tag{5}$$

Equation (5) acts such that for positive misfit, i.e. $a_e > a_s$, the rotation is in the opposite direction to the vicinal angle, and vice versa. We note that this is equivalent to an earlier description of misorientation (Pond et al. (1987)), in which the vicinal surface was considered to consist of terraces of low-index surface separated by steps. Misfit on the terraces was assumed to be relaxed, and eq.(5) obtained using an array of interfacial line defects at steps according to the classification of Pond (1988). Equation (5) has met with some success in describing misorientations of semicoherent epilayers in several cases (Aindow (1989)), indicating that dislocation populations were not greatly affected by the vicinal offcut for the vicinal angles studied (up to 2°). However, there are notable examples which do not conform to this simple model, e.g. Olsen & Smith (1975). These discrepancies must therefore arise from a difference in the populations of dislocations with Burgers vector components normal to the interface, which in turn implies a difference in dislocation nucleation and/or multiplication rates on the various glide systems, related to the stress in a vicinal epilayer. We now consider the possible reasons for this difference, using the Hornstra and Bartels model to indicate the main influences on dislocation nucleation in a coherent-semicoherent transition.

As the strain energy of an epilayer of given thickness is independent of interfacial orientation, it is clear that the individual glide systems must be considered. As a first approximation, we therefore take the glide force per unit length on a given dislocation F_l as a rough comparative measure of its nucleation rate. The glide force per unit length on a dislocation with Burgers vector $\underline{b}$ gliding on a plane $\underline{m}$ is given by

$$F_l = (\Sigma\underline{m})\bullet\underline{b} \qquad\qquad (6)$$

where Σ is the stress tensor. Applying Hornstra and Bartel's model to $In_{0.25}Ga_{0.75}As$ on InP as an example of a typical elastically anisotropic misfitted semiconductor epilayer, ($\epsilon_{\parallel} \approx -2\%$), the value of F_l may be calculated for any of the 12 possible glide systems and any given interface.

A plot of F_l is given for $\underline{b}=1/2\{1\bar{1}0\}$ on (111) and (11$\bar{1}$), for interfaces with the zone axis [0$\bar{1}$1] in figure (2). (Note that the range of interfacial orientations required to obtain all

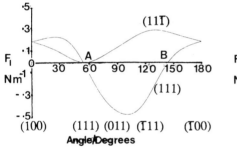

Figure 2. Variation of glide force with interfacial orientation for the $1/2[1\bar{1}0]$ (111) / (11$\bar{1}$) glide systems.

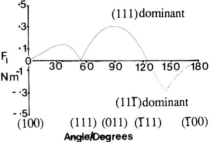

Figure 3. Difference in glide forces on a $\underline{b}=1/2[1\bar{1}0]$ dislocation for the (111) / (11$\bar{1}$) glide planes.

values of F_l is only 180° due to the centrosymmetric nature of the stress tensor.) It can be seen that the points where $F_l=0$, A and B, result from the geometry of the system, as a dislocation must have some misfit relieving effect through glide to experience a force from coherency strains. At point A, the (111) glide plane is parallel to the interfacial plane, and glide of the dislocation produces no misfit relieving effect. The (11$\bar{1}$) plane is positioned such that a $1/2[1\bar{1}0]$ dislocation is restricted to perfect screw orientation if it lies in the interface, again relieving no misfit. At point B, the (111) glide plane is perpendicular to the ($\bar{2}$11) interface, and a $1/2[1\bar{1}0]$ dislocation on this plane is restricted to a screw component parallel to the interface, once more relieving no misfit. The maxima and minima of figure 2 are *not* determined by geometry, and the positions and magnitudes are variable with respect to the material's elastic constants. Figure (3) shows the difference between the glide forces of Figure 2, i.e. $|F_l(111)| - |F_l(11\bar{1})|$. The (111) plane is dominant throughout most of the range of θ_v, reaching peaks at (0$\bar{1}$1) and (211), (although not geometrically constrained to do so). The curve passes through zero at interfacial orientations where the two glide systems are equivalent by the crystal's symmetry, and when the dislocation experiences no glide forces at (111). It can be seen from Figure 3 that if there is no barrier to cross slip, $1/2[1\bar{1}0]$

dislocations are expected to lie preferentially on the (111) glide plane for interfaces from (100) to ($\bar{1}$11), and on the (11$\bar{1}$) glide plane from ($\bar{1}$11) to ($\bar{1}$00). The $1/2[10\bar{1}](111)$ and $1/2[10\bar{1}](1\bar{1}\bar{1})$ glide systems are equivalent to the $1/2[1\bar{1}0](111)$ and $1/2[1\bar{1}0](1\bar{1}\bar{1})$ glide systems respectively throughout the range of θ_v used here, as they are related by a ($01\bar{1}$) mirror which is not broken by any rotation about the [$01\bar{1}$] axis: Figures 2 and 3 therefore also apply to $1/2[10\bar{1}]$ dislocations. A similar plot of F_l against interfacial orientation for the $1/2[110](1\bar{1}\bar{1})/(1\bar{1}1)$ and $1/2[101](1\bar{1}\bar{1})/(\bar{1}11)$ glide systems is shown in Figure 4. It is easily seen that a positive rotation on Figure 2 is equivalent to a negative rotation on Figure 4. This now covers 8 of the 12 glide systems available. The two glide systems for dislocations with $\underline{b}$ = $1/2[0\bar{1}1]$, i.e. $1/2[0\bar{1}1](111)$ and $1/2[0\bar{1}1](111)$, must always have F_l =O) as the dislocation can only lie in the interface in perfect screw orientation. The remaining two glide systems of $\underline{b}$=$1/2[011](11\bar{1})$ and $1/2[011](1\bar{1}1)$ are equivalent under all rotations about [$0\bar{1}1$], and are shown in Figure 5. Once more, the zero points are geometrically constrained to interfaces where $\underline{b}$ is either completely perpendicular to the interface or the dislocation must lie in perfect screw orientation.

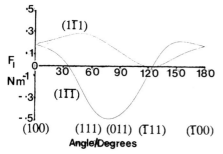

Figure 4. Variation of glide force with interfacial orientation for the $1/2[110]$ (1$\bar{1}\bar{1}$) / (1$\bar{1}$1) glide systems.

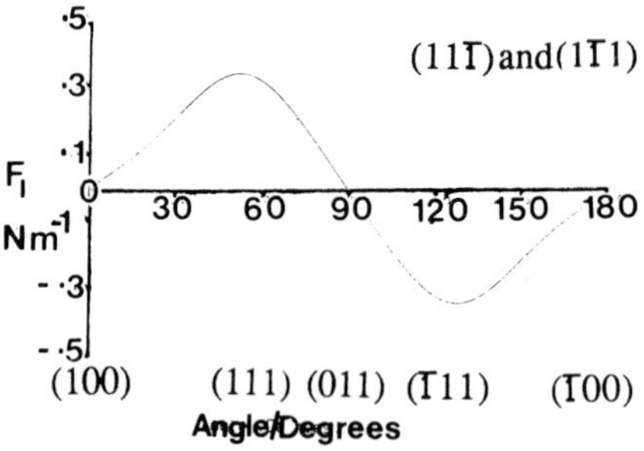

Figure 5. Variation of glide force with interfacial orientation for the $1/2[011]$ (11$\bar{1}$) / (1$\bar{1}$1) glide systems.

3.0 APPLICATIONS TO MISORIENTATION EFFECTS

3.1 Vicinal (100) Epilayers

For an exact (100) epilayer, 8 of the 12 glide systems have equal and non-zero values of F_L. It would therefore be expected that the only difference in dislocation populations would arise from the α/β and glide/shuffle nature of the dislocations as seen by Kavanagh et al. (1988). However, this will not give rise to a misorientation since sufficient Burgers vectors are available to allow cancellation of components normal to the interface. Now, for a (100) epilayer offcut by θ_v towards [011], the symmetry in the forces F_l is broken: this is illustrated in Table 1 for 5° and 10° offcuts. Now, assuming the nucleation and multiplication of

m	(111)		(1$\bar{1}\bar{1}$)		(11$\bar{1}$)/(1$\bar{1}$1)	
b	$1/2[1\bar{1}0]$	$1/2[0\bar{1}1]$	$1/2[110]$	$1/2[1\bar{1}0]$	$1/2[011]$	$1/2[101]$
θ_v	$1/2[10\bar{1}]$		$1/2[101]$	$1/2[10\bar{1}]$		$1/2[110]$
0°	0.1922	0.0	0.1922	0.1922	0.0	0.1922
5°	0.2036	0.0	0.1797	0.1799	0.2039	0.2038
10°	0.2140	0.0	0.1682	0.1663	0.0488	0.2151

Table 1.   Values of F_l for the 12 $1/2<110>\{111\}$ glide systems for a coherent (100) In$_{0.25}$Ga$_{0.75}$As epilayer offcut towards [011].

dislocations during the coherent-semicoherent transition is sensitive to F_l, 4 glide systems will be operative on 3 glide planes in vicinal epilayers, compared to 8 glide systems on 4 glide planes in exact (100) epilayers. It has been suggested by Ahearn and Uppal (1987) that fewer dislocation-pinning interactions will occur in vicinal (100) epilayers where this reduction of operative glide systems occurs, leading to fewer dislocation tangles within the epilayer, and a more efficient relief of misfit. A further consequence of symmetry breaking is that dislocations on the (111) plane experience a greater force per unit length than those on (1T̄T̄). It will therefore be likely that the density of dislocations gliding on (111) exceeds that of (1T̄T̄) due to preferential nucleation. This will tend to reduce the misorientation described by Equation 5, as can be seen in Figure 1, if glide plane A is (1T̄T̄), glide plane B is (111) and [0T̄11] is out of the paper. It has been noted that there is a tendency for this to occur in studies of rough (100) surfaces on a local scale (Feuillet et al (1987), indicating that the boundary conditions imposed by the Hornstra and Bartels model are not critical in some cases.

3.2 Vicinal (011) Epilayers

In an exact (011) epilayer, 8 of the 12 glide systems have non-zero values of F_1 as shown in Table 2. Of these, four are dominant, with F_1 over three times the value of that for the minority four, and it would be expected that stress relief would mainly occur on the 'majority' glide systems, through cross slip of dislocations from the minority to the majority planes. Thus, only two glide planes, (111) and (1T̄T̄), would contain most of the dislocation population. For vicinal (011) surfaces offcut towards [100], symmetry is once more broken, and dislocations on the (1T̄T̄) plane will experience a greater force than those on (111), although both planes are still dominant. The action of glide on the two planes (111) (1T̄T̄) will result in an increase in tilt, as can be seen in Figure 1 if glide plane A is (111), glide plane B is (1T̄T̄), and [0T̄1] is into the paper.

m	(111)		(1T̄T̄)	(11T̄)/(1T̄1)		
b	1/2[1T̄0]	1/2[0T̄1]	1/2[110]	1/2[1T̄0]	1/2[101]	1/2[011]
θv	1/2[10T̄]		1/2[101]	1/2[10T̄]	1/2[110]	
0°	-0.4661	0.0	-0.4661	0.1350	0.1350	0.0
5°	-0.4131	0.0	-0.4680	0.1075	0.1625	0.0549
10°	-0.3690	0.0	-0.4780	0.0805	0.1895	0.1090

Table 2. Values of F_1 for the 12 $1/2<110>\{111\}$ glide systems for a coherent (011) $In_{0.25}Ga_{0.75}As$ epilayer offcut towards [100].

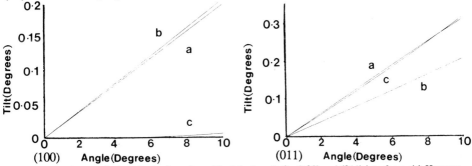

Figures 6 & 7. Variation of misorientation with vicinal angle for (100) and (011) interfaces. (a)-Hornstra and Bartels model (eqn.3.), (b)-Semicoherent interface with equal populations of 60° dislocations (eqn.5.), (c)-Semicoherent interface with unequal populations of 60° dislocations (eqn.6).

3.3 Misorientations in semicoherent vicinal (100) and (011) epilayers

The dependence of misorientation will not, in general, be a simple function of the differences in F_1. However, we may illustrate the principles of the model by taking the density of dislocations to be proportional to the differences in F_1.

This allows a modification of equation 5:

$$\text{Tan}\,(\theta^{t}m') = [F_{l_1}\sin(\varnothing+\theta_v) - F_{l_2}\sin(\varnothing-\theta_v)]\epsilon_{\#}/[F_{l_1}\cos(\varnothing+\theta_v) + F_{l_2}\cos(\varnothing-\theta_v)]$$

(6)

Again neglecting the contribution of dislocations on the (1Ī1) and (11Ī) planes. This function is shown in Figures 5 and 6 for vicinal (100) and (011) epilayers: misorientations $\theta^e m$ and $\theta^t m$ are also shown.

4.   Discussion and Conclusions

The results obtained above are subject to the assumptions inherent in the Hornstra and Bartels model. The most important of these with regard to large misfit epitaxy is that the epilayer is of uniform thickness and of infinite extent, and so the model can only be applied with confidence to layer by layer growth. Island growth modes, for example GaAs on silicon, will have substantially different boundary conditions and stress relief mechanisms.

Another important consideration is the image forces experienced by a dislocation loop as it grows in size to form a section of 'misfit' dislocation. This will vary according to the glide system, being a maximum when the glide plane is perpendicular to the free surface.

Furthermore, the variation of elastic constants and misfit with temperature should be considered for any given system, and the values appropriate to the growth conditions used. Any predictions must also take account of differential thermal contractions if the epilayer is grown at an elevated temperature.

To summarise, it has been shown that vicinal (100) epilayers offcut towards [011] will tend to become less misoriented as stress relief occurs, although the reduction of tilt will depend to a large extent on the sensitivity of the glide systems to the differences in F_l. Vicinal (011) epilayers offcut towards (100) will tend to maintain or increase their misorientation as stress relief occurs, but it is unclear how much positive feedback will result. Finally, it must be emphasised that many material and geometrical considerations have been neglected which may prove crucial to a deeper understanding of the processes involved, and that this work must be regarded as an initial foray into the subject.

ACKNOWLEDGEMENTS

It is a pleasure to acknowledge helpful discussions with Prof. D.J. Bacon and N.P. Huxford in the course of this work.

REFERENCES

Aindow M 1989 PhD Thesis (Liverpool University, UK)
Ahearn J S and Uppal P 1987 Mat. Res. Soc. Symp. Proc. **91** 167
Feuillet G, Di Cioccio L and Millon A 1987 Inst. Physics Conf.Series **87**(2) 135
Hornstra J and Bartels W J 1978 J. Cryst. Growth **44** 513
Igarishi O 1976 Jap. J. Appl. Phys. **15** 1435
Kavanagh K L, Capano M A, Hobbs L W, Maree P M J , Schaff W, Mayer J W,
    Petit D, Woodall J M, Stroscio J A, Feenstra R M and Barbour J C 1988
    J. Appl. Phys. **64** 4843
Nagai H 1974 J. Appl. Phys. **45**(a) 3789
Olsen G H and Smith R T 1975 Phys. Stat. Sol. (a) **31** 739
Pond R C 1988 in "Dislocations in Solids" ed. F R N Nabarro (Amsterdam:
    North-Holland)
Pond R C, Aindow M, Dineen C and Peters T 1987 Inst. Phys. Conf. Ser. **87** 181

Inst. Phys. Conf. Ser. No 104: Chapter 5
Paper presented at Int. Symp. on Struct. Prop. Disloc. Semicond., Oxford, 1989

461

On the homogeneous nucleation of dislocations during integrated circuit processing

J Vanhellemont , H Bender and C Claeys
Interuniversity Micro-Electronics Center (IMEC), Kapeldreef 75,B-3030 Leuven, Belgium

ABSTRACT: The homogeneous nucleation of dislocations in silicon by the direct condensation of self-interstitials is discussed. A two-step nucleation mechanism is proposed that allows to understand both low and high temperature observations.

1. INTRODUCTION

Extended lattice defects in substrates used for silicon processing can have a detrimental influence on the electrical performance and reliability of integrated circuits. Since the beginning of the microelectronics era the study of the nucleation and annihilation mechanisms of processing induced defects has been very intense. Excellent reviews on the generation and properties of substrate defects in integrated circuits can be found e.g. in Kolbesen and Strunk 1985 and in Hu 1986. Each increase of the wafer size and decrease of the component size, requiring novel processing techniques, has introduced new defect related problems resulting in a continuing interest in processing induced defects.

As the generation of substrate defects has to be avoided as much as possible a good insight in the nucleation and propagation mechanisms of dislocations is essential for the process development engineer. More and more evidence exists that the behaviour of intrinsic and extrinsic point defects is the driving force for most of the observed extended defect phenomena. The dominant extended defect type in silicon is extrinsic i.e. it is formed by the condensation of self-interstitials.

2. NUCLEATION OF DISLOCATIONS DURING SILICON PROCESSING

Two types of defect nucleation have to be considered: firstly heterogeneous nucleation which occurs by activating dislocation sources (e.g. grown-in dislocations or precipitates) which are present at the start of the processing step and secondly homogeneous nucleation which occurs by the direct precipitation of self-interstitials in cases where no other sources with a lower activation energy are available. The present paper focuses mainly on homogeneous nucleation of dislocations.

2.1. Heterogeneous Nucleation

In low quality substrates with a relatively high grown-in dislocation density ($>10^8$ m^{-2}), plastic deformation occurs mainly by the glide, interaction and multiplication of the grown-in "generator" defects. As dislocation glide occurs at low stress levels, the macroscopic yield stress of such materials will be low. Pinning effects due to the formation of a Cottrell atmosphere around the generator dislocations will improve the strength of the wafer. This is the main reason why high oxygen content Czochralski material was preferred in the early days of integrated circuit processing.

A similar plastic deformation related to the substrate quality can occur when "generator" dislocations are formed by the high stress fields around precipitates (metallic or SiO_x) during heat treatments. This type of yielding is also considered as heterogeneous.

2.2. Homogeneous Nucleation

In material with a low density of grown-in dislocations as is the case for the nowadays commercially available silicon substrates, deformation will mainly occur after a first nucleation by direct condensation of self-interstitials resulting in the formation of "generator" dislocations which can subsequently act as multiplication sources. This homogeneous nucleation requires much higher stress levels as the glide force exerted by the externally applied stress field must exceed the line tension force. This line tension force can be very large for the dislocation nucleus as it is inversely proportional to the radius of curvature of the dislocation.

To explain the homogeneous nucleation of thin film edge induced dislocations in silicon substrates Vanhellemont *et al* (1987) proposed a model in which the first defect nucleus is formed by the direct precipitation of self-interstitials in a highly stressed area. This nucleus can only grow by the absorption of self-interstitials (climb) until the critical size for growth by glide is reached. The total climb force F_n will thus be the main parameter determining the yield stress. F_n is the sum of the external climb force due to mechanical stresses and of the internal climb force (or chemical force) which is due to the supersaturation of self-interstitials. Under constant mechanical stress the chemical force (or equivalently the supersaturation of self-interstitials) will determine the yield point (Vanhellemont and Claeys 1988a). Typical examples of homogeneously nucleated lattice defects observed in silicon treated at different temperatures are shown in Figures 1 and 2.

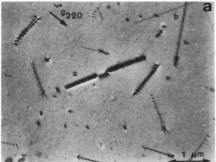

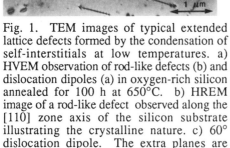
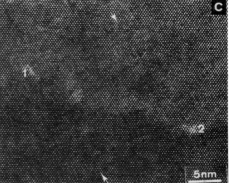

Fig. 1. TEM images of typical extended lattice defects formed by the condensation of self-interstitials at low temperatures. a) HVEM observation of rod-like defects (b) and dislocation dipoles (a) in oxygen-rich silicon annealed for 100 h at 650°C. b) HREM image of a rod-like defect observed along the [110] zone axis of the silicon substrate illustrating the crystalline nature. c) 60° dislocation dipole. The extra planes are indicated by the arrows.

The most important parameters determining the homogeneous nucleation of extended lattice defects during processing are thus the presence of high stresses and cooling rates and a supersaturation of intrinsic and extrinsic point defects.

Preferential sites for defect nucleation will be localised stress fields which occur e.g. near the edges of many of the thin film structures, near precipitates, below ion implanted areas and at the rim and the centre of the wafer during cooling or heating cycles. The cooling rate, e.g. during pulling out of the wafer from the furnace, is also an important parameter as the rate at which supersaturated self-interstitials annihilate during cooling will determine the yield stress level. The self-interstitial concentration will be considerably higher during rapid cooling which can easily lead e.g. to the creation of high densities of glide dislocations moving inwards from the rim of the wafer leading to a macroscopically observable bowing of the wafer (warpage).

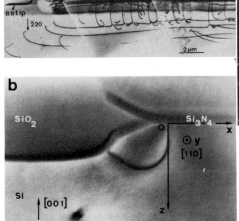
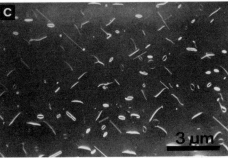

Fig. 2. Examples of homogeneously nucleated extended defects in silicon after high temperature treatments. a-b) 60° dislocations nucleated in silicon at Si₃N₄ film edges during a local oxidation process. c) Weak-beam HVEM image of circular dislocation loops and dipoles observed in ion implanted and annealed silicon (courtesy A. De Veirman).

Some of the most important sources of intrinsic point defects during integrated circuit processing are:

- Crystal growth: Strong indications exist that the increase of the wafer size is accompanied by an increase of the grown-in concentration C_I^0 of self-interstitials. The value of C_I^0 has a large impact on the precipitation kinetics of e.g. the interstitial oxygen and thus also the variation of the yield stress with time and temperature.
- Oxygen precipitation: The precipitation of the excess of interstitial oxygen occurs with a volume increase which is partially released by the emission of self-interstitials. The high stress levels close to the precipitates together with the decrease of the yield stress due to the high supersaturation of self-interstitials around the precipitate leads to the nucleation of prismatic loops and stacking faults.
- Ion implantation: The introduction of extrinsic point defects by high energy implantation leads to the creation of a large number of Frenkel pairs of which the largest part recombines during the initial stages of a thermal anneal leaving an excess of self-interstitials of the same order of magnitude as the implanted dose.
- Film growth (silicidation, oxidation,...): film growth with the consumption of part of the silicon substrate can occur as well with the emission of self-interstitials as with the emission of vacancies. In the last case a shrinking of extrinsic defects can be observed.

3. THE HOMOGENEOUS NUCLEATION MECHANISM

3.1 Low Temperatures (<850°C)

At low (<850°C) anneal temperatures the critical glide force (or frictional force) for dislocation glide is extremely large so that nucleation of dislocations by a glide mechanism is very unlikely. As at these temperatures the anisotropy of the elastic constants of the silicon matrix is larger than at higher temperatures, the formation of rod-like defects (RLD) and dislocations lying along <110> directions requires the minimum of energy as also the line tension is minimised.

Based on computer simulations of HREM images, initially the defects of Figure 1b were identified as ribbons of coesite, a high pressure phase of silicon oxide (Bourret 1984, Bender 1984). Recently however, strong evidence was presented that the defects consist of hexagonal silicon and are thus formed by self-interstitial precipitation (Bourret 1987, Reiche and Breitenstein 1987, Bender and Vanhellemont 1988). A possible nucleation and growth mechanism for the RLD is (Bender and Vanhellemont 1988): precipitation of self-interstitials occurs in {113} habit planes by a mechanism described by Tan (1981). Further precipitation of self-interstitials leads to a coarsening of the {113} defects and to a tendency towards a {100} habit plane. The presence of a large uniaxial compressive stress in the habit plane of the defect causes a shear transformation into hexagonal silicon. The hexagonal silicon rods can be considered as the precursors of the 60° and 90° dislocation dipoles which are observed in the same material (Reiche *et al* 1988), and also of the dislocation loops and stacking faults which are e.g. observed in ion implanted material after higher temperature anneals. A transformation of the RLD into 60° or 90° dislocation dipoles or stacking faults will occur when this is energetically favourable.

Two beam diffraction contrast TEM studies have shown that the displacement field connected with the RLD has no screw component . The strain field and thus also the strain energy of the RLD is thus equivalent with that of a pure edge dislocation dipole with Burgers vector $b_{rod} = k[100]$ in the coordinate system of Figure 3a where the x-axis lies along the Burgers vector in the glide plane (Bourret et al 1984). k varies between 0.18a and 0.33a.
The self energy E per unit length of a dislocation dipole consisting of two straight edge dislocations with Burgers vector $\mathbf{b}$ ($= b[\sin\phi,0,\cos\phi]$) is given by:

$$E = \frac{\mu b^2}{2\pi(1-\nu)} (1 - \nu\cos^2\phi) \ln \frac{r}{r_0} . \qquad (1)$$

ϕ is the angle between the Burgers vector and the dislocation line, $\mathbf{r}$ the vector connecting the cores of the two dislocation segments, r_0 the nucleation size of the {113} defect and μ and ν are the shear modulus and the Poisson ratio, respectively. The attracting glide force on both dipole dislocation segments is minimised for $\theta = \pi/2$ indicating that the {001} habit planes are the equilibrium planes for larger glide dislocation dipoles (Weertman and Weertman 1964). The attracting force on both dislocation segments is inversely proportional to r so that it is very large at the moment of nucleation of the RLD. The driving force for their nucleation must be the chemical force which has an internal component due to the supersaturation of self-interstitials and possibly also an external component due to externally applied stress fields.

The energy of a 90° dipole is obtained by substituting b by $a/\sqrt{2}$ and ϕ by $\pi/2$. It is energetically favourable for the rod-like defect to transform in a 90° dipole when $b_{rod} = a/\sqrt{2} = 0.38$ nm which is considerably larger than the experimentally observed largest value of $b_{rod} = 0.18$ nm (Bender 1984). This theoretical value of b_{rod} is however the upper limit as the strain energy of the rod and the self-energy of hexagonal silicon, which is slightly higher than that of silicon with the diamond structure, is not taken into account. Comparing the self-energy of a dipole consisting of 60° dislocations (E_{dip60})with a 90° dipole one obtains: $E_{dip60} = (1 - 0.25\nu)E_{dip90}$. The energy of a 90° dipole is thus about 10% higher than of a 60° dipole which explains why the majority of experimentally observed dipoles is of the 60° type.

Expressing conservation of self-interstitials allows to obtain a more realistic value for b_{rod} at which transformation into a perfect dislocation dipole will occur. The number of silicon atoms N in a dislocation dipole is given by the Burgers vector component perpendicular to the dipole plane times the dipole width. For a (001) rod-like defect and a (001) 60° dislocation dipole this yields $N_{rod} = \rho_{Si}r_{rod}b_{rod}$ and $N_{dip} = \rho_{Si}r_{dip}a/\sqrt{2}\cos(34.5°)\sqrt{3}/2$, with ρ_{Si} the atomic density of silicon which is assumed equal for the hexagonal and diamond cubic phase. Assuming at the moment of transformation $N_{rod} = N_{dip}$ and $r_{rod}=r_{dip}$ one obtains $b_{rod} = 0.16$ nm which gives the lower limit for the transformation of a rod in a dipole and is very close to the observed average value of $b_{rod} = a/3$ (= 0.18 nm) in material which contained both rod like defects, dislocation dipoles and mixed defects (Bender 1984). When the transformation of RLD into dipole occurs for a b_{rod} value large than this lower limit r_{dip} will be larger than r_{rod}

after the transformation as observed in Figure 1a. After nucleation of the dipole which is no longer sessile this will grow further by a combined climb and glide mechanism and will tend towards a near {001} or {110} habit plane to minimise the attracting glide force. The total attracting force between the two dipole dislocation segments increases with decreasing distance between them and creates an uniaxial compressive stress along the <110> direction in case of a (001) habit plane. This compressive stress induces the shear transformation of the silicon lattice into hexagonal silicon in between the two dislocations.

The described low temperature nucleation mechanism of 60 and 90° dislocation dipoles is very similar to the one proposed for film edge induced dislocation nucleation: first a precursor of the dislocation loop is formed by precipitation of excess self-interstitials. This small nucleus can only grow by the absorption of self-interstitials (=climb) until a critical size is reached at which a transformation towards glide dislocations is favoured.

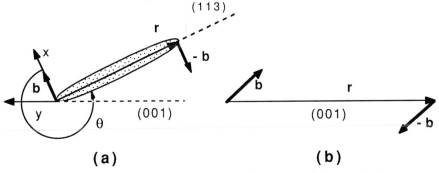

(a)   **(b)**

Fig. 3. Schematical representation of the growth of a rod-like defect nucleating in a {113} plane (a) and during growth tending towards a {001} plane and finally transforming to a 60° or 90° dislocation dipole (b).

3.2 High Temperatures (>850°C)

Here also the nucleation of dislocation loops is only possible starting from a nucleus which can grow by climb until (for a glide dislocation) it reaches the critical size when further growth by glide can take over. As the dislocation loops grow faster and to a larger size than at low temperatures the attracting force between the segments decreases allowing the loop to take a circular shape which contains a maximum of the self-interstitials with a minimal growth energy. At these higher temperatures the nucleation of partial dislocations (Frank stacking faults) is also energetically possible. To minimise their climb energy for a maximal number of precipitated self-interstitials the dislocation loops will minimise their circumference by taking a circular shape. An expression for the critical radius r_c of a dislocation loop can be obtained by following a similar reasoning as for the formation of spherical precipitates (Vanhellemont and Claeys 1987). The change of free energy dG at temperature T for an infinitesimal increase dr is given by

$$dG = \sigma 2\pi r dr - 2\pi r dr b k T \ln \frac{C_I}{C_I^*} + 2\pi dr\, \mathbf{L} = 0 \ , \text{ with } \mathbf{L} \approx \frac{\mu b^2}{4\pi} \ln \frac{r}{5b} \ .$$

C_I^* and C_I are the thermal equilibrium concentration and the actual concentration of self-interstitials respectively. L is the line tension and k the Boltzmann constant. The critical radius r_c for loop growth is obtained by posing dG = 0, or

$$r_c = \frac{\mathbf{L}}{bkT\ln \dfrac{C_I}{C_I^*} - \sigma} \ . \tag{2}$$

Equation (2) allows to describe stacking fault growth and shrinkage due to the evolution of C_I and also to understand the growth of larger dislocation loops at the expense of smaller ones which is observed during annealing of ion implanted silicon.

3.3 Multiplication Mechanisms

Macroscopic plastic deformation will only occur when high densities of dislocations are formed. The basic mechanism for multiplication in silicon is cross glide of dislocations in their cross glide plane. This leads to the formation of Frank Read and Frank Read-like sources of dislocations leading to a rapid dislocation multiplication which is followed by the formation of dense networks by interaction of dislocations with different Burgers vectors leading to a hardening of the substrate. Especially in case of <110> aligned stress fields such multiplication mechanisms are easily triggered once the yield stress for homogeneous nucleation is exceeded, making these directions not very suitable for device edges in integrated circuit processing.

In Figure 4 plan view HVEM micrographs illustrate the multiplication of 60° dislocations starting from a homogeneously nucleated small dislocation half loop at <110> silicon nitride film edges during a local oxidation process. The extended defect in Figure 4b is one single dislocation formed by a Frank Read like source (Vanhellemont J and Claeys C 1988b).

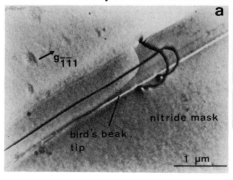
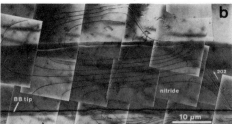

Fig. 4. HVEM micrographs illustrating dislocation multiplication by a cross-glide mechanism at <110> oriented thin film edges on (001) (a) and (111) silicon substrates (b).

REFERENCES

Bender H 1984 *Phys. Stat. Sol. (a)* **86** 245
Bender H and Vanhellemont J 1988 *Phys. Stat. Sol. (a)* **107** 455
Bourret A, Thibault-Desseaux J and Seidman D N 1984 *J. Appl. Phys.* **55** 825
Bourret A 1987 *Inst. Phys. Conf. Ser.* **87** 39
Hu S M, in *Semiconductor Silicon 1986*, Huff HR, Abe T and Kolbesen B, eds,(Pennington, N J: The Electrochem. Soc.) 722
Kolbesen BO and Strunk HP 1985 *VLSI Electronics: Microstructure Science* , Silicon Materials, Huff HR, Volume Ed., Vol 12 (New York: Academic) pp. 143-222
Reiche M and Breitenstein O 1987 *Phys. Stat. Sol. (a)* **101** K97
Reiche M, Reichel J and Nitzsche W 1988 *Phys. Stat. Sol. (a)* **107** 851
Tan T Y 1981 *Phil. Mag. A.*44 101
Vanhellemont J, Amelinckx S and Claeys C 1987 *J. Appl. Phys.* **61** 2176
Vanhellemont J and Claeys C 1987 *J. Appl. Phys.* **62** 3960
Vanhellemont J and Claeys C 1988a *J. Electrochem.Soc.* **135** 1509
Vanhellemont J and Claeys C 1988b *J. Appl. Phys.* **63** 5703
Weertman J and Weertman J R 1964 *Elementary Dislocation Theory* (New York: The Macmillan Company)

Author Index

Subject Index†

† Page numbers refer to the first pages of papers in which citations occur.